**Neutrons and Synchrotron Radiation in Engineering Materials Science**

*Edited by*
*Walter Reimers, Anke Rita Pyzalla, Andreas Schreyer, and Helmut Clemens*

## *Further Reading*

U. Krupp

**Fatigue Crack Propagation in Metals and Alloys**
**Microstructural Aspects and Modelling Concepts**

2007
ISBN 978-3-527-31537-6

M. Birkholz

**Thin Film Analysis by X-Ray Scattering**

2006
ISBN 978-3-527-31052-4

G. De With

**Structure, Deformation, and Integrity of Materials**
**Volume I: Fundamentals and Elasticity**
**Volume II: Plasticity, Visco-elasticity, and Fracture**

2006
ISBN 978-3-527-31426-3

V. Levitin

**High Temperature Strain of Metals and Alloys**
**Physical Fundamentals**

2006
ISBN 978-3-527-31338-9

# Neutrons and Synchrotron Radiation in Engineering Materials Science

From Fundamentals to Material and Component Characterization

*Edited by*
*Walter Reimers, Anke Rita Pyzalla, Andreas Schreyer, and Helmut Clemens*

WILEY-VCH Verlag GmbH & Co. KGaA

**The Editors**

***Prof. Dr. Walter Reimers***
Technical University Berlin
Institute for Material Science and Technologies
Metallic Materials
Ernst-Reuter-Platz 1
10587 Berlin
Germany

***Prof. Dr. Anke Rita Pyzalla***
Max-Planck-Institut für Eisenforschung GmbH
Max-Planck-Strasse 1
40237 Düsseldorf
Germany

***Prof. Dr. Andreas K. Schreyer***
GKSS Research Center
Institute of Materials Research
Max-Planck-Strasse 1
21502 Geesthacht
Germany

***Prof. Dr. Helmut Clemens***
University of Leoben
Department of Physical Metallurgy and Materials Testing
Franz-Josef-Strasse 18
8700 Leoben
Austria



**Library of Congress Card No.:** applied for

**British Library Cataloguing-in-Publication Data**
A catalogue record for this book is available from the British Library.

**Bibliographic information published by the Deutsche Nationalbibliothek**
Die Deutsche Nationalbibliothek lists this publication in the Deutsche Nationalbibliografie; detailed bibliographic data are available in the Internet at http://dnb.d-nb.de.



Printed in the Federal Republic of Germany
Printed on acid-free paper

**Typesetting** Asco Typesetters, Hong Kong
**Printing** betz-druck GmbH, Darmstadt
**Binding** Litges & Dopf GmbH, Heppenheim

**ISBN** 978-3-527-31533-8

# Contents



*Neutrons and Synchrotron Radiation in Engineering Materials Science: From Fundamentals to Material and Component Characterization*
Edited by Walter Reimers, Anke Rita Pyzalla, Andreas Schreyer, and Helmut Clemens

ISBN: 978-3-527-31533-8

# Preface

Neutron and photon sources offer unique possibilities by complementary use of their radiation for structure analyses of advanced engineering materials. Neutrons and photons delivered by a spallation source or a reactor, respectively, a synchrotron radiation source, provide information about materials microstructures in the near-surface region and in the bulk of components nondestructively. Compared to conventional laboratory X-rays the spatial resolution and in-depth information achievable can be increased by up to several orders of magnitude by using synchrotron radiation and neutrons. The new possibilities for microstructure analyses for advanced materials and multimaterial systems meet with increasing demands from the materials engineering point of view. In engineering materials science, the establishment and refinement of relationships between microstructure parameters and macroscopic properties requires information on different length and time scales, both covering several orders of magnitude.

The use of neutrons and synchrotron radiation has been the domain of physicists and other natural scientists for many years. The development of experimental techniques suitable for the characterization of the often complex engineering materials and components as well as the availability of dedicated instruments for material science applications at reactors, spallation sources, and electron storage rings today render the use of neutrons and synchrotron radiation attractive also for material scientists and engineers working in universities, research institutes, and industry. The predominant experimental techniques used are diffraction, small-angle scattering, and tomography. Analyzing the number of publications listed in the search engine, SCOPUS, the scientific community's increasing interest in these techniques becomes apparent (see Figure). A more detailed view on the number of publications where scientists used neutrons or synchrotron radiation for residual stress analyses using diffraction, for small-angle scattering, as well as for tomography reveals the continuing and even increasing importance of neutrons in these scientific fields as well as the more recent strong impact of synchrotron radiation.

The idea to write this book is a result of the discussions with lecturers and attendees of the first autumn school, "Engineering Material Science with Neutrons and Synchrotron Radiation", which took place in the conference center "Haus am Schüberg" in Ammersbek near Hamburg, Germany, from 10th to 14th of

*Neutrons and Synchrotron Radiation in Engineering Materials Science: From Fundamentals to Material and Component Characterization*
Edited by Walter Reimers, Anke Rita Pyzalla, Andreas Schreyer, and Helmut Clemens

ISBN: 978-3-527-31533-8

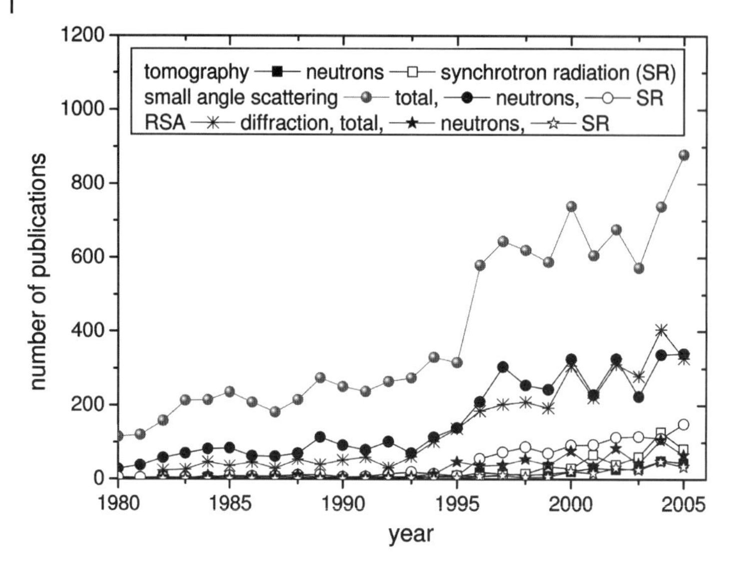


Number of publications where X-rays, synchrotron radiation, and neutrons were employed for residual stress analyses (RSA) by diffraction, for small-angle scattering and for tomography. The number of publications was determined using the search engine SCOPUS (www.scopus.com) in October 2006 using the keywords, residual stress analyses + diffraction, small-angle scattering and tomography combined with the keywords synchrotron and neutron.

October 2005. This autumn school was designed to provide a systematic overview of this field to students from all over Europe. It was organized within the Virtual Institute "Photon and Neutron Research on Advanced Engineering Materials (VI-PNAM)" in co-operation with the Montanuniversität Leoben (Austria). The virtual institute VI-PNAM is a collaboration of the Technical Universities of Berlin (Germany), Vienna (Austria), Dresden (Germany), Clausthal (Germany), the Max-Planck-Institut für Eisenforschung GmbH, and the three German research centers HMI (Berlin), GKSS (Geesthacht), and DESY (Hamburg), which has been funded by the Helmholtz-Gemeinschaft (HGF), which is their roof organization. The lectures given in the autumn school obviously inspired the students for future research both at neutron and synchrotron radiation sources: while more than 50% and about 40% of the students had not used neutrons respectively synchrotron radiation before, after the autumn school nearly all of them intend to perform experiments both with neutrons and synchrotron radiation in future.

This book is intended both as a reference for the attendees of the autumn school and as an introduction for those interested, but not yet experts in the field. In order to provide the information necessary for readers with a background in natural science, material science, and engineering disciplines with basics of engineering material science and the properties and generation of neutrons and synchrotron radiation, these topics are briefly summarized in Part I "General" of this book. The following Part II comprises detailed information about state-of-the art diffraction, scattering and imaging methods for the use of neutrons and synchrotron radiation for engineering materials' characterization. New and emerging methods in this scientific field are presented in Part III. Part IV gives examples for industrial applications of neutrons and synchrotron radiation.

We hope that this textbook will serve as a brief introduction to newcomers in the field and that it will foster the interest of students, scientists, researchers, and practitioners in industry in using neutrons and synchrotron radiation for pursuing their research and development interests.

The editors thank all authors for their enthusiasm to contribute to this book. In addition, we are extremely grateful for the help of all those co-workers in our institutes and departments who helped us preparing the manuscripts, managing the review process of the chapters and the communication between editors and authors as well as between editors and publishers. For their continuing strong contributions to these tasks we would like to express our strong gratitude especially to Mr. Dimas Souza and Mr. Benjamin Breitbach and in particular to Ms. Adelheid Adrian of Max-Planck-Institut für Eisenforschung GmbH (Düsseldorf). Last but not least we would like to thank the publishers, most notably Dr. Nicola Oberbeckmann-Winter and Dr. Jörn Ritterbusch, for their great support during writing and editing this book.

*Walter Reimers*
*Anke Rita Pyzalla*
*Andreas Schreyer*
*Helmut Clemens*

# List of Contributors

***Funda S. Bayraktar***
GKSS Research Center
Institute of Materials Research
Max-Planck-Strasse 1
21502 Geesthacht
Germany

***Felix Beckmann***
GKSS Research Center
Institute for Materials Research
Notkestrasse 85
22607 Hamburg
Germany

***Heinz-Günter Brokmeier***
Clausthal University of Technology
Institute of Materials Science and Engineering
Department TEXMAT
Agricolastrasse 6
38678 Clausthal-Zellerfeld
Germany

***Ulrike Cihak***
University of Leoben
Department of Physical Metallurgy and Materials Testing
Franz-Josef-Strasse 18
8700 Leoben
Austria

***Helmut Clemens***
University of Leoben
Department of Physical Metallurgy and Materials Testing
Franz-Josef-Strasse 18
8700 Leoben
Austria

***Christoph Genzel***
Hahn-Meitner Institute
c/o BESSY
Albert-Einstein-Strasse 15
12489 Berlin
Germany

***Günter Goerigk***
Forschungszentrum Jülich
Außenstelle JCNS-FRM II
Forschungsneutronenquelle
Hein-Maier-Leibniz
85748 Garching
Germany

***Astrid Haibel***
GKSS Research Center
Institute of Materials Research
Department Research with Neutrons and Synchrotron Radiation
Notkestrasse 85
22607 Hamburg
Germany

*Neutrons and Synchrotron Radiation in Engineering Materials Science: From Fundamentals to Material and Component Characterization*
Edited by Walter Reimers, Anke Rita Pyzalla, Andreas Schreyer, and Helmut Clemens

ISBN: 978-3-527-31533-8

***Kristoffer Haldrup***
University of Copenhagen
Niels Bohr Institute
Center for Molecular Movies
H.C. Ørsted Institute
Universitetsparken 5
2100 Copenhagen
Denmark

***Gene E. Ice***
Oak Ridge National Laboratory
P.O. Box 2008
Oak Ridge, TN 37831-6118
USA

***Augusta Isaac***
Max-Planck-Institut für
Eisenforschung GmbH
Max-Planck-Strasse 1
40237 Düsseldorf
Germany

***Wolfgang Knop***
GKSS Research Center
Max-Planck-Strasse 1
21502 Geesthacht
Germany

***Mustafa Koçak***
GKSS Research Center
Institute of Materials Research
Max-Planck-Strasse 1
21502 Geesthacht
Germany

***Michael Lohmann***
Deutsches Elektronen
Synchrotron
HASYLAB
Notkestrasse 85
22603 Hamburg
Germany

***Wolfgang Ludwig***
Riso National Laboratory
Materials Research Department
Frederiksborgvej 399
4000 Roskilde
Denmark

***René Valéry Martins***
GKSS Research Center
Institute of Materials Research
Notkestrasse 85
22607 Hamburg
Germany

***Henning Friis Poulsen***
Riso National Laboratory
Materials Research Department
Frederiksborgvej 399
4000 Roskilde
Denmark

***Philipp Klaus Pranzas***
GKSS Research Center
Institute of Materials Research
Max-Planck-Strasse 1
21502 Geesthacht
Germany

***Anke Rita Pyzalla***
Max-Planck-Institut für
Eisenforschung GmbH
Max-Planck-Strasse 1
40237 Düsseldorf
Germany

***Walter Reimers***
Technical University Berlin
Institute for Material Science
and Technologies
Metallic Materials
Ernst-Reuter-Platz 1
10587 Berlin
Germany

***Andre Rothkirch***
Deutsches Elektronen
Synchrotron
HASYLAB
Notkestrasse 85
22670 Hamburg
Germany

***Javier Roberto Santisteban***
CONICET and Instituto Balseiro
Centro Atómico Bariloche
San Carlos de Bariloche – 8400
Argentina

***Christina Scheu***
University of Leoben
Department of Physical
Metallurgy and Materials Testing
Franz-Josef-Strasse 18
8700 Leoben
Austria

***Søren Schmidt***
Riso National Laboratory
Materials Research Department
Frederiksborgvej 399
4000 Roskilde
Denmark

***Peter Schreiner***
GKSS Research Center
Zentralabteilung
Forschungsreaktor
Max-Planck-Strasse 1
21502 Geesthacht
Germany

***Andreas Schreyer***
GKSS Research Center
Institute of Materials Research
Max-Planck-Strasse 1
21502 Geesthacht
Germany

***Peter Staron***
GKSS Research Center
Institute of Materials Research
Max-Planck-Strasse 1
21502 Geesthacht
Germany

***Martin Stockinger***
Böhler Schmiedetechnik
GmbH & Co KG
Mariazellerstrasse 25
8605 Kapfenberg
Austria

***Wolfgang Treimer***
University of Applied Sciences Berlin
Department II Physics, Mathematics,
Chemistry
Luxemburgerstrasse 10
13353 Berlin
Germany
and
Hahn-Meitner Institute
Dept. SF1
Glienicker Strasse 100
14109 Berlin
Germany

***Rolf Treusch***
Deutsches Elektronen
Synchrotron
HASYLAB
Notkestrasse 85
22603 Hamburg
Germany

***John A. Wert***†
Danish Technical University Risø
National Laboratory
Frederiksborgvej 399
4000 Roskilde
Denmark

***Thomas Wroblewski***
Deutsches Elektronen
Synchrotron
HASYLAB
Notkestrasse 85
22670 Hamburg
Germany

***Sang-Bong Yi***
Clausthal University of Technology
Institute of Materials Science
and Engineering
Agricolastrasse 6
38678 Clausthal-Zellerfeld
Germany

# Part I
# General

# 1
# Microstructure and Properties of Engineering Materials

*Helmut Clemens and Christina Scheu*

## 1.1
## Introduction

In general, engineering materials are grouped into four basic classifications: metals, ceramics, polymers, and semiconductors. Whereas semiconductors represent exclusively functional materials, the remaining three – depending on their application – can be assigned either to the group of structural or functional materials. Independent of to which group they belong, the important properties of solid materials depend on the geometrical atomic arrangement and also the type of bonding that exists between the constituent atoms. The three types of primary or chemical bonds which are found in engineering materials – covalent, ionic and metallic – and the main contributions to the individual groups are shown in Figure 1.1. Metals and their alloys possess primarily metallic bonding; semiconductors have mainly covalent bonds, whereas many ceramics exhibit a mix-

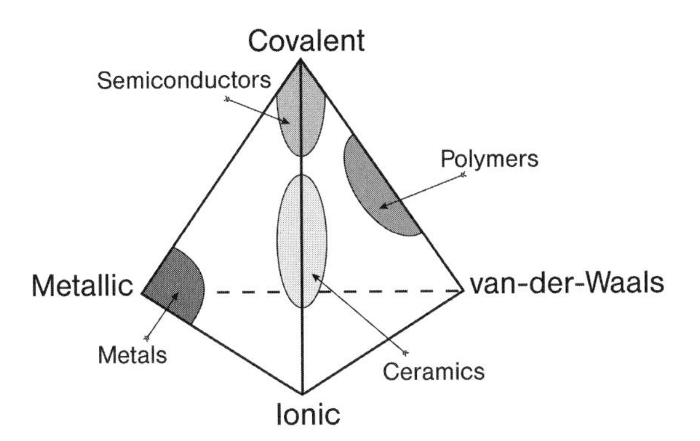


**Figure 1.1** Bonding behavior present in different groups of engineering materials (after [2]). Covalent, ionic, and metallic bonds represent strong primary bonds, whereas van der Waals attraction is a weak secondary bond.

*Neutrons and Synchrotron Radiation in Engineering Materials Science: From Fundamentals to Material and Component Characterization*
Edited by Walter Reimers, Anke Rita Pyzalla, Andreas Schreyer, and Helmut Clemens

ISBN: 978-3-527-31533-8

ture of covalent and ionic bonding. In engineering polymers weak secondary forces of attraction (van der Waals forces) exist between the extended covalently bound hydrocarbon chains (Figure 1.1). In general, the nature of bonding depends on the electronic structure of the constituent atoms forming the solid and arises from the tendency of atoms to obtain stable electron configurations.

The structure of engineering materials relates to the arrangement of its internal components. On an atomic level, structure is understood as the organization of atoms relative to each other. In crystalline materials the atoms are arranged in periodically repeating arrays which are termed crystal or lattice structures. Metals, for instance, have particularly simple crystal structures: (a) face-centered cubic (fcc), (b) body-centered cubic (bcc), (c) hexagonal closed-packed (hcp), and (d) tetragonal. Many metals and their alloys exist in more than one crystal structure depending on temperature and composition, but, in most cases, the transitions are between these four crystal structures. In contrast, semiconductors usually crystallize either in the diamond structure (silicon, germanium) or often in the zincblende structure (e.g., gallium arsenide).

The next larger structural level is the microscopic level. Here, large groups of atomic arrangements are considered as components of the microstructure which determines most of the properties of the material. The microstructure of engineering materials is described by identification of the grain size, types of phases present and description of their structure, shape and size distributions. In addition, two-dimensional defects such as grain boundaries and heterophase interfaces, one-dimensional defects such as dislocations, and zero-dimensional defects such as point defects are important microstructural features which often control the resulting properties.

In this introductory chapter the microstructure of engineering materials is explained with focus on structural metallic materials, showing a polycrystalline multiphase assembly. The most important microstructure parameters are pre-

**Table 1.1** Influence of atomic arrangement and microstructure on the properties of engineering metallic materials. Control of the atomic arrangement and the microstructure is possible through processes such as casting, powder metallurgy, working, and heat treatment (after [13]).

| Property | Influence of atomic arrangements and atomic defects | Influence of microstructure |
|---|---|---|
| Mechanical (e.g., strength and ductility) | Strong | Strong |
| Electrical, magnetic and thermal (e.g., resistivity, magnetization, conductivity) | Moderate to strong | Slight to strong |
| Chemical (e.g., corrosive resistance, catalytic potential) | Slight | Slight to moderate |

sented and their influence on mechanical properties is briefly discussed. Table 1.1 roughly summarizes the influence of atomic arrangement, atomic defects, and microstructure upon the properties of metallic materials. In addition, the most important methods for microstructural characterization on a nanometer and micrometer scale will be outlined in this introductory chapter with emphasis on analytical electron microscopy. At the end of the chapter a selection of textbooks and journal articles is listed which might be helpful for the reader to deepen his understanding of the microstructure and properties of engineering materials [1–15] as well as of methods used for microstructural characterization [6, 16–23].

## 1.2 Microstructure

Figure 1.2 shows schematically the microstructure of a polycrystalline multiphase metallic material. For a comprehensive and better understanding of the following explanations, Table 1.2 lists the typical mole fractions and size ranges of the individual microstructural features and Table 1.3 describes their most important characteristics and their influence on various properties.

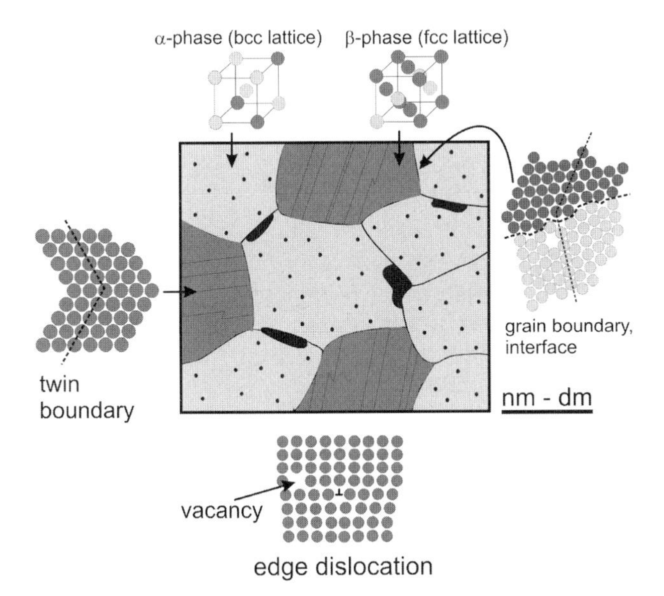


**Figure 1.2** Schematic microstructure of a polycrystalline multiphase metallic material. The microstructure consists of $\alpha$ and $\beta$ grains showing bcc and fcc crystal structures. Within the grains the existence of atomic defects is indicated (vacancies, dislocations, twin boundaries). The grains are separated by grain (phase) boundaries. On some grain boundaries large precipitates are visible. Within one type of grains nanometer-sized particles are present. For further explanations see text and refer to Tables 1.2 and 1.3.

**Table 1.2** Microstructure of engineering metallic materials: constituents and their concentration and size ranges.

| Microstructural constituents | Range |
|---|---|
| Vacancies | Equilibrium concentration (mole fraction): $10^{-15}$ (room temperature)–$10^{-4}$ (near the melting point) |
| Dislocations | Density[a]: $10^{10}$ (annealed)–$10^{16}$ $m^{-2}$ (heavily cold-worked) |
| Grains | Size: nm–dm |
| Subgrains/domains | Size: nm–μm |
| Alloying elements | Concentration (mole fraction): ppm–50% |
| Second phases | Volume fraction: 0–70% |
| Particles (precipitates, dispersoids) | Size: nm–μm; Volume fraction: 0–70% |

a) Dislocation density: the number of dislocations that intersect a unit area of a random surface section; alternatively: the total dislocation length per unit volume of material.

**Table 1.3** The role of microstructural constituents in engineering metallic materials.

| Microstructural constituent | Dependent on/characteristics (selection) | Responsible for (examples) |
|---|---|---|
| Vacancies | Temperature, deformation | Hardening at low temperatures; diffusion processes at elevated temperatures; diffusional creep |
| Dislocations | Deformation, temperature, recovery and recrystallization processes; at elevated temperatures edge dislocations may climb and leave their slip planes | Plastic deformation; strength is controlled by their number and motion; driving force for recrystallization; dislocation ceep |
| Stacking faults | Crystal structure, alloying | Mobility of dislocations, e.g., climb of edge dislocations and cross-slip of screw dislocations is hampered |
| Mechanical twins | Stacking fault energy, deformation, temperature | Additional deformation mechanism at low temperatures and/or high strain rates |
| Subgrains/ domains | Deformation, temperature, stacking fault energy/ordered crystal structure; antiphase boundary energy | Work hardening, creep, creation of antiphase boundaries |

**Table 1.3** *(continued)*

| Microstructural constituent | Dependent on/characteristics (selection) | Responsible for (examples) |
|---|---|---|
| Grain boundaries | Lattice orientation between neighboring grains; subdivision in small-angle, medium-angle and high-angle grain boundaries | Work hardening by acting as barriers to slip from one grain to the next; segregation site of impurity atoms |
| Phase boundaries | Alloy system, composition, phase stability at elevated temperatures | Strengthening effects, e.g., in duplex or multiphase steels |
| Grains | Alloy system, type of nucleation, processing, deformation, heat treatment, recrystallization | Strengthening (see grain boundaries) but ductility is maintained; grain boundary sliding at elevated temperatures (creep, superplasticity) |
| Annealing twins | Stacking fault energy; characteristic of face-centered cubic materials exhibiting a low stacking fault energy | Lowering of total boundary energy during grain growth |
| Precipitates/ dispersoids | Alloy system, composition, heat treatment, processing; the interface between particle and matrix can be coherent, semicoherent or incoherent | Increase in strength by the interaction of moving dislocation; dislocations can loop, cut through or cross-slip the particles at ambient temperatures; at elevated temperatures the dislocations can surmount the particles by climb processes |
| Phase arrangement (e.g., eutectics, duplex, dual phase) | Alloy system, composition, processing, heat treatment | Positive: control of mechanical and thermo-physical properties; negative: embrittlement in case of brittle phases situated at grain boundaries |

The schematic microstructure shown in Figure 1.2 consists of grains of two different phases. The phases differ in their crystal structures (fcc and bcc) and their chemical compositions. As indicated in the depicted crystal structures each phase forms a solid solution. A solid solution represents a homogeneous crystalline phase that contains two or more chemical species. Both substitutional and interstitial solid solutions are possible, such as nickel and chromium in iron (e.g., austenitic steels) and carbon in iron (e.g., heat treatable steels), respectively. The solubility of a metal, i.e., its alloying behavior, depends on the atomic size factor (difference in size between solute and solvent atom), the electrochemical effect (the higher the difference in electronegativity, the higher the tendency for the elements to form intermetallic phases rather than extensive solid solutions), and the relative electron valency (a metal of higher electron valency is more likely to dissolve to a large extent in one of lower electron valency than vice versa).

### 1.2.1 Crystal Defects

The grains in a microstructure represent individual crystals within the polycrystalline material (Figure 1.2). Within each grain, atoms are regularly arranged according to the basic crystal structure but also a variety of imperfections, termed crystal defects, may occur. These defects are point defects (vacancies, interstitial atoms), line defects (dislocations), planar defects (stacking faults, twin boundaries), and volume defects (voids, cavities). Of particular interest are dislocations, because plastic deformation mainly corresponds to the motion of dislocation in response to an applied shear stress (see Chapters 19 and 20). In contrast, hindering of dislocation movement is the basic concept of all strengthening mechanisms (see Section 1.3). Dislocations are subdivided into edge and screw dislocations. At temperatures where no thermally activated diffusion processes take place edge dislocations are confined to their slip planes, whereas screw dislocations can change their slip planes rather easily by cross-slip processes. A schematic drawing of an edge dislocation is shown in Figure 1.2. An edge dislocation is a linear crystalline defect associated with the lattice distortion produced in the vicinity of the end of an extra half-plane of atoms within a crystal. Depending on processing history and/or mechanical loading subgrains or cell structure, separated by dislocation networks or tangles, can be formed within the grains. In grains showing an ordered crystal structure (e.g., intermetallic phases) domain structures may appear. The individual domains are separated by antiphase boundaries. The corresponding energy is referred to as antiphase boundary energy.

### 1.2.2 Grain (Phase) Boundaries and Twins

The size of the grains depends on materials processing and heat treatments and can be adjusted in a wide range. In most technical relevant structural metallic materials, such as steels, aluminium alloys and titanium alloys, the grain size is in the range of several ten micrometers. In contrast, in nanostructured functional materials, e.g., superhard coatings with high wear resistance, a grain size in the range of few nanometers is required. The grains as shown in Figure 1.2 are separated by grain (phase) boundaries. In general, grain (phase) boundaries are interfaces which separate two adjoining grains (phases) having different crystallographic orientations and in the case of phases different crystal structures and/or chemical compositions. Within the boundary region, which can have a width of one to several atomic distances, an atomic mismatch due to the transition from the crystalline orientation of one grain to that of an adjacent one can occur. Depending on the structure one can distinguish between high-angle grain boundaries, small-angle grain boundaries, etc. Because the atoms are differently coordinated and/or bonded along grain boundaries, there is an interfacial or grain boundary energy associated with them. The magnitude of this energy is a function of the degree of misorientation between the grains, being larger for high-angle grain boundaries although some energy cusps can occur for special grain boundaries. Simple small-angle grain boundaries can be described by dislocation arrangements. A twin boundary as shown in Figure 1.2 is a special type of grain boundary. Atoms of one side of the boundary are located in mirror image positions of the atoms of the other side. Twins result from atomic displacements that are produced from an external stress state (mechanical or deformation twins) and also during annealing heat treatments subsequent to deformation (annealing twins). The formation of twins is closely related to the stacking fault energy of the material. In general, low stacking fault energy facilitates twinning as can be seen in the high density of annealing twins in fcc metals and their alloys, such as copper, $\alpha$-brass and austenitic steels. The positive effect of deformation twinning on strain hardening and deformability, for example, is exploited in TWIP (twinning-induced plasticity) steels.

### 1.2.3
### Precipitates and Dispersions

In many structural engineering metal materials precipitates occur. In Figure 1.2 two types of precipitates are drawn schematically: few large ones at grain boundaries and a large number of small particles homogeneously dispersed within individual grains. In many alloys, e.g., steels or nickel-based alloys, these precipitates are carbides or intermetallic phases. Their influence on mechanical properties primarily depends on volume fraction, size, distribution, type of precipitate, and arrangement in the microstructure. Large precipitates along grain boundaries as shown in Figure 1.2 can either have a positive or negative effect on the properties. For example, in nickel-based superalloys precipitates are generated at grain boundaries by means of a special heat treatment in order to minimize grain boundary sliding at high service temperatures. However, such a phase arrangement can also lead to serious embrittlement as observed in steels containing non-metallic inclusions or cementite films along grain boundaries. Nanometer-sized particles of a second phase which are uniformly dispersed within the grains provide the most versatile strengthening mechanism for metallic materials in addition to solid solution strengthening (see Section 1.3). There are different ways to produce extremely fine particles in a metallic matrix: a variety of metallic alloy systems have been developed for which so-called precipitation heat treatments are employed to precipitate a new phase from a supersaturated solid solution. Examples of engineering alloys that are hardened by precipitation treatments include aluminium–copper (e.g., Duraluminium or Dural), nickel–aluminium (e.g., nickel-based superalloys), and some ferrous alloys (e.g., maraging and tool steels). A common feature of these nanometer-sized particles, which usually precipitate in the form of metastable phases, is their coherency with the matrix in the early stages of precipitation. However, during exposure at service temperatures these particles may change their chemistry and are prone to coarsening. Very often this process is accompanied by loss of coherency; thus a semicoherent or an incoherent interface between particle and matrix is formed. As a consequence the initial hardening mechanism is altered, leading to a decrease in strength.

Another way to strengthen metals and their alloys is to produce a uniform dispersion of several volume fractions of extremely small particles of a very hard and inert material. The dispersed phase may be either metallic or nonmetallic and they usually do not show coherency with the matrix. Examples are oxide dispersion strengthened (ODS) superalloys: hard nanometer-sized $Y_2O_3$ particles are mechanically alloyed into the matrix powder and consolidated and processed by powder metallurgical techniques. The dispersion strengthening effect is often

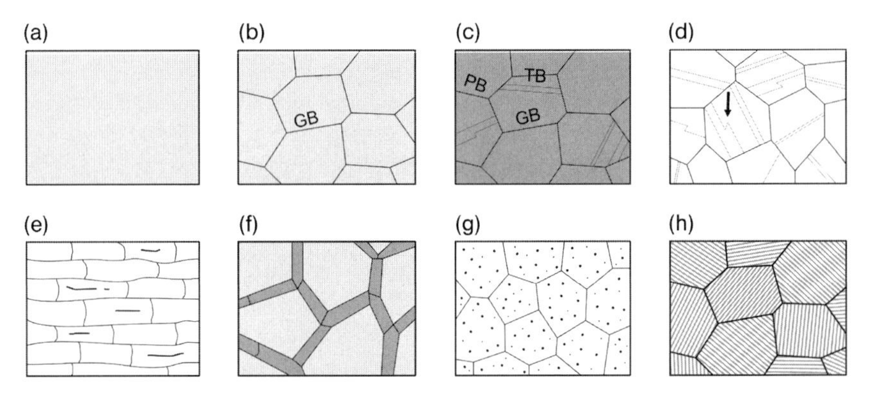


**Figure 1.3** Schematic drawings of different microstructures: (a) Single crystal: crystalline solid for which the periodic and repeated atomic pattern extends throughout the entire sample without interruption. The properties depend strongly on the orientation of the crystal. Example: single-crystal made of nickel-based superalloys. (b) Polycrystalline single-phase material. The individual grains differ in their crystallographic orientations and are separated by grain boundaries (GB). Example: $\alpha$-iron (ferrite) with body-centered cubic (bcc) lattice structure. (c) Two-phase material. The phases differ in chemical composition and crystal structure. The grains are separated either by phase boundaries (PB) or GBs. One phase, most probably a phase possessing a face-centered cubic (fcc) lattice structure, shows the appearance of annealing twins. TB denotes a coherent twin boundary. Example: $\alpha+\beta$-brass, consisting of $\alpha$-grains (fcc) and $\beta$-grains (bcc). (d) Single-phase material exhibiting a large number of annealing twins; arrow: incoherent TB. Example: $\alpha$-brass (fcc), austenitic stainless steel. (e) Deformed grains with elongated inclusions. Example: ferritic steel with non metallic inclusions after rolling to sheet. Due to rolling the sheet can exhibit a marked texture which may be reflected in anisotropic mechanical properties. (f) Two-phase material, where one of the phases is situated along GBs. Example: pearlitic steel with proeutectoid ferrite on GBs. (g) Polycrystalline material with precipitates. Example: nickel-based superalloy containing $\gamma'$-$Ni_3Al$ precipitates. (h) Two-phase material after eutectoid transformation which represents the outcome of a diffusion-controlled reaction. The grains consist of alternating layers (or lamellae) of the constituting phases. The mechanical properties, e.g., the yield strength, depend primarily on the lamellae spacing. Example: pearlitic steels.

technologically more difficult to realize, however, the strengthening effect is retained at elevated temperatures and for extended service times. This is a direct effect of the inertness of the extremely fine particles, leading to a high resistance against particle growth and re-dissolution effects.

The previous explanation was focused on the various microstructural constituents which range from atomic dimensions to mesoscopic scale. In engineering metallic materials these constituents appear in a great variety of arrangements which in turn determine many of their properties (see Tables 1.1 to 1.3). In Figure 1.3 a schematic drawing of different microstructures is given along with references to structural metallic materials which are widely used. For completeness it should be mentioned that metals and their alloys which have undergone a severe amount of deformation, as in rolling, forging or wire drawing, will develop a preferred orientation or deformation texture, in which certain crystallographic planes within the deformed grains tend to align themselves in a preferred manner with respect to the direction of maximum strain. A recrystallization heat treatment, conducted on a cold-worked metal, can produce a preferred orientation which is different to that existing in the deformed material. This type of texture is termed annealing or recrystallization texture (see Chapters 3 and 20).

As examples for the described microstructures, Figure 1.4 displays images of a pearlitic steel and the nanostructure of a superhard $TiB_2$ coating. The grain size of the pearlitic steel is about 10 μm, whereas the grain size of the $TiB_2$ coating is below 5 nm. Today's advanced engineering metallic materials represent a combination of both features. For example, nickel-based superalloys, some aluminium alloys, and iron-based tool steels possess a "conventional" matrix with regard to grain size. The matrix, however, is hardened and strengthened by nanometer-sized and uniformly dispersed particles that precipitate from a supersaturated solid solution.

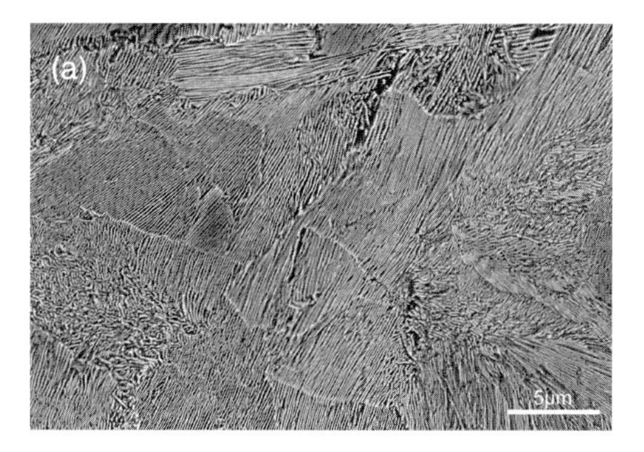

**Figure 1.4** (a) "Conventional microstructure" of a pearlitic steel (scanning electron microscope image) and (b) "advanced nanostructure" of a superhard $TiB_2$ coating (high-resolution transmission electron microscope image) [15]. The grain size of the pearlitic steel is about 10 μm, whereas the grain size of the $TiB_2$ coating is below 5 nm.

## 1.3 Microstructure and Properties

In the previous section it was pointed out that the properties of engineering metallic materials depend on the atomic arrangement, the prevailing crystal defects as well as the arrangement and morphology of the constituting phases/particles (see Figure 1.2 and Tables 1.1 to 1.3). In the following the influence of microstructural parameters on mechanical strength will be discussed. In general, the strength of a metal is controlled by the number and motion of dislocations. The stress required to move dislocations, the Peierls–Nabarro stress, is relatively low in pure metals. Consequently, in order to strengthen metals one must restrict the motion of dislocations by either generating internal stresses that oppose their motion, or by placing particles in their path that require them either to cut or to loop the particles. Figure 1.5 summarizes the basic strengthening mechanisms for metallic materials at low ($T < 0.3T_M$) and high ($T > 0.3T_M$) temperatures. $T_M$ is the melting point (in Kelvin) of the metal or alloy under consideration. Practically, there are four major strengthening mechanisms which will be outlined in the following: (1) work (dislocation density) hardening, (2) strengthening by grain size reduction, (3) solid solution strengthening, and (4) strengthening by particles.

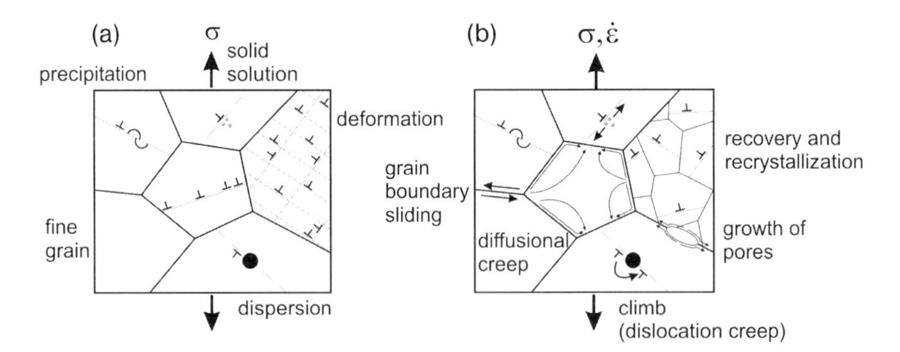


**Figure 1.5** Basic strengthening mechanisms for metallic materials at low ($T < 0.3\ T_M$) and high ($T > 0.3\ T_M$) temperatures. $T_M$ is the melting point in Kelvin. (a) At low temperatures the (yield) strength of a material is controlled by dislocation density (work hardening or strain hardening), grain size (grain boundary strengthening), concentration and size of alloying atoms (solid solution strengthening), and size and volume fraction of particles (precipitation or dispersion strengthening). (b) At high temperatures thermally activated processes and creep determine the occurring strength. For example, a high dislocation density is reduced by recovery and recrystallization. Fine-grained microstructures lead to high diffusion creep rates and pronounced grain boundary sliding. Particles, which are effective barriers to dislocations at low temperatures, are surmounted by climb processes (dislocation creep). Depending on the loading conditions, pores nucleate and grow at grain boundaries leading to micro- and macrostructural damage and consequently to a reduced lifetime.

(1) The work hardening phenomenon can be explained on the basis of dislocation–dislocation strain field interactions. Plastic deformation during cold working produces an increase in the number of dislocations (Table 1.2). As a consequence, the mean distance between individual dislocations decreases. On average, dislocation–dislocation strain interactions are repulsive. The net result is that the motion of a dislocation is hindered by the presence of other dislocations. As the dislocation density increases with increasing degree of deformation, the resistance to dislocation motion becomes more pronounced.

(2) The yield strength of a metal is almost universally observed to increase as the grain size decreases. The experimental data virtually always show a linear relationship between yield strength and the reciprocal value of the square root of the grain diameter. The strengthening effect produced by grain size reduction results from blockage of dislocations by grain boundaries. Therefore, a fine-grained material is stronger than one that is coarse grained, since the former has a greater total grain boundary area to obstruct dislocation motion. Two reasons can be given why grain boundaries act as barriers to dislocation motion during plastic deformation: firstly, grains are of different crystallographic orientations. If a dislocation passes from one grain to another it will have to change its direction of motion. This process becomes more difficult as the misorientation between the grains increases. Secondly, the atomic disorder within a grain boundary region results in a discontinuity of slip planes from one grain into the other. Boundaries between two different phases are also barriers to dislocations. Such a behavior is utilized in the strengthening of complex multi-phase metallic materials.

(3) Solid solution hardening is another effective technique to strengthen and harden metals. When a solute atom (alloying atom) dissolves in a solid metal it may act as an atomic-sized obstacle to dislocation motion. The strengthening effect depends on the nature of the interaction of the dislocation with the solute atoms. Usually, two general interactions are considered, one of a chemical nature and the other of an elastic nature. The difference in chemical bonding between solute atoms and solvent atoms is reflected in the difference in their elastic shear moduli. This difference gives rise to a change in the dislocation–atom interaction. If the solute atom has a different size than the matrix atoms, then a misfit strain field is produced around the solute atom that interacts with the strain field of the dislocations.

(4) Particles may be introduced into the matrix either by precipitation or powder metallurgical approaches (see Section 1.2). These particles will interact with the dislocations causing the dislocations either to cut through the particles or to loop them. It should be noted that particle cutting is restricted to particles which are coherent or at least semi-coherent to the matrix. The degree of strengthening resulting from nanometer-sized particles depends on their distribution in the matrix (see Chapter 13). In addition to the shape, the second-phase dispersion can be described by specifying volume fraction, average particle diameter and mean interparticle spacing.

At elevated temperatures ($T > 0.3T_M$) the microstructure may become thermally unstable (Figure 1.5b) and thermally activated processes such as diffusion and creep start determining the strength of the material. For example, a hardened cold-worked material can lose its strength due to recovery and recrystallization (see Chapter 19). Fine-grained materials which show good strength properties at ambient temperatures are prone to diffusional creep and pronounced grain boundary sliding. Furthermore, particles which are effective barriers to dislocations at low temperatures can be surmounted by diffusion-assisted climb processes (dislocation creep).

As a conclusion, Table 1.4 summarizes the discussed basic strengthening mechanisms and assesses their effect at low and high temperatures. It is worth mentioning here that the mechanical properties might not show the same size dependence when the grain sizes or the materials dimension reach the nanometer regime. This is most likely related to the difficulty to generate and move dislocations in these materials and ongoing research works are addressing this problem. The interested reader might find an introduction to this research field in [24–28].

## 1.4 Microstructural Characterization

As has been outlined in the previous sections the microstructure has major influence on the properties of engineering materials, and the most relevant microstructural features are summarized in Table 1.5. Usually, a combination of different characterization methods has to be applied to obtain the necessary information, and the present section is devoted to this topic. However, it will be only a rough guideline which methods can be applied to assess a specific microstructural feature but it will not be exhaustive. For a detailed description of the operating modes of the different methods the reader is referred to literature [6, 16–23] and to specific chapters of this book.

**Table 1.4** Basic strengthening mechanisms for engineering metallic materials and their assessment.

| | Dislocations | Grain boundaries | Solute atoms | Particles | Transformation | Anisotropy | |
|---|---|---|---|---|---|---|---|
| Strengthening mechanism | Work hardening | Grain boundary hardening | Solid-solution hardening | Precipitation/ dispersion hardening | Transformation hardening, e.g., martensitic transformation | Crystal (structure, texture) | Microstructure (fiber composites; directed grains; duplex micro-structure) |
| Scaling parameters | Dislocation density | Grain size | Concentration of solute atoms | Particle size and volume fraction | Quenching rate, alloy composition | Crystal orientation, intensity of texture | Strength, orientation and volume fraction of fibers; grain orientation; deformation behavior of constituent phases |
| Assessment (after [14]): | | | | | | | |
| Low-temperature strength | ++ | ++ | + | ++ | ++ | + | + |
| High-temperature strength (creep) | ±[a] | – | + | ++ | ±[b] | + | + |

Impact on yield strength:
++ increases strongly
+ increases
– decreases
a) + if dislocations are pinned by stable particles; – if recrystallization takes place during high-temperature application (creep)
b) + if martensitic structure is maintained

**Table 1.5** Information on microstructure needed.

- Grain/subgrain/domain size
- Crystal structure and chemistry of grains and particles
- Preferred grain orientation (texture)
- Three-dimensional arrangement of phases
- Phase transitions (onset, temperatures...)
- Size, shape and volume fraction of particles (precipitates, dispersoids)
- Structure and type of appearing interfaces, segregation to interfaces
- Types of defects and defect density (pores, cracks...)
- Vacancy concentration and dislocation density
- Local/residual stresses
- Microstructural evolution during deformation and/or thermal treatment
- Nucleation and growth processes

The most frequently used characterization techniques for studying the microstructure of engineering materials are light optical microscopy, electron and ion beam microscopy and corresponding analytical measurements as well as X-ray, neutron, and electron diffraction experiments. All of these methods are based on the elastic or inelastic interaction of a probe (visible light, electrons, ions, X-rays, neutrons) with the material under investigation giving rise to a scattered intensity of the initial beam and to the generation of secondary signals (photons, electrons, ions). Each method allows the access of microstructural features on different length scales as indicated in Figure 1.6 and, except for X-ray and neutron diffraction techniques which will be addressed in the remaining book chapters, the others will be shortly described in the following.

Light optical microscopy (LOM) is the common method which is employed to determine the grain size of engineering materials. In addition, the size and distribution of larger inclusions and precipitates can be investigated. However, due to a resolution limit in the order of the wavelength of light (i.e., around 500 nm) it is not suitable to investigate nanocrystalline materials or sub-μm precipitates. In addition, no information on the chemical composition or crystal structure of the individual phases can be obtained.

Scanning electron microscopy (SEM) and focused ion beam microscopy (FIB) enable us to study grain and precipitate sizes as well as their arrangement with a spatial resolution in the order of several ten nanometers. The resolution in the image is thereby mainly governed by the beam size. Often these microscopes are equipped with analytical tools to perform energy-dispersive X-ray spectroscopy (EDX) and wavelength dispersive X-ray spectroscopy (WDX) (in the SEM) or secondary ion mass spectroscopy (SIMS) (in the FIB). These analytical methods can be used to determine the chemical composition of different phases. Since the interaction volume of the incident electron beam within the sample is much larger than the beam size (which can be as small as a few nanometers), semiquantitative EDX or WDX measurements can only be done with a resolution of about 1 μm.

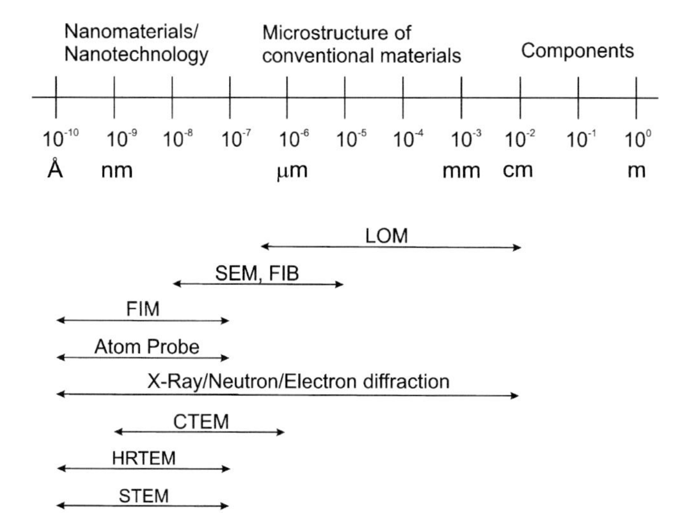


**Figure 1.6** Length scale which is covered in engineering materials, ranging from the atomic/nano scale to the dimensions of large components. Several characterization methods are listed as well as their resolution limits. LOM: light optical microscopy; SEM: scanning electron microscopy; IM: ion microscopy; FIM: field ion microscopy; CTEM: conventional transmission electron microscopy; HRTEM: high resolution transmission electron microscopy; STEM: scanning transmission electron microscopy.

In principle, for SIMS a sub-µm lateral resolution can be achieved; however, in practice this resolution is often not obtained due to insufficient counting statistics. Thus, even if the size distribution of small particles can be determined, the classification of the corresponding particle types (regarding e.g., chemical composition) is not possible and requires the use of an additional characterization method. Modern SEM are often equipped with an electron back scatter diffraction (EBSD) detector which allows to investigate the crystal structure of the occurring phases and their preferred orientation (texture) within the sample surface. The spatial resolution is in the range of 50 nm.

The crystal structure of sub-µm-sized particles and precipitates can be studied by transmission electron microscopy (TEM) using electron diffraction experiments. These studies can also be conducted to determine the orientation relationship between different phases or to show the presence of special grain boundaries such as twin boundaries. With the help of conventional TEM images (bright-field, dark-field, weak-beam) microstructural features such as dislocation densities, antiphase boundaries, grain/subgrain/domain sizes, particle shape, size, and distribution can be addressed. The spatial resolution for conventional TEM investigations is in the order of some nanometers.

Analytical TEM measurements such as EDX and electron energy-loss spectroscopy (EELS) allow to determine the chemical composition of individual phases, particles or at interfaces. The spatial resolution of these methods depends strongly on the beam size, and for modern TEM with a scanning unit (STEM) a resolution of about 1 nm is achieved for EDX and $\geq 0.1$ nm for EELS measurements. The main reason for the differences in resolution is thereby determined by a larger specimen thickness (and thus stronger effect of beam broadening within the sample) for EDX measurements (to obtain a better signal-to-noise ratio in the data) and due to the detection geometries.

The EELS measurements can not only be used to determine the chemical composition of the investigated region, but also to get insight into the electronic structure. This is obtained by analyzing the electron energy-loss near-edge structure (ELNES) which is associated with each element specific ionization edge and which contains information on, e.g., bonding characteristics and nominal oxidation states of the probed atoms. In addition, studying the extended energy-loss fine structure (EXELFS), which occurs around 50 eV above the ionization edge onset, allows to obtain information on the radial distribution function of the atoms. The valence loss region with its characteristic plasmon features at an energy loss of around 15–25 eV can be used to investigate the optical properties of the materials by a subsequent Kramers–Kronig analysis. However, due to the nature of the excitation process these latter measurements can only be done with a spatial resolution of a few nanometers.

High-resolution TEM (HRTEM) and so-called *Z*-contrast images (*Z* stands for the atomic number) using a STEM allow to study the atomic structure of interfaces or the crystal structure of nm-sized precipitates. The HRTEM image formation can be described with the help of Abbe's theory, and the image can be understood as an interference pattern of different diffracted beams. For the imaging a parallel beam is used, and the whole interference pattern is detected simultaneously. In contrast, for a *Z*-contrast image a convergent electron beam is used and scanned over the sample. At each position of the beam, the intensity of electrons scattered in large scattering angles is detected and the image is formed serial point by point. The detected signal is roughly proportional to the square of the atomic number. The *Z*-contrast image can be understood as a convolution of the specimen function (atomic columns) with the electron beam function. The resolution limit of both methods is mainly governed by the spherical aberration ($C_S$) of the electron lenses, i.e., of the objective lens which is most important for the imaging in HRTEM and of the condenser lens which is responsible for the electron beam size in the case for *Z*-contrast imaging in the STEM. Recent developments of $C_S$ correctors allow obtaining, for both methods, a spatial resolution of $\leq 0.1$ nm.

It is important mentioning that all TEM images reveal a two-dimensional projection of the three-dimensional sample. This can cause problems, e.g., if particle distributions are investigated, and thus complementary methods have to be applied. In addition, problems can occur if the sample thickness is too large since then small particles embedded in the matrix might become invisible. Dislocation densities can only be estimated up to $10^{15}$ m$^{-2}$, and other methods have then to be applied. Also the TEM specimen preparation has to be taken into account, and care has to be taken not to change the original structure or at least to minimize possible damaging effects. In addition, only a limited specimen volume is analyzed in TEM, and statistically evaluations of microstructural features are time consuming. Therefore, integral methods which probe the features over a large sample volume and which are nondestructive (regarding the sample preparation) such as X-ray and neutron scattering should be performed in addition.

Another method to image lattice defects such as dislocations and grain boundaries at an atomic level is the field-ion microscope (FIM). A positive voltage is applied to a fine tip of the material of interest which leads to the ionization of an imaging gas (e.g., He, Ne). The ions of the imaging gas are then radially accelerated to a fluorescent screen which is on a negative potential. The image represents the geometry of the atomic arrangement of the terraces of the tip. Particles of a second phase might lead to a different ionization behavior of the imaging gas and thus appear differently. If the applied electrical field is high enough, the atoms of the tip can be ionized themselves and leave the tip in radial directions. With the help of a time-of-flight mass spectrometer the ions can be identified, which is the basic working principle of an atom probe. With suitable detectors, a three-dimensional image of the tip can be obtained. This method is especially suitable to study the initial stages of precipitations or to determine the chemical composition of impurity elements at defects such as dislocations or interfaces. The tip preparation of samples containing defects can be rather time consuming, but using a FIB can help to produce a needle from the area of interest. However, not all materials can withstand the high electrical field and, as for the TEM investigations, the analyzed sample volume is rather small. Again, complementary methods are required to access the microstructural features governing the properties of engineering materials.

Despite the methods described so far, a variety of other imaging characterization techniques exist such as scanning probe microscopes, e.g., scanning tunneling microscope and atomic force microscope [23]. These methods are helpful to get insight into the surface structure of engineering materials down to the atomic level, but information on, e.g., surface stresses on a larger scale are not easy to address. Since material scientists are generally interested to obtain all the information listed in Table 1.5 with a high statistical relevance, diffraction techniques are the right choice for microstructural characterization – if possible always linked to complementary methods such as the ones mentioned in this chapter. The following chapters will provide the basic background in the underlying physics of X-ray and neutron diffraction. In addition, the experimental set-ups used for the measurements are explained and fundamental descriptions of data treatment and analysis are given.

## References


1 W. D. Callister 1997, *Materials Science and Engineering – An Introduction*, Wiley, New York, Weinheim.
2 J. F. Shackelford 2005, *Introduction to Materials Science for Engineers*, Pearson Education, New Jersey.
3 A. Tetelman, C. R. Barrett, W. D. Nix 2005, *The Principles of Engineering Materials*, Prentice-Hall, Englewood Cliffs, NJ.
4 G. Weidmann, P. Lewis, N. Reid 1990, *Structural Materials*, Butterworth, London.
5 G. Gottstein 2001, *Physikalische Grundlagen der Materialkunde*, Springer, Berlin.
6 P. Haasen 1984, *Physikalische Metallkunde*, Springer, Berlin.
7 R. W. Cahn, P. Haasen, E. J. Kramer (eds.) 2005, *Materials Science and Technology*, vols. 2a/2b, 6/7, 15/16, Wiley-VCH, Weinheim.
8 D. A. Porter, K. E. Easterling 2001, *Transformations in Metals and Alloys*, Nelson Thornes, Cheltenham.
9 D. Hull, D. J. Bacon 2001, *Introduction to Dislocations*, Butterworth-Heinemann, Oxford.
10 G. E. Dieter 1988, *Mechanical Metallurgy*, McGraw-Hill, London.
11 R. E. Smallman, R. J. Bishop 1999, *Modern Physical Metallurgy & Materials Engineering*, Butterworth-Heinemann, Oxford.
12 T. H. Courtney 1990, *Mechanical Behavior of Materials*, McGraw-Hill, London.
13 J. D. Verhoeven 1975, *Fundamentals of Physical Metallurgy*, Wiley, New York.
14 E. Hornbogen 1974, in *High-Temperature Materials in Gas Turbines*, eds. P. R. Sahm, M. O. Speidel, Elsevier, Amsterdam, pp. 187–205.
15 P. H. Mayrhofer, C. Mitterer, and H. Clemens 2005, Self-organized nanostructures in hard ceramic coatings, *Adv. Eng. Mater.*, 7, 1071–1082.
16 David Brandon and Wayne D. Kaplan 1999, *Microstructural Characterization of Materials*, Wiley, West Sussex, England.
17 L. Reimer 1998, *Scanning Electron Microscopy*, Springer Series in Optical Sciences, 2nd edn, Springer, Berlin.
18 D. B. Williams and C. B. Carter 1996, *Transmission Electron Microscopy*, vol. 1–4, Plenum, New York.
19 B. Fultz and J. M. Howe 2001, *Transmission Electron Microscopy and Diffractometry of Materials*, Springer, Berlin.
20 L. Reimer 1997, *Transmission Electron Microscopy*, Springer Series in Optical Sciences, 4th edn, Springer, Berlin.
21 M. K. Miller, A. Cerezo, M. G. Hetherington and G. D. W. Smith 1996, *Atom Probe Field Ion Microscopy*, Oxford University Press, Oxford.
22 K. Hono 2002, Nanoscale microstructural analysis of metallic materials by atom probe field ion microscopy, *Prog. Mater. Sci.*, **46** (6), 621–729.
23 C. R. Brundle, C. A. Evans, and S. Wilson 1992, Encyclopedia of Materials Characterization – Surfaces, Interfaces, *Thin Films*, Butterworth-Heinemann, Stoneham.
24 W.D. Nix 1989, Mechanical properties of thin films, *Metall. Trans. A*, **20** (11), 2217–2245.
25 H. Gleiter 2000, Nanostructured materials: basic concepts and microstructure, *Acta Mater.*, **48** (1), 1–29.
26 E. Arzt, G. Dehm, P. Gumbsch, O. Kraft, and D. Weiss 2001, Interface controlled plasticity in metals: dispersion hardening and thin film deformation, *Prog. Mater. Sci.*, **46** (3–4), 283–307.
27 L. B. Freund and S. Suresh 2004, *Thin Film Materials: Stress, Defect Formation and Surface Evolution*, Cambridge University Press, Cambridge.
28 G. Dehm, C. Motz, C. Scheu, H. Clemens, P. Mayrhofer, and C. Mitterer 2006, Mechanical size-effects in miniaturized and bulk materials, *Adv. Eng. Mater.*, **8**, 1033–1045.

# 2
# Internal Stresses in Engineering Materials

*Anke Rita Pyzalla*

## 2.1
## Definition

### 2.1.1
### Stress Tensor, Strain Tensor, and Elasticity Tensor

#### 2.1.1.1 Stress Tensor

In all materials, workpieces, and components, every (infinitesimal) small volume element is subject to forces exerted by its surroundings onto the surface of the volume element. The surface of the volume element can be described by surface areas, mathematically equivalent to planes, whose spatial orientation is given by their normal vector $\vec{n}$

$$\vec{n} = \begin{pmatrix} n_1 \\ n_2 \\ n_3 \end{pmatrix} \tag{2.1}$$

The force $\vec{F}$ acting on the volume element is

$$\vec{F} = \begin{pmatrix} F_1 \\ F_2 \\ F_3 \end{pmatrix} \tag{2.2}$$

The stress tensor $\tilde{\sigma}$ which has the components $\sigma_{ij}$ acting on a cubic volume element results from the forces exerted on each orthogonal section plane $A$ of the volume element:

$$\tilde{\sigma} \cdot \vec{n} = \frac{\vec{F}}{A}, \quad \sigma_{ij} n_j = \frac{F_i}{A} = t_i \tag{2.3}$$

where $\vec{t}$ is called the stress vector (Figure 2.1) [1].

*Neutrons and Synchrotron Radiation in Engineering Materials Science: From Fundamentals to Material and Component Characterization*
Edited by Walter Reimers, Anke Rita Pyzalla, Andreas Schreyer, and Helmut Clemens

ISBN: 978-3-527-31533-8

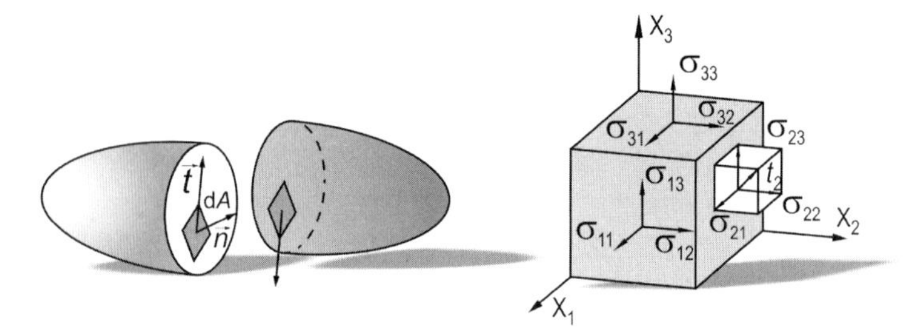


**Figure 2.1** Stress vector and stress tensor [1].

The diagonal components $\sigma_{kk}$ of the stress tensor, which act in the normal direction onto the cubic volume element in Figure 2.1, are often referred to as normal stresses. The stress tensor components $\sigma_{ij}$, $i \neq j$, which act parallel to the surface elements are referred to as shear stresses. The stress tensor $\tilde{\sigma}$ is a symmetric second-order tensor

$$\tilde{\sigma} = \begin{pmatrix} \sigma_{11} & \sigma_{12} & \sigma_{13} \\ \sigma_{12} & \sigma_{22} & \sigma_{23} \\ \sigma_{13} & \sigma_{23} & \sigma_{33} \end{pmatrix} \tag{2.4}$$

The stress tensor can be transposed to any other reference system using the tensor transformation laws. It can, therefore, be diagonalized ($\sigma_{ij} = 0$ if $i \neq j$) by a principal axis tensor transformation. In the principal axis system only normal stresses act in the direction of the principal axes and there are no shear stresses between them.

#### 2.1.1.2 **Strain Tensor**

The external forces exerted on a material, workpiece, or the component result in displacements $\vec{u}(\vec{x})$. The components of the displacement vector $\vec{u}$ in the direction of the coordinates $x_1$, $x_2$, and $x_3$ (for simplicity we assume a Cartesian coordinate system in the following paragraphs) are defined by:

$$\vec{u} = \begin{pmatrix} u_1 \\ u_2 \\ u_3 \end{pmatrix} \tag{2.5}$$

The deformation associated with the displacement is described by the symmetric second-order strain tensor $\tilde{\varepsilon}$:

$$\tilde{\varepsilon} = \begin{pmatrix} \varepsilon_{11} & \varepsilon_{12} & \varepsilon_{13} \\ \varepsilon_{12} & \varepsilon_{22} & \varepsilon_{23} \\ \varepsilon_{13} & \varepsilon_{23} & \varepsilon_{33} \end{pmatrix} \tag{2.6}$$

Within the limit of "small" deformations (this is nearly always the case for metals and other engineering materials but, not for, e.g., rubber-like materials) the relation between deformations and strain tensor components is given by

$$\varepsilon_{ij} = \frac{1}{2}\left(\frac{\partial u_i}{\partial x_j} + \frac{\partial u_j}{\partial x_i}\right) = \frac{1}{2}(u_{i,j} + u_{j,i}) \tag{2.7}$$

In the case of pure elastic deformation the relation between strain and stress tensor is given by Hooke's law:

$$\tilde{\sigma} = \tilde{C}\tilde{\varepsilon} \tag{2.8}$$

The elasticity tensor $\tilde{C}$ is a fourth-order tensor and its components are $C_{ijkl}$. The symmetry of the strain and the stress tensor results in, e.g.,

$$C_{ijkl} = C_{klij} \tag{2.9}$$

$\tilde{C}$ in general has 21 different components. $\tilde{C}$ is also referred to as the stiffness tensor; its inverse tensor $\tilde{S}$ is the compliance tensor.

For cubic crystals the elasticity tensor has only two independent components, $c_{12}$ and $c_{44}$:

$$\tilde{C} = \begin{pmatrix} c_{12}+2c_{44} & c_{12} & c_{12} & 0 & 0 & 0 \\ c_{12} & c_{12}+2c_{44} & c_{12} & 0 & 0 & 0 \\ c_{12} & c_{12} & c_{12}+2c_{44} & 0 & 0 & 0 \\ 0 & 0 & 0 & c_{44} & 0 & 0 \\ 0 & 0 & 0 & 0 & c_{44} & 0 \\ 0 & 0 & 0 & 0 & 0 & c_{44} \end{pmatrix} \tag{2.10}$$

In the case of isotropy, Hooke's law can be simplified to the expressions

$$\sigma_{ij} = \frac{E}{1+\nu}\varepsilon_{ij} - \frac{\nu E}{(1+\nu)(1-2\nu)}\delta_{ij}\varepsilon_{kk} \tag{2.11}$$

$$\varepsilon_{ij} = \frac{1+\nu}{E}\sigma_{ij} - \delta_{ij}\frac{\nu}{E}\sigma_{kk} \tag{2.12}$$

$\delta$ is the Kronecker $\delta$-function ($\delta=1$ for $i=j$, $\delta=0$ for $i \neq j$), $E$ is Young's modulus, and $\nu$ is Poisson's ratio of the material.

## 2.1.2
## Definitions, Residual Stresses

### 2.1.2.1 Stress Equilibrium

Residual stress is the stress, which remains in a body that is stationary and at equilibrium with its surroundings [2]. Residual stresses, thus, are mechanical stresses present in a workpiece or a component, which is not subject to external forces, a momentum or temperature gradient. Residual stresses $\tilde{\sigma}$ are by definition self-equilibrating stresses. At the surface of the material, respectively, the component, the residual stresses must fulfill the equilibrium condition

$$\tilde{\sigma} \cdot \vec{n} = 0, \quad \sigma_{ij} n_j = 0 \tag{2.13}$$

In each point of the material, respectively, the component equilibrium means

$$\frac{\partial \sigma_{ij}}{\partial x_j} = 0 \tag{2.14}$$

In a Cartesian coordinate system, the equilibrium conditions thus are [3]

$$\begin{aligned}
&\frac{\partial \sigma_{11}}{\partial x_1} + \frac{\partial \sigma_{12}}{\partial x_2} + \frac{\partial \sigma_{13}}{\partial x_3} = 0 \\
&\frac{\partial \sigma_{12}}{\partial x_1} + \frac{\partial \sigma_{22}}{\partial x_2} + \frac{\partial \sigma_{23}}{\partial x_3} = 0 \\
&\frac{\partial \sigma_{13}}{\partial x_1} + \frac{\partial \sigma_{23}}{\partial x_2} + \frac{\partial \sigma_{33}}{\partial x_3} = 0
\end{aligned} \tag{2.15}$$

In a cylinder coordinate system, Eq. (2.14) becomes [3]

$$\begin{aligned}
&\frac{\partial \sigma_{rr}}{\partial r} + \frac{1}{r}\frac{\partial \sigma_{r\varphi}}{\partial \varphi} + \frac{\partial \sigma_{rz}}{\partial z} + \frac{1}{r}(\sigma_{rr} - \sigma_{\varphi\varphi}) = 0 \\
&\frac{\partial \sigma_{r\varphi}}{\partial r} + \frac{1}{r}\frac{\partial \sigma_{\varphi\varphi}}{\partial \varphi} + \frac{\partial \sigma_{\varphi z}}{\partial z} + \frac{2}{r}\sigma_{r\varphi} = 0 \\
&\frac{\partial \sigma_{rz}}{\partial r} + \frac{1}{r}\frac{\partial \sigma_{\varphi z}}{\partial \varphi} + \frac{\partial \sigma_{zz}}{\partial z} + \frac{1}{r}\sigma_{rz} = 0
\end{aligned} \tag{2.16}$$

$r$ is the cylinder radius, $\phi$ the azimuth, and $z$ is the coordinate in direction of the cylinder longitudinal axis.

In case of flat parts such as sheets or, e.g., in coatings due to their small expansion in one direction (e.g., the $x_3$-direction) as compared to the other two directions $(x_1, x_2)$, often stresses in the $x_3$-direction can be assumed to be negligible $(\sigma_{13} = \sigma_{23} = \sigma_{33} = 0)$, this is called the "plane stress" condition. In the case of "plane stress" Eq. (2.14) simplifies to [3]

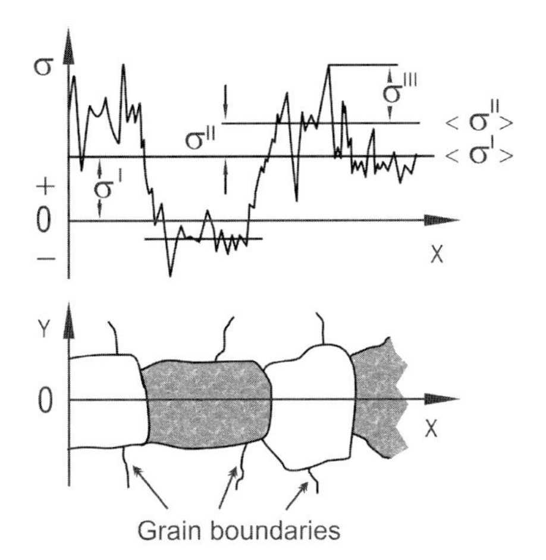


**Figure 2.2** Definition of type I, type II, and type III residual stresses in a single phase material [5].

$$\frac{\partial \sigma_{xx}}{\partial x} + \frac{\partial \sigma_{xy}}{\partial y} = 0 \quad \text{and} \quad \frac{\partial \sigma_{xy}}{\partial x} + \frac{\partial \sigma_{yy}}{\partial y} = 0 \tag{2.17}$$

Hooke's law can be expressed simply by

$$\sigma_{11} = \frac{E}{1 - \nu^2}(\varepsilon_{11} + \nu \varepsilon_{22}) \quad \text{and} \quad \sigma_{22} = \frac{E}{1 - \nu^2}(\varepsilon_{22} + \nu \varepsilon_{11}) \tag{2.18}$$

The stress equilibrium conditions imply that tensile residual stresses in a certain direction within one part of a body are always balanced by matching compressive residual stresses in another part. Thus, the residual stress state of a workpiece or a component can never be expressed by a single residual stress tensor (or even a single residual stress value), but only by a three-dimensional residual stress distribution [4]. This also implies the presence of residual stress gradients, since the residual stress distribution has to be (mathematically) continuous. Strong residual stress gradients are often present in the near-surface area of components, because the residual stress component normal to the surface due to the equilibrium condition (2.13) needs to vanish but stress continuity has to be observed in the bulk material (Figure 2.2).

2.1.2.2 **Residual Macro- and Microstresses**

Residual stresses can be categorized by various schemes, e.g., regarding the manufacturing method producing the residual stresses (e.g., welding residual stresses, shot-peening residual stresses), the method by which they are measured, and their cause (e.g., temperature or deformation inhomogeneities). The most common categorization of residual stresses introduced by [5, 6] when dealing with diffraction methods is based on the length scale. This categorization distinguishes between type I, type II, and type III residual stresses (Figure 2.2). Type I residual stresses are also referred to as residual macrostresses, type II and type III residual stresses are commonly denominated as residual microstresses, e.g., [7].

The residual stress distribution in a material, respectively, a component is the sum of type I, type II, and type III residual stresses:

$$\tilde{\sigma}(\vec{x}) = \tilde{\sigma}^{\mathrm{I}}(\vec{x}) + \tilde{\sigma}^{\mathrm{II}}(\vec{x}) + \tilde{\sigma}^{\mathrm{III}}(\vec{x}), \quad \int_V \tilde{\sigma}(\vec{x})\,\mathrm{d}V = 0 \tag{2.19}$$

*Type I residual stresses* $\tilde{\sigma}^{\mathrm{I}}(\vec{x})$ represent the average residual stresses $\tilde{\sigma}(\vec{x})$ acting within all phases and crystallites in the gauge volume V.

$$\tilde{\sigma}^{\mathrm{I}}(\vec{x}) = \frac{1}{V}\int \tilde{\sigma}(\vec{x})\,\mathrm{d}V \tag{2.20}$$

The gauge volume must be large enough to represent the macroscopic material (e.g., contain a sufficient number of crystallites and all phases present). Releasing $\tilde{\sigma}^{\mathrm{I}}(\vec{x})$ causes macroscopic shape changes of the material, respectively, the component.

*Type II residual stresses* $\tilde{\sigma}^{\mathrm{II}}(\vec{x})$ describe the mean deviation from the macroscopic residual stress level $\tilde{\sigma}^{\mathrm{I}}(\vec{x})$ of an individual crystallite (single phase material)

$$\tilde{\sigma}^{\mathrm{II}}(\vec{x}) = \frac{1}{V}\int [\tilde{\sigma}(\vec{x}) - \tilde{\sigma}^{\mathrm{I}}(\vec{x})]\,\mathrm{d}V \tag{2.21}$$

In a multiphase material type II residual stresses are taken as the average residual stresses $\langle\tilde{\sigma}^{\mathrm{II}}(\vec{x})\rangle^{\alpha}$ of the crystallites belonging to a phase $\alpha$ or as the average residual stresses of those crystallites of the phase $\alpha$ contributing to the measurement:

$$\langle\tilde{\sigma}^{\mathrm{II}}(\vec{x})\rangle^{\alpha} = \frac{1}{V^{\alpha}}\int [\tilde{\sigma}(\vec{x}) - \tilde{\sigma}^{\mathrm{I}}(\vec{x})]\,\mathrm{d}V^{\alpha} \tag{2.22}$$

Release of the type II residual stresses may result in macroscopic distortions.

*Type III residual stresses*, $\tilde{\sigma}^{\mathrm{III}}(\vec{x})$ represent the local deviation of the residual stresses within an individual crystallite from its average residual stress (variation on the atomic scale). Thus, the average type III residual stress $\tilde{\sigma}^{\mathrm{III}}(\vec{x})$ of a crystallite is zero by definition. Release of type III residual stresses does not result in macroscopic distortions.

Since the categorization of type I, type II, and type III residual stresses is based on their length scale and not their magnitude, e.g., type III residual stresses may in some cases be as detrimental to a component's lifetime as type I residual stresses (Figure 2.3) [5].

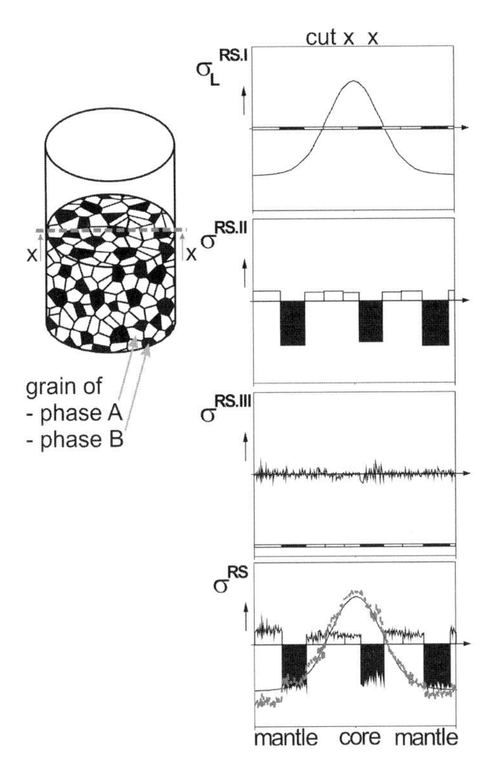


**Figure 2.3** Definition of type I, type II, and type III residual stresses in a multiphase material (see e.g., [5]).

## 2.2
## Origin of Residual Macro- and Microstresses

Residual stresses originate from the misfit between different regions [2]. These misfits result from gradients in deformation or temperature or both deformation and temperature across the component. The most common sources of residual stresses in a component are manufacturing processes, but, residual stresses can also result from temperature or deformation gradients exerted under service conditions.

Type I residual stresses $\tilde{\sigma}^{\mathrm{I}}(\vec{x})$ result from long range strain incompatibilities introduced, e.g., by strain or temperature gradients in a manufacturing process. These strain inhomogeneities arise, e.g., due to friction between the semi-finished component and the die for instance in cold rolling or extrusion [8]. In welding processes, temperature gradients across the welds cause misfits in thermal expansion and strength and, thus, residual macrostresses, e.g., [9].

While the distribution and magnitude of type I residual stresses often can be controlled by modifying the process parameters of a manufacturing process, this

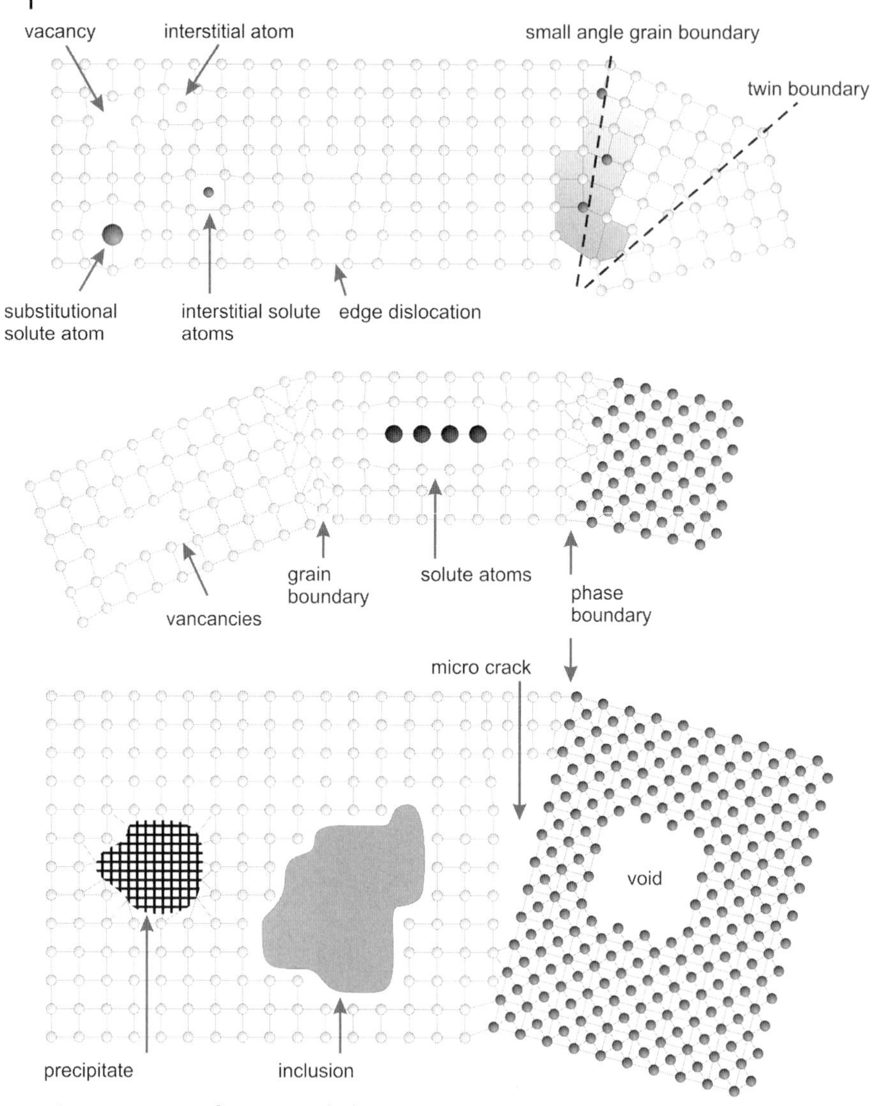


**Figure 2.4** Origin of type III residual stresses [5].

in general is not possible for type II and type III residual stresses. Type II residual stresses arise for instance due to deformation misfits between neighboring grains and due to temperature or deformation induced misfits between the phases of a multiphase material. Type III residual stresses are caused, e.g., by voids, solute atoms, or dislocations in the crystal lattice (Figure 2.4) [5].

For about two decades research efforts have revealed the elementary processes of residual stress formation in manufacturing processes, some of the results will be outlined in the next sections.

### 2.2.1
### Residual Stress Formation in Primary Forming Processes

Primary forming by processes, such as casting, sintering, and isostatic pressing, usually occurs at elevated temperatures, the ingot or semifinished product then is cooled down to room temperature. In large cast or sintered parts, temperature and density gradients occur, which can lead to gradients in the thermal expansion and, thus, to residual strains and stresses. The formation of residual stresses during primary forming by hot isostatic pressing of a two-phase material is a well-defined problem that can be understood relatively easily and, thus, will be used as an example here:

"When a child squeezes snow to make a snowball, he is using the technique of hot-isostatic pressing (inelegantly, but universally known as "HIP"ing)" [10] is a very descriptive explanation of the HIP process. HIP is used to produce precisely shaped – often near net-shape – bodies from metal and/or ceramic powders. The powder is packed into a sheet metal pre-form, which is then evacuated and inserted into a pressure vessel. Within the pressure vessel the preform is heated and simultaneously subjected to a high pressure, often by argon gas. Depending on the process parameters, several mechanisms contribute to the densification of the powders, they involve diffusional redistribution of matter, plastic flow and/or power law creep, e.g., [10].

HIP is one of the predominant techniques for producing particle reinforced wear resistant metal matrix composites (PMMCs). These PMMCs consist of a metal matrix with up to 30% coarse (several μm diameter) ceramic hard phase particles. During cooling from the HIP temperature (about 1100 °C) residual stresses emerge due to differences in the thermal expansion of the metal matrix and the ceramic hard phase particles (Figure 2.5).

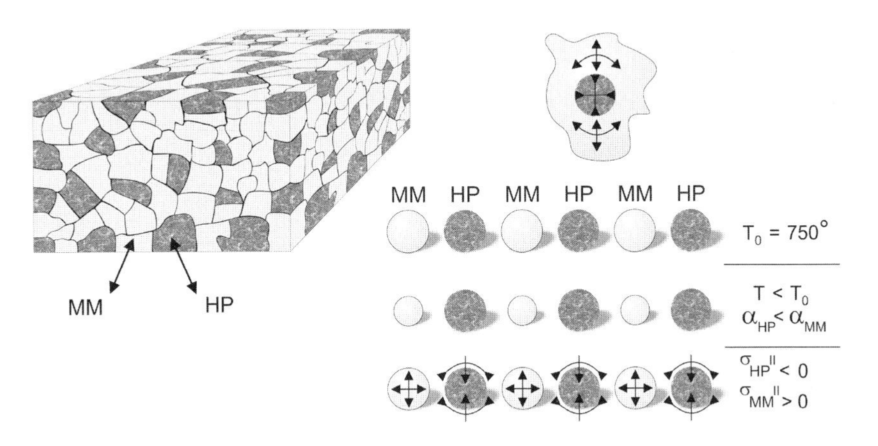


**Figure 2.5** Residual stress formation during cooling of a composite. The dark particles could be, e.g., ceramic hard particles (HP), which have a lower thermal expansion coefficient than the metal matrix (MM, "white" grains).

A very simple model, using a parallel or a series connection between metal matrix and ceramic particles after cooling yields compressive residual microstresses in the ceramic particles (due to their lower thermal expansion), while tensile residual microstresses develop in the metal matrix. A more detailed analysis of the residual stresses in the metal matrix using a model of a spherical ceramic particle embedded in an infinite metal matrix reveals tensile residual microstresses parallel, but, compressive residual microstresses perpendicular to the interface to the ceramic particle.

### 2.2.2 Residual Stress Formation in Heat Treatment Processes

Heat treatment processes usually lead to residual stress formation during cooling the component to room temperature because of temperature gradients and/or volume changes induced by phase transformations. Among the technically most important heat treatment processes are hardening and annealing. The following sections therefore will concentrate on residual stress formation during these processes, starting with the less complex residual stress formation during annealing and simplifying the complex geometry of technical components, e.g., camshafts or crankshafts by a cylinder model (Figure 2.6) [11].

#### 2.2.2.1 Residual Stresses in a Material without Phase Transformation (Pure Cooling Residual Stresses)

When immersing a cylinder lengthwise into a cooling medium, e.g., water, oil or a salt bath, during the first seconds of cooling the temperature in the mantle decreases faster than in the core of the cylinder. Thus, volume shrinkage (due to thermal expansion) is stronger in the mantle than in the cylinder core. Because the mantle aims at shrinking more than the core permits, tensile stresses evolve in the mantle which are balanced by compressive stresses in the core. The differences in the volume changes and, thus, the stresses increase with increasing temperature difference of mantle and core (until point 1 in Figure 2.6 is reached).

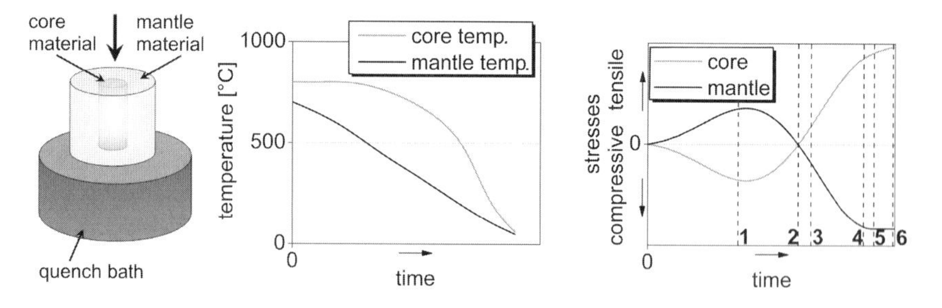


**Figure 2.6** Temperature and stress development of a material without phase transformation during water cooling [14].

After reaching the maximum temperature difference between mantle and core, where tensile stresses exist in the mantle, respectively, compressive stresses in the core, further cooling results in a decrease of the temperature difference in the mantle and core. The macroscopic stresses of the cylinder therefore decrease. Once the temperature decrease in the core becomes faster than in the cylinder mantle (point 2 in Figure 2.6), the material in the core aims at shrinking faster than the mantle. Thus, tensile stresses evolve in the core and compressive stresses develop in the cylinder mantle. The magnitude of the stresses increases with decreasing temperature. The residual stress level after cooling to room temperature is limited by the strength of the material; once the stresses reach the yield point (point 3 in Figure 2.6), further stress increase is limited by hardening of the material and the increase of yield strength with decreasing temperature (points 4 and 5 in Figure 2.6).

The stresses evolving due to temperature gradients between mantle and core during cooling the cylinder are macroscopic stresses. If the stresses formed exceed the yield stress of the cylinder and inhomogeneous plastic deformations occur, residual stresses remain after cooling. The distribution and magnitude of these residual stresses depend strongly on the cooling process (e.g., medium, immersion speed), the geometry of the cylinder (diameter, length-to-diameter ratio) and (the temperature dependent!) material properties (e.g., thermal diffusivity, thermal expansion coefficient, yield strength) (Figure 2.6).

#### 2.2.2.2 **Residual Stresses in a Material with Phase Transformation**

In the particular case of steels residual stress formation processes have to take volume changes due to phase transformations into account. These volume changes depend on the phases formed and the carbon content of the steel (Table 2.1) [12].

The transformation of pearlite into austenite during heating decreases the volume, during cooling the transformation of austenite into martensite, bainite, and pearlite increases the volume. The strains induced by the volume increase during martensitic phase transformation, in most cases, are larger than the thermal strains generated during cooling a component. Thus, phase transformations significantly alter the residual stress state after thermal treatments.

**Table 2.1** Phase transformation induced volume changes during heat treatment processes of steels, e.g., phase transformation from austenite into martensite of a steel C45 with 0.45%C, results in a volume increase of $(4.64 - 0.53 * 0.45) = 4.4\%$.

| Phase transformation | Volume change (%C) |
|---|---|
| Pearlite → austenite | $-4.64 + 2.21$ |
| Austenite → martensite | $4.64 - 0.53$ |
| Austenite → lower bainite | $4.64 - 1.43$ |
| Austenite → upper bainite or pearlite | $4.64 - 2.21$ |

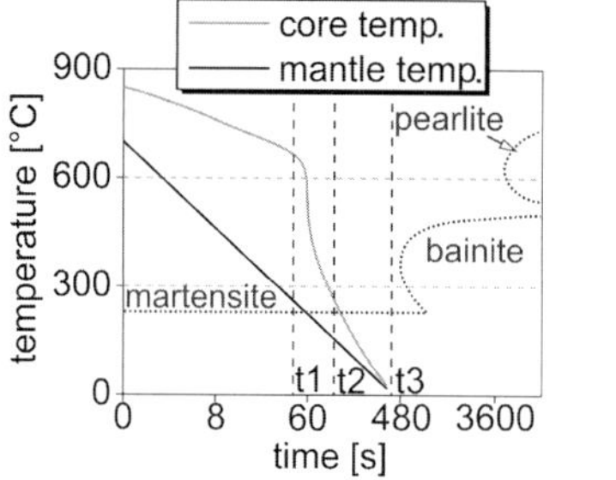


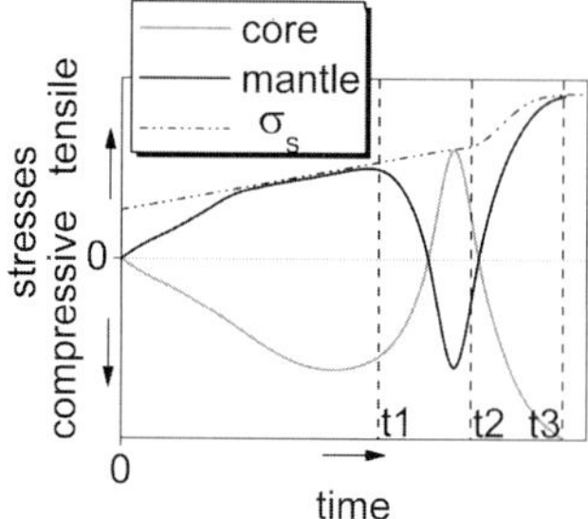


**Figure 2.7** Residual stresses in mantle and core of a through-hardening ∅100 mm steel cylinder after water cooling [13, 14].

In the following a long axle-like component is again simplified by a cylinder model, first a through-hardened cylinder, then a surface hardened cylinder will be considered:

*Through-hardening component:* Hardening of the steel cylinder necessitates rapid cooling of high-temperature austenite. During cooling the cylinder mantle cools faster than the core, thus, as long as both the mantle and the core remain austenitic, tensile stresses develop in the mantle, which are balanced by compressive stresses in the cylinder core (Figure 2.7) [13, 14]. At a time $t_1$ the mantle reaches the martensite start temperature $M_S$ and the austenite → martensite transformation starts. Due to martensite formation, the volume of the cylinder mantle increases and, consequently, the tensile stresses in the mantle decrease. As soon as the difference of strains due to temperature differences are outweighed by the volume increase in the mantle, compressive stresses develop in the mantle and tensile stresses form in the core of the cylinder. At time $t_2$ the temperature in the cylinder core reaches $M_S$ and martensite formation leads to a volume increase within the cylinder core. Thus, the tensile stresses in the core and the balancing compressive stresses in the mantle decrease. Since at this time due to the low temperature, the mantle is rigid (yield strength increases with decreasing temperature), expansion of the core is limited, leading to compressive stresses in the core and tensile stresses in the cylinder mantle. In through-hardening components, thus, the residual stress state induced by martensitic hardening usually is the contrary of the residual stress state resulting solely from temperature gradients in the component.

#### Partially Hardened Component

In the case of a nonthrough hardening component stress formation during cooling starts again with the development of tensile stresses in the mantle and compressive stresses in the cylinder core (Figure 2.8) [13, 14]. Reaching the time $t_1$ austenite → pearlite transformation starts within the cylinder core. The compres-

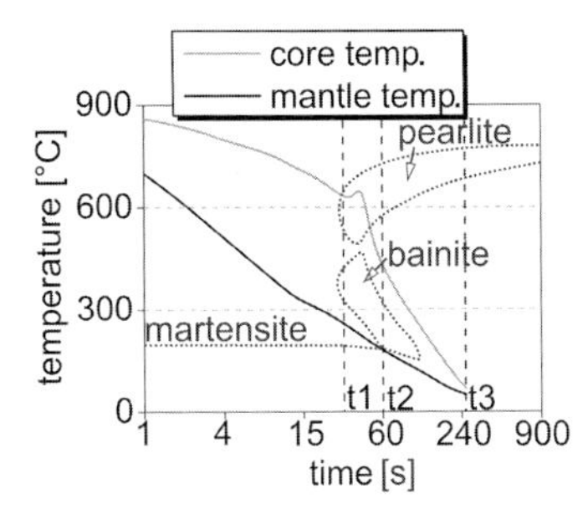


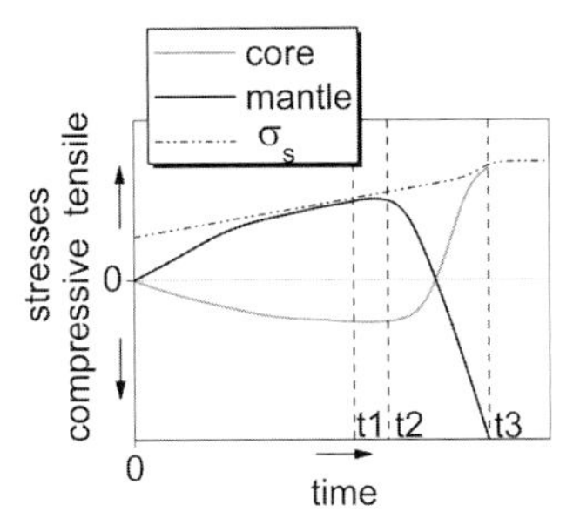


**Figure 2.8** Residual stresses in mantle and core of a surface hardening ∅100 mm steel cylinder after water cooling [13, 14].

sive strain induced by the volume increase caused by phase transformation and the compressive stresses due to cooling add to each other. As soon as the time $t_2$ is reached, martensite formation of the cylinder mantle starts and the volume of the cylinder mantle increases. As a consequence of the volume increase, the tensile stresses in the mantle and the compressive stresses in the cylinder core decrease and after cooling ($t_3$) the cylinder mantle is under compressive residual stresses, the core contains the balancing tensile residual stresses. Thus, in surface-hardening components the residual stress state after cooling is similar to the residual stress state of a component experiencing only thermal stresses and opposite to the residual stress state of a through-hardened component.

The residual stress state of components, which are neither through-hardened nor only hardened within a thin surface layer, depends on a number of factors, such as the component dimensions, the phase transformation behavior of the material, the (temperature dependent) mechanical strength of the material, its martensite start temperature, and the cooling medium.

#### 2.2.2.3 Residual Stress Formation in Surface Hardening Processes (Nitriding, Carbo-Nitriding, and Case Hardening)

In order to increase strength and wear resistance in the near-surface region, steel parts can be subjected to nitriding, carbo-nitriding and case hardening treatments. In principle the elementary processes of residual stress formation in case of hardened components are similar to those discussed for the surface-hardened components. However, due to its different chemical composition the near-surface region must be regarded as a material different from that of the core. In the case of hardening, the carbon content of the cylinder mantle is higher as compared to the core, thus, the phase transformation behavior of mantle and core differ significantly (Figure 2.9) [14, 15]. The $M_s$ temperature of the mantle of case-hardened components is, therefore, substantially lower than in the core. During cooling of case hardening workpieces martensite transformation and volume increase in the cylinder mantle, thus, occurs usually later than the phase transformation in

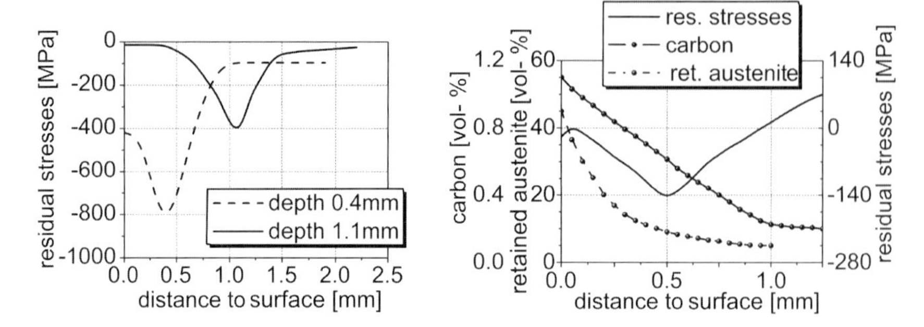


**Figure 2.9** Relation between carbon, retained austenite volume fraction, and residual stresses in a case hardened component (distance to surface) [14, 15].

the core, resulting in compressive residual stresses in the cylinder mantle and tensile residual stresses in the cylinder core. Besides decreasing the $M_s$ temperature the high carbon content in the cylinder mantle implies that usually $M_f$ is not reached, thus, the cylinder mantle besides martensite also contains retained austenite. The retained austenite volume fraction due to the gradient in carbon content is highest and, thus, the martensite volume fraction is lowest at the cylinder surface and decreases toward the interface between the case hardened zone and the parent material. Because of this gradient in phase composition maximum compressive residual stresses are present in case hardened components at the interface between the hardened zone and parent material.

Residual stresses formed during nitriding are in most cases comparatively small compressive residual stresses in the near-surface zone balanced by tensile residual stresses in the bulk [14, 16, 17], due to nitriding temperatures being lower than case hardening ones. Besides temperature gradients also the volume increase due to nitrogen diffusion and nitride precipitation in the near-surface zone contributes to the compressive stresses. Because the steels are hardened and annealed previous to nitriding, phase transformations usually do not occur [18].

### 2.2.3
### Residual Stress Formation in Forming Processes

Forming processes necessitate plastic deformation of a workpiece or semifinished product. Due to the unavoidable inhomogeneity of plastic deformation, forming processes create residual stresses. In addition to residual macrostresses in forming processes with strong plastic deformation, residual microstresses of the same magnitude also evolve due to locally very high defect densities or due to texture formation. In order to introduce the processes of residual stress formation in forming processes, in the following section, deep-rolling and cold extrusion are taken as examples.

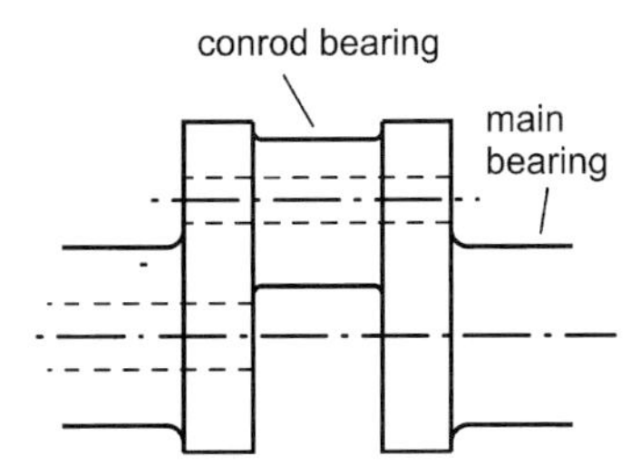


**Figure 2.10** Crankshaft.

### 2.2.3.1 Deep-Rolling Residual Stresses

Deep-rolling is a well-established industrial process for deformation hardening of axial symmetric components such as axles, bolts, and threads. Deep-rolling substantially improves the fatigue strength of dynamically loaded parts [19–22], e.g., by deep-rolling of the shoulder transient between jowl and bearing (Figure 2.10) the bending fatigue strength of crankshafts could be improved by up to 200% (Figure 2.10) [23].

During deep-rolling, the rolls exert Hertzian pressure onto the surface of the workpiece. The Hertzian pressure and friction forces lead to a triaxial stress state in the near-surface area (Figure 2.11) [24]. In those regions of the near-surface zone, where the equivalent stress exceed the yield stress, the material deforms plastically. In deep-rolling processes the near-surface zone is usually stretched during the process. If the yield strength is exceeded compressive residual stresses remain. Thus, after loading compressive residual stresses remain in the near-surface zone, which are balanced by tensile residual stresses in the bulk material.

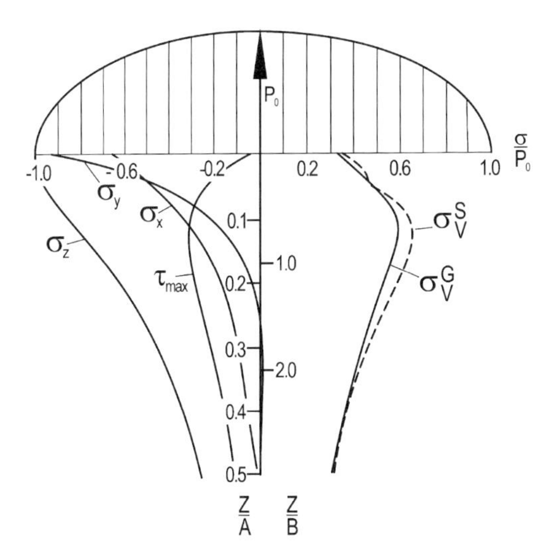


**Figure 2.11** Stresses due to Hertzian pressure (see e.g., [24]).

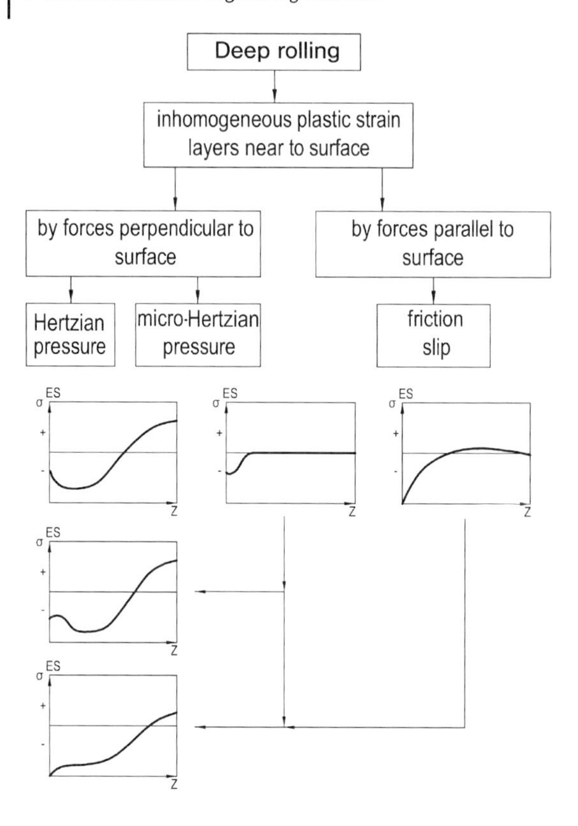


**Figure 2.12** Residual stresses introduced by deep-rolling (adapted from [24]).

While deformation due to tangential forces, e.g., due to friction between the tool and the workpiece, leads to compressive stress maxima at the surface of the workpiece, the maximum stress and deformation caused by Hertzian pressure is not directly at, but, below the surface (Figure 2.12).

The depth of the residual stress maximum after deep-rolling, thus, depends on the process parameters, material behavior, and geometry of tool and workpiece (Figure 2.12): for instance an increase in rolling force increases the distance between the depth of the residual stress maximum and the surface of the workpiece [24]. Besides the depth and the magnitude of the residual stresses also surface roughness and hardness need to be taken into account as criteria for choosing an optimum rolling force. An excess in rolling force results in an increase of surface roughness and, thus, a decrease of fatigue strength. The hardness distribution in the near-surface zone correlates with the residual stress distribution because of increasing lattice defect formation with increasing plastic deformation (Figure 2.13).

For an example, for residual stress measurement in deep-rolled components and an optimization of the process parameters, see Chapter 24.

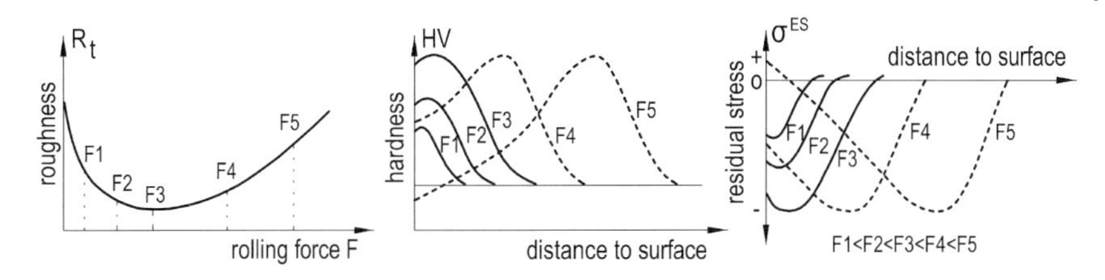


**Figure 2.13** Relation between rolling force and surface roughness, hardness, and residual stresses in deep-rolling (adapted from [24]).

#### 2.2.3.2 Cold Extrusion Residual Stresses

Cold forward extrusion (Figure 2.14) is industrially used for manufacturing bolts, which will be further processed resulting, e.g., into screws or thread rods (Figure 2.14).

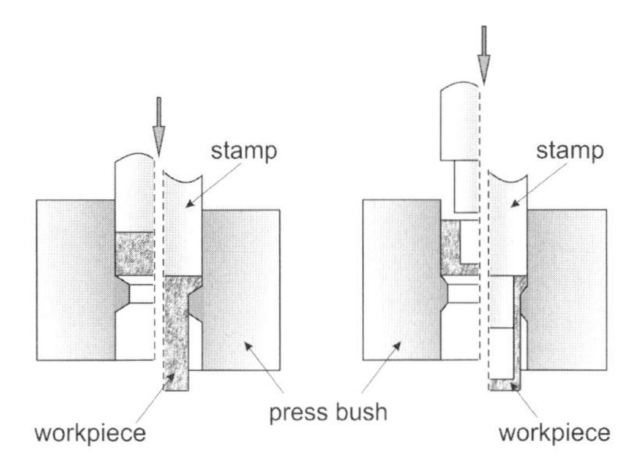


**Figure 2.14** Scheme of a forward extrusion process of a full and a hollow workpiece.

Both in forward extrusion processes of full (Figure 2.14) and hollow workpieces, the workpiece is heavily plastically deformed. Due to deformation obstruction at the shoulder of the die, the material flow at the outer surface is slower and more inhomogeneous than in the bulk of the workpiece. Therefore, within the bulk of the workpiece the grains are homogeneously stretched, whereas at the outer surface of the samples the grains are first compressed and later stretched, while passing the transient radius of the die [25–27]. Thus, tensile residual stresses remain in near-surface zone, while the bulk of the workpiece is under compressive residual stresses (Figure 2.15) [27].

The heavy plastic deformation of the workpiece during the cold extrusion process leads to fiber texture formation. The changes of grain orientation in fiber texture evolution result in strain incompatibilities between neighboring crystallites. As a consequence the residual stress state of a crystallite depends on its crystallographic orientation (Figure 2.16). These residual microstresses between

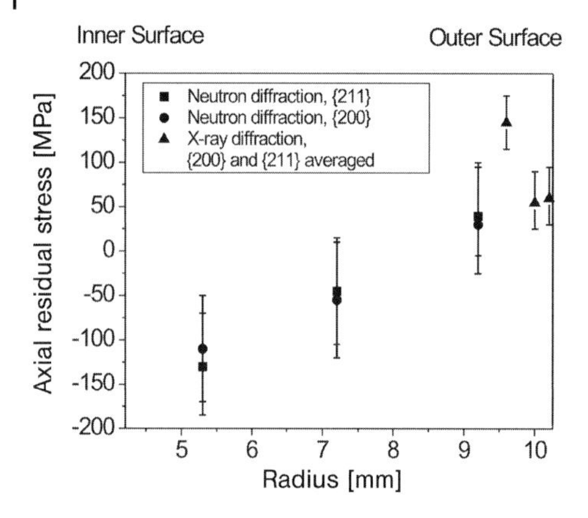


**Figure 2.15** Axial residual macrostress in hollow forward extruded steel samples [27].

different oriented crystallites usually are called intergranular stresses. The intergranular microstresses in different crystallite groups (crystallites of similar orientation) depend on the amount of plastic deformation. Since these intergranular residual stresses can reach the magnitude of the residual macrostress, the formation of the intergranular stresses during plastic deformation has been extensively investigated for uniaxial [28, 29], and multiaxial deformation (Figure 2.16) [27, 30].

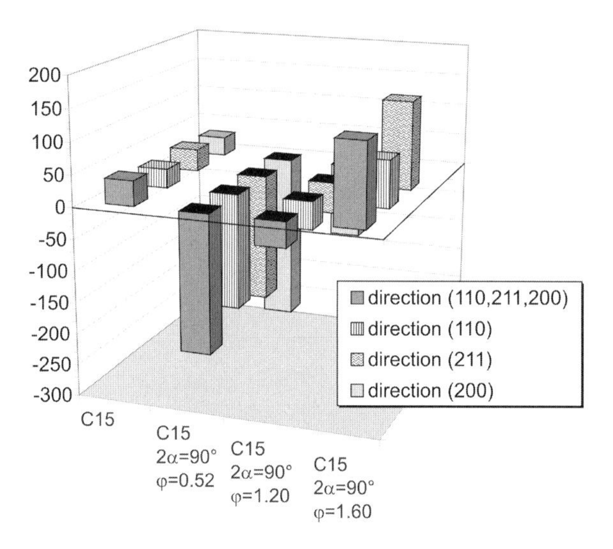


**Figure 2.16** Dependence of the difference of the principal residual stresses in axial and radial direction on the orientation of the crystallite group [27].

## 2.2.4
## Residual Stress Formation in Metal Cutting Manufacturing

In chip-forming metal cutting processes, such as drilling, turning, machining, and grinding, the dimensions and the surface topography of the workpiece and, therefore, its residual macrostress state are altered. The main mechanisms of residual stress formation in metal cutting processes depend on the material, its heat treatment (e.g., hardened or annealed steel), the machining process, and the process parameters. During chip formation the near-surface zone of the workpiece experiences elastic and plastic deformations, thermal loading, and may undergo phase transformations [31]. Elastic and plastic deformations are caused by the mechanical forces the die exerts on the workpiece. Thermal loading is the result of plastic deformations and of friction, e.g., between the die and the workpiece. Because of the decrease of yield strength with temperature, additional plastic deformation can result from heat production. Phase magnitude of transformations may occur during heating as well as during rapid cooling because of the heat flow from the rather thin near-surface zone into the cold surrounding workpiece (Figure 2.17) [31].

Plastic deformation introduced by squeezing the material in the near-surface zone will extend the surface area, thus, introducing compressive residual stresses in the near-surface zone and tensile residual stresses in the bulk material. In contrast thermal loading results in a hot near-surface zone compared to the bulk, thus, in materials without phase transformations tensile residual stresses form in the near-surface zone, which are balanced by compressive residual stresses in

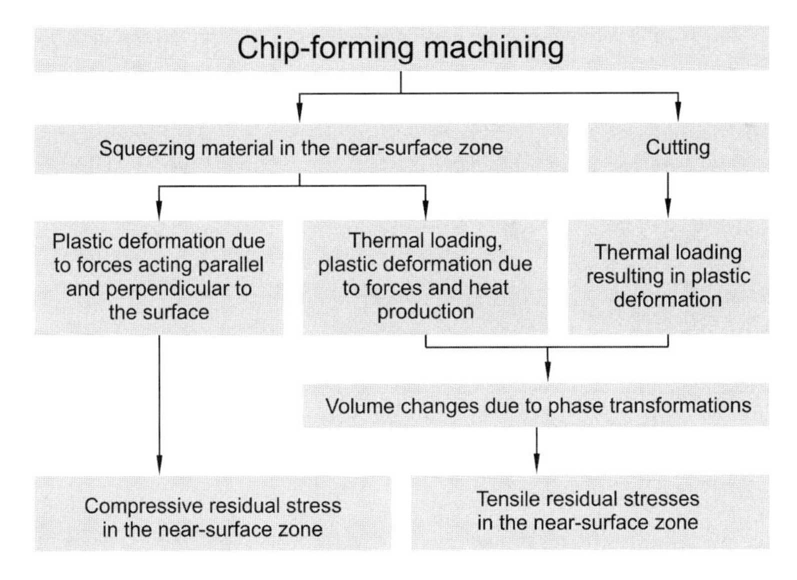


**Figure 2.17** Mechanisms of residual stress formation in chip forming processes (adapted from [31]).

the bulk. The resulting residual stress state after machining is, therefore, determined by the superposition of the residual stresses introduced by deformation and by temperature. In the near-surface region predominant deformation causes compressive residual stresses, whereas a predominant thermal loading result in tensile residual stresses (Figure 2.17). In materials undergoing phase transformations, e.g., in steels, also strains induced by the phase transformation need to be considered. Since only the surface zone of the workpiece hardens, in analogy to surface hardening processes (see Section 2.2.2), compressive residual stresses will form in the near-surface region due to the volume increase caused by martensitic transformation.

#### 2.2.4.1 Grinding Residual Stresses

Grinding often is the last step in a manufacturing process of a component. Surface integrity after grinding, thus, has a strong influence on component behavior under service conditions, e.g., low cycle (LCF) and high cycle (HCF) fatigue behavior, e.g., [24].

Abrasive particles on the surface of a grinding wheel perform the cutting process. Depending on the workpiece material (e.g., steel or ceramics), the grinding wheel (e.g., type of abrasive and binder, dressing), and the parameters of the grinding process (e.g., process type, depth of cut, feed rate, cooling medium) strong plastic deformations and high temperatures may occur [24]. Because the surface zone influenced by the grinding process usually is rather thin, the residual stress state after grinding is a plane stress state (see Section 2.1.2); however, particularly in the case of ground ceramics and ground hardened steels also triaxial residual stress states and steep residual stress gradients have been observed, e.g., [32–34] (Figure 2.18).

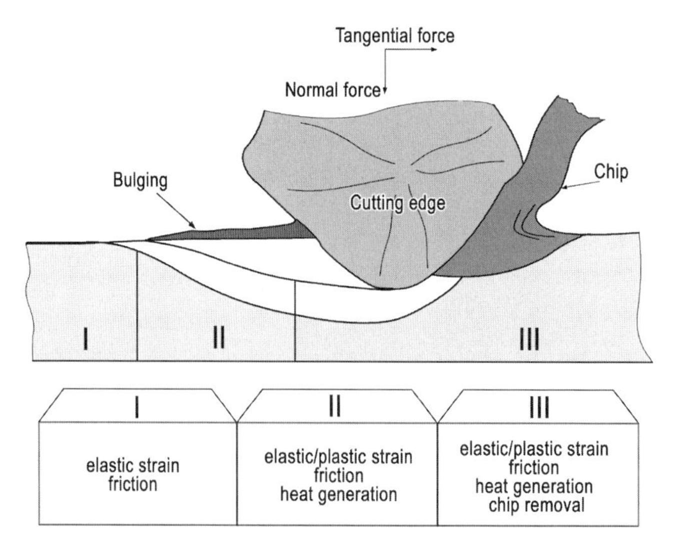


**Figure 2.18** Elementary processes in chip formation in grinding processes [24].

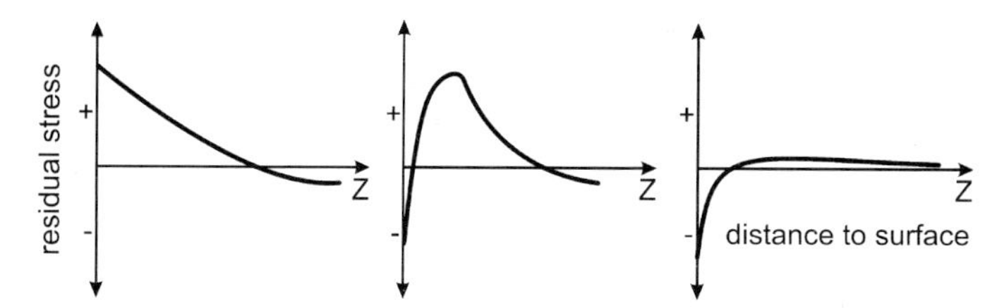


**Figure 2.19** Typical residual stress distributions in the near-surface zone of ground materials.

A detailed view on chip formation in grinding shows that first the cutting edge of an abrasive grain touches the workpiece at a low incidence angle, producing elastic deformations and heat because of friction between the abrasive grain and the workpiece (Figure 2.18). Increasing forces result in plastic deformation and temperature increase due to increase of friction. As a result of temperature increase, the yield strength of the workpiece material decreases and plastic deformation becomes easier. As a consequence of changes in workpiece surface topography, e.g., local bulges form at the workpiece surface. Local shear band formation starts, contributing strongly to heat formation and preparing chip cutting. The cutting process ends by elastic or elastoplastic deformation, while the abrasive looses contact with the workpiece [24]. The macroscopic interaction between grinding wheel and workpiece occurs in a very similar way [24].

The residual stress state after grinding depends on whether plastic deformation or temperature effects prevail (Figure 2.19a, b). Predominant temperature effects result in tensile residual stresses at the workpiece surface [24, 35]. Inhomogeneous plastic deformations and negligible temperature effects lead to compressive residual stresses at the surface and tensile residual stresses beneath

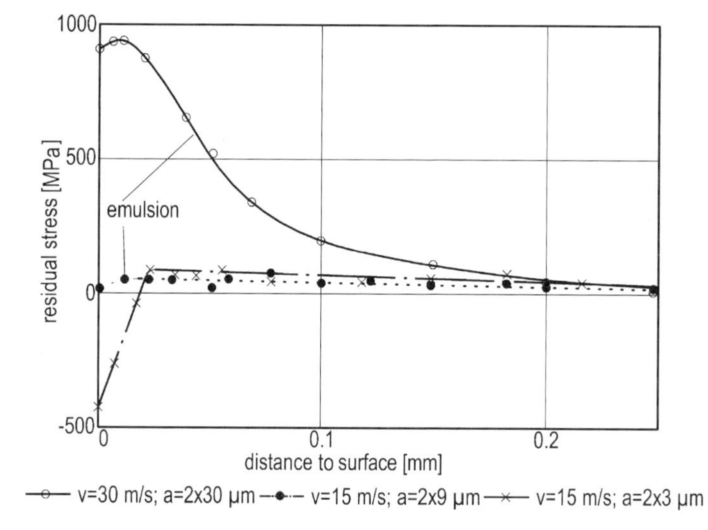


**Figure 2.20** Dependence of near-surface residual stresses on grinding conditions (adapted from [24]).

the surface. A combination of both temperature effects and inhomogeneous plastic deformation often results in compressive residual stresses at the surface and a wide near-surface zone containing lower tensile residual stresses (Figure 2.19c).

An assessment of the influence of the grinding process on the workpiece residual stress state revealed that the material removal rate (workpiece speed × depth of cut) does not relate to the residual stress state of the material in a unique way. Instead, individual process parameters of the grinding process need to be taken into account. For instance, increasing workpiece speed for instance diminishes local heating and, thus, favors compressive residual stresses at the surface (Figure 2.20). Increasing the depth of cut on the other hand increases heat input and, therefore, induces surface tensile residual stresses. The selection of grinding fluids is of particular importance both with respect to their cooling and lubrication capacity, details are given, e.g., [31, 36].

### 2.2.5
### Residual Stress Formation in Joining Processes

Joining by substance-to-substance-bonding includes the process groups welding, brazing, soldering, and gluing. Differentiation between welding and brazing, respectively, soldering is usually based on the melting of the parent material during the welding process, while only the filler melts in brazing and soldering processes. Due to the high industrial and commercial relevance of welding, the number of welding processes has been steeply increasing for the last hundred years. In the following section residual stress formation in two fusion welding techniques using plasma or electric arcs is considered in detail.

In fusion welding residual stresses are caused by inhomogeneous temperature fields. In the weld seam temperatures will be significantly higher than in the parent material, thus, the weld seam during cooling shrinks, but, is hindered by the surrounding colder parent material (Figure 2.21) [37, 38].

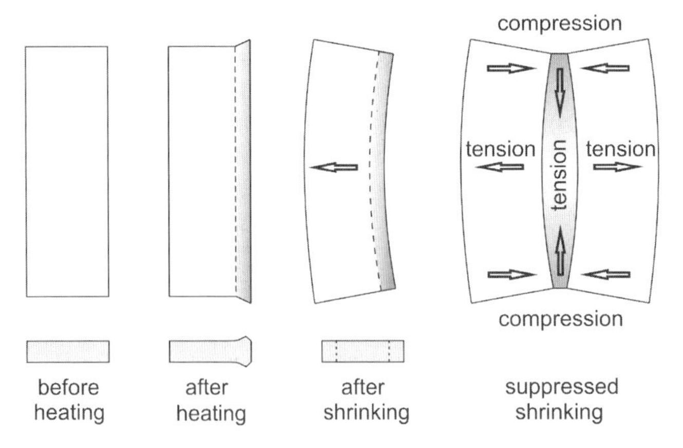


**Figure 2.21** Residual stress formation in thin sheet fusion welds [37, 38].

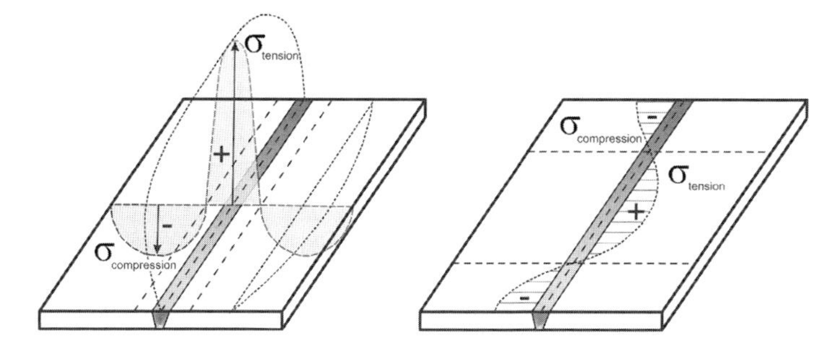


**Figure 2.22** Residual stresses in thin sheet welds of a material without phase transformation.

Residual stress distribution in welds of materials without phase transformation are, therefore, characterized in the longitudinal joint direction (i.e., along the weld line) by tensile residual stresses in the weld seam and heat affected zone (HAZ) of the parent material (next to the weld seam) and balancing compressive residual stresses in the parent material (Figure 2.22) [38].

In case of thin sheets and single pass fusion welding, the welding residual stress state often is a plane stress state, whereas in case of thick sheets or profiles the welding residual stress state usually is triaxial. With respect to the weld behavior under load, e.g., fatigue loading, the biaxial, respectively, triaxial tensile residual stresses in the center of the weld are critical (Figure 2.22).

In the case of fusion welding of steel with austenite → martensite (or bainite) phase transformation, martensite forms in the weld seam and the HAZ. As a

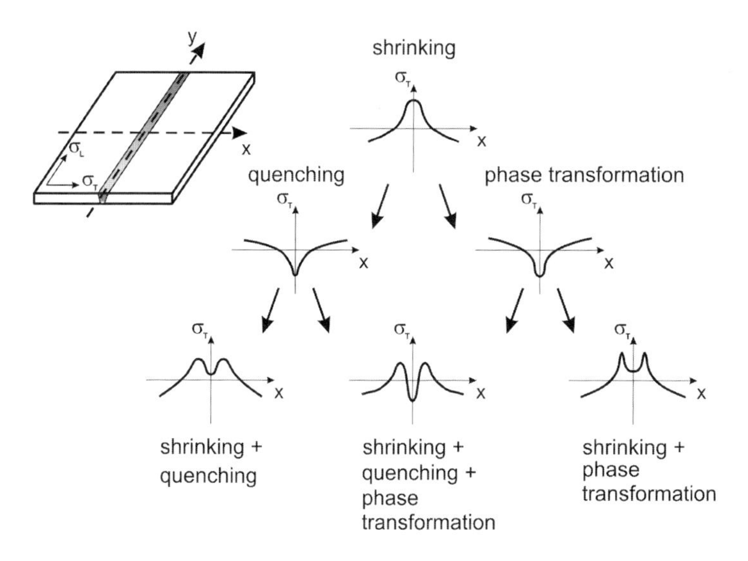


**Figure 2.23** Residual stress formation in a thin sheet weld of a steel with phase transformation.

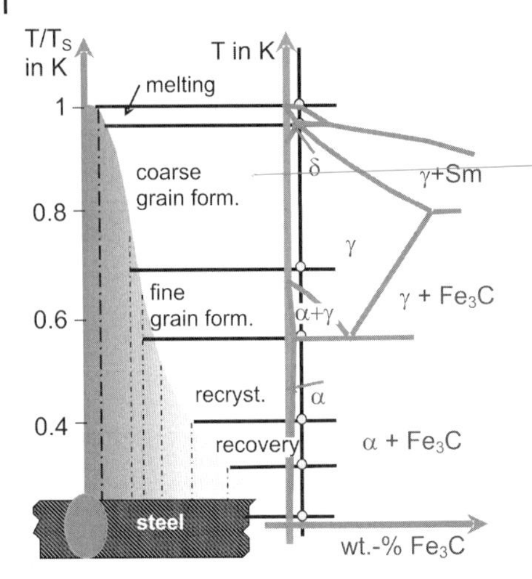


**Figure 2.24** Relation between temperature distribution and physical processes in a steel weld with phase transformation.

consequence, the volume in the weld seam and the HAZ increases and usually at room temperature compressive residual stresses are present in the weld seam (Figure 2.23) and the HAZ (except for materials with very high $M_s$ temperature). These compressive residual stresses will be higher, the lower the phase transformation temperature of the material is, because the yield strength increases with increasing bainite/martensite volume fraction and decreasing temperature [39].

Besides martensitic phase transformation other physical processes, such as recovery and recrystallization processes in case of cold rolled sheets or precipitation aging in the HAZ, affect the residual stress state and the mechanical properties of the weld (Figure 2.24).

The distribution and the magnitude of the welding residual stresses in both materials with and without phase transformation depend on the material (e.g., temperature-dependent thermal expansion coefficient and mechanical properties) and the process parameters (e.g., heat input, heat source, welding speed, weld geometry, environment temperature) of the welding process (Figure 2.25) [39].

The restraint within the weld can be described by the ratio between the extension of the hot zone and the cold parent material and (in case of materials with phase transformation) by the ratio between the transforming and the nontransforming zone [39]. Thus, comparing the residual stress distribution of a weld produced by conventional gas welding and a weld of similar geometry produced by tungsten inert gas (TIG) welding, residual stresses of higher magnitude are expected in the TIG weld due to the higher and more focused heat input during TIG welding.

A reduction in magnitude of high tensile residual stresses in welded joints can be performed either by preventive means (Figure 2.26) [39] or by postweld heat treatments. The most effective method of reducing the magnitude of welding re-

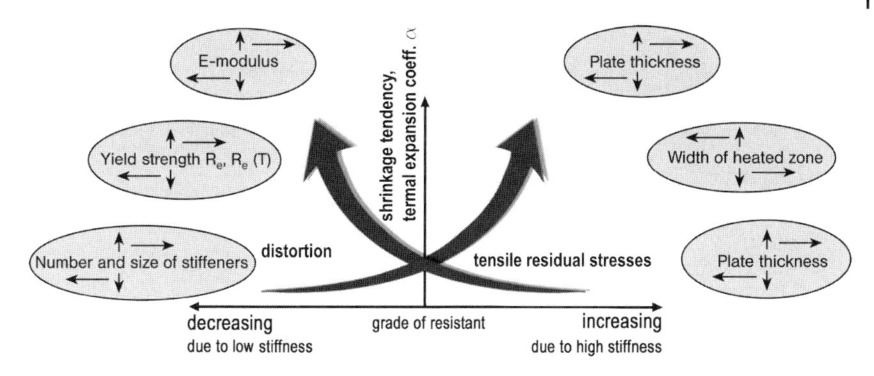


**Figure 2.25** Relation between grade of restraint, shrinkage tendency, stiffness, and residual stresses in welds [39].

sidual stresses is the reduction in stiffness of the construction [39]. A lower stiffness, however, may result in unacceptable distortions of the welded structure.

Most effective in reducing the magnitude of the residual stresses in a welded structure is a postweld heat treatment (stress relief annealing), which is more effective the higher the annealing temperature is [39]. The annealing temperature, however, is limited by phase transformations or other significant microstructure changes, such as grain coarsening. In the case of dissimilar material welds, postweld heat treatments even may be detrimental due to the thermal expansion mismatch of the dissimilar parent materials [40].

Besides construction welds, repair welds are also introduced into components and structures by routine inspection either to remedy initial fabrication defects found in castings or welds or to rectify in-service degradation of components and thereby extending their lifetime [41]. Omitting postweld heat treatments,

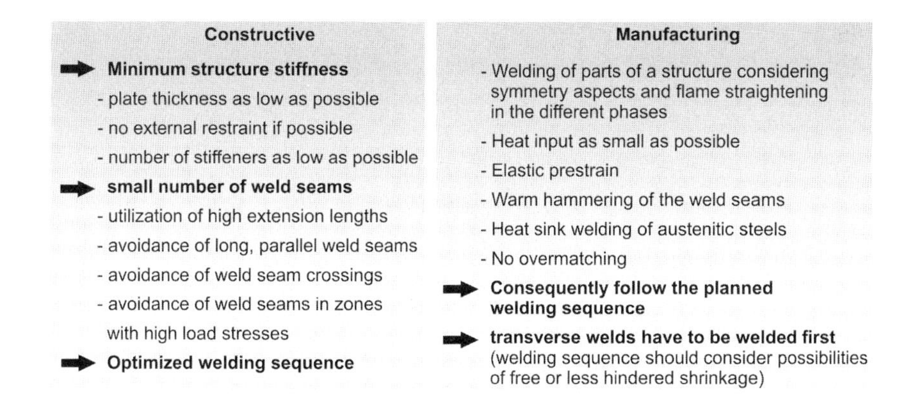


**Figure 2.26** Preventive possibilities for avoiding high tensile residual stresses and their consequences in welded structures [39].

repair welds can play an important role in many subsequent component failures due to the high residual stresses associated with the repair process. For instance, in case of steam leak at a nonstress relieved pipe-work repair weld, both the magnitude and multiaxial nature of the residual stress field was instrumental in driving creep damage leading to reheat cracking [41, 42].

### 2.2.6
### Residual Stress Formation in Coatings

Thin films and coatings are applied to increase, e.g., the wear resistance [43], high temperature corrosion resistance [44], resistance against wet corrosion and biocompatibility [45] of a substrate. Multilayer thin films are extensively used in the microelectronics industry because of their superior magnetic properties [46] and become increasingly used in X-ray diffraction [47].

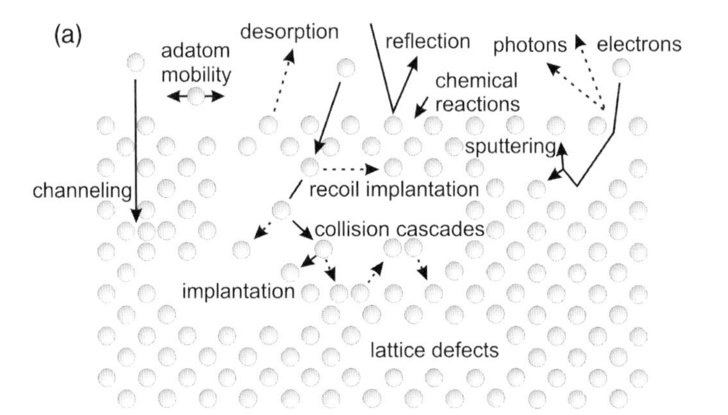


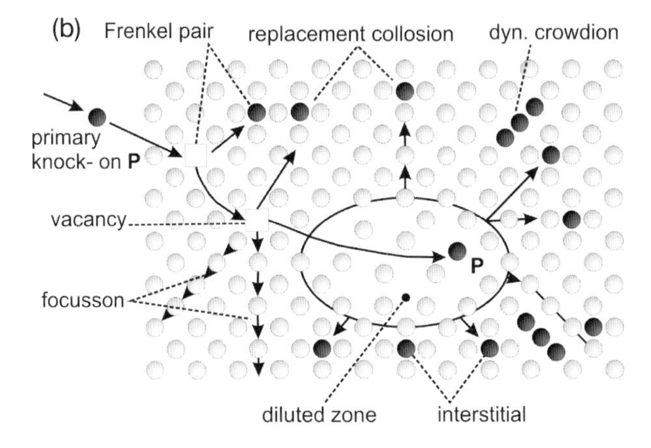


**Figure 2.27** (a) Effects of an ion bombardment on a growing film [49, 50] and in the 10–1000 eV range commonly employed in thin film processing. (b) Schematic view of possible lattice defects created by an impinging energetic atom, primary knock-on atom (P) [43, 51].

The methods for applying coatings are manifold, including, e.g., thermal spraying, thermally activated chemical vapor deposition (CVD), and plasma-assisted physical vapor deposition (PVD) techniques. In coatings produced by CVD residual stresses form during cooling from deposition temperature due to the difference in thermal expansion of the coating and the substrate. In addition to these extrinsic residual stresses, PVD coatings undergo so-called intrinsic residual stresses due to the ion bombardment during the PVD process (Figure 2.27) [43, 48–51].

These incoming ions or knock-on atoms will remove atoms from their lattice positions (Figure 2.27). Collision cascades result in a rearrangement of the lattice atoms. The impinging particles may also be physically implanted and surface species may be recoil-implanted into the subsurface lattice. Residual interstitials and vacancies are formed, which can, in turn, lead to an increased density of extended defects and form one-, two- (e.g., grain and sub-grain boundaries, twins) or three-dimensional defects (e.g., voids, cracks or particles) [43].

The microstructure of PVD coatings depends strongly on deposition parameters. Thornton [52] developed a "zone model", which describes the dependence of the microstructure on argon partial pressure, respectively, temperature for thinner PVD deposited layers based on a correlation of the sputter particle direction, shadowing effects and surface, respectively, volume diffusion. As a consequence of the deposition process the layer possess a microstructure gradient, the maximum of defect density usually is at the interface between substrate and coating; as a consequence residual macro- and microstress gradients are present (Figure 2.28) [53].

The magnitude of compressive residual macrostresses in hard coatings can reach up to several GPa [26]. The steepness of the residual stress gradients within the coating depends on the process parameters. Often steep residual stress gradients of up to several hundred MPa per micrometer coating thickness are encountered (Figure 2.29). Residual microstresses in thin films also strongly depend on the deposition process, because it determines the defect density and distribution (e.g. [53]). By annealing thin films relaxation processes, such

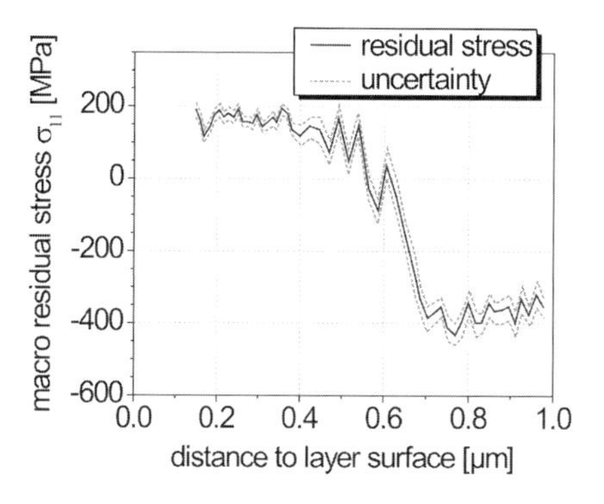


**Figure 2.28** Residual macrostress distribution within a PVD sputtered CoFe single layer [53].

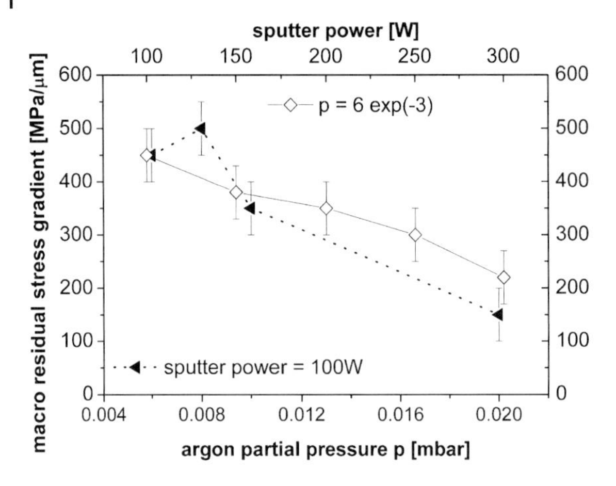


**Figure 2.29** Dependence on macro-internal stress gradients in CoFe-single layers on sputter power and argon partial pressure [53].

as plastic flow, creep, and microcracking, can substantially reduce the residual macro- and microstresses [54].

## 2.3 Relevance

The basis for assessing the influence of residual stresses on the strength and lifetime of a component is the superposition principle, which means that in a first approximation a linear superposition (addition) of load and residual stress distributions represent the total stress distribution within a workpiece or a component. Residual stresses, therefore, have influence on the strength, the fatigue behavior, corrosion resistance (in particular resistance against stress corrosion cracking), magnetic properties, etc. Workpiece or component failure in most cases occurs due to the onset of plastic deformation or fracture when subjected to tensile loads. The near-surface area is critical, e.g., because of notch effects. Therefore, usually compressive residual stresses in the near-surface zone are beneficial with respect to workpiece or component lifetime. Thus, in large cast or forged technical components, such as generator shafts, pressure vessels, terms of delivery often include specifications regarding maximum residual stress values and permissible residual stress distributions [55]. In component design known residual stress distributions can be included in the design criteria (e.g., the effective stress), whereas unknown residual stress distributions are incorporated into lifetime predictions by sweeping assumptions, conservative application of statistics or by an increase in safety factors [55, 56].

The relevance of residual stress distributions measured or calculated can only be assessed taking its context into account [55], e.g., material, microstructure, heat treatment, mechanical properties (brittle, respectively, ductile state), surface roughness, loading condition, environment (corrosive media, high temperature, extreme low temperature, etc.), and the method used for measurement or calculation. For instance in case of a brittle steel workpiece, the residual stress state needs to be considered both regarding the onset of plastic deformation and fracture (at not much higher plastic strains). In contrast, in case of a ductile workpiece, residual stresses need to be considered with respect to the onset of plastic deformation, but, their influence on fracture is not pronounced [57].

In the following sections the relevance of residual stresses on workpiece and component behavior under external loading will be explained in some more detail using technically relevant examples.

### 2.3.1 Failure due to Residual Stress Formation or Residual Stress Relief Induced by Temperature Changes

Depending on the yield strength of a material stresses formed during cooling will either result in residual stresses or distortion after cooling (see Section 2.2). Once thermal stresses are higher than the yield strength, distortion induced by plastic deformation results in size and shape alterations and, thus, may reduce the component to scrap. Chatterjee-Fischer [14] compared the distortions of two 200 mm × 200 mm × 20 mm-sized sheets of a plain carbon steel (0.1% C) and an austenitic stainless steel X5CrNi18-8 after cooling from 920 °C in water

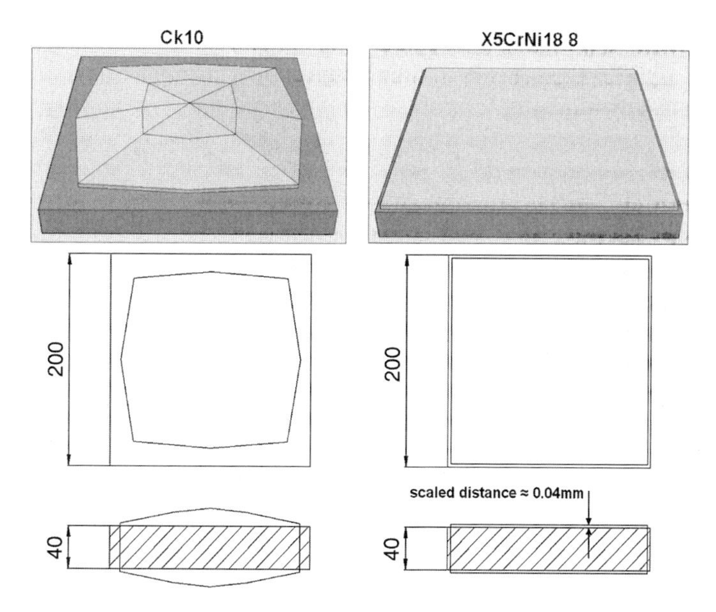


**Figure 2.30** Distortions introduced by quenching of steel sheets [14].

(Figure 2.30). The sheet cross-section after cooling depicts well the tendency of the component to reach a spherical shape. Due to the higher yield strength of the stainless steel compared to the plain carbon steel, the distortions introduced are significantly smaller, thus, residual stresses in the X5CrNi18-8 steel sheet will be significantly higher than in the plain steel sheet.

In case of materials with phase transformation, the residual stress state of long components, such as shafts or axles (see Section 2.2), formed after cooling usually consists of compressive residual stresses in the core and tensile residual stresses in the mantle. When the tensile residual stresses in the mantle exceed the strength of the material, cracks usually occur in longitudinal direction of shafts or axles [58].

In order to prevent crack formation, large surface hardening tools are usually cooled in air before quenching, while through-hardening tools are instantly quenched. Thus, a further increase of tensile residual stresses in the near-surface zone due to the martensitic transformation can be avoided [58].

Failure due to stress formation may occur also during rapid heating. Due to the faster temperature increase in the mantle compared to the core of the cylindrical component, the mantle will try to expand faster than the core, but, its expansion is hindered by the more rigid core. Thus, the mantle experiences compressive stresses, while tensile stresses evolve in the core of the component. Particularly in case of large dies used in continuous casting, tensile stresses introduced by fast heating processes may cause crack formation in the core of dies, e.g., in high alloyed steels with low thermal conductivity and low toughness [14].

### 2.3.2 Influence of Residual Stresses on Component Failure Under Static and Dynamic Mechanical Loads

Compressive residual stresses can stop fatigue cracks. In a fatigue experiment using a notched specimen a defined compressive residual stress field was produced in front of the growing crack by a hardness indentation (Figure 2.31). During cyclic loading the crack growth rate initially increases with increasing crack length. As soon as the crack tip reaches the compressive residual stress field, the crack growth rate decreases and the crack is finally arrested [59].

A very simple but impressive example of the influence of residual stresses on fatigue lifetime gives the SN (Wöhler-) curves obtained on hardened and ground low-carbon (C45) steels (Figure 2.32). Tensile residual stresses introduced by the grinding process yield a significantly lower fatigue strength and endurance limit compared to an almost stress-free sample, while compressive residual stresses in the near-surface zone increased both the fatigue strength and the endurance limit [59].

The highest benefit of compressive residual stresses appears for hardened steel, also quenched and tempered steel shows an increase of bending fatigue strength with increasing compressive residual stresses. In contrast, the bending fatigue strength of the same steel in normalized condition benefits only very slightly

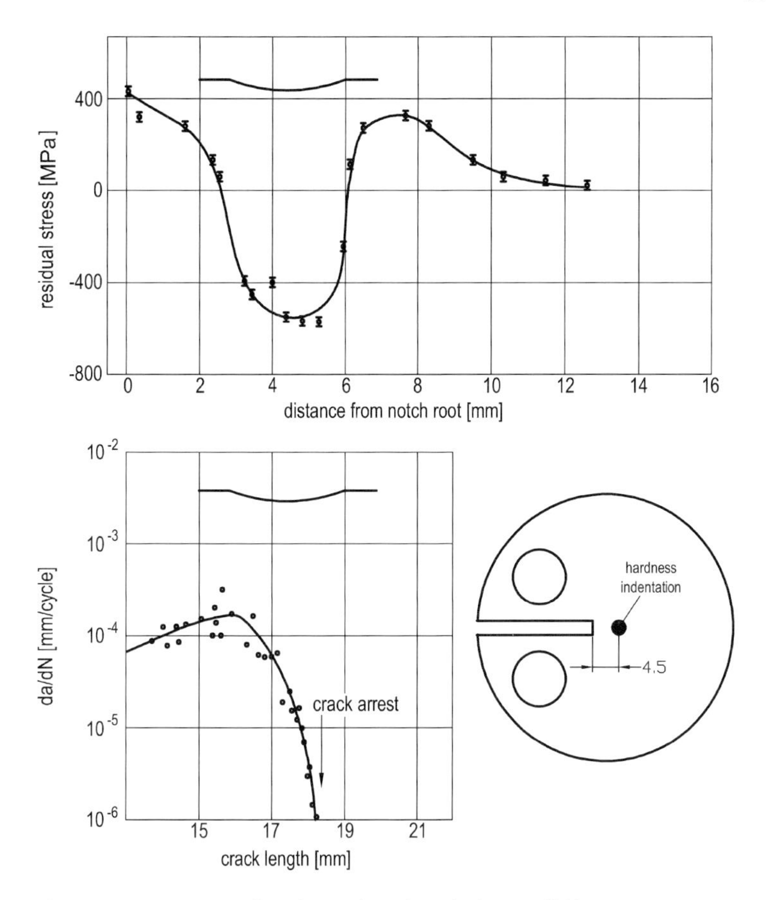


**Figure 2.31** Interaction of crack growth and residual stress field.

from compressive surface stresses due to its higher ductility. In the opposite case of tensile residual stresses, the hardened steel is the most susceptible to premature failure under fatigue loading (Figure 2.33).

Based on many of these experiments, Starker et al. [60] and Kloos et al. [61] proposed to consider the influence of residual stresses in the design of components exposed to fatigue loading by adding in the Haigh diagram the residual stress value measured to the mean fatigue strength. The local failure condition for single step cyclically loaded components then is given by

$$\sigma_{\mathrm{eq},a} = R_{tf}\left[1 - \frac{(\sigma_{1,m} + \sigma_1^{\mathrm{RS}}) + (\sigma_{2,m} + \sigma_2^{\mathrm{RS}})}{R_m}\right] \tag{2.23}$$

$\sigma_{\mathrm{eq},a}$ equivalent stress amplitude according to the Von Mises failure criterion

$\sigma_{1,m}$, $\sigma_{2,m}$ principal mean stresses

$\sigma_1^{RS}$, $\sigma_2^{RS}$ principal residual stresses
$R_{tf}$ tension compression fatigue strength
$R_m$ tensile strength

The increase in component lifetime due to the introduction of compressive residual stresses is one of main motivations of mechanical surface treatments, such as deep-rolling and shot-peening.

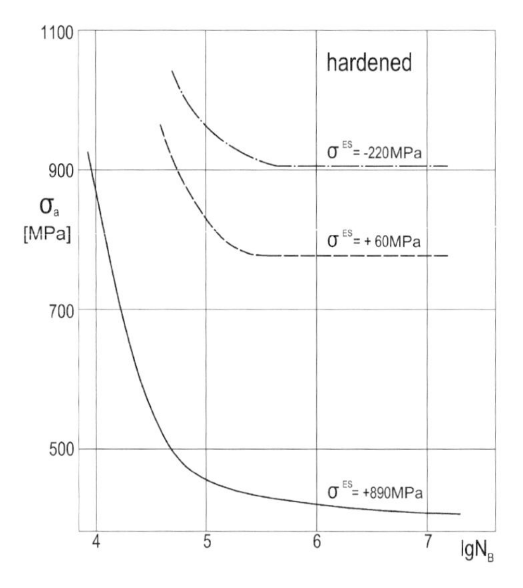


**Figure 2.32** Influence of surface residual stresses on fatigue strength (Quelle), $N_B$ is the number of loading cycles, $\sigma_a$ is the stress amplitude.

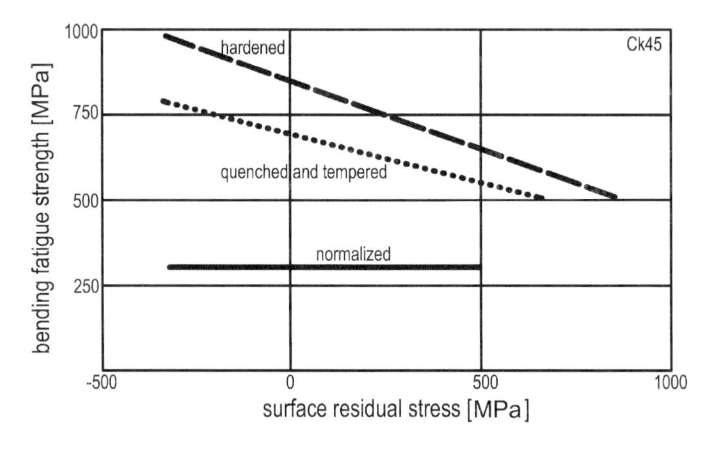


**Figure 2.33** Influence of heat treatment and surface residual stresses on bending fatigue strength [31].

### 2.3.3
### Influence of Residual Stresses on Component Failure in Corrosive Environments

An influence of residual stresses on component failure in corrosive environments is in general affirmed, however, detailed and quantitative experimental investigations are still sparse. This is probably due to the interdependence of corrosion resistance, microstructure, and, e.g., heat treatment and the residual stress state, which complicates the derivation of a simple relation between residual stresses and characteristic values describing the corrosion sensitivity.

For duplex stainless steels recent investigations revealed that the local stress gradient was the key parameter in pit initiation and that the local average stress was the parameter governing the transition from metastable to stable pitting [62]. In the case of uniform corrosion of martensitic tool steels, tensile residual surface stresses introduced local maxima into the current-density potential curve [63]. Also electrochemical measurements for quantifying intergranular corrosion of an aluminum alloy revealed a significantly higher breakdown potential in zones with residual compressive stress [64].

Residual stresses, thus, can have a strong influence on corrosion processes, however, corrosion processes can also exert a strong influence on the residual stress state of a component. Amirat et al. [65] showed by experimental and numerical investigations the redistribution of residual stresses in a pipeline tube caused by corrosion and the effect of the residual stress redistribution on the reliability and lifetime of the tube.

### 2.3.4
### Influence of Residual Stresses on Wear

Residual stresses can both directly and indirectly affect the wear rate of components (Figure 2.34).

Until recently consensus existed that because of the very high contact stresses in wear processes, the effect of the residual stress state of the first and the second body (counterbody) in a tribo-system on the wear process appeared negligible. The magnitude of residual stresses experimentally determined on worn surfaces, e.g., railway rails [66] and laser hardened steels [63, 67, 68], however, are in the same order of magnitude as the stresses generated by Hertzian pressure and friction in wear processes (Figure 2.34). These experimental results indicate that in some tribo-systems significant interaction between residual stresses and wear rate exist. This will be explained in more detail on the example of a nitrogen alloyed tool steel X30CrMoV15-1 + 0.3%N that was laser surface hardened in order to increase its wear resistance, while maintaining corrosion resistance. The amount of martensite and retained austenite generated in the near-surface region depend on the maximum temperature reached during the laser hardening process and the heating ratio. High maximum surface temperatures result in high retained austenite volume fractions in the near-surface zone due to nitride dissolution in the steel matrix. During the wear test the surface hardened zone

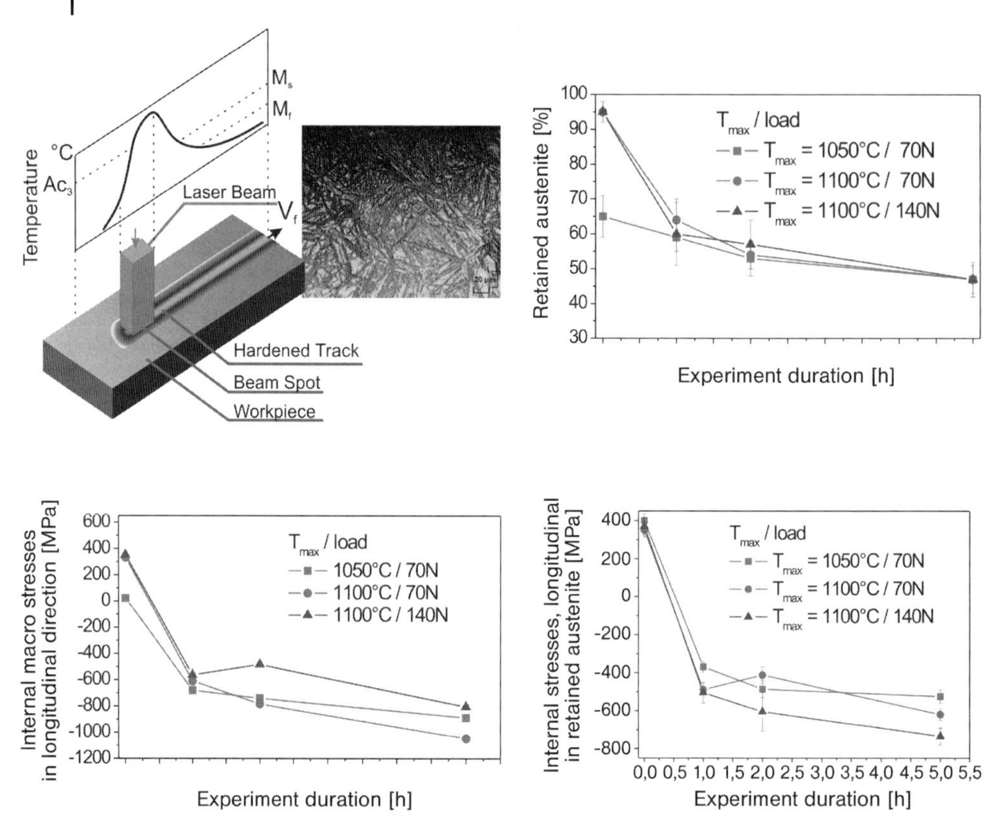


**Figure 2.34** Development of retained austenite volume fraction and residual stresses during wear of a laser surface hardened martensitic steel.

is subjected to Hertzian pressure and friction, which lead to an expansion of the near-surface layer. Thus, compressive residual stresses develop during unloading (similar to the mechanism of residual stress formation discusses for deep-rolling). In addition, compressive residual stresses form due to the transformation of part of the retained austenite into martensite. As a consequence, the near-surface zone contains compressive residual macrostresses and the phase specific residual stresses in the retained austenite are also compressive. The high compressive macro- and microstresses stabilize the retained austenite, because its transformation into martensite would generate additional compressive residual stresses due to the volume expansion.

An indirect influence of residual stresses on wear rate occurs through their interaction with the damage induced by the wear process. James et al. [56] present an example of a cutter head gear box casing, where a fatigue crack initiated at local fretting damage and preferentially followed the local residual stress field.

## References


1 D. Gross, *Bruchmechanik 1*, Springer, Berlin 1992.
2 P.J. Withers, H.D.K.H. Badeshia, *Mater. Sci. Technol.* 2001, **17**, 355–365.
3 H. Eschenauer, W. Schnell: *Elastizitätstheorie I*, BI Wissenschaftsverlag, Zürich, 1986.
4 A. Pyzalla, W. Reimers, Study of stress gradients using synchrotron X-ray diffraction, in: M. Fitzpatrick, A. Lodini (eds.), *Analysis of Residual Stress by Diffraction using Neutron and Synchrotron Radiation*, Taylor and Francis, London, 2003, pp. 219–232.
5 E. Macherauch, H. Wohlfahrt, U. Wolfstieg, *Härterei Technische Mitteilungen (HTM)* 1973, **28**, 201–211.
6 H. Behnken: Some basic relations to the stress analysis using diffraction methods, in: V. Hauk (eds.), *Structural and Residual Stress Analysis by Nondestructive Methods*, Elsevier, Amsterdam, 1997, pp. 39–65.
7 I.C. Noyan, J.B. Cohen, *Residual Stress – Measurement by Diffraction and Interpretation*, Springer, New York, 1987.
8 A. Pyzalla, W. Reimers, *Textures Microstruct.* 1999, **33**, 291–301.
9 M.B. Prime, T. Gnäupel-Herold, J.A. Baumann, R.J. Lederich, D.M. Bowden, R.J. Sebring, *Acta Mater.* 2006, **54**, 4013–4021.
10 A.S. Helle, K.E. Easterling, M.F. Ashby, *Acta Metal.* 1985, **33**, 2163–2174.
11 H. Buchholtz, H. Bühler, *Archiv für das Eisenhüttenwesen* 1933, **6**, 335–340.
12 B. Lement, *Amer. Soc. Metals*, Metals Park, Ohio, 1959.
13 A. Rose, *Härterei Technische Mitteilungen (HTM)* 1966, **21**, 1–6.
14 R. Chatterjee-Fischer, *HTM* 1973, **28**, 276–288.
15 D.P. Koistinen, *Trans. Amer. Soc. Metals* 1958, **50**, 227–241.
16 R. Glocker, H. Hasenmaier, *Z. Metallkunde* 19949, **40**, 182–186.
17 M. Koch, *Z. Werkstofftechnik* 1973, **4**, 81–85.
18 A. da Rocha, T. Hirsch, *Mat. Sci. Eng. A* 2005, **395** (1–2), 195–207.
19 K. Roettger, G. Wilcke, S. Mader, *Materialwissenschaft und Werkstofftechnik* 2005, **36** (6), 270–274.
20 L. Wagner, *Mat. Sci. Eng. A*, 1999, **263** (2), 210–216.
21 I. Altenberger, U. Noster, B.L. Boyce, J.O. Peters, B. Scholtes, R.O. Ritchie, *Mat. Sci. Forum* 2002, 404–407, 457–462.
22 W. Zinn, B. Scholtes, *J. Mater. Eng. Perform.* 1999, **8** (2), 145–151.
23 H.M. Mayer, C. Achmus, A. Pyzalla, W. Reimers, *Mat. Sci. Forum* 2000, **347**, 340–345.
24 B. Scholtes, *Eigenspannungen in mechanisch randschicht-verformten Werkstoffzuständen*, DGM Informationsgesellschaft Verlag, Oberursel, 1991 and Habilitationsschrift, TU Karlsruhe, 1990.
25 A.E. Tekkaya, *Ermittlung von Eigenspannungen in der Kaltmassivumformung*, Thesis, Universität Stuttgart, 1986.
26 C. Genzel, W. Reimers, R. Malek, K. Pöhlandt, *Mat. Sci. Eng.* 1996, **A205**, 79.
27 A. Pyzalla, W. Reimers, K. Pöhlandt, *Residual Stresses and Texture Evolution in Cold Extrusion of Full and Hollow Steel Bodies*, in: T. Ericsson, M. Odén, A. Andersson (eds.), *Proc. ICRS 5, 16–18 June 1997*, Linkoeping, Schweden, pp. 58–63.
28 J.W.L. Pang, T.M. Holden, J.S. Wright, T.E. Mason, *Acta Mater.* **48** (2000), 1131–1140.
29 A.M. Korsunsky, K.E. James, M.R. Daymond, *Eng. Fracture Mech.* **71** (2004), 805–812.
30 R.V. Martins, U. Lienert, L. Margulies, A. Pyzalla, *Mat. Sci. Eng. A* 2005, **402** (2005), 278–287.
31 B. Scholtes, Adv. in Surface Treatments, *Technol. Appl. Eff.* 1987, **4**, 59–71.
32 S. Immelmann, E. Welle, W. Reimers, *Mat. Sci. and Eng. A* 1997, **238**, 287–292.
33 B. Eigenmann, B. Scholtes, E. Macherauch, Materialwissenschaft und Werkstofftechnik 1990, **21**, 257–265.
34 H. Ruppersberg, I. Detemple, *Mat. Sci. Eng. A* 1993, **161**, 41–44.
35 M.J. Balart, A. Bouzina, L. Edwards, M.E. Fitzpatrick, *Mat. Sci. Eng. A* 2004, **367**, 132–142.

**36** E. Brinksmeier, E. Schneider, W.A. Theiner, H.K. Tönshoff, *Annals CIRP* 1984, **33/2**, 489–509.

**37** R. Malisius, *Schrumpfungen, Spannungen und Risse beim Schweißen*, DVS-Verlag, Düsseldorf, 1977.

**38** U. Boerse, D. Werner, H. Wirtz, *Das Verhalten der Stähle beim Schweißen*, Tl.1,2 Crundlagen, DVS-Verlag, Duisburg, 1995, pp. 2000.

**39** T. Nitschke-Pagel, H. Wohlfahrt, *Mat. Sci. Forum* 2002, 404–407, 215–224.

**40** M.J. Balart, A. Bouzina, L. Edwards, M.E. Fitzpatrick, *Mat. Sci. Eng. A* 2004, **367**, 132–142.

**41** P.J. Bouchard, D. George, J.R. Santisteban, G. Bruno, M. Dutta, L. Edwards, E. Kingston, D.J. Smith, *Int. J. Pressure Vessels Piping* 2005, **82** (4), 299–310.

**42** J. Dunn, J. MacGuigan, R.L. McLean, L. Miles, R.A. Stevens, *Investigation and repair of a leak at a high temperature stainless steel butt weld, Proceedings of International Conference on Integrity of High Temperature Welds*, Nottingham, 1998, pp. 241–258.

**43** P.H. Mayrhofer, C. Mitterer, L. Hultman, H. Clemens, *Prog. Mater. Sci.* 2006, **51**, 1032–1114.

**44** S. Kuroda, T.W. Clyne, *Thin Solid Films* 1991, **200** (1), 49–66.

**45** S. Tadano, M. Todoh, J.-I. Shibano, T. Ukai, *JSME: Int. J., Series A: Mech. Mater. Eng.* 1997, **40** (3), 328–335.

**46** P. Grünberg, M.B. Brodsky, *Phys. Rev. Letters* 1986, **57** (19), 2442–2445.

**47** M. Schuster, H. Goebel, L. Bruegemann, D. Bahr, F. Burgaezy, C. Michaelsen, M. Stoermer, P. Ricardo, R. Dietsch, T. Holz, H. Maid, Laterally graded multilayer optics for X-ray analysis, *Proc. SPIE – Int. Soc. Optic. Eng.* 1999, **3767**, 183–193.

**48** A.J. Perry, J.A. Sue, P.J. Martin, *Practical measurement of the residual stress in coatings, Surface and Coatings Technology, First Australian-USA Workshop on Critical Issues in High Performance Wear Resistant Films, 1996, Vol. 81, Issue 1*, pp. 17–28.

**49** D.M. Mattox, *J. Vac. Sci. Technol.* 1989, **A7** (3), 1105–1114.

**50** W. Ensinger, *Nucl. Instr. Meth. Phys. Res. B* 1997, 127–128, 796–808.

**51** R.W. Cahn, P. Haasen, *Physical Metallurgy*, 4th edition, University Press, Cambridge, 1996.

**52** J.A. Thornton, *J. Vac. Sci. Technol.* 1986, **A4** (6), 3059–3065.

**53** S. Dieter, A. Pyzalla, A. Bauer, N. Schell, J. McCord, K. Seemann, N. Wanderka, W. Reimers, *Z. Metallkde* 2004, **95**, 163–175.

**54** P.J. Withers, H.D.K.H. Badeshia, *Mater. Sci. Technol.* 2001, **17**, 366–375.

**55** V. Hauk, Eigenspannungen, ihre Bedeutung für Wissenschaft und Technik, in: *Eigenspannungen–Entstehung–Messung–Bewertung-*, E. Macherauch, V. Hauk (eds.), Deutsche Gesellschaft für Metallkunde e.V., Oberursel 1983, 9–48.

**56** M.N. James, D.J. Hughes, Z. Chen, H. Lombard, D.G. Hattingh, D. Asquith, J.R. Yates, P.J. Webster, *Eng. Failure Anal.* **14** (2004), 384–395.

**57** E. Macherauch, K.H. Kloos, *HTM Beiheft Last- und Eigenspannungen*, 1982, pp. 175–190.

**58** H. Bühler, E. Scheil, *Archiv für das Eisenhüttenwesen* 1933, **6**, 283–288.

**59** E. Macherauch, K.H. Kloos, Origin, measurement and evaluation of residual stresses, in: *Eigenspannungen, Entstehung, Messung, Bewertung*, E. Macherauch, V. Hauk (eds.), Deutsche Gesellschaft für Metallkunde e.V., Oberursel 1986, 3–26.

**60** P. Starker, H. Wohlfahrt, E. Macherauch, *Archiv Eisenhüttenwes* 1980, **51**, 439–445.

**61** K.H. Kloos, P.K. Braisch, *HTM* 1982, **37**, 83–86.

**62** V. Vignal, N. Mary, R. Oltra, J. Peultier, *J. Electrochem. Soc.* 2006, **153**, B352–B357.

**63** A. Pyzalla, C. Bohne, M. Heitkemper, A. Fischer, *Mater. Corrosion* 2001, **52**, 99–105.

**64** X. Liu, G.S. Frankel, *Corrosion Sci.* 2006, **48**, 3309–3329.

**65** A. Amirat, A. Mohamed-Chateauneuf, K. Chaoui, *Int. J. Pressure Vessels Piping* 2006, **83**, 107–117.

**66** E. Wild, L. Wang, B. Hasse, T. Wroblewski, G. Goerigk, A. Pyzalla, *Wear* 2003, **254**, 876–883.

**67** M. Heitkemper, A. Fischer, C. Bohne, A. Pyzalla, *Wear* 2001, **250**, 477–484.

**68** M. Heitkemper, C. Bohne, A. Pyzalla, A. Fischer, *Int. J. Fatigue* 2003, **25**, 101–106.

# 3
# Texture and Texture Analysis in Engineering Materials

*Heinz-Günter Brokmeier and Sang-Bong Yi*

## 3.1
## Introduction

The crystallographic texture describes the orientation distribution of the crystallites in a polycrystalline sample and is one of the basic parameters to characterize polycrystalline materials. General introductions in texture analysis can be found in the textbooks by Wassermann and Grewen [1], Bunge [2], Kocks et al. [3], and Bunge and Esling [4]. Texture analysis is a statistical method so that a sufficient number of crystallites given by their crystal coordinate system $K_A$ (Figure 3.1) are needed. Firstly, the orientation distribution has to be related to a sample coordinate system $K_B$ using the orientation $g$ of all grains. The sample coordinate system can be given by the deformation mode or by the shape of the sample. In Figure 3.1, the sample coordinate system is given for a rolling process (RD – rolling direction, TD – transverse direction, and ND – normal direction). Similar to the rolling geometry, geologists define a sample coordinate system by the foliation (equivalent to the rolling plane) and the lineation (equivalent to RD). The resulting orientation distribution function $f(g)$ gives the volume fraction d$V$ of grains in an orientation increment dg. For a detailed description of different ways to define the rotation $g$ by Euler angles, by $(h\ k\ l)[u\ v\ w]$-values, or by the orientation matrix, see Bunge [2].

Secondly, the orientation distribution may depend on the location of the volume element inside a compact sample, so that a texture gradient is observed. Typical examples are more or less all semifinished products. Consequently, the crystal orientation has to be analyzed not only related to the sample coordinate system but also to the position $x$, $y$, $z$ in the sample, which is called orientation stereology [5].

Thirdly, the measured crystallographic texture documents the present state of the material. Any thermomechanical treatment or aging can change the texture, so that during the treatment the texture changes with time as a function of process parameters. For a detailed understanding of such processes, it is of great interest to follow the texture changes during the process. Therefore, recently

*Neutrons and Synchrotron Radiation in Engineering Materials Science: From Fundamentals to Material and Component Characterization*
Edited by Walter Reimers, Anke Rita Pyzalla, Andreas Schreyer, and Helmut Clemens

ISBN: 978-3-527-31533-8

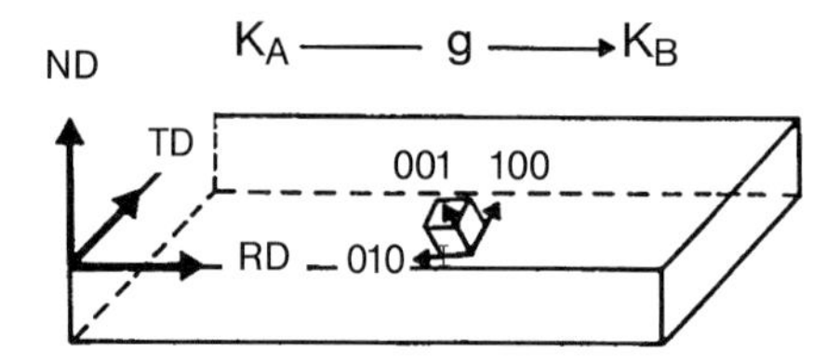


**Figure 3.1** Definition of the orientation g of an individual grain. $K_A$ is the coordinate system of a crystallite, $K_B$ is the sample coordinate system with the labeled directions RD (rolling direction), TD (transverse direction), and ND (normal direction).

developed methods for *in situ* experiments [6–8] need fast scanning possibilities adapted to the reaction speed of, e.g., recrystallization or loading.

Fourthly, also the orientation stereology is a function of the external conditions acting on the material that can change with time, see for instance, nucleation processes. However, automatic scanning of crystallite orientations over a sufficient number of individual grains requires a certain amount of time and, thus, the use of this method for the investigation of time-dependent processes where, e.g., the temperature changes with time, is limited. In Table 3.1, a summary of multidimensional texture analyses with up to nine dimensions is given.

Due to the crystallographic structure, material properties of single crystals are anisotropic, so that the crystallographic texture also describes the anisotropy of the material properties of a polycrystal. Consequently, the anisotropy in the material properties of a polycrystalline-textured material (like, e.g., Young's modulus, electric conductivity, thermal expansion, plastic anisotropy, or magnetic behavior) is between single crystal anisotropy and isotropic behavior of a texture-free powder (Figure 3.2). Thus, the knowledge of texture is essential for under-

**Table 3.1** Summary of different texture definitions.

| Dimension | Experiment |
|---|---|
| Two-dimensional texture | Pole figure, inverse pole figure |
| Three-dimensional texture | Orientation distribution function (ODF) |
| Six-dimensional texture | Orientation stereology<br>Crystal orientation as a function of $x$, $y$, $z$ |
| Six-dimensional texture | Orientation distribution function (ODF) as a function of temperature, time, and mechanical loading |
| Nine-dimensional texture | Orientation stereology, as a function of temperature, time, and pressure |

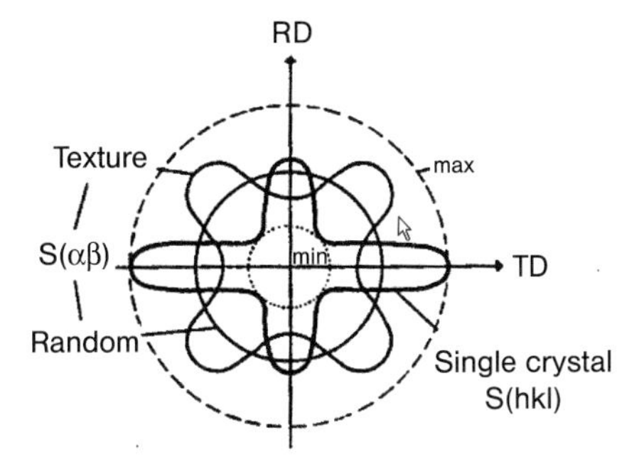


**Figure 3.2** Single crystal and polycrystal anisotropy.

standing the properties of technical polycrystalline materials. Here, one has to distinguish on one hand the texture type, which is the qualitative information, from the texture sharpness on the other hand, which gives also quantitative information. A well-known example of the influence of crystallographic texture on the deformation behavior is shown in Figure 3.3, where deep drawing produces an uneven rim of an Al can show the appearance of four ears.

In the last decades, there have been significant advances in texture measuring techniques. These originate not only from technical progresses in instrumentation, computer, and detector technologies, but also from the increasing request for understanding the mechanism of texture evolution which appears unavoidably in thermomechanical processing of almost all polycrystalline materials, such as metals, ceramics, or rocks. The need for efficient use of materials in industrial processes increased the interest in quantitative texture analysis as an effective tool for understanding materials behavior. In this sense, the directional anisotropy of physical and mechanical properties of textured materials is the driving force for advances in quantitative texture analysis. These advances go hand in hand with progress in numerical methods and software development based, e.g., on the series expansion [2, 9] and the WIMV [10, 11] algorithms for the calculation of orientation distribution functions from measured pole figures. Current topics in

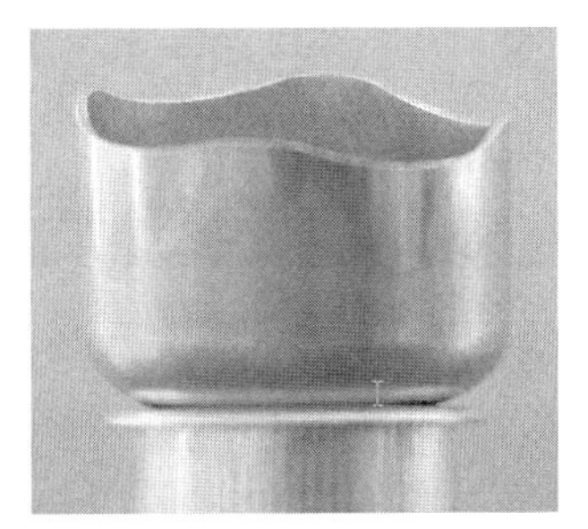

**Figure 3.3** Al deep drawing cup showing the earing effect.

**Table 3.2** Penetration power of different radiations as half value thickness $I/I_o = 50\%$.

| Metal | X-ray Cu Kα 1.54 Å | s-ray 0.124 Å–100 keV | S-ray 0.062 Å–200 keV | n-ray 1.00 Å |
|---|---|---|---|---|
| Mg | 0.0140 cm | 3.40 cm | 4.20 cm | 6.10 cm |
| Al | 0.0053 cm | 1.50 cm | 2.70 cm | 7.67 cm |
| Cu | 0.0015 cm | 0.20 cm | 0.50 cm | 0.85 cm |
| Ti | 0.0011 cm | 0.60 cm | 1.30 cm | 1.61 cm |
| Pb | 0.0003 cm | 0.03 cm | 0.04 cm | 2.10 cm |

texture analysis are, e.g., prediction of anisotropic behavior, optimization of textures for industrial applications, and new aspects in texture research like modeling and simulation of polycrystalline plasticity.

In this chapter, texture-measuring techniques that nowadays are widely used will be introduced and compared to each other. By comparing each method, the necessities for the use of large-scale facilities in texture analysis and the importance of selecting the adequate equipment for user demands will be emphasized. An important issue is the penetration depth because it largely determines the experimental technique to be used for a specific problem. In Table 3.2, penetration depths of different types of radiation are collected. Finally, characteristics of diffraction methods will be presented with selected examples.

## 3.2 Pole Figures

Pole figures are used for the graphical representation of textures. A pole figure is a projection of directions on a plane. Often the stereographic projection is used, which is shown in Figure 3.4a. A direction is defined by a line from the center of a sphere ($O$) to a point on the surface of the sphere ($A$). The direction can be, e.g., the normal of a set of lattice planes. When the point A is observed from the point $P$, the point of intersection $B$ marks the stereographic projection of the direction $\overline{\mathrm{OA}}$ on a plane. Directions determined in this way can be drawn into a polar grid (Figure 3.4b). When the intensity distribution gained from the measurement of a lattice reflection with indices $h\ k\ l$ where the sample is rotated around two axes (angles $\alpha$ and $\beta$) is drawn into such a polar grid using contour lines, it is called a pole figure. The diffracted intensity is proportional to the fraction of crystallites within the tested volume oriented in the specific direction.

Three angles are required for defining the orientation of a crystallite in relation to the sample axes, where different sets of angles are possible. A common choice is the set of Euler angles $\varphi_1$, $\phi$, and $\varphi_2$ for the consecutive rotations around the

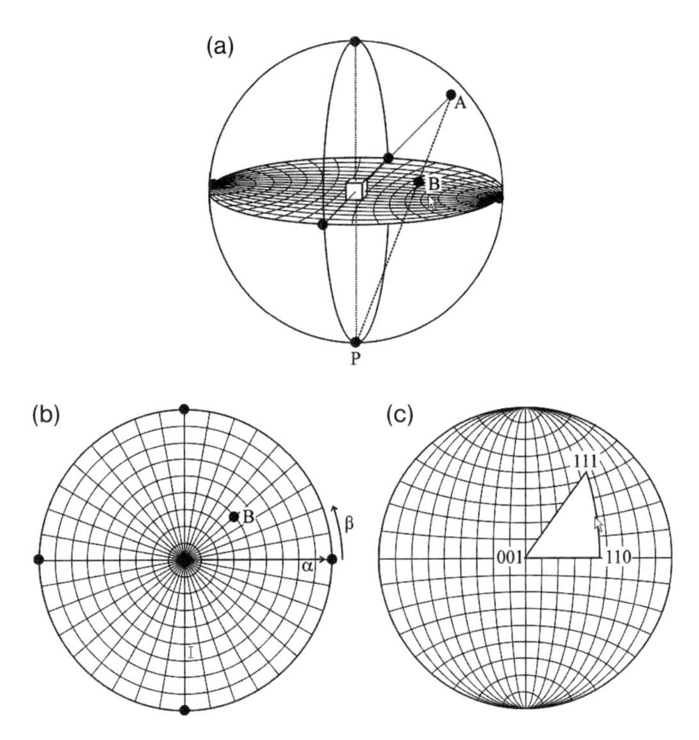


**Figure 3.4** (a) Stereographic projection (see the text). (b) Polar grid in which pole figures are drawn. (c) Part of a polar grid used for inverse pole figures in cubic materials.

*z*-axis, *x*-axis, and again *z*-axis [2]. Mathematically, texture can be described by the orientation distribution function (ODF) $f(g)$ with the orientation $g(\varphi_1, \phi, \varphi_2)$. Since the diffracted intensity does not depend on the rotation angle around the scattering vector, two or more pole figures have to be determined for a calculation of the ODF, where the required number of pole figures and suitable choices of $h\ k\ l$ depend on the crystal symmetry and on the sample symmetry. Orientation distribution functions are usually plotted for a set of different values of $\varphi_1$ (e.g., every 5°), for each $\varphi_1$, the distribution is drawn as a contour plot with the orthogonal axes $\phi$, $\varphi_2$ [2]. A very practical description of the orientation distribution function, which can be calculated from $\varphi_1$, $\phi$, $\varphi_2$, is the lattice plane $(h\ k\ l)$ parallel to the rolling direction and the lattice direction $[u\ v\ w]$ parallel to the rolling direction.

A representation of the crystal orientation with respect to a sample orientation, e.g., a wire axis or rolling or transverse direction in a rolled sheet, is called inverse pole figure. Only a part of the polar grid is required for such a representation, depending on the crystal symmetry of the material (Figure 3.4c). It has to be noticed that any sample orientation has its own inverse pole figure.

## 3.3
## Texture Measurements on Laboratory Scale

### 3.3.1
### X-ray Diffraction

As a standard laboratory method, X-ray diffraction is the most commonly used for texture measurements. The essential set-up of an X-ray texture diffractometer consists of an X-ray tube, a detector, and a four-circle goniometer that is called Eulerian cradle. Two methods of pole figure measurement are generally used, reflection and transmission geometry (Figure 3.5). The sample is rotated around two axes, tilting ($\chi$ in reflection and $\omega$ in transmission) and rotation ($\phi$ in reflection and $\chi$ in transmission) at fixed detector position ($2\theta$) for a certain crystallographic plane. Because X-rays are strongly absorbed by matter, the transmission geometry is applicable to thin foils or materials with low absorption. Figure 3.6 represents the pole figure grid of $5 \times 5$ in tilting and rotation, and the pole figure coverage at one sample position by using different type of detectors. More than one pole figure can be measured simultaneously by using a one-dimensional position sensitive detector (line detector) covering more than one reflection. In addition, several measuring points in each pole figure can be covered simultaneously by using a two-dimensional position sensitive detector (area detector).

Because of the low penetration power and small beam size in laboratory X-ray diffraction, when compared to those for neutrons (Table 3.2), poor grain statistics are achieved in coarse-grained materials. To increase the grain statistics, sample oscillation over some millimeters or more is used during the measurement under the assumption of homogeneous texture in the oscillated area. Another way of raising the statistics is to add the pole figures measured at more than one position (or sample). Due to the strong defocusing effect and the diffraction geometry, complete pole figures cannot be measured by laboratory X-ray diffraction. In reflection geometry, the pole figure measurement is carried out in general up to $\chi = 70^\circ$ or at most $\chi = 85^\circ$. On the contrary, the peripheral part of the pole figure is covered in transmission geometry. In transmission, pole figures are measured up to $\omega = 55^\circ$ or at most $\omega = 60^\circ$. Diverse methods are used to obtain complete

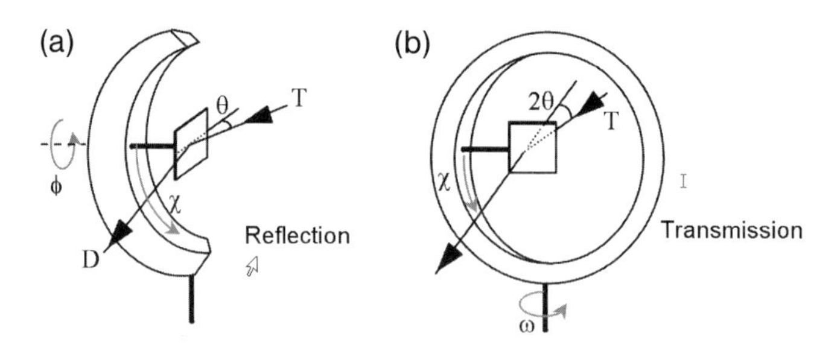


**Figure 3.5** Diffraction geometry in pole figure measurements: (a) in reflection and (b) in transmission.

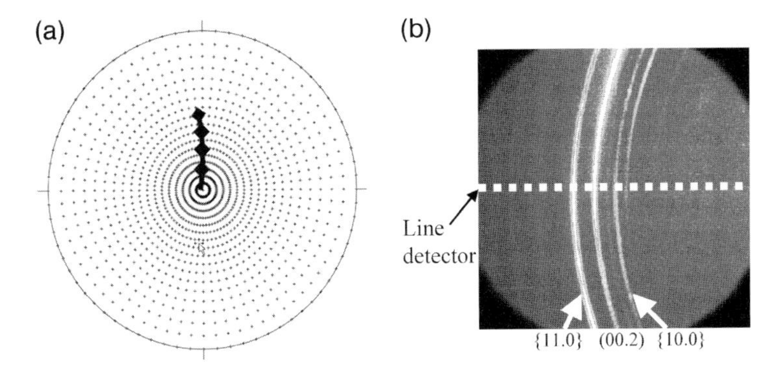


**Figure 3.6** (a) Pole figure measuring grid with 5 × 5 in tilting and rotation. Each grid point is measured individually by a point or line detector. (b) Diffraction image of extruded Mg on an area detector. The pole figure coverage by one diffraction line on this image, e.g., (0 0 2), is marked in (a) with a black line.

pole figures, e.g., recalculation of the pole figure from the orientation distribution function (ODF), combination of transmission with reflection method, or the use of oblique samples [12].

However, the last two methods are seldom used due to difficulties in measurement and sample preparation, while the ODF calculation has become a standard procedure in X-ray texture analysis. It is often found in practical measurements that some diffraction peaks are located very close to each other or totally overlapped, especially in polyphase samples and/or materials with low crystal symmetry. In such cases, the overlapped peaks are hardly to be used for pole figure measurements unless mathematical peak fitting can be employed to separate the peaks. Despite of the above limitations, laboratory X-ray diffraction technique is the most widely used for texture analysis, because of the easiness in machine usage, low cost, and relatively easy sample preparation.

### 3.3.2 Electron Diffraction

Orientation determination by electron diffraction is accomplished by indexing of the so-called Kikuchi patterns or spot diffraction patterns, using, e.g., selected area diffraction (SAD) or small-angle convergent beam electron diffraction (SCBED). Figure 3.7 illustrates the formation of the Kikuchi line and SCBED spot pattern. The Kikuchi patterns in SEM and TEM are formed in similar way. The only difference is that in SEM the backscatter electrons have an opposite direction to the incident beam, while in TEM transmitted electrons form the Kikuchi pattern. When an electron beam is directed onto a crystalline sample, electrons are scattered from lattice planes according to Bragg's law. These inelastically scattered electrons emerge along diffraction cones having an opening

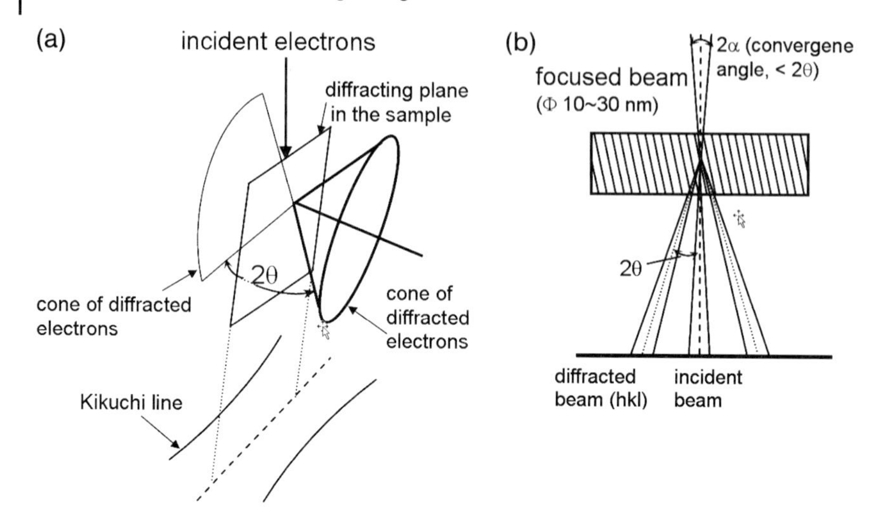


**Figure 3.7** (a) Kikuchi line formation in TEM. (b) Spot pattern in TEM–SCBED, Kossel–Möllenstedt pattern in the case of $2\alpha < 2\theta$ [15].

angle of $180^\circ - 2\theta$. When the cones of diffracted electrons intersect a detecting screen, they form Kikuchi lines on the screen. The usage of SCBED in TEM results in significant improvement in spatial resolution of below 10 nm, while SAD delivers average information from about 0.5 μm [13]. Once the electron beam is focused to a smaller convergence angle ($2\alpha$) than the diffraction angle ($2\theta$), the nonoverlapped spot pattern by elastic diffraction of electrons is observed on the TEM screen, called Kossel–Möllenstedt pattern. For a detailed description of electron diffraction, see Refs. [14, 15]. With the aid of the current developments in electron microscopy, computer techniques, and automated pattern indexing, e.g., Kikuchi pattern indexing based on the Hough transformation [15] or SAD spot pattern indexing by template matching [16], electron diffraction has become an important tool for microstructure and texture analysis. The real power of electron diffraction in materials analysis lies in the simultaneous determination of both orientation and microstructure with high angular and lateral resolution at defined sample positions. Also, energy dispersive X-ray chemical analysis (EDX) can be carried out simultaneously with orientation mapping. Based on the above possibilities, texture components can be separated with respect to microstructural elements, e.g., chemical composition, grain size, phase, position, etc. Another important feature of electron diffraction is the possibility of direct texture measurement, i.e., the three-dimensional orientation at the measuring point is determined from indexing the diffraction patterns. Therefore, the results from electron diffraction can be presented directly in the orientation space without additional mathematical calculations, which are necessary in X-ray and neutron measurements. Based on this feature, possible errors occurring during mathematical treatments can be avoided in electron diffraction. However, long-time and large-scaled mapping is necessary to reach satisfactory statistics for determining the global texture, which is hardly to get in TEM. The diffraction quality is very sensitive to sample conditions, e.g., sample surface flatness, grain size, dislocation density, electric conductivity, etc. In electron backscatter diffraction (EBSD) in SEM, the electrons penetrate only some tens of nanometers such that the pattern quality is even more sensitive than in TEM. It is generally recommended to use high-beam currents for getting a high signal-to-noise ratio in SEM–EBSD and also to apply electropolishing for preventing sample damages during preparation.

## 3.4 Texture Measurements at Large-Scale Facilities

### 3.4.1 Neutron Diffraction

The advantages of neutron diffraction lie in the combination of high penetration depth (at some centimeter level) for almost all materials (Table 3.2) and the relatively large beam cross-sections. Thus, the typical sample geometry is completely different from laboratory X-ray and EBSD samples (Figure 3.8). Moreover, complete pole figures are obtained without special sample preparation, and measured pole figures can be used for ODF calculation after background intensity corrections. These characteristics make neutron diffraction a powerful tool for

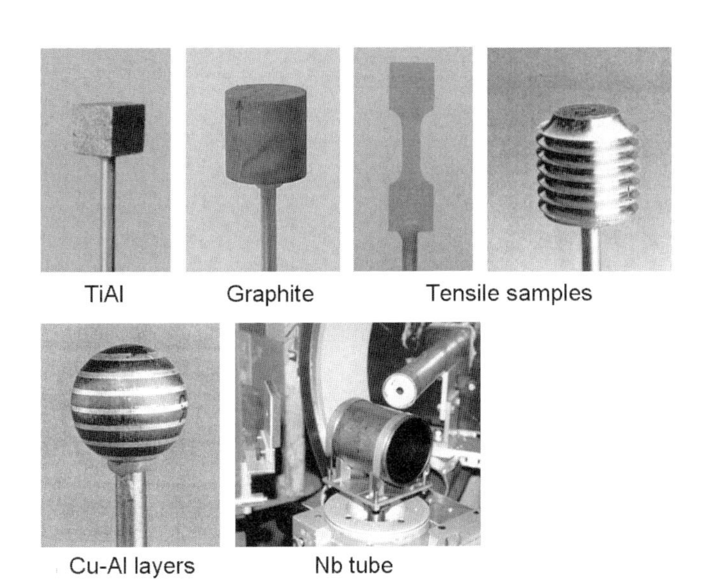


**Figure 3.8** Various sample geometries for neutron texture measurements. The sample volume varies from some $mm^3$ to $cm^3$, depending on the sample material.

texture analysis, e.g., in geological materials as well as in metallic samples having a complex shape or a large volume. Due to the low absorption for neutrons, satisfactory grain statistics are easily obtained for coarse-grained materials. *In situ* pole figure measurements are possible under various sample environments. The basic principle of pole figure measurement with neutrons is analogous with that of laboratory X-ray reflection. However, longer counting times are often necessary because of the low neutron flux from a reactor. Because of long counting times and possible flux variations depending on reactor conditions, the flux-controlled measuring scheme is generally employed, i.e., constant neutron flux at a measuring point, not constant measuring time.

Figure 3.9 shows a typical set-up for pole figure measurements using a monochromatic neutron beam. Since thermal neutrons from a reactor cover a wide range in wavelength, in general 0.5–4 Å, a bunch of single crystals is employed for getting a monochromatic beam. The single crystals are located on a plate having a slight curvature such that there is a focused beam at the sample position.

The different wavelengths are easily selected by changing the monochromator take-off angle. Like for X-ray diffraction, line and area detectors are also applicable in neutron diffraction for recording more than one measuring points simultaneously (Figure 3.6). Another way to cover several reflections simultaneously is use of the time-of-flight (TOF) method with neutrons. The wavelength of a neutron is inverse proportional to its velocity so that a polychromatic neutron beam can be analyzed with regard to the flight time of the neutrons. In TOF measurements, the diffraction angle ($\theta$ in Bragg's law) remains constant, hence a lattice spacing ($d$ in Bragg's law) profile is achieved based on the wavelength pattern. A long flight path of neutrons is necessary for high resolution in time-of-flight analysis. Since the whole diffraction profile, according to the wavelength range, is measured at once, for example, 0.5–4 Å of wavelength range corresponds to the $d$-spacing of 0.35–2.83 Å at fixed $\theta = 45°$, the TOF technique is well suited for low crystal symmetric as well as for multiphase samples. The wide range of the measured diffraction profile can be directly utilized for the whole diffraction profile analysis, e.g., Rietveld refinement-based texture analysis and structure analysis

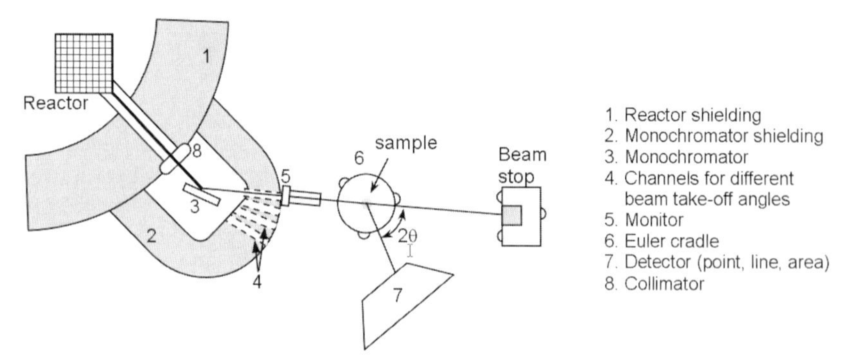


**Figure 3.9** Neutron texture diffractometer TEX-2, GKSS Research Centre, Germany.

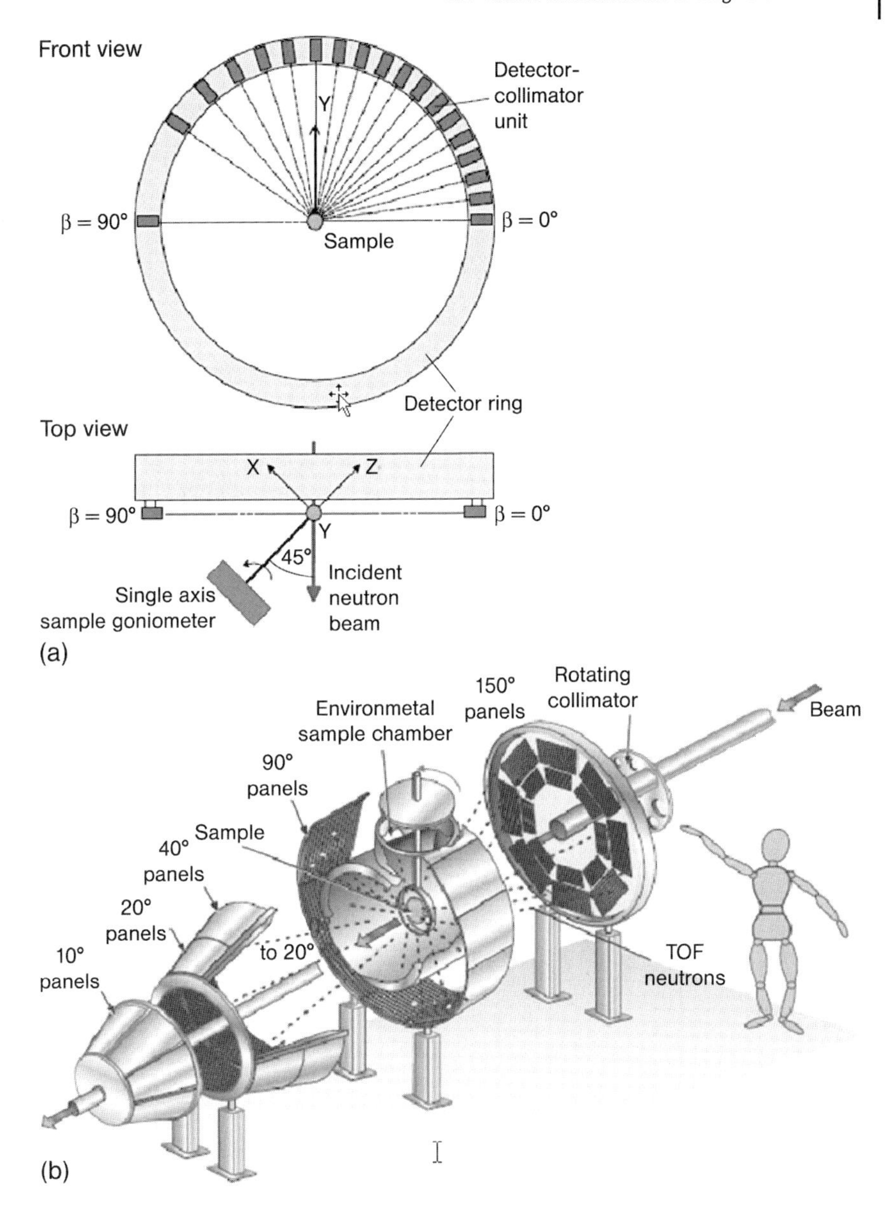


**Figure 3.10** Examples of TOF diffractometers with many detectors. (a) SKAT in Dubna, Russia [17]. (b) HIPPO in Los Alamos, USA [18].

[10, 16]. An example of texture analysis using Rietveld refinement will be given in Section 3.4.1.1. With the installation of many detectors at a TOF beam line, complete pole figures can be measured by rotation around one sample axis only. This capability also allows *in situ* texture analysis. Figure 3.10 shows two different

TOF diffractometers equipped with many detectors, the SKAT diffractometer with 19 detectors at the IBR2 Dubna Research Centre in Russia [18] and the HIPPO diffractometer with the total of 1360 detectors at the Los Alamos Laboratory in the USA [19].

Neutrons are a powerful probe for texture analysis not only in the case of coarse-grained material or for measurement of complete pole figures or weak textures, but also for nondestructive texture analysis in semifinished products, which shall be demonstrated in the following example.

#### 3.4.1.1 Texture of Semifinished Products

For texture investigations of semifinished products, there are special requirements for the experimental facilities. On the one hand, in most cases texture varies over the sample volume, which requires determination of the texture at different locations within a semifinished product. On the other hand, often a whole set of different material characterizations have to be applied to identical samples (e.g., qualitative and quantitative phase analysis, stress determination, and texture investigations). In the ideal case, texture analyses are performed nondestructively so that other experimental tests (e.g., a tensile test) can be applied afterward to exactly the same samples. However, a large number of texture investigations are still done destructively by sectioning a large sample due to the limited availability of nondestructive methods. In all these cases, the standard techniques described in this chapter are in use. Nondestructive measurements are nonstandard techniques, so that only a few investigations have been carried out up to now. In all the cases, for avoiding artifacts due to sample preparation, particularly on the sample surface, a beam with a high penetration power is needed (Table 3.2).

In the case of nondestructive investigations, the sample shape influences the texture measurement due to a changing illuminated volume and changing absorption during sample rotation and tilting. In some cases, one has to work with incomplete pole figures. In such an extreme case, even just a few characteristic points in the pole figure can be sufficient to calculate an ODF, however, only with a very limited resolution. Following this idea, a very practical and fast method can be utilized using 60 keV X-rays for characterizing the deep drawing behavior of 2 mm thick steel strips and a few millimeters thick aluminum strips during sheet production [20]. Such fast and simple methods are very effective for quality insurance purposes in industrial production lines. Nevertheless, to get more precise quantitative texture information, a set of complete pole figures is necessary.

For relatively simple sample geometries like tensile test samples or long cylindrical samples most corrections (constant volume, constant absorption) have only a minor influence on the result or can be neglected [21]. But more complicated is the nondestructive measurement of textures in semifinished products like tubes using the reflection method and an absorption correction [22], which shall be shown with the following examples. A Nb-tube of 140 mm length, 84 mm diameter, and 3 mm wall thickness shown in Figure 3.11 was investi-

**Figure 3.11** Photograph of a Nb tube mounted in the texture diffractometer TEX-2 at GKSS.

gated with respect to texture anisotropy around the perimeter. Seamless cavities for the manufacturing of accelerator units for a linear collider are produced from such tubes by hydroforming. A texture variation around the perimeter of the tube can lead to uneven cavity formation during the hydroforming step and to a rough inner surface of the tube making such cavities rejects. Local textures averaged over the wall thickness could be obtained from a whole tube with neutron diffraction. This was possible by using 4 mm wide slits on both sides, incoming and outgoing beam. It could be shown that this texture variation was on the one hand due to the texture of the ingot the tube was produced from, and on the other hand to the low deformation degree during tube manufacturing [22].

Seamless cavities made of a Cu–Nb composite are other candidates for the manufacturing of accelerator units for a linear collider. In this case, a composite Cu–Nb tube is the preproduct for hydroforming, which is produced by coextrusion of a Cu and a Nb-tube, or by explosive bonding of a Cu and a Nb-tube [23]. In both the cases, a homogeneous texture around the perimeter and sufficient material flow is required. Besides the type of initial materials, such as the type of the used Cu alloy and the purity of Nb, the production process itself is of great interest. Figure 3.12a shows a Cu–Nb ring, 140 mm in diameter and 15 mm wide, cut from a coextruded tube with additional flow forming to get a tube wall of 4 mm thickness (1 mm Nb and 3 mm Cu). Three pole figures of each phase allow the ODF calculation of the Nb and the Cu part of the composed tube (Figure 3.12b). The ODF section $\varphi_1 = 0°$ with $\Phi$ from 0° to 90° and $\varphi_2$ from 0° to 90° shows the typical BCC-texture with a strong $\alpha$-fiber at $\varphi_2$ 45° and $\Phi$ from 0° to 90°. The anisotropy of the Nb texture around the perimeter is shown for three points in Figure 3.12c with different $\alpha$-fibers for three positions [23]. These investigations were carried out at the neutron texture diffractometer TEX-2 (FRG-1, Geesthacht).

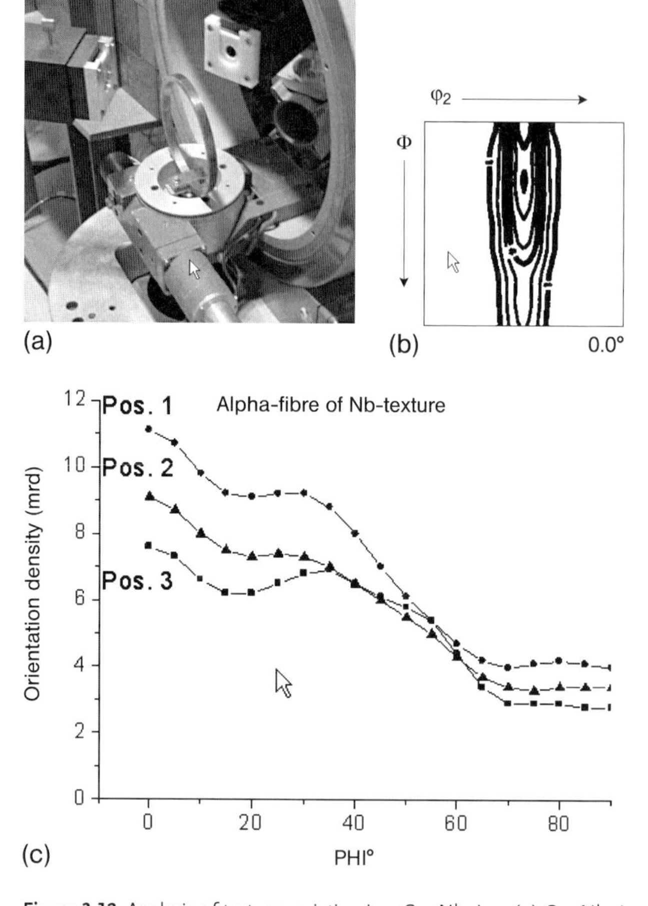


**Figure 3.12** Analysis of texture variation in a Cu–Nb ring. (a) Cu–Nb ring mounted in a Eulerian cradle. (b) ODF section for $\varphi_1 = 0^\circ$ of the Nb texture. (c) $\alpha$-Fibers of the Nb texture for three positions along the perimeter of the tube.

### 3.4.2
### Texture Analysis Using Synchrotron X-rays

Synchrotron X-rays are attractive for texture analysis because of high spatial resolution combined with excellent brilliance and high penetration depth for many materials in the same order as for thermal neutrons. The high spatial resolution results from the highly intense X-ray beam with possible sizes down to a few micrometers. In addition, the divergence of synchrotron X-rays is extremely small so that high orientation resolutions can also be achieved (<0.01). On the basis of the high photon flux, a few percent sample transmission is already enough for a satisfactory intensity at the detector. For this reason, texture investigations of heavy elements in relatively large volumes are possible in transmission geometry.

By employing an area detector in transmission, full Debye–Scherrer rings are recorded simultaneously such that significant reduction of the measuring time and high angular resolution are obtained. In fact, the texture measurements utilizing synchrotron X-rays began with the advent of area detectors in the mid-1990s. The combination of synchrotron X-rays with area detectors has opened new fields in texture analysis. Texture analysis at localized volumes with sizes from micrometers to centimeters is now possible in a nondestructive way. Though neutron diffraction allows nondestructive texture analysis, its spatial resolution is limited to the order of millimeters. Moreover, synchrotron X-rays offer the possibility for *in situ* measurements under various sample environments, due to the fast measuring time together with high penetration power and small wavelength. As with the neutron TOF technique, sample revolution along one axis is enough to cover full pole figures, which offers more freedom for sample environment devices.

The principle of pole figure measurement using synchrotron X-rays is similar to that of laboratory X-ray measurement that was introduced already in the beginning of the 20th century using X-ray films as the only available area detector at that time. The main difference is the wavelength, which has a strong influence on the scanning routine (Figure 3.13). Depending on the research interests and machine availabilities, various measuring methods have been developed by many working groups. For the detailed description of other methods, see the following references: stationary and moving detector modes [24, 25], employing a conical slit system using monochromatic beam [26], and energy dispersive method using white beam. In this chapter, the texture measurement using monochromatic beams and area detectors will be introduced, so-called stationary detector mode, as a basic measuring technique available in almost all synchrotron X-ray beam lines without employing special slit systems. Figure 3.13 illustrates the basic beam line set-up for texture measurement at the synchrotron beam line BW5 at Hasylab in Hamburg, Germany. The primary white beam from the wiggler enters the experimental station; the desired wavelength is selected using a mono-

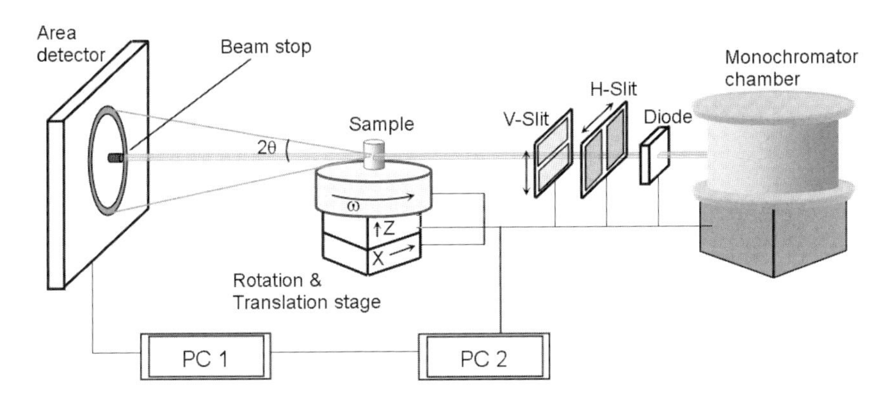


**Figure 3.13** Set-up for texture measurements at beam line BW5 at Hasylab, Germany. PC 1 is responsible for reading data from the area detector; PC 2 controls all motors.

chromator located in a vacuum chamber. The beam size is controlled by motorized horizontal and vertical slits. Because a synchrotron beam decays gradually with time and, thus, periodical reinjection of accelerated particles into the storage ring is necessary, the incident beam intensity is monitored at a gas-counter (diode) for the correction of measured intensity (detector). The sample is positioned on a $\omega$-rotation table mounted on a $XZ$-stage which allows sample translations in sample height ($Z$) and in the perpendicular direction to the incident beam ($X$). For a complete texture measurement, the sample is rotated around the $\omega$-axis with a fixed step size. At each $\omega$-position, the sample is irradiated and the occurring Debye–Scherrer rings are collected at the area detector. Since the intensity along the Debye–Scherrer rings is proportional to the pole density on the orientation sphere, the textured sample shows an intensity variation with the angle $\gamma$ along the ring (Figure 3.14). A Debye–Scherrer ring recorded at a certain sample rotation $\omega$ covers a circle in the corresponding $\{h\ k\ l\}$ pole figure and the full pole figure coverage is completed by the $\omega$-rotation. As shown in Figure 3.14, there is a blind area in the pole figure, which originates from the diffraction geometry. Due to the low wavelength, however, this unmeasurable area, which corresponds to the diffraction angle $\theta$, appears only in a small area. In case of using an energy of 100 keV (wavelength of 0.12 Å) the diffraction rings of most interest locate at $\theta$ ranges less than 3°.

The task of commonly available programs is the ODF calculation (series expansion and WIMV method), more correctly, the pole figure inversion and normalization, meaning that measured pole figure presented in the $\{\omega, \gamma\}$ grid has to be interpolated into a regular $\{\alpha, \beta\}$ grid. In the following sections, some selected examples of texture analysis using synchrotron X-rays are given.

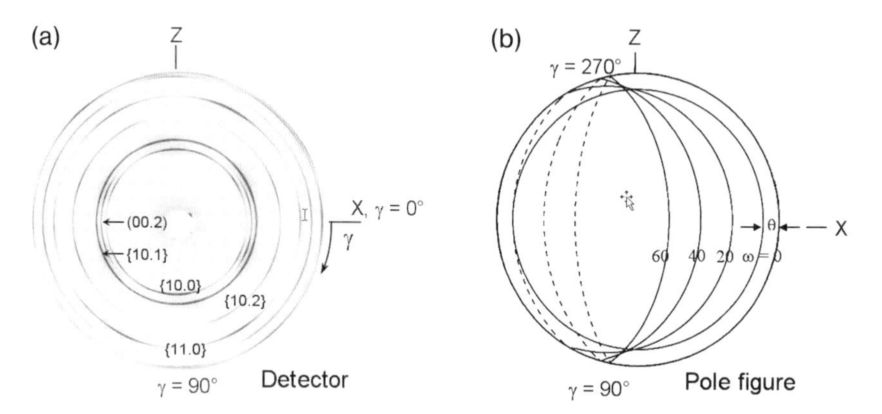


**Figure 3.14** (a) Diffraction image from strongly textured Mg on an area detector (transmission method). (b) Pole figure coverage by Debye–Scherrer rings obtained at different $\omega$-positions. Solid lines in the pole figure are corresponding to Debye–Scherrer rings within the range from $\gamma = 90°$ to 270°. For emphasizing the blind area effect, $\theta = 10°$ and equal area projection are applied to this pole figure.

#### 3.4.2.1 Local Texture Measurement in an Extruded Mg Rod

Texture heterogeneity in a round extruded Mg rod, which shows a different texture in core and mantel, was examined by a synchrotron beam having a cross-section of 1 mm × 1 mm [27]. The rod with a diameter of 14 mm was produced by extrusion at 300 °C of precompacted Mg powder (∅74 mm). The sample for the texture measurement with a size of 3 mm × 5 mm × 14 mm was cut with the long edge along a rod diameter (Figure 3.15a). A stronger ⟨1 0 0⟩ fiber texture component parallel to the extrusion direction is observed in the mantel (sample Mg1–0) than in core (sample Mg1–4) of the extruded rod (Figure 3.15b). Looking on the (0 0 2) pole figure the texture variation can be seen as a shift of the (0 0 2) pole. Due to the high brilliance of the synchrotron X-ray, which corresponds to a small pole figure window, the detection of the orientation pole shift in a very small range is also possible. As shown in Figure 3.15c, this pole shift goes from 90° (Mg1–0) to 85° (Mg 1–4). Because of the high texture symmetry, the results are represented by intensity variation along the equator on the (0 0 2) pole figure, so-called linear pole figure.

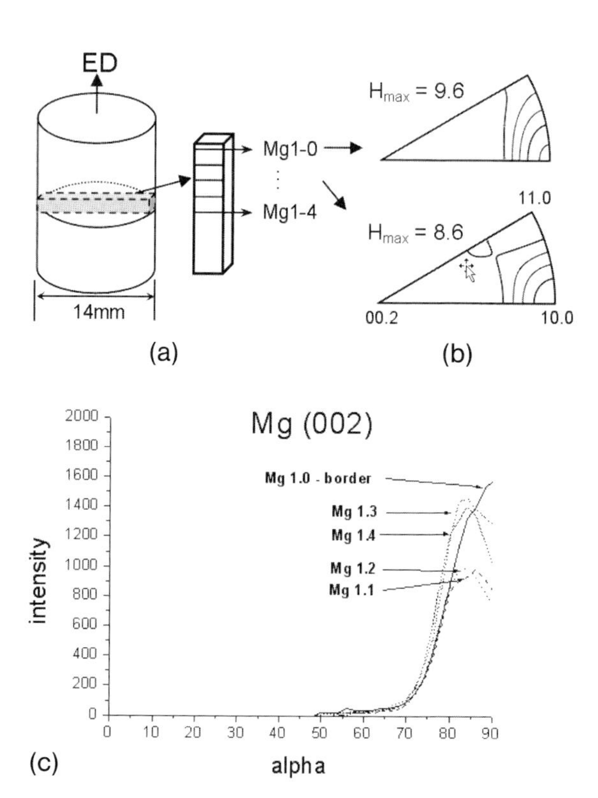


Figure 3.15 (a) Samples for texture measurements cut from an extruded Mg rod (ED stands for the extrusion direction). (b) Inverse pole figures in ED for mantle and core of the rod. (c) Intensity variation along the equator on the (0 0 2) pole figures.

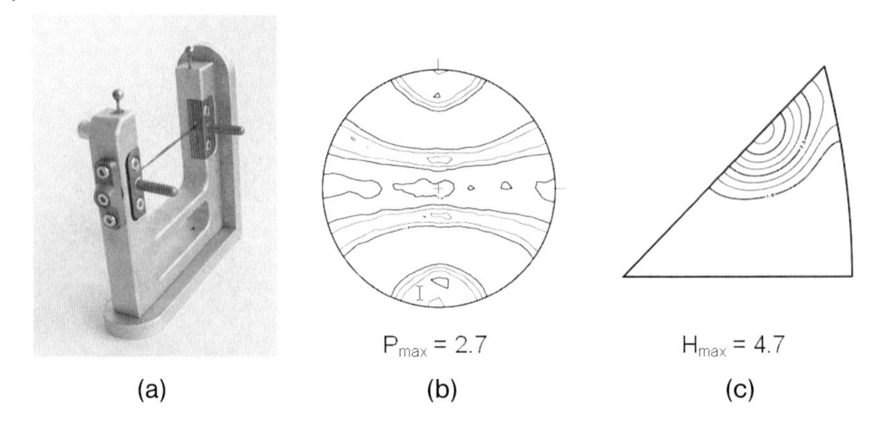


**Figure 3.16** (a) Wire sample holder with ten softly stretched Cu wires. (b) The average of six measured (1 1 1) pole figures; the wire direction is in the center of the pole figure. (c) Inverse pole figure for the wire direction.

#### 3.4.2.2 Global Texture in Cu Wire

The determination of the global texture in a Cu wire with a thickness of 0.12 mm is shown in this example [28]. Texture measurement in such a thin wire using other techniques, e.g., electron and neutron diffractions, suffers from grain statistics and low diffraction intensity, respectively. Also, a conventional X-ray measurement needs an intensity correction according to the wire geometry [29]. By employing a specially designed wire holder (Figure 3.16a), which extends slightly the wire bundle, the global texture of the thin wire was successfully obtained. For increasing grain statistics, the measurements were carried out at six different sample positions. Thereafter, the measured pole figures were averaged into one pole figure. The averaged (1 1 1) pole figure (Figure 3.16b) and the inverse pole figure in the wire direction (Figure 3.16c) show a strong $\langle 1\ 1\ 2 \rangle$ fiber texture component parallel to the wire direction.

#### 3.4.2.3 *In situ* Texture Measurement at Elevated Temperatures

Texture transformation, i.e., texture inheritance during phase transformation, was investigated by heating of a steel sample [30]. The texture of plain steel (carbon content: 0.2 mass%) sample, having BCC crystal structure ($\alpha$-ferrite), was measured at room temperature. Thereafter, the sample was heated until the phase transformation to FCC structure ($\gamma$-austenite) was completed. The texture measurement of the $\gamma$-austenite was carried out at elevated temperatures. Figure 3.17a shows the vacuum furnace used for the *in situ* texture measurement. The furnace can be installed on the $\omega$-rotation table and causes no additional diffraction peaks. Because of grain growth effects, fast measuring time is basically necessary for this *in situ* measurement. For reducing the measuring time, the $\omega$-rotation was conducted up to 45° with 5° step, which covers only a quarter of

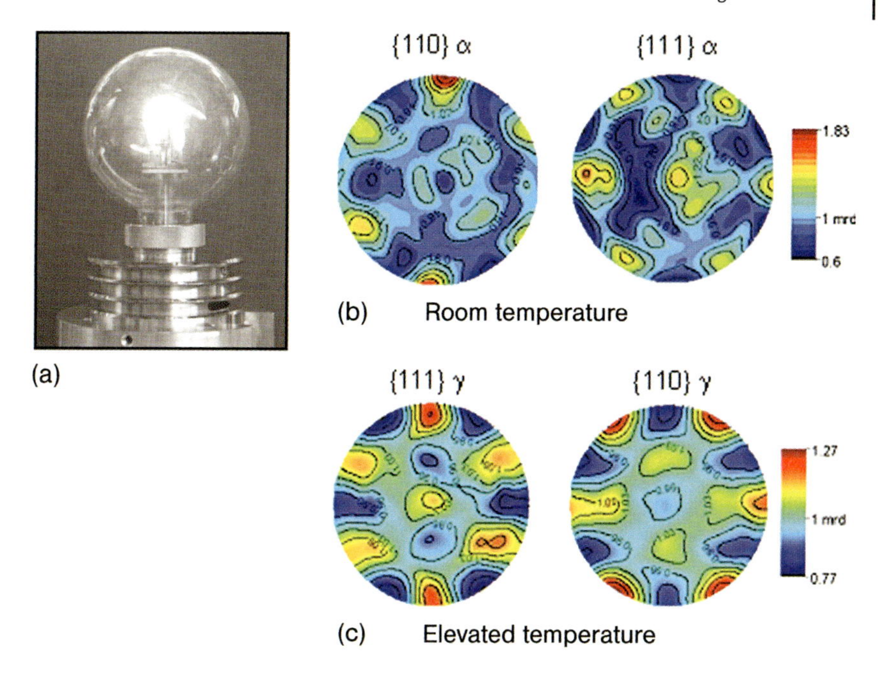


**Figure 3.17** (a) Vacuum furnace with glass wall and graphite heating element. Pole figures of (b) $\alpha$-ferrite and (c) $\gamma$-austenite as calculated using the MAUD software.

the pole figure. For extracting the complete pole figures from the restricted data, the Rietveld texture analysis program MAUD [10], which is based on whole diffraction profile analysis, was employed. The calculated pole figures of the $\alpha$-ferrite and the $\gamma$-austenite are shown in Figure 3.17b and c, respectively. The texture results match well to the Kurdjumov–Sachs orientation relationship for BCC ↔ FCC transformation: $(1\ 1\ 1)_{FCC} \| (0\ 1\ 1)_{BCC}$ and $[1\ 0\ 1]_{FCC} \| [1\ 1\ 1]_{BCC}$.

#### 3.4.2.4 *In situ* Texture Measurement Under Loading

*In situ* measurements guarantee to study the texture evolution without the influence originating from the usage of many different samples. In fact, the mechanical properties and the texture evolution are strongly dependent on the initial texture and microstructure, and this dependency, for example, becomes more pronounced in magnesium alloys having a HCP crystal structure. The texture variation was investigated during uniaxial compression of Mg alloy AZ31 [31]. AZ31 is a Mg-based grade, alloyed with Al and Zn. A universal testing machine (UTM), by which a sample can be loaded up to 20 kN, was installed on the $\omega$-rotation table (Figure 3.18). During the irradiation of the sample for texture measurement, the loading was stopped, but not released, and the sample was rotated together with the UTM around the $\omega$-axis. Figure 3.18b presents the texture vari-

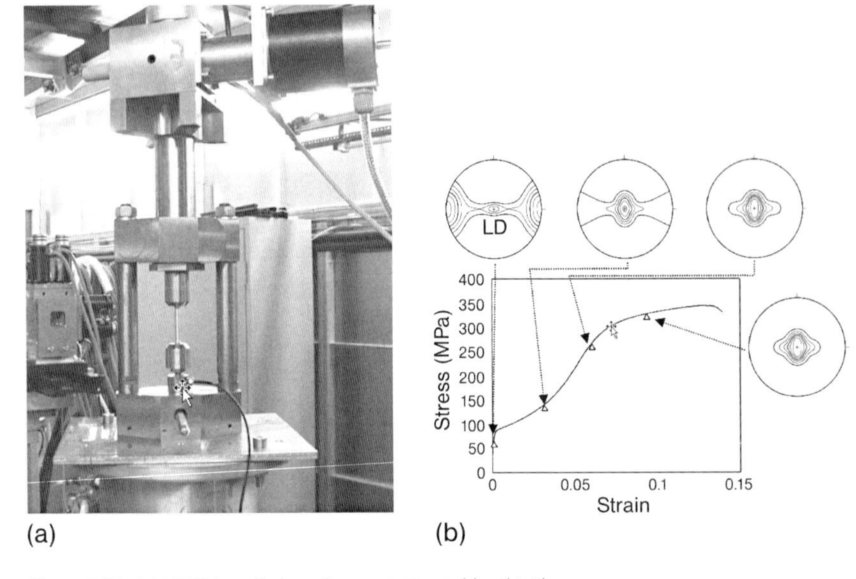


**Figure 3.18** (a) UTM installed on the $\omega$-rotation table. (b) Flow curve during compression of AZ31 and (0 0 2) pole figures at different strains. The loading direction (LD) is in the center of the pole figure.

ation in terms of (0 0 2) pole figures and the stress–strain curve during compression. By interpreting the measured texture results together with the results of a numerical simulation of texture development the anisotropic behavior and the activity of the potential deformation modes in a Mg alloy could be analyzed as a function of initial texture and deformation degree.

## References

1 G. Wassermann, J. Grewen, *Texturen metallischer Werkstoffe*, 2. Auflage, Springer, Berlin, Göttingen, Heidelberg, 1962.
2 H.J. Bunge, *Texture Analysis in Materials Science*, Cuvillier Verlag, Göttingen, 1993.
3 U.F. Kocks, C.N. Tomé, H.-R. Wenk, *Texture and Anisotropy*, Cambridge University Press, Cambridge, 1998.
4 H.J. Bunge, C. Esling, *Quantitative Texture Analysis*, DGM-Informationsverlag, Oberursel, 1986.
5 H.J. Bunge, R.A. Schwarzer, Orientation stereology–a new branch in texture research, *Adv. Eng. Mater.* 2001, **3**, 25–39.
6 H.-G. Brokmeier, S.B. Yi, N.J. Park, J. Homeyer, In-situ texture analysis using hard X-rays, *Solid State Phen.* 2005, **105**, 55–60.
7 S.C. Vogel, D. Bhattacharyya, G.B. Viswanathan, D.J. Williams, H.L. Fraser, Phase transformation textures in Ti–6Al–4V alloy, *Mater. Sci. Forum* 2005, **495**, 681–686.
8 H.-G. Brokmeier, In-situ texture analysis under applied load rays, in: *Advanced Materials* 2005 eds: M. Farooque, S.A. Rizvi, J.A. Mirza, KRL Rawalpindi, Pakistan 2007, 292–301.
9 M. Dahms, H.J. Bunge, The iterative series-expansion method for quantitative texture anaysis, *J. Appl. Crystallogr.* 1989, **22**, 439–447.

10 S. Matthies, G.W. Vinel, On the reproduction of the orientation distribution function of texturized samples from reduced pole figures using the conception of a conditional ghost correction *Phys. Status Solidi* 1982, **B112**, 111–114.

11 L. Lutterotti, http://www.ing.unitn.it/~maud/.

12 H.J. Bunge, *Experimental Techniques of Texture Analysis*, DGM Informationsgesellschaft Verlag, 1985.

13 S. Zaefferer, Investigation of the correlation between texture and microstructure on a submicrometer scale in the TEM, *Adv. Eng. Mater.* 2003, **5**, 745–752.

14 D.B. Williams, C.B. Carter, *Transmission Electron Microscopy: A Textbook for Materials Science*, Plenum Press, New York, 1996.

15 A.J. Schwartz, M. Kumar, B.L. Adams, *Electron Backscatter Diffraction in Materials Science*, Kluwer Dordrecht, 2000.

16 E.R. Rauch, A. Duft, Orienation maps derived from TEM diffraction patterns collected with an external CCD camera, *Mater. Sci. Forum* 2005, 495–497, 197–202.

17 L. Lutterotti, S. Matthies, H.-R. Wenk, A.J. Schultz, J.W. Richardson, Combined texture and structure analysis of deformed limestone from time-of-flight neutron diffraction spectra, *J. Appl. Phys.* 1997, **81**, 594–600.

18 K. Ullemeyer, P. Spalthoff, J. Heinitz, N.N. Isakov, A.N. Nikitin, K. Weber, The SKAT texture diffractometer at the pulsed reactor IBR-2 at Bubna: experimental layout and first measurements, *Nucl. Instrum. Methods Phys. Res.* 1998, **412**, 80–88.

19 S.C. Vogel, C. Hartig, R.B. Von Dreele, H.-R. Wenk, D.J. Williams, Quantitative texture measurements using neutron time-of-flight diffraction, *Mater. Sci. Forum* 2005, 495–497, 107–112.

20 H.-J. Kopineck, H. Otten, in: *Advances and Application of Quantitative Texture Analysis*, eds. H. J. Bunge, C. Esling, DGM-Informationsgesellschaft, Oberursel, 1991, pp. 153–165.

21 H.-G. Brokmeier, B. Schwebke, J. Homeyer, *Texture investigation on an engine shaft*, *Hasylab Annual Report* 2005, 541–542.

22 H.-G. Brokmeier, W. Singer, H. Kaiser, Neutron diffraction–a tool to optimize processing of niobium tubes, *Appl. Phys.* 2002, **A74**, s1704–s1706.

23 W. Ye, H.-G. Brokmeier, V. Singer, Nondestructive texture analysis of a co-extruded Cu–Nb tube, *Mater. Sci. Forum* 2002, **408**, 185–189.

24 H.R. Wenk, S. Grigull, Synchrotron texture analysis with area detectors, *J. Appl. Crystallogr.* 2003, **36**, 1040–1049.

25 H.J. Bunge, L. Wcislak, H. Klein, U. Garbe, J.R. Schneider, Texture and microstructure analysis with high-energy synchrotron radiation, *Adv. Eng. Mater.* 2002, **4**, 300–305.

26 R.V. Martins, U. Lienert, L. Margulies, A. Pyzalla, Determination of the radial crystallite microstrain distribution within an $AlMg_3$ torsion sample using monochromatic synchrotron radiation, *Mater. Sci. Eng.* 2005, **A402**, 278–287.

27 H.-G. Brokmeier, A. Günther, S.B. Yi, W. Ye, Investigation of local textures in extruded magnesium by synchrotron radiation, *Adv. X-ray Anal.* 2003, **46**, 151–156.

28 H.-G. Brokmeier, B. Weiss, S.B. Yi, W. Ye, K.D. Liss, T. Lippmann, Texture determination of thin Cu-wires by synchrotron radiation, *Mater. Sci. Forum* 2005, 495–497, 131–136.

29 T. Montesin, J.J. Heinzmann, A. Vadon, Absorption corrections for X-ray texture measurement of any shape sample, *Textures Microstruct.* 1991, **14**, 567–572.

30 H.-G. Brokmeier, S.B. Yi, B. Schwebke, J. Homeyer, *In situ* analysis of crystallographic texture using high-energy X-rays, *Z. Kristallogr.*, in press.

31 S.B. Yi, C.H.J. Davies, H.-G. Brokmeier, R.E. Bolmaro, K.U. Kainer, J. Homeyer, Deformation and texture evolution in AZ31 magnesium alloy during uniaxial loading, *Acta Mater.* 2006, **54**, 549–562.

# 4
# Physical Properties of Photons and Neutrons

*Andreas Schreyer*

## 4.1
## Introduction

Photons in the X-ray wavelength regime and neutrons share a number of physical properties which make both probes a unique tool for the investigation of materials.

Both probes are available with wavelengths on the order of atomic distances, their strength of interaction with matter is of similar magnitude, and both probes penetrate into matter from μm to many cm deep. From the interference patterns (e.g., Bragg peaks) generated, when X-rays or neutrons are scattered from matter, essential information on the structure of materials on the atomic level can be obtained in a destruction-free way.

An important distinction from electrons, which are also frequently used for scattering experiments, is the fact that electrons interact more strongly with matter due to the Coulomb interaction and penetrate less deeply into it. Therefore, electrons are only used for the study of surfaces, which is not the focus of the present book. Here we are especially interested in the capability of X-rays and neutrons to unravel the structure within materials and components in a destruction-free way, i.e., to penetrate deep into samples.

Some basic physical properties of X-rays and neutrons are listed in Table 4.1. Both particles are very different in nature: for example, the neutron has a mass, its energy depends on this mass and its velocity, and it is described by the Schrödinger equation. The X-ray photon, on the other hand, has no mass, always moves at the velocity of light, and is described by the Maxwell equations. As a consequence, neutrons of 1 Å wavelength have an energy in the meV regime, whereas the energy of X-ray photons is in the keV range. It is not surprising that the physics of the interaction of X-ray photons and neutrons with matter is also very different. However, as will be shown in the following, the formalism for describing this interaction and the strength of the interaction are so similar that the same theories can be used to describe the scattering of photons and neutrons from matter.

*Neutrons and Synchrotron Radiation in Engineering Materials Science: From Fundamentals to Material and Component Characterization*
Edited by Walter Reimers, Anke Rita Pyzalla, Andreas Schreyer, and Helmut Clemens

ISBN: 978-3-527-31533-8

**Table 4.1** Basic physical properties of neutrons and photons. The equations of motions are the Schrödinger equation (neutrons) and the Maxwell equations (photons), respectively, yielding the listed relations for momentum and energy. Some useful formulae and typical values for the energy $E$, wavelength $\lambda$ and velocity $v$ are shown as well (after [8]).

| **Neutrons** | | **X-rays (photons)** |
|---|---|---|
| Three quarks and gluons | | Point-shaped |
| $m_n = 1.675 \times 10^{-27}$ kg $\approx 1$ u | Mass | 0 |
| 0 | Charge | 0 |
| $\frac{1}{2}$ | Spin | 1 |
| $\mu_n = \gamma \cdot \mu_N$; $\gamma = -1.91$, $\mu_N = 5 \times 10^{-27}$ J T$^{-1}$ | Magnetic dipole moment | 0 |
| *Schrödinger equations* | *Wave* | *Maxwell equations* |
| *(Particle wave)* | | *(Electromagnetic wave)* |
| $\mathbf{p} = m \cdot \mathbf{v} = \hbar \mathbf{k};\ p = \frac{h}{\lambda},$ | Momentum | $\mathbf{p} = \hbar \mathbf{k}; p = \frac{h}{\lambda}$ |
| $E = \frac{1}{2} m \cdot v^2 = \frac{\hbar^2 k^2}{2m} =: k_B T$ | Energy | $E = h \cdot \nu = \frac{hc}{\lambda}$ |
| $\lambda[\text{Å}] = \frac{4000}{v[m/s]}$ | *Useful data* | |
| $E[\text{meV}] = \frac{81.8}{\lambda^2[\text{Å}^2]} = 0.086 T[K]$ | | $E[\text{keV}] = \frac{12.4}{\lambda[\text{Å}]}$ |
| "Thermal" neutrons (300 K) | *Example* | |
| $E = 25$ meV | | |
| $\lambda = 1.8$ Å | | $\lambda = 1$ Å $\equiv E = 12.4$ keV |
| $v = 2200$ m s$^{-1}$ | | |

### 4.2 Interaction of X-Ray Photons and Neutrons with Individual Atoms

A simplified picture of the process of interaction between X-ray photons or neutrons is visualized in Figure 4.1. A plane wave of wavevector $|\mathbf{k}| = 2\pi/\lambda$ (wavelength $\lambda$) along $z$ of amplitude

$$\psi_0 = e^{ikz} \tag{4.1}$$

interacts with an individual scatterer of point dimension, generating a spherical wave

$$\psi_S = b/re^{ikr} \tag{4.2}$$

centered around the scatterer. The amplitude of the scattered spherical wave at the location $r$ is determined by the parameter $b$ which has the dimension of a length; thus $b$ is called the scattering length. In general, $b$ is a complex number, where a positive real part describes a repulsive interaction, a negative real part an attractive interaction, and the imaginary part the absorption.

### 4.2.1 Neutrons

Let us first consider the scattering length for the scattering of neutrons from an individual atom. Being nuclear particles neutrons interact via the so-called strong interaction with the particles in the nucleus of the atom, namely with an ensemble of protons and neutrons, the numbers of which are given by the order number $Z$ and the mass number $A$ of the atom. Due to the extremely short range of the strong interaction and the small size of the nucleus (both on the order of fm $= 10^{-15}$ m) compared to the wavelength of the neutron (on the order of Å $= 10^{-10}$ m) the scattering process is actually well described by Figure 4.1, i.e., by assuming a scatterer of point dimension.

The values of the neutron scattering length $b_n$ turn out to be on the order of fm, i.e., the scattering is weak. $b_n$ depends sensitively on the order number $Z$ and the mass number $A$ of the scattering atom, since $Z$ and $A$ determine the number of protons and neutrons in the nucleus, respectively.

Since $b_n$ varies unsystematically throughout the periodic table a large scattering contrast between elements with similar $Z$ or between different isotopes (same $Z$, different $A$) can be readily obtained with neutrons (see Figure 4.2). This is not

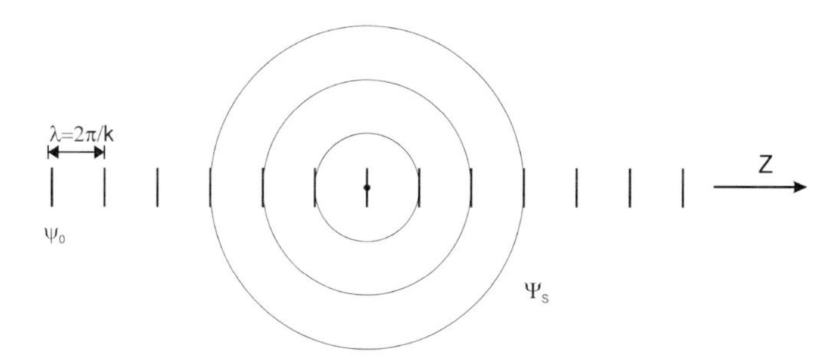


**Figure 4.1** Schematic view of the interaction between a neutron or X-ray photon represented by a plane wave $\psi_0$ of wavelength $\lambda$ with a point-like particle creating a spherical wave $\psi_S$.

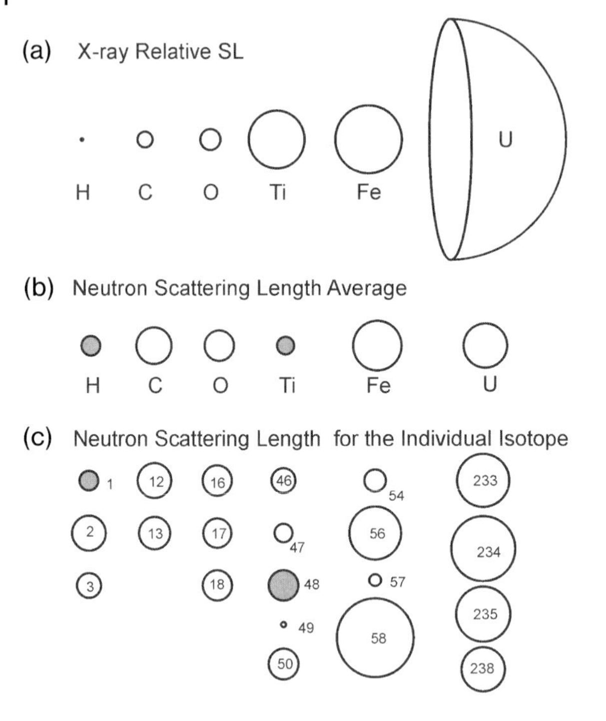


**Figure 4.2** Scattering lengths for X-rays (a) and neutrons (b), averaged over the natural isotope distribution and (c) for each isotope of the respective element) represented by circles whose radii are proportional to the scattering lengths. The circles representing the neutron scattering lengths have been multiplied by a factor of 2 compared to the X-ray scattering lengths. Shaded circles represent negative values (after [1]).

possible with X-rays. As opposed to the case of X-rays $b_n$ are negative for a few isotopes.

In the neutron wavelength regime considered here, absorption of neutrons by most isotopes is very weak leading to a very small imaginary part of the scattering length for these cases. For the problems discussed in the present book, this property is the most significant advantage of neutrons since this allows the penetration of especially large samples (see, e.g., Chapters 7 and 8).

Since in general a sample will contain different isotopes of an element the disorder in $b_n$ will lead to a so-called incoherent scattering contribution. In addition, in cases where the spin $i$ of a scattering nucleus is not zero this leads to an additional interaction with the dipole moment of the neutron. Since the nuclear spins are disordered above temperatures in the mK range, i.e., for all purposes of this book, this effect will lead to an additional spin-disorder-induced incoherent scattering. Therefore, in the case of neutrons, effective coherent and incoherent scattering lengths have to be distinguished carefully.

However, the interaction of neutrons with atoms is not always limited to the nucleus. In the case of magnetic materials, the magnetic dipole moment of the neutron leads to an additional interaction with any magnetic moment in the shells of the atoms, which is of similar magnitude as the nuclear interaction. For this reason neutrons are the classic probe of magnetic structures and magnetism is a main scientific field in neutron scattering. However, since the study of magnetism is not the main scope of this book, we do not provide any details on the magnetic interaction here.

A table of the coherent and incoherent scattering lengths including a comparison with X-ray scattering lengths can be found in chapter 2 of [1]. An in-depth discussion of the neutron nuclear scattering length is given in the article by Sears in [2]. If not stated otherwise $b_n$ denotes the coherent nuclear scattering length in the remainder of this book.

### 4.2.2 X-Rays

In the case of X-rays, on the other hand, the interaction takes place with the electronic shells of atoms or the electronic bandstructure which evolves when atoms form crystals. To begin with, let us consider the interaction of a linearly polarized photon plane wave with an individual electron (see Figure 4.1). Being an electromagnetic wave, the oscillating electric-field component of the plane wave, defined to be parallel to $x$, will couple to the charge of the electron and cause its oscillation along $x$. This oscillation of the electron in turn generates an electromagnetic dipole field, which is considered to be the scattered wave. Quantitatively, this scattered wave is described by a dipolar field instead of the simpler spherical wave introduced above. In Figure 4.3 this dipolar field in the $x$–$z$ plane around the oscillating electron is shown schematically.

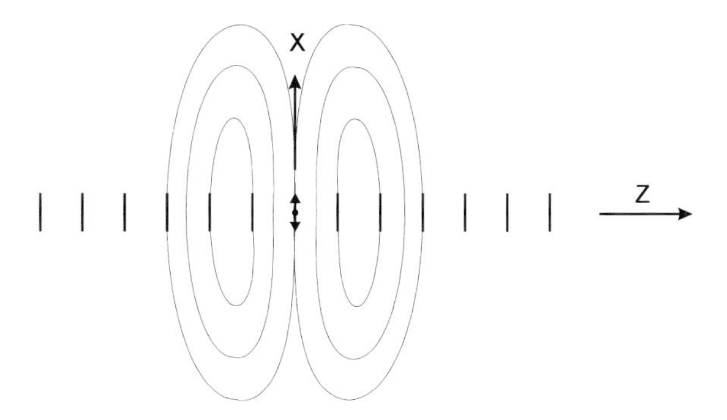


**Figure 4.3** Schematic representation of the interaction of an electromagnetic plane wave propagating along $z$ with a free electron. The electric field oscillating along $x$ causes an oscillation of the electron along $x$ which creates a dipolar field. In the orthogonal direction ($y$–$z$ plane) Figure 1 holds.

A very important consequence is that no intensity is emitted along the oscillating direction of the electron, i.e., along $x$. Furthermore, in the $x$–$z$ plane, the scattered intensity strongly depends on the scattering angle (see Figure 4.3), whereas due to the characteristics of the dipolar field in the $y$–$z$ plane Figure 4.1 holds, i.e., the scattered intensity is independent of the scattering angle.

Since synchrotron sources generate linearly polarized light, it is essential to take care of the principles outlined above. In general, this is achieved by making sure that the scattering experiment is performed in the $y$–$z$ plane to avoid the anisotropy of the scattering shown in Figure 4.3. Since due to their size storage rings are horizontal, producing horizontally polarized light, the scattering angle is usually measured in the vertical direction ($y$–$z$ plane). In cases, where the scattering angle deviates from the $y$–$z$ plane (e.g., when large two-dimensional detectors are used) a correction factor has to be applied, which accounts for the form of the dipolar field in the $x$–$z$ plane [3].

It is actually quite straightforward to calculate the scattering length of the scattering process introduced above [3, 4]. The scattering length of this so-called Thomson scattering from a free electron turns out to be

$$r_e = e^2/m_e c^2 \tag{4.3}$$

where $e$ is the electron charge, $m_e$ is the electron mass, and $c$ is the velocity of light. This Thomson scattering length $r_e$ has the value 2.82 fm, i.e., it has the same magnitude as the neutron scattering length $b_n$.

In the next step we have to take care of the fact that we are interested in the scattering from all electrons in the atomic shells. Since these have a dimension on the order of Å, i.e., five orders of magnitude larger than $r_e$, the assumption of a point-like scatterer does not hold for the scattering of X-rays from atoms. To account for this effect, an integral over the scattering contributions from all electrons within the shell accounting for the phase difference due to their different location within the shell has to be performed. This integral over the volume of the shell is the so-called form factor

$$f_e = \int \rho_e(r) e^{iqr}\, dv \tag{4.4}$$

with the electron density $\rho_e$, and the scattering vector

$$\mathbf{q} = \mathbf{k}_f - \mathbf{k}_i \tag{4.5}$$

where $\mathbf{k}_i$ and $\mathbf{k}_f$ are the incident and final wavevectors before and after the scattering process, respectively (see Figure 4.4) [3, 4]. From simple geometrical considerations it can be shown that

$$|\mathbf{q}| = 4\pi/\lambda \sin\theta \tag{4.6}$$

where $2\theta$ is the scattering angle between $\mathbf{k}_i$ and $\mathbf{k}_f$. The scattering vector $\mathbf{q}$ is the fundamental quantity which is used to describe any scattering process.

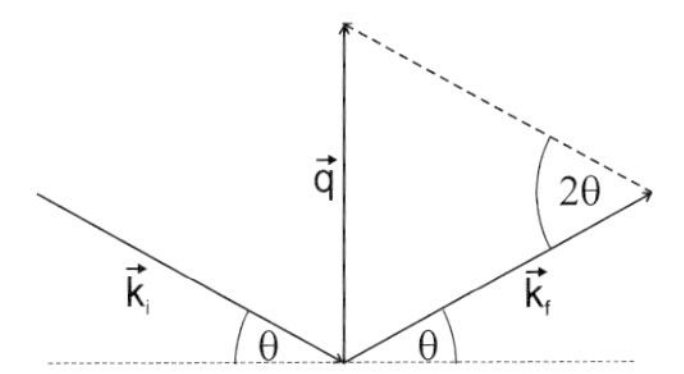


**Figure 4.4** Geometrical construction of the scattering triangle defined by $\mathbf{q} = \mathbf{k}_f - \mathbf{k}_i$ with the scattering angle $\theta$.

For small $\mathbf{q}$, i.e., in the small angle regime $f_e \approx Z$, where $Z$ is the number of electrons (order number) of the atom. With increasing $\mathbf{q}$ a rapid decay of the form factor occurs (see Figure 4.5).

The complete scattering length for X-rays is given by

$$b_X = r_e(f_e + \Delta f_e' + i\Delta f_e'') \tag{4.7}$$

where $\Delta f_e'$ and $\Delta f_e''$ are the real and imaginary parts of the dispersion correction near resonances, where the incoming X-rays can excite core electrons of the atom. The scattering contrast between elements with similar $Z$ can therefore be enhanced significantly by tuning the X-ray wavelength to a resonance of one element inducing a major dispersion correction.

The imaginary part of $b_X$ describes the absorption and is related to the more common length absorption coefficient $\mu_l$ by

$$\mu_1 = 2\lambda n_0 r_e \Delta f_e'' \tag{4.8}$$

with the atomic number density $n_0$ [5]. Compared to neutrons the absorption of X-rays is significantly higher. According to Eq. (4.8) it can, however, be strongly reduced by reducing $\lambda$, i.e., by increasing the X-ray energy accordingly. This

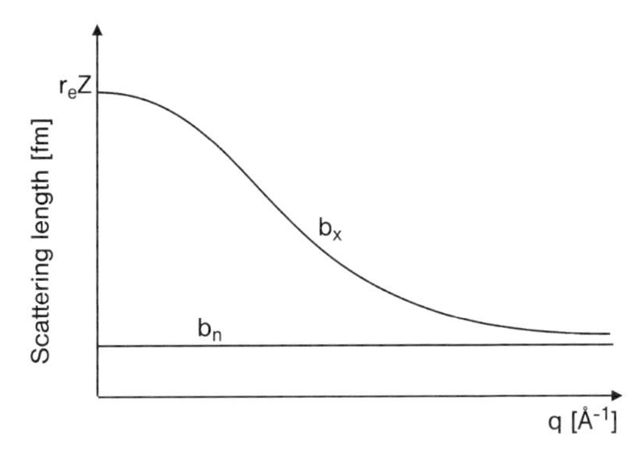


**Figure 4.5** Comparison of the $q$ dependence of the scattering lengths for X-rays and neutrons. Whereas $b_X$ strongly depends on $q$, $b_n$ is constant.

scheme is commonly used in materials research (see Chapters 9 and 10.1). In addition, absorption is the most important contrast mechanism for imaging methods, which are of increasing significance in materials research (see Chapters 15–17).

The values of the form factors and the dispersion corrections are tabulated in the International Tables [6]. An excerpt can be found, e.g., in [3].

In addition to the Thomson scattering discussed so far, a very much weaker spin-dependent contribution to the scattering of X-rays from magnetic materials exists, which can be enhanced using resonance effects similar to those leading to the dispersion correction [4]. As for the case of neutrons, we do not elaborate on this contribution to the scattering here.

In Figure 4.5 a qualitative comparison of the $q$-dependence of $b_X$ and $b_n$ is shown for the case of a negligible dispersion correction. Whereas $b_n$ is constant $b_X$ increases significantly for small $q$ and can be larger than $b_n$ for large $Z$.

Another useful comparison is made in Figure 4.2, where $b_n$ and $b_X$ are plotted as a function of the order number $Z$. Whereas $b_X$ increases continuously with $Z$, $b_n$ varies irregularly and also depends on the mass number $A$. The latter property actually allows a variation of $b_n$ by isotope substitution. This technique is very important for the investigation of hydrogenous materials, where the exchange of hydrogen (negative scattering length, see Figure 4.2) against deuterium (large positive scattering length) allows an isotope labeling.

## 4.3
## Scattering of X-Ray Photons and Neutrons from Ensembles of Atoms

We have seen that $b_n$ and $b_X$ are of similar magnitude and quite small. This has the important consequence that the probability that one photon or neutron is scattered more than once by an ensemble of atoms is small as well, i.e., multiple scattering effects can often be neglected. Based on this assumption the scattering from an ensemble of atoms can be calculated similar to the case of the form factor introduced above. Instead of an integral, a discrete sum over the scattering contributions from all atoms accounting for the phase difference due to their different location within the ensemble is performed. The amplitude of the scattered wave then is

$$A(\mathbf{q}) = \sum b e^{iqr} \tag{4.9}$$

where the sum runs over all atoms with the scattering length $b_n$ or $b_X$, respectively. The scattered intensity $I(\mathbf{q})$ is then obtained by multiplying with its complex conjugate

$$I(\mathbf{q}) = A(\mathbf{q})A^*(\mathbf{q}) = \left|\sum b e^{iqr}\right|^2 \tag{4.10}$$

For the present book the specific case of scattering from a crystal, i.e., from a regular periodic three-dimensional array of atoms is most relevant. It can be shown [1, 3, 4, 7] that for an infinite rigid crystal

$$I(\mathbf{q}) \sim \sum_{\mathrm{G}} \delta(\mathbf{q}-\mathbf{G})|F(\mathbf{G})|^2 \tag{4.11}$$

with the structure factor

$$|F(\mathbf{G})| = \sum_{\mathrm{m}} b_{\mathrm{m}} \mathrm{e}^{\mathrm{i}\mathbf{G}\boldsymbol{\rho}} \tag{4.12}$$

where $\boldsymbol{\rho}$ are the positions of the atoms within a unit cell describing the crystal. The sum runs over all potentially different atoms within this unit cell. The so-called reciprocal lattice vectors $\mathbf{G}$ are related to the lattice via the lattice spacings $d_{hkl}$:

$$|\mathbf{G}_{hkl}| = 2\pi/d_{hkl} \tag{4.13}$$

According to Eq. (4.11) the scattered intensity $I(\mathbf{q})$ has strong interference maxima ($\delta$-functions), so-called Bragg peaks, whenever

$$\mathbf{q} = \mathbf{G} \tag{4.14}$$

where $\mathbf{G}$ is directly related to the lattice spacings $d_{hkl}$ via Eq. (4.13). For details, e.g., on the definition of the unit cell and the Miller indices *hkl*, which describe the lattice planes in a crystal, see, e.g., [1, 3, 4, 7]. The structure factor (Eq. 4.12) contains information on the symmetry of the unit cell and determines the relative intensity of the Bragg peaks. A very important consequence is that for certain reflections the structure factor can become zero, implying that certain reflections do not occur for a given unit cell. In [3] the structure factors are calculated for the most common crystal structures.

From Eqs. (4.6), (4.13) and (4.14) the well-known Bragg equation is obtained as

$$n\lambda = 2d_{hkl} \sin\theta \tag{4.15}$$

From the measured $I(\mathbf{q})$ the crystal structure can be obtained including the position of all atoms (phase analysis), variations in $d_{hkl}$ as a function of location within the sample can be measured (strain analysis, see Chapters 6 to 8), or the orientation distribution of crystallites in a sample can be determined (texture analysis, see Chapter 3).

For a real nonrigid crystal at finite temperatures Debye Waller factors must be included to account for the temperature-induced deviation of the atoms from their lattice positions [1].

In practice, crystals are neither infinitely large nor perfect. Instead of infinitely sharp Bragg peaks ($\delta$-functions, Eq. 4.11) the finite number of coherently scattering planes leads to peaks of, e.g., Gaussian or Lorenzian shape, depending on the kind of disorder. The width of the peak is a measure for the number of coherently scattering planes. A common approximation for the determination of the crystal size is the Debye Scherrer formula given, e.g., in [3].

The formalism outlined above can also be modified to describe scattering from noncrystalline matter, as detailed, e.g., in Chapter 12 for small angle scattering.

The scattering theory introduced here is called the kinematical scattering theory. As mentioned above it neglects multiple scattering. In large perfect crystals, however, multiple scattering can occur even for small scattering lengths since the photon or neutron has a nonnegligible probability to be scattered more than once. The reason is that a scattered neutron/photon still fulfils the Bragg condition, thus its probability to be scattered more than once increases with the quality and size of the crystal. Consequently, a second scattering process can occur, leading to a reduction of the scattered beam intensity resulting from the first scattering process. This effect is called extinction. Therefore, extinction corrections have to be applied in these cases within the kinematical theory [5].

Alternatively, the dynamical scattering theory can be used, which takes multiple scattering into account [3, 4]. It is based on the solution of the Schrödinger/Maxwell equation(s) for the given sample for neutrons/X-rays, respectively (see Table 4.1). Unlike the kinematical theory, the dynamical theory also takes into account optical effects like refraction and total reflection which can become dominant for small scattering vectors $\mathbf{q}$, when surfaces or interfaces are involved [4, 5].

For the purpose of the present book, which deals with the investigation of polycrystalline materials, the simpler kinematical scattering theory is sufficient in general. As outlined above, the same formalisms apply for neutrons and X-rays, making a complementary use of both probes easy for the user.

## Acknowledgment

The drawing of the figures by Danica Solina is gratefully acknowledged.

## References


1 G. E. Bacon, *Neutron Diffraction*, Clarendon Press, Oxford, 1975.

2 K. Sköldand D. L. Price(eds.), *Neutron Scattering in Methods of Experimental Physics*, Part A, vol. 23, Academic Press, Orlando, 1986.

3 B. E. Warren, *X-Ray Diffraction*, Dover, New York, 1990.

4 J. Als-Nielsen and Des McMorrow, *Elements of Modern X-Ray Physics*, Wiley, New York, 2001.

5 Varley F. Sears, *Neutron Optics*, Oxford University Press, Oxford, 1989.

6 International Tables for X-Ray Crystallography, International Union of Crystallography, Vol. C, 2004. For more details see http://www.iucn.org.

7 Stephen W. Lovesey, *Theory of Neutron Scattering from Condensed Matter*, vols. 1 and 2, Clarendon Press, Oxford, 1987.

8 Magnetische Schichtsysteme in Forschung und Anwendung, Hrsg. Institut für Festkörperforschung des Forschungszentrums Jülich GmbH.

# 5
# Radiation Sources

## 5.1
## Generation and Properties of Neutrons

*Wolfgang Knop, Philipp Klaus Pranzas, and Peter Schreiner*

### 5.1.1
### Introduction

Neutron scattering is one of the most effective ways for the study of microstructure and dynamics of condensed matter. A wide scope of problems, ranging from physics, chemistry, biology, medicine, environmental research, geology, polymers, and material science can be investigated with neutrons. For this reason, a high neutron flux $\Phi_{th} > 10^{13}$ n/s cm$^2$ with different spectra including cold, thermal, and hot neutrons is required from the neutron supplier/operator. Aside from the important neutron scattering techniques, nondiffractive methods like imaging techniques and isotope production can also be applied, which are of increasing relevance for industrial applications.

### 5.1.2
### Generation of Neutrons

For the supply of high neutron fluxes $>10^{13}$ n/s cm$^2$ with a high availability worldwide two production methods are at the disposal: research reactors and spallation sources [1].

#### 5.1.2.1 Research Reactors

Here the neutrons are produced by the nuclear fission process [2] as schematically shown in Figure 5.1.1.

The Uranium-235 nucleus has a high fission cross-section for thermal neutrons, thus U-235 breaks into lighter elements and liberates 2 to 3 neutrons for every fissioned element (average value 2.5).

The fast neutrons generated during the nuclear fission have a energy within the MeV range. The deceleration of these fast neutrons into the thermal energy

*Neutrons and Synchrotron Radiation in Engineering Materials Science: From Fundamentals to Material and Component Characterization*
Edited by Walter Reimers, Anke Rita Pyzalla, Andreas Schreyer, and Helmut Clemens

ISBN: 978-3-527-31533-8

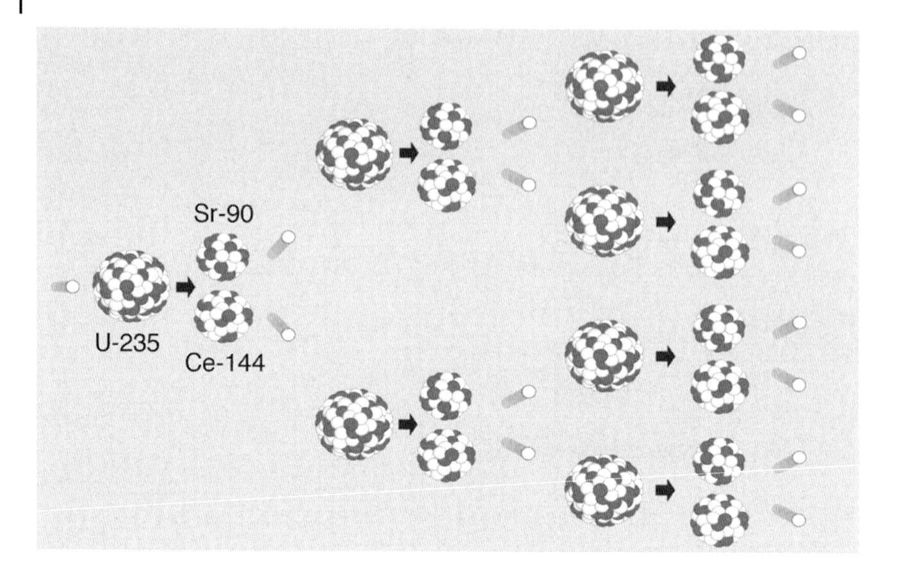


Figure 5.1.1 The fission process in a nuclear reactor [3].

region (meV) results via impacts with moderator atoms (Figure 5.1.2). This is necessary to maintain the fission process and to exploit them for experiments.

A good moderator medium should have a small neutron absorptive capacity as well as a high probability of scattering (deceleration by impacts). In Table 5.1.1 the characteristics of typical moderator materials are presented.

From this it becomes clear that deuterium and beryllium are very good moderator materials (small absorption coefficients, strong scattering effect). Hydrogen (or actually water) possesses a high scattering length, however, a higher neutron absorption coefficient. Therefore, heavy water reactors or beryllium encased cores are mainly used for moderation.

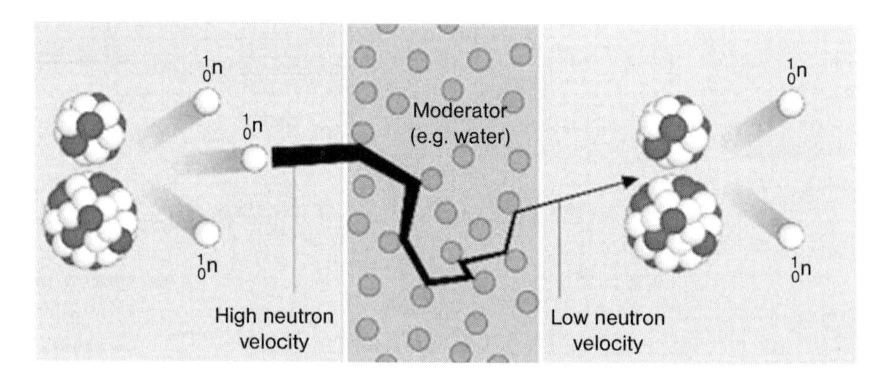


Figure 5.1.2 Moderation of neutrons [3].

**Table 5.1.1** Characteristics of moderator materials.

| Element | Absorption ($cm^{-24}$) | Scattering ($cm^{-24}$) |
|---|---|---|
| H | 0.33 | 82 |
| D | 0.00052 | 7.64 |
| Be | 0.0077 | 7.63 |

An example for the principal layout of a research reactor is shown in Figure 5.1.3.

Figure 5.1.3 shows the new German high flux neutron source Heinz Maier–Leibnitz (FRM II) in Garching near to Munich, Germany, with a nominal thermal power of 20 MW. The compact core (one fuel element) is in a central cooling duct and is cooled with light $H_2O$. According to Table 5.1.1, hydrogen (H) has a high absorption coefficient. Therefore, a moderator tank with heavy water ($D_2O$) is built around the cooling duct. For $D_2O$ the loss of neutrons is significantly smaller compared to $H_2O$. Therefore, more neutrons are available for the beam tubes at the experiments.

Another example is the Materials Testing Reactor (MTR)-type swimming pool research reactor FRG-1 with a thermal power of 5 MW, operated by the GKSS-Research Centre Geesthacht GmbH. In order to make available a high neutron flux for the neutron scattering experiments seven radial tubes (see Figure 5.1.4) and one tangential beam tube are positioned at one side of the core. The tangential beam tube has a very low background as it does not have a direct view to the core. For the reduction of measurable background from gamma radiation and fast neutrons at the experimental positions, a system of curved neutron guides is installed at three of the radial beam tubes.

The beam tubes (AlMg3) are approx. 310 cm long with a diameter of either 15 or 20 cm. One tube has been extended to 35 cm diameter in parts to accommodate the neutron guide system. At its inner end this tube is facing the cold neutron source (CNS). The CNS is used to decelerate neutrons starting with a speed of about 2.2 km/s down to 0.5 km/s, the so-called "cold neutrons."

#### 5.1.2.2 Spallation Sources

The principle process which takes place in a spallation source [5] is schematically shown in Figure 5.1.5: Protons, generated in a large accelerator facility dislodge one or two nucleons with high energy from a nucleus through direct impact and transfer a large amount of energy into the entire volume of the nucleus. This energy is then released from nucleus by evaporating up to 10 spallated neutrons compared to 2 or 3 neutrons for each fusion process.

Independently of the kind of the generation of neutrons (nuclear fission, spallation) the facilities for both reactions are similar in principle. The emerging fast

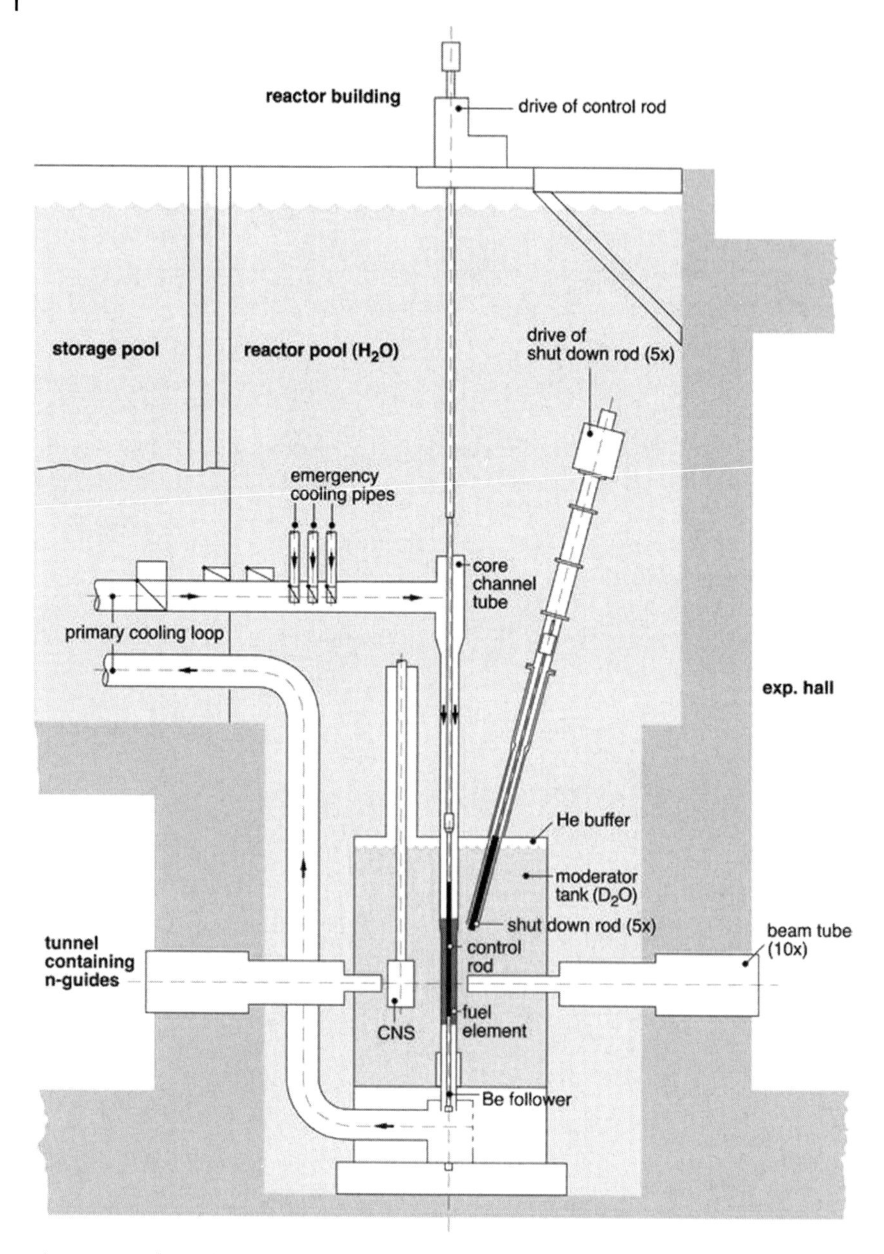


**Figure 5.1.3** The High Flux Reactor FRM II in Garching near to Munich, Germany [4].

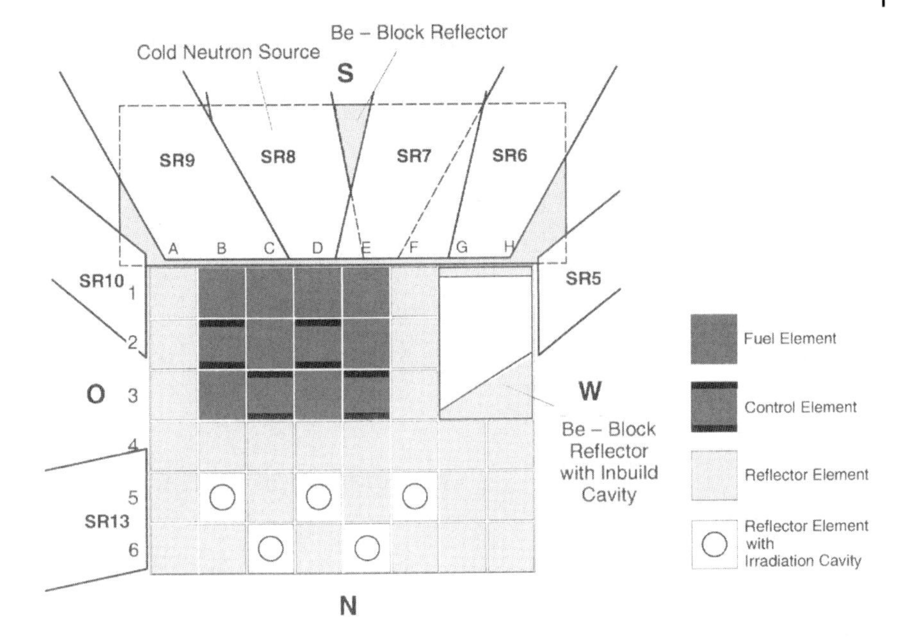


**Figure 5.1.4** FRG-1 compact core.

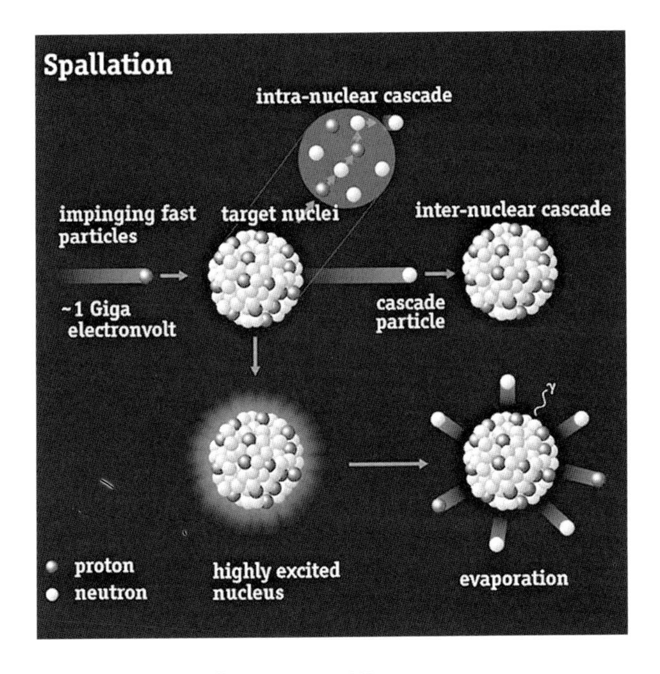


**Figure 5.1.5** The spallation process [6].

**Table 5.1.2** Neutrons needed for experimental facilities.

Cold, hot and thermal neutrons

- The fast neutrons generated during nucleus fission exhibit energy of about one million electron volt – too high for practical use.
- Thus, they are forced to collide with a special material (moderator) and lose energy down to useable energy levels
- For scientific experiments we need
  - hot neutrons (0.1–0.5 eV) moderated at 2000 °C
  - thermal neutrons (0.01–0.1 eV) moderated at 40 °C
  - cold neutrons (0–0.01 eV) moderated at −250 °C

neutrons must be moderated (Figure 5.1.2) in order to use them for different experiments.

For these experimental facilities neutrons with different spectra (wavelengths) are needed (Table 5.1.2). Corresponding moderator vessels are foreseen at neutron sources as described above by way of example for the cold neutron source of the FRG-1.

## 5.1.3 Instrumentation

Neutrons are used as radiation source for structural investigations in a large number of different scattering instruments. As an example, the experimental hall of the FRG-1 at GKSS is shown in Figure 5.1.6.

The instrumentation of the Geesthacht Neutron Facility (GeNF) is focused on materials science research due to a strong research effort of GKSS in this field. GeNF offers the possibility to study structural properties of materials over a wide size range from the atomic scale up to micrometers [7]. About half of the instruments – marked with a blue label – use cold neutrons from the CNS, the other half thermal neutrons. At two diffractometers Ares-2 and FSS residual stresses are analyzed in metal alloys and welds. At the instrument Tex-2 texture in metals and geological samples is studied. Structures with sizes between 1 nm up to the micrometer scale are investigated at the small-angle scattering instruments SANS-1 and SANS-2 and at the double crystal diffractometer (DCD) using ultrasmall angles. Magnetic and nonmagnetic nanostructures are studied at the reflectometers, NeRo and PNR, as well as at the diffractometer, POLDI. Nondestructive static and dynamic imaging is performed at the neutron radiography and tomography facility, GENRA-3. An additional application is the irradiation of samples (ICI) for universities, industry and research centers, e.g., for the production of radionuclides. An up-to-date overview of neutron facilities worldwide is available in [8].

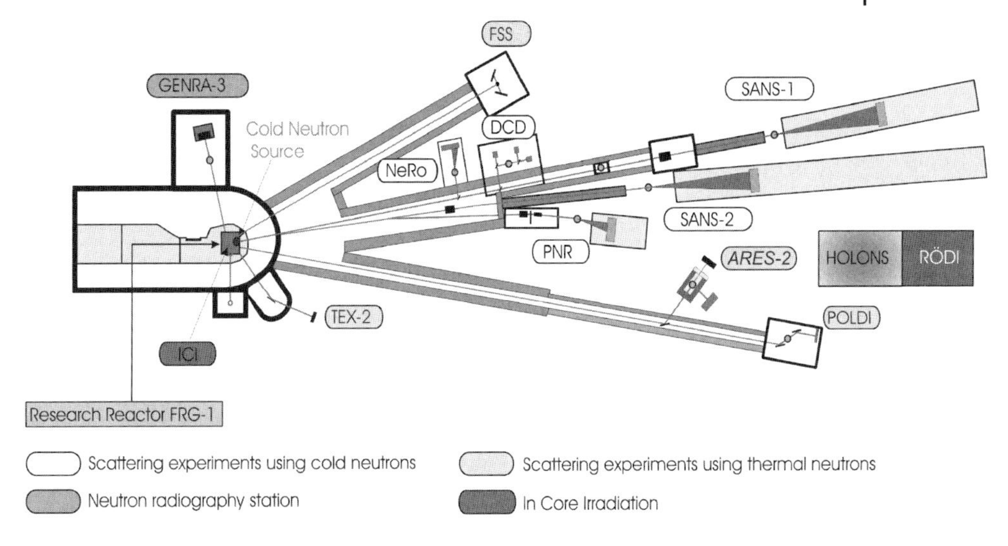


**Figure 5.1.6** Instrumentation in the experimental hall of the FRG-1 [7].

## References


1 Neutron Data Booklet, A.J. Dianoux and G. Lander(editors), Institute Laue Langevin, available via http://www.ill.fr.
2 S. Glasstone and M.C. Edlund, *Elements of Nuclear Reactor Theory*, Van Nostrad, Princeton, NJ, 1952.
3 Basiswissen Kernenergie, Martin Volkmer, Informationskreis Kernenergie; Berlin, Germany, available from (http://www.kernenergie.de/r2/documentpool/de/Gut_zu_wissen/Materialen/Downloads/018basiswissen2006.pdf.)
4 http://www.frm2.tum.de.
5 J. M. Carpenter, Pulsed Spallation Sources for Slow Neutrons, *Nucl. Instr. Meth.* 91–113 (1977).
6 http://neutron.neutron-eu.net/n_ess.
7 http://genf.gkss.de.
8 http://www.neutron.anl.gov/.


## 5.2 Production and Properties of Synchrotron Radiation

*Rolf Treusch*

### 5.2.1 Introduction

It is well known that charged particles emit electromagnetic radiation when accelerated. One example from classical electrodynamics is the accelerated (and decelerated) motion of electrons in an antenna yielding the well-known $\cos^2$

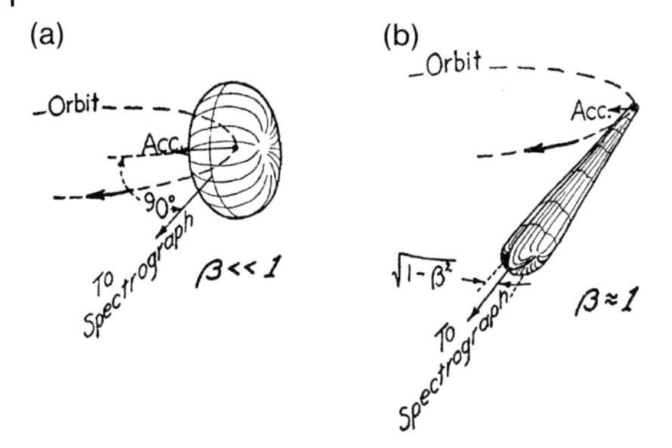


**Figure 5.2.1** Angular distribution of radiation emitted from charged particles on a circular orbit: (a) nonrelativistic velocities, (b) relativistic case (from [6]).

Θ-distribution of the Hertz dipole (see, e.g., [1]). The same effect also forms the basis for the today widely used synchrotron radiation. Synchrotron radiation (in the following mostly "SR" for brevity) is generated in particle accelerators when charged particles are accelerated to relativistic velocities (with a Lorentz factor $\gamma$ of typically a few 100 to several 1000) and deflected in magnetic fields, e.g., in order to keep them on a circular orbit. Due to their high ratio of charge to mass ($q/m$), SR is particularly pronounced for electrons and positrons. Synchrotron radiation got its name because it was first observed at the General Electric 70 MeV Synchrotron in Schenectady, New York in 1947 [2].

More than 100 years ago, at the end of the 19th century, Larmor already worked out the theoretical description of the radiation of electrons on a circular orbit [3]. In contrast to the Hertz dipole, motion and acceleration vectors are not parallel, but perpendicular here. The centripetal force which keeps the electron on its orbit by accelerating it toward the center is typically the Lorentz force of a dipole magnet. Larmor's work dealt with slow, nonrelativistic electrons and yielded a doughnut-like pattern for the spatial distribution of the radiation (Figure 5.2.1a). For the transition to the case of relativistic motion, basically a Lorentz transformation has to be applied to Larmor's classical treatment. As worked out by Ivanenko and Pomeranchuk [4] and Schwinger [5], this Lorentz transformation results in a strongly forward directed radiation cone with a typical opening angle of $1/\gamma$ in the range of a few mrad down to some tenths of a mrad (Figure 5.2.1b).

In the first particle accelerators, which were mainly used for nuclear and high-energy physics, SR was a rather undesirable byproduct, since it significantly affected the machine performance. The radiation power per particle is

$$P \simeq \frac{2e^2c}{3\rho^2} \cdot \beta^4 \cdot \left(\frac{E}{mc^2}\right)^4 = \frac{2e^2c}{3\rho^2} \cdot \beta^4 \cdot \gamma^4 \tag{5.2.1}$$

where $E$ is the particle energy, $e$ its charge, $mc^2$ is the particle rest mass, $\beta = v/c$ is the speed of the particles with respect to the speed of light, and $\rho$ is the orbit radius,

$$\rho(\mathrm{m}) = 3.34 \frac{E(\mathrm{GeV})}{B(\mathrm{T})} \tag{5.2.2}$$

It is evident that for proton accelerators this radiation loss is in most cases negligible, since – due to the much higher mass – it is about $10^{13}$ times smaller than for electrons or positrons. For the latter two, the radiative energy loss per turn is (in practical units)

$$I_{\mathrm{tot}}[W] \simeq 88.5 \frac{E^4(\mathrm{GeV})}{\rho(\mathrm{m})} j(\mathrm{mA}) \tag{5.2.3}$$

where $j$ is the circulating current. For today's machine energies of a few GeV and currents of some 100 mA it typically is in the range of several 100 kW to few megawatts and hence, apart from the more technical problems in accelerator design, a substantial part of the electricity bill. Scaling with $E^4/\rho$ it also sets a "natural" limit to the upscaling in energy for particle physics accelerators which have today, e.g., with the 27 km circumference of LEP at CERN, already reached enormous dimensions. For much higher energies, it is obviously advisable to scale up the radius $\rho$ to infinity, i.e., use a linear accelerator in order to avoid the production of synchrotron radiation. Whereas in the sixties and seventies of the last century, SR has been mostly used in a "parasitic" mode at the particle physics accelerators, the growing demand during the last 40 years resulted in a continuously growing number of facilities ("storage rings," see Section 5.2.3) which are exclusively dedicated to the generation of SR. Presently, about 50 facilities around the world [7] serve thousands of users in a wide variety of fields in fundamental and applied research, ranging from physics, biology, chemistry across crystallography, materials and geological sciences to medical applications. SR is well suited for all these fields since it possesses the following outstanding properties:

1. continuous spectrum from the infrared to X-rays,
2. high intensities,
3. small source size,
4. collimation in forward direction,
5. high degree of linear polarization in the orbit plane,
6. elliptical or circular polarization above and below the orbit plane (for bending magnets),
7. well-defined, pulsed time structure (short electron/positron bunches in machine),
8. quantitatively known characteristics,
9. clean (ultra high vacuum) environment.

These properties will now be detailed (additional background information, also on beamline optics and experimental techniques can be found in [8–12]):

### 5.2.2
### Properties of Synchrotron Radiation

(1) *Spectrum*:
Due to the strong forward collimation of the radiation and the circular motion of the electron/positron bunches in the storage ring (see Section 5.2.3), it very much resembles a sweeping search light. An external, stationary observer who looks tangentially onto the circular orbit only sees a short flash of radiation every time the bunches pass by. Due to the shortness of the pulses, one observes besides the rotation frequency $\omega_o$ also its higher harmonics up to a typical frequency of

$$\omega_{\text{typ}} \simeq \frac{3\pi c\gamma^3}{2\rho} = \frac{3\pi}{2}\omega_o\gamma^3 \tag{5.2.4}$$

Since the rotation frequency $\omega_o$ is not exactly constant due to so-called betatron and synchrotron oscillations around the nominal orbit (see, e.g., [13]), the infrared harmonics are already smeared out to a continuum. For bending magnet radiation, the spectrum as a function of photon energy can be displayed in "universal curves" (Figure 5.2.2) with one characteristic parameter, the critical energy

$$\varepsilon_c(\text{keV}) = 0.665 \cdot E^2(\text{GeV}) \cdot B_0(\text{T}) \tag{5.2.5}$$

which is defined such that 50% of the total radiation energy is emitted above and below $\varepsilon_c$, respectively. $B_o$ is the magnetic field of the bending magnet which is typically around 1 T.

The continuous spectrum of synchrotron radiation allows to tune the photon energy to the experimental demands using an appropriate monochromator, e.g., a grating or crystals. Figure 5.2.3 illustrates the length scales corresponding to

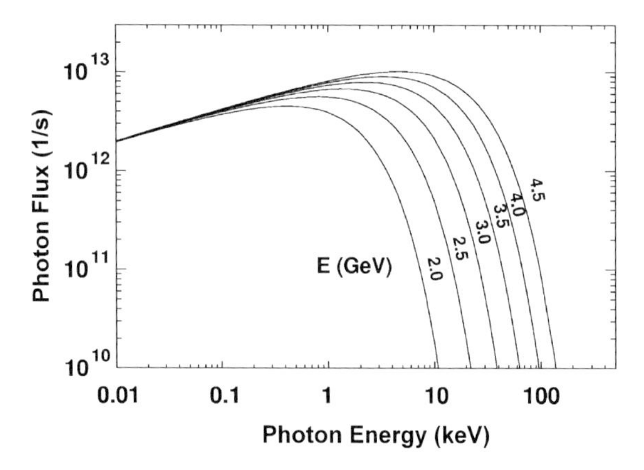


**Figure 5.2.2** "Universal curves" for the spectral distribution of photon flux as a function of machine energy.

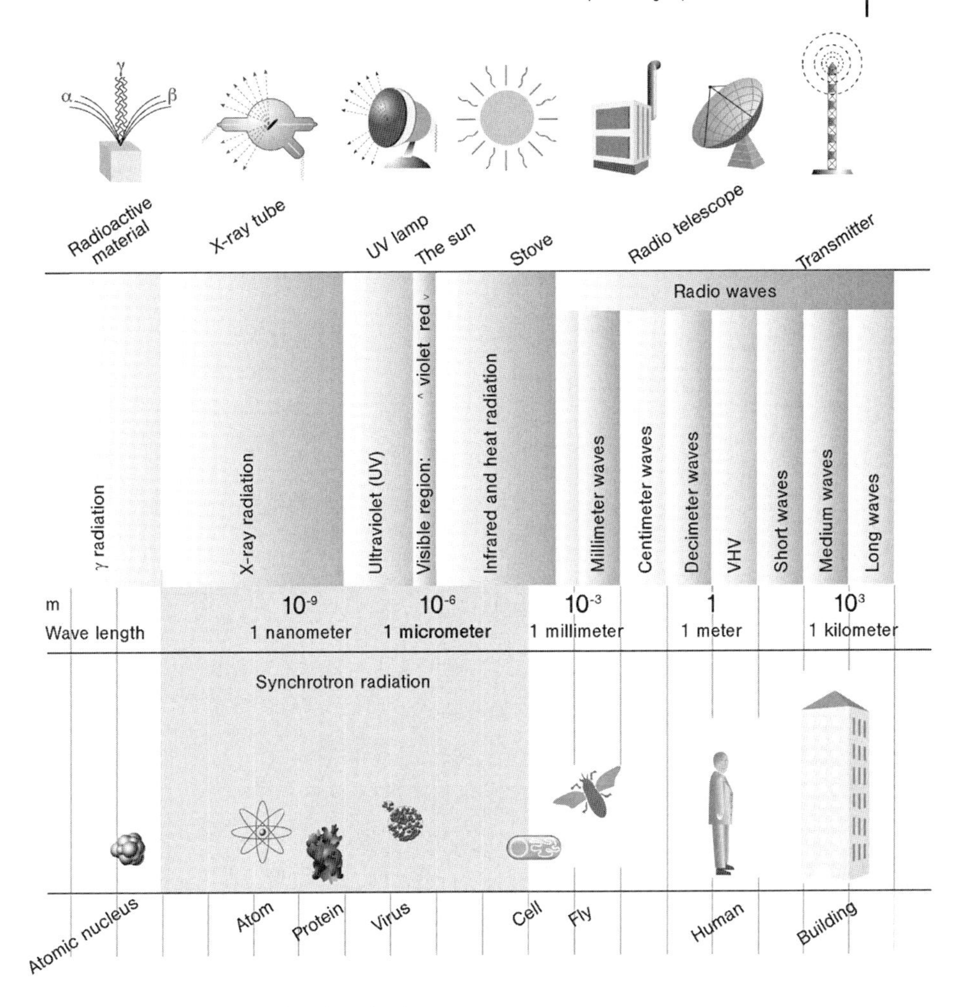


**Figure 5.2.3** Length and wavelength scales from macrosopic across microscopic toward nuclear dimensions.

typical wavelength ranges of today's SR sources. Storage rings with energies below 1 GeV are best suited for the infrared (IR) to ultraviolet (UV) part of the spectrum, whereas multi-GeV storage rings are required for the X-ray range with wavelengths around 1 Å and access to corresponding atomic length scales for scientific investigations.

(2) *Intensities*:
Usually, intensities of SR sources are compared in terms of brilliance, which is the number of photons per time interval (s), per source size ($mm^2$), per opening angle ($mrad^2$) and per 0.1% spectral bandwidth. Figure 5.2.4 compares a few

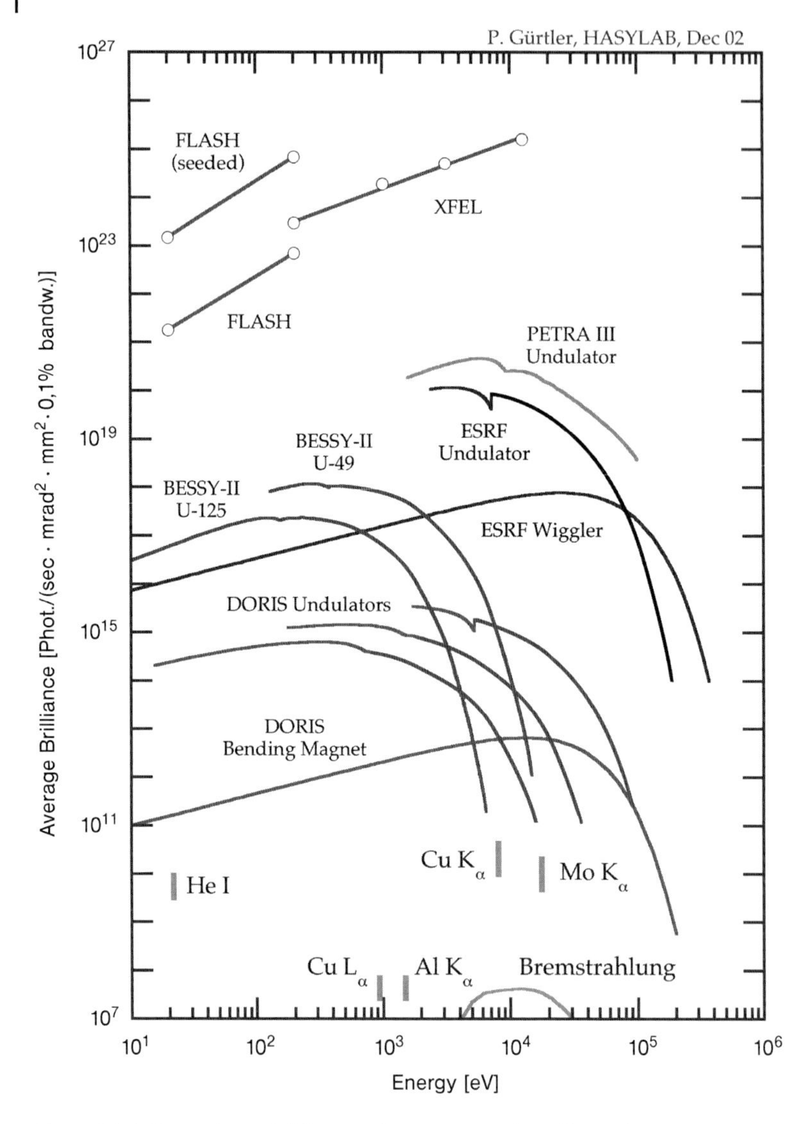


**Figure 5.2.4** Comparison of different radiation sources from X-ray tubes across synchrotron radiation storage rings toward the free electron lasers (FELs) at DESY (FLASH and the future XFEL) in terms of average brilliance (see units at left) as a function of photon energy. Note that the practical conversion from energy to wavelength is $\lambda$ (Å) = 12.398/$E$ (keV).

typical examples of SR sources (from DORIS and PETRA at DESY, the ESRF in Grenoble and BESSY II in Berlin) to other sources, such as X-ray tubes, He lamps and above all of that, the newest technological developments – free electron lasers (FELs). Compared to X-ray tubes with rotating anodes, SR is superior by five to ten orders of magnitude which, besides the general advantage of continuous tunability, allows to get better signals in shorter time with less sample volume required and/or better energy resolution and spatial resolution. A short outlook on free electron lasers which yield another five orders of magnitude enhancement of average brilliance will be given at the end of this chapter.

(3/4) *Source size and collimation*:
Typical electron beam diameters and hence SR source sizes of modern storage rings are in the range of 100 μm. This small size along with the high degree of collimation which has already been illustrated in the introduction of this chapter (Figure 5.2.1) makes the SR beam well suited for focusing onto small samples or for space resolved techniques. Beam sizes of a few 10 μm at the sample can easily be achieved and even submicron focusing is feasible with the appropriate optics.

(5/6) *Polarization*:
A typical example for the polarization properties of SR is depicted in Figure 5.2.5. Shown are the horizontal ($\sigma$), vertical ($\pi$), and total polarization component of bending magnet radiation for different wavelengths as a function of the elevation angle $\psi$ relative to the orbital plane of the storage ring. In the orbit plane, the

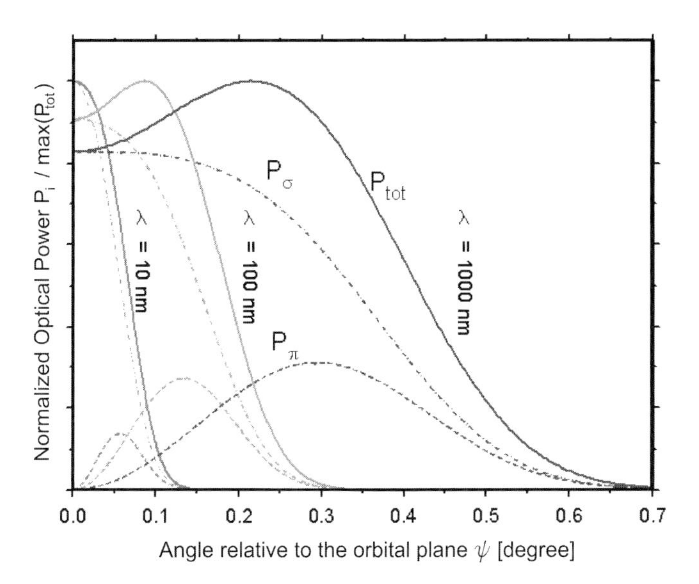


**Figure 5.2.5** Example of angular distribution and polarization of synchrotron radiation from a bending magnet as function of emitted wavelength (http://physics.nist.gov/MajResFac/SURF/SURF/sr.html).

beam is only linearly (horizontally) polarized and with growing elevation angle it becomes elliptically polarized and finally obtains full circular polarization. The longer the wavelength (the lower the photon energy) the wider is the angular distribution.

Note that the polarization properties of radiation from (symmetric) wigglers and undulators are different, as will be discussed later.

(7) *Time structure*:
In synchrotron radiation storage rings, from one up to several hundred electron bunches circulate at almost the speed of light. A ring circumference of, e.g., 300 m implies that each individual bunch runs around at 1 MHz repetition rate and all bunches together yield SR pulse repetition rates of one up to several 100 MHz. This pulsed structure of SR is illustrated in Figure 5.2.6. The typical spacing between the pulses (depending on the storage ring and its specific operation mode/filling pattern) ranges from 2 ns to 1 μs and the typical pulse length (electron/positron bunch length divided by $c$) from 10 to 100 ps.

(8) *Quantitative description of SR*:
Some examples with formulas are given in the text, for additional information the reader is referred to [8–12].

(9) *Clean environment*:
Since the storage rings are operated under ultra high vacuum (UHV, typically $10^{-9}$ mbar) conditions and since no debris is produced – as, e.g., from X-ray laser targets – one can directly connect UHV chambers, e.g., for surface science experiments.

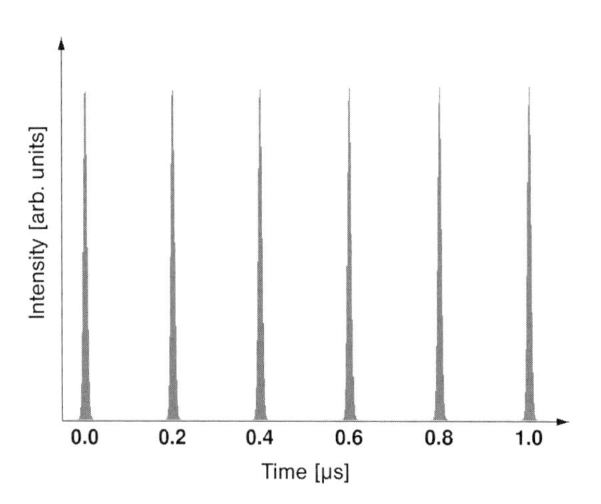


**Figure 5.2.6** Illustration of the pulsed time structure of synchrotron radiation. For details see the text.

## 5.2.3
## Sources of Synchrotron Radiation

Today's machines solely dedicated to the production of synchrotron radiation are mostly so-called storage rings where the particles are kept on a closed orbit at constant energy. A few of the basic ingredients of a synchrotron radiation storage ring are sketched in Figure 5.2.7. Electrons or positrons, respectively, are running around on a closed orbit at 99,99...% the speed of light and emit synchrotron radiation when they pass bending magnets or special magnetic insertion devices. Since a substantial amount of kinetic energy of the particles is lost via SR, the bunches have to be reaccelerated every turn running through accelerating radio frequency cavities.

During the last 20 years, advanced magnetic insertion devices for SR, called wigglers and undulators, became more and more the key components of SR storage rings. Wigglers and undulators are periodic magnetic multipole structures of typically a few meters length which are put into straight sections of the storage rings. The repetitive left and right turns of the electron/positron bunches result in no net deviation from the original direction but in a significant enhancement of the emitted SR as will be outlined below. A simplified overview of the difference between the sources of SR is given in Figure 5.2.8 and will be discussed in detail in the following section.

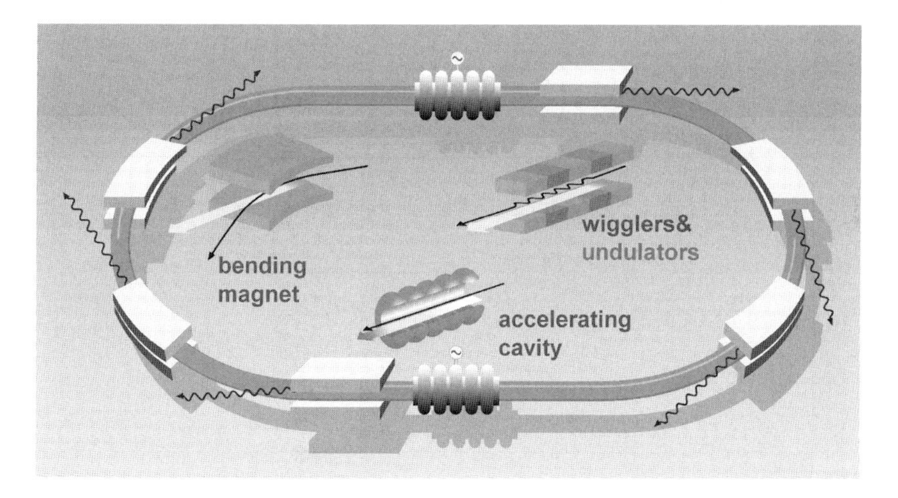


**Figure 5.2.7** Sketch of a synchrotron radiation storage ring where either electron or positron bunches are traveling through a closed system of evacuated tubes with dipole magnets to force the particles onto a closed orbit. Energy losses due to synchrotron radiation are compensated in radio-frequency cavities which reaccelerate the particles. Note that this is a very simplified picture, in particular omitting all other magnets, such as quadrupoles and sextupoles and all the beam diagnostics around the ring.

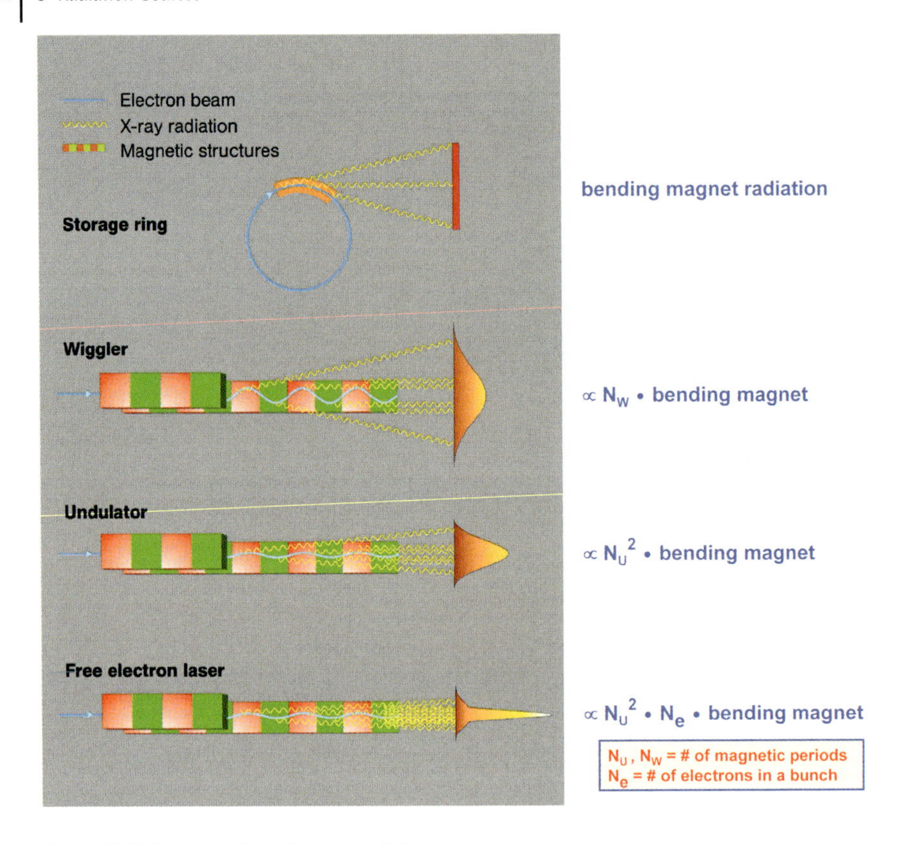


**Figure 5.2.8** Sources of synchrotron radiation.

#### 5.2.3.1 Bending Magnets

For a *bending magnet*, the horizontal fan of radiation corresponds roughly to the net deflection angle which is typically in the order of a few up to several ten degrees. This means that even several beamlines can collect radiation (typically a few mrad horizontally each) from different points of the bent trajectory. The characteristics of synchrotron radiation from a bending magnet can be exactly calculated via the Schwinger equation [5]

$$I(\lambda, \psi) = \frac{27e^2c}{32\pi^3\rho^3} \cdot \left(\frac{\lambda_c}{\lambda}\right)^4 \cdot \gamma^8 \cdot (1 + (\gamma\psi)^2)^2 \cdot \left(K_{2/3}^2(\xi) + \frac{(\gamma\psi)^2}{1 + (\gamma\psi)^2} \cdot K_{1/3}^2(\xi)\right) \tag{5.2.6}$$

with $\rho$, the bending radius of the trajectory according to Eq. (5.2.2), and

$$K_X(\xi) \tag{5.2.7}$$

modified Bessel functions, where the first term with $K_{2/3}$ corresponds to radiation polarized parallel to the orbit plane and the second term with $K_{1/3}$ corresponds to the radiation component polarized perpendicular to the orbit plane with

$$\xi = \frac{\lambda_c}{2\lambda} \cdot \left(1 + (\gamma\psi)^2\right)^{3/2} \tag{5.2.8}$$

where $\lambda_c$ is the critical wavelength, which can be derived from the critical energy described before as

$$\lambda_c(\text{Å}) = \frac{12.398}{\varepsilon_c(\text{keV})} = 5.59 \frac{\rho(\text{m})}{E(\text{GeV})^3} \tag{5.2.9}$$

$\gamma$ is the usual Lorentz parameter and $\psi$ again the elevation angle from the orbit plane.

#### 5.2.3.2 Wigglers and Undulators

As already mentioned above, wigglers and undulators are special magnetic multipole devices with alternating magnetic field, where the electrons/positrons are forced on a usually sinusoidal trajectory (Figure 5.2.9) in order to enhance the radiation output.

A wiggler with its $N$ magnetic periods is basically a superposition of a certain number $2N$ of bending magnet sources with a bending radius given by the strength of the alternating magnetic field. Wigglers and undulators are described by the wiggler parameter

$$K := \alpha \cdot \gamma = \frac{eB_0\lambda_0}{2\pi m_e c} \tag{5.2.10}$$

which gives the ratio of deflection angle $\alpha$ of the particles in the $B$-field with respect to opening cone $1/\gamma$ of the emitted radiation. $\lambda_0$ is the period length of the magnetic field and $B_0$ its peak value. In practical units

$$K = 0.934 \cdot \lambda_0(\text{cm}) \cdot B_0(T) \tag{5.2.11}$$

If $K \gg 1$, one speaks of a *wiggler*. Here, the intensity enhancement is about $2N$-fold compared to a single dipole of same strength and the radiation is emitted in forward direction with typical horizontal opening angles in the mrad range.

If $K \lesssim 1$, the device is called an *undulator*. In this case, it is not just a pure superposition of $2N$ independently radiating bending magnet sources. Since the net deflection is similar or smaller than the typical SR opening cone, radiation emitted from different poles can interfere constructively. An undulator can be

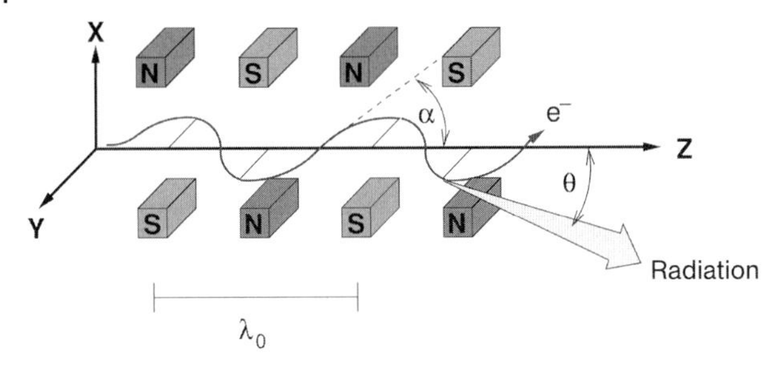


**Figure 5.2.9** Principle of the periodic motion of electrons in the magnetic field of a wiggler/undulator along with the characteristic parameters as used in the text.

treated in a simple geometrical picture where one just calculates the path difference between the photons emitted from consecutive poles. This set-up very much resembles the emission of radiation from a $2N$-fold slit and it is hence not surprising that the intensity enhancement of an undulator is not just $2N$-fold but scales with $N^2$ and results in certain resonance wavelengths of undulator emission given by

$$\lambda_i = \frac{\lambda_0}{2i\gamma^2} \cdot \left(1 + \frac{K^2}{2} + \gamma^2\Theta^2\right) \tag{5.2.12}$$

where $\lambda_i$ is the wavelength of the $i$-th harmonic, $\lambda_0$ the period length of the magnetic field and $\Theta$ the observation angle with respect to the forward direction.

To obtain the energy and angular distribution of wiggler/undulator radiation, one has to put the (sinusoidal) motion of electrons in the periodical magnetic field of the wiggler/undulator in the far field approximation of radiation equation

$$\frac{dI(\omega)}{d\Omega} = \frac{e^2\omega^2}{4\pi^2 c} \left| \int_{-\infty}^{\infty} \vec{n} \times (\vec{n} \times \vec{\beta}) \exp(i\omega(t - \vec{n} \cdot \vec{r}(t)/c))\, dt \right|^2 \tag{5.2.13}$$

(Eq. (14.67) from [1]). A detailed treatment of this procedure and the results can, e.g., be found in [14]. Figure 5.2.10 shows examples of a wiggler spectrum and an undulator spectrum with the characteristic "undulator lines." At the right, the fluorescence caused by a white (= full spectrum) wiggler beam traveling through air is depicted.

This also illustrates that SR can be really intense, with up to several kilowatts of radiation power in a "white," non-monochromatized wiggler beam and power densities up to the order of kW mm$^{-2}$. This heatload is quite a challenge, advanced beamline optics, beamstops and filters, etc. have to deal with, e.g., via special cooling designs or adaptive (distortion compensating) optics.

Note that for wigglers and undulators, since the trajectory has an equal number of left and right turns, left- and right-handed circular polarization above and

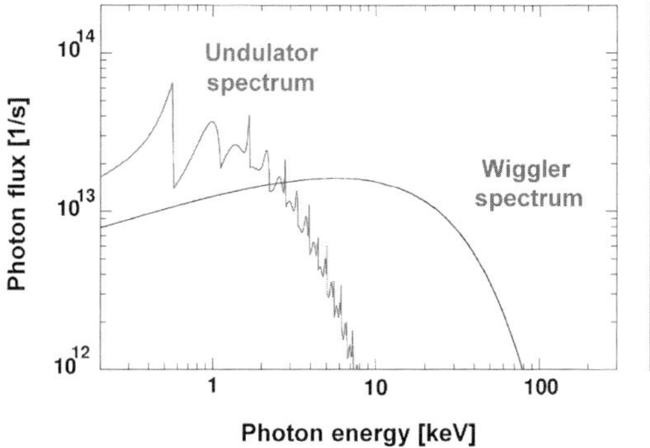


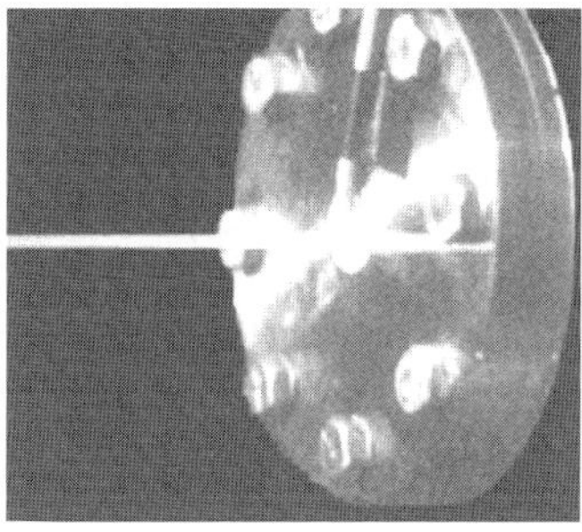

**Figure 5.2.10** Typical (angle-integrated) wiggler and undulator spectra (left) along with the trace of a "white" (= non-monochromatized) wiggler beam producing intense visible fluorescence while traveling through air (from ID11 at the ESRF, Grenoble).

below the orbit plane, respectively, cancel out. On the other hand, insertion devices with full polarization control, such as helical undulators or asymmetric wigglers are also available today [15].

## 5.2.4 Outlook: Free Electron Lasers

What will be presented here is mainly a short excerpt from [16]. More background information for the interested reader can, e.g., be found in [17].

The basic principle of the free electron laser can be described within the standard picture for the generation of synchrotron radiation: while traveling with relativistic velocity ($v \simeq c$, $\gamma \simeq 10^2$–$10^5$) through the undulator, the electrons are accelerated in the direction transverse to their propagation due to the Lorentz force introduced by the magnetic field. They propagate along a sinusoidal path and emit SR in a narrow cone in the forward direction. In the undulator, the deflection of the electrons from the forward direction is comparable to the opening angle of the synchrotron radiation cone. Thus the radiation generated by the electrons while traveling along the individual magnetic periods overlaps (as described before). This interference effect is reflected in the formula for the wavelength $\lambda$ of the first harmonic of the spontaneous, on-axis ($\Theta = 0$) undulator emission (cf. also Eq. (5.2.12))

$$\lambda = \frac{\lambda_0}{2\gamma^2}\left(1 + \frac{K^2}{2}\right) \tag{5.2.14}$$

The interference condition basically means that, while traveling along one period of the undulator, the electrons slip by one radiation wavelength with respect to the (faster) electromagnetic field. This is one of the prerequisites for the SASE (self-amplified spontaneous emission) process of the FEL. To obtain an exponential amplification of the spontaneous SR emission present in any undulator, some additional criteria have to be met: One has to guarantee a good electron beam quality and a sufficient overlap between radiation pulse and electron bunch, which generates it, along the undulator. To achieve that, one needs a low emittance, low energy spread electron beam with an extremely high charge density in conjunction with a very precise magnetic field and accurate beam steering through a long undulator. Note that electron bunches of the high quality that is necessary to get lasing in the vacuum-ultraviolet (VUV) or X-ray range cannot be produced and transported in a storage ring. It needs a state-of-the-art linear accelerator, e.g., the one built in TESLA technology that is used for FLASH (*F*ree Electron *Las*er in *H*amburg) at DESY [18].[1)]

Oscillating through the undulator, the electron bunch then interacts with its own electromagnetic field created via spontaneous undulator emission. Depending on the relative phase between radiation and electron oscillation, electrons experience either a deceleration or acceleration: electrons that are in phase with the electromagnetic wave are retarded while those with opposite phase gain energy. Through this interaction a longitudinal fine structure, the so-called microbunching, is established which in turn amplifies the electromagnetic field (Figure 5.2.11).

The longitudinal distribution of electrons in the bunch is "cut" into equidistant slices with a separation corresponding to the wavelength $\lambda$ of the emitted radiation which causes the modulation. More and more electrons begin to radiate in phase, which results in an increasingly coherent superposition of the radiation emitted from the microbunched electrons. The more intense the electromagnetic field gets, the more pronounced the longitudinal density modulation of the electron bunch and vice versa. In the beginning – without microbunching – all $N_e$ electrons in a bunch ($N_e \geq 10^9$) can be treated as individually radiating charges with the power of the spontaneous emission $\propto N_e$ (this is the usual SR from an undulator!). With complete microbunching, all electrons radiate almost in phase. They basically behave like a single point charge $N_e \cdot e$. This leads to a radiation power $\propto N_e^2$ (see also Figure 5.2.8) and thus an amplification of many orders of magnitude with respect to the spontaneous emission of the undulator. Due to the progressing microbunching, the radiation power of such a SASE FEL grows exponentially with the distance $z$ along the undulator (Figure 5.2.11, bottom).

The peak brilliance of such SASE FELs surpasses the spontaneous undulator radiation from today's state-of-the-art synchrotron radiation facilities by about eight or more orders of magnitude while the average brilliance is about five orders of magnitude higher (see Figure 5.2.4). The peak brilliance is the brilliance scaled to the length of a single pulse while average brilliance is normalized to seconds at the highest possible repetition rate.

1 At the time of writing, lasing down to a shortest wavelength of 13 nm in the EUV/soft X-ray range has been obtained at FLASH with pulse energies of about 100 μJ and pulse lengths of about 10–20 fs.

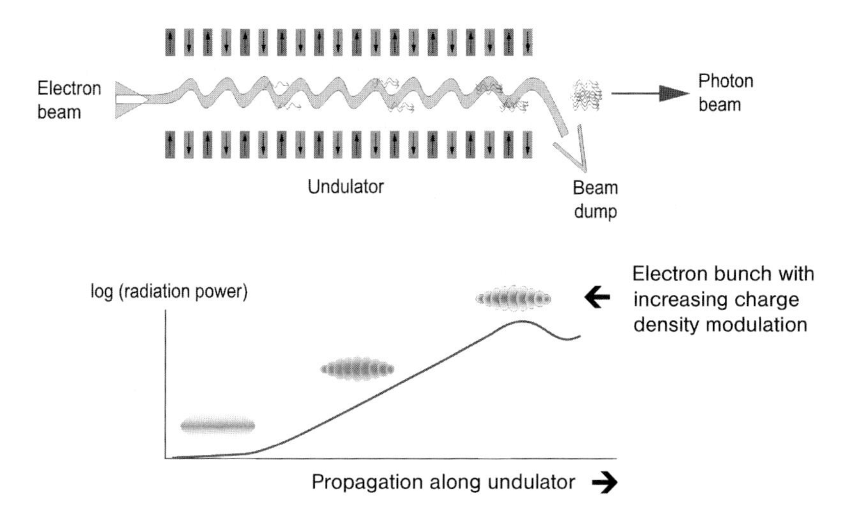


**Figure 5.2.11** Sketch of the self-amplification of spontaneous emission (SASE) in an undulator resulting from the interaction of the electrons with the synchrotron radiation they emit. In the lower part of the figure the longitudinal density modulation (microbunching) of the electron bunch is shown together with the resulting exponential growth of the radiation power along the undulator.

In summary, with the present FLASH facility and future XFELs around the world, in comparison to a state of the art synchrotron radiation source (spontaneous undulator radiation) one obtains

- $10^8 \times$ more peak brilliance,
- short pulses ($\approx$100 fs versus 100 ps),
- full transverse coherence,
- partial/almost full temporal coherence.

This opens the door to a completely new class of experiments, ranging from nonlinear X-ray physics, across femto(second)chemistry to plasma physics and biological (protein) dynamics. FELs will give scientists, for example, insight into hitherto unknown properties of materials.

### 5.2.5
### Summary

Today's synchrotron radiation sources as well as the upcoming FELs provide unique capabilities for a wide range of applications from physics and material sciences across geology, environmental sciences, chemistry and biology toward medical and industrial applications. With their unique combination of properties such as the wide wavelength spectrum (respectively tunability) from the infrared to the X-ray range, a tightly collimated beam, high intensities and short pulses, these radiation sources allow us to study the subtleties of the structure of matter down to atomic length scales and with the FELs even at femtosecond time scales.

## References


1 J.D. Jackson, *Classical Electrodynamics*, Wiley, New York, 1962.
2 F. R. Elder, A. M. Gurewitsch, R. V. Langmuir, H. C. Pollock, Radiation from electrons in a synchrotron, *Phys. Rev.* 1947, **71**, 829–830.
3 J. Larmor, On the theory of the magnetic influence on spectra and on the radiation from moving ions, *Phil. Mag.* 1897, **44**, 503–512.
4 D. Ivanenko, I. Pomeranchuk, On the maximal energy attainable in a betatron, *Phys. Rev.* 1944, **65**, 343.
5 J. Schwinger, Electron radiation in high energy accelerators, *Phys. Rev.* 1946, **70**, 798–799; J. Schwinger, On the classical radiation of accelerated electrons, *Phys. Rev.* 1949, **75**, 1912–1925.
6 D. H. Tomboulian, P. L. Hartman, Spectral and angular distribution of ultraviolet radiation from the 300 Mev Cornell synchrotron, *Phys. Rev.* 1956, **102**, 1423–1447.
7 See, e.g., http://hasylab.desy.de/facilities/sr_and_fel_labs/sr_labs/index_eng.html.
8 *Handbook on Synchrotron Radiation*, vol. 1 (Ed.: E.E. Koch), North-Holland, Amsterdam, 1983.
9 *Handbook on Synchrotron Radiation*, vol. 2 (Ed.: G.V. Marr), North-Holland, Amsterdam, 1987.
10 *Handbook on Synchrotron Radiation*, vol. 3 (Eds.: G.S. Brown and D.E. Moncton), North-Holland, Amsterdam, 1991.
11 *Handbook on Synchrotron Radiation*, vol. 1 (Eds.: S. Ebashi, M. Koch and E. Rubinstein), North-Holland, Amsterdam, 1991.
12 *Synchrotron Radiation – Techniques and Applications* (Ed.: C. Kunz), Springer, Berlin, 1979.
13 S. Krinski et al., chapter 3 in [8], vol. 1a.
14 S. Krinski et al., chapter 2 in [8], vol. 1a.
15 J.A. Clarke, *The Science and Technology of Undulators and Wigglers*, Oxford Series on Synchrotron Radiation, vol. 4, Oxford University Press, Oxford, 2004.
16 *TESLA Technical Design Report, Part V* (Eds.: G. Materlik and Th. Tschentscher), DESY Report 2001-011/TESLA-FEL 2001-05, Hamburg, 2001. The introduction to the SASE FELs is also available at http://hasylab.desy.de/facilities/sr_and_fel_basics/fel_basics/index_eng.html.
17 A list of references is given on the FLASH user information pages at http://hasylab.desy.de/facilities/flash/publications/selected_publications/basic_papers_and_books/index_eng.html.
18 (a) See, e.g., http://hasylab.desy.de/facilities/flash/index_eng.html. (b) V. Ayvazyan et al., First operation of a free-electron laser generating GW power radiation at 32 nm wavelength, *Eur. Phys. J. D* 2006, **37**, 297–303; for the European XFEL project see http://xfel.desy.de.

# Part II
# Methods

# 6
# Introduction to Diffraction Methods for Internal Stress Analyses

*Walter Reimers*

## 6.1
## General Aspects

Diffraction methods are widely used for internal stress analyses due to the following advantages:

- Diffraction methods can be applied nondestructively, which is of importance, e.g., for inspection purposes, for following the development of residual stresses over the lifetime of components and also for residual stress analysis in expensive components.
- Diffraction methods can be used for crystalline materials which represent the major part of engineering materials.
- Diffraction methods are phase sensitive for enabling the determination of phase specific stresses.
- Diffraction methods allow in most cases the determination of the absolute values and the sign of the stress tensor components. Therefore, diffraction methods are often needed for the calibration of other methods, such as ultrasonic or magnetic measurement techniques.

Beyond these general aspects for all diffraction methods the use of synchrotron X-rays and neutrons for residual stress analysis offers additional advantages:

- Due to the often weak absorption of neutrons and high-energy photons (see Chapter 4), these radiations enable the extension of the information depth from the surface to the bulk of materials and components. Since residual stresses are balanced in a component (see Chapter 2), the stress states at the surface and in the bulk are different in most cases.

*Neutrons and Synchrotron Radiation in Engineering Materials Science: From Fundamentals to Material and Component Characterization*
Edited by Walter Reimers, Anke Rita Pyzalla, Andreas Schreyer, and Helmut Clemens

ISBN: 978-3-527-31533-8

- The possibilities of defining not only the wavelength/energy of the neutron or photon beam but also its size give manifold possibilities of defining the gauge volume under study. This is an important advantage for analyzing technical components with complex geometries and these possibilities are often needed for the investigation of spatially inhomogeneous residual stress distributions.
- The high photon flux delivered by third-generation synchrotron sources has stimulated the field of *in situ* studies. Here the short data acquisition times enable the investigation of time-dependent processes such as, e.g., creep, phase transformation stresses, and the formation of welding residual stresses.

## 6.2 Principles of Diffraction Methods

The diffraction condition for X-rays and neutrons on crystalline materials can be described by the Laue equations or more simply by the Bragg equation:

$$n\lambda = 2d(hkl)\sin\theta(hkl) \tag{6.1}$$

Figure 6.1 illustrates geometrically the Bragg equation and the variables.

So using the experimental set-up shown in Figure 6.2, a diffractogram of polycrystalline materials can be obtained (Figure 6.3).

The measured intensity distributions of the reflections *hkl* have to be evaluated. Therefore, numerous peak fitting programs are available which yield the reflection position $2\theta(hkl)$, the integral intensity of the reflection, and the full width

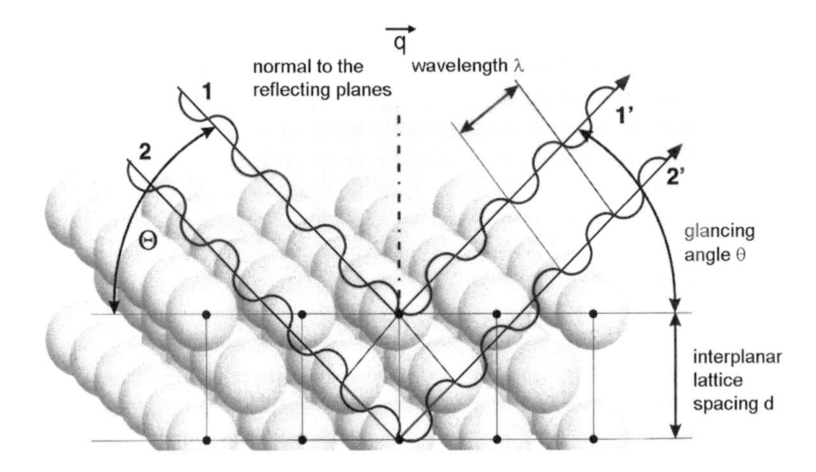


**Figure 6.1** Reflection of X-rays or neutrons at lattice planes *hkl* for illustrating the Bragg equation.

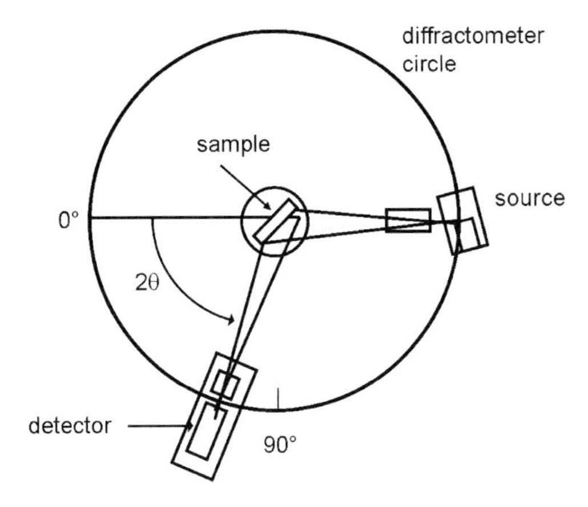


**Figure 6.2** Two-circle diffractometer.

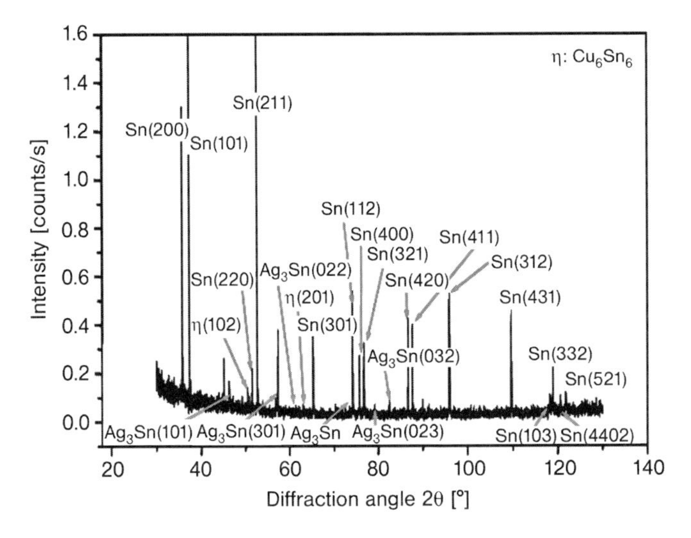


**Figure 6.3** Typical diffractogram (Sn-Ag-Cu alloy), X-ray diffractogram, CoK$\alpha$ radiation.

of the half maximum (FWHM) of the reflection. These data deliver different basic information about the material state:

- Using the Bragg equation, the interplanar lattice spacings $d(hkl)$ can be calculated from the reflection positions $2\theta(hkl)$. The *hkl* values obtained are the fingerprint of the crystalline phases present in the material under study and the identification is done using the JCPDS data files [1].

- The reflection positions $2\theta(hkl)$ may be shifted due to the presence of internal stresses. Mostly this effect is limited to $\Delta 2\theta(hkl) < 1^\circ$ and therefore does not affect the phase identification. The sensitivity of the precise reflection position $2\theta(hkl)$ to internal stresses is the basis for internal stress determination by diffraction methods. The reflection position shift is caused by the change of the interplanar lattice distances due to the presence of strains and stresses.
- The relative intensities of the reflections *hkl* of one crystalline phase give hints to the orientation distribution of the reflecting crystallites. If these intensity ratios are in agreement with the data given in the JCPDS data file, a random distribution of the crystallites is present. If the deviations between the experimentally observed intensity ratios and the JCPDS intensity ratios are significant, then preferred orientation of the crystallites (texture) has to be suspected (see Chapter 3 for further details).
- The relative intensities of the reflections *hkl* of the two or more crystalline phases can be used for the calculation of the crystalline phase volume fractions. The most prominent example in the field of engineering materials is the determination of retained austenite in steels [2].
- The FWHM of the reflection profiles depends on the experimental conditions, especially the divergence of the beam, but also on the material state under study. So small grain sizes and microstrains lead to a broadening of the reflection profiles. However, grain size effects and microstrains have different dependences on $\theta$, so that they can be separated [3]. Beyond the more or less qualitative information about the material state when evaluating the FWHM of the reflection profiles, there are also methods [4] available, which perform full profile fittings of the whole diffractogram. Due to the amount of data included this way, additional information about microstructural defects like twin density, stacking fault, and dislocation densities can be obtained. However, this topic is beyond the scope of this book and readers are referred to specialized literature (see, e.g., [5]).

## 6.3 Principles of Strain Determination by Diffraction Methods

The crystallites forming the polycrystalline solid aggregate act as strain gauges for the analysis of internal stresses using diffraction methods. So the presence of tensile stresses leads to an expansion of the interplanar lattice spacings and compressive stresses to a compression of the interplanar lattice spacings in the direction of the acting stresses. The strain of an interplanar lattice spacing is defined as follows:

$$\varepsilon = \frac{d - d_0}{d_0} \tag{6.2}$$

with $d_0$ the interplanar lattice distance of the stress-free material.

From the Bragg equation it can then be obtained:

$$\Delta\theta = -\varepsilon \tan\theta \tag{6.3}$$

So the analysis of reflection shifts allows the determination of strain values with sign and magnitude.

Due to the tensorial character of strains and stresses, in most cases diffraction measurements in different sample directions are necessary. Therefore, a minimum of two Cartesian systems has to be defined:

- *Sample system S*: the $S_3$-axis is the normal to the sample surface and the $S_1$- and $S_2$-axes are often given by the sample itself, e.g., parallel and perpendicular to a weld.
- *Laboratory system L*: the $L_3$-axis is given by the measuring direction, which is the normal to the reflecting planes *hkl*. The $L_2$-axis is parallel to the sample surface and the $L_1$-axis is given by the cross product of the vectors $\vec{L}_2 \times \vec{L}_3$.

These two Cartesian systems and the azimuth angle $\psi$ and the pole angle $\varphi$ are shown in Figure 6.4.

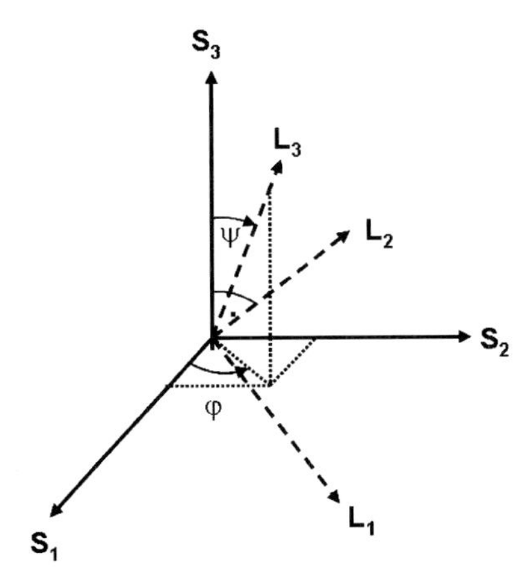


**Figure 6.4** Orientation of the laboratory system relative to the sample system.

The transformation matrix $\omega_{ij}$ between the sample system and laboratory system is given by

$$\omega_{ij} = \begin{pmatrix} \cos\varphi\cos\psi & \sin\varphi\cos\psi & -\sin\psi \\ -\sin\varphi & \cos\varphi & 0 \\ \cos\varphi\sin\psi & \sin\varphi\sin\psi & \cos\psi \end{pmatrix} \tag{6.4}$$

Since the $L_3$-axis is defined by the measuring direction $\varphi$, $\psi$, the measurement delivers the $\varepsilon'_{33}$ component, where the prime indicates the laboratory system. $\varepsilon'_{33}$ is obtained from the transformation of the tensor $\tilde{\varepsilon}$ from the sample to the laboratory system:

$$\begin{aligned} \varepsilon'_{33} = \varepsilon'_{\varphi,\psi} &= \omega_{3k}\omega_{3l}\varepsilon_{kl} \\ &= \varepsilon_{11}\cos^2\varphi\sin^2\psi + \varepsilon_{22}\sin^2\varphi\sin^2\psi + \varepsilon_{33}\cos^2\psi \\ &\quad + \varepsilon_{12}\sin 2\varphi\sin^2\psi + \varepsilon_{13}\cos\varphi\sin 2\psi + \varepsilon_{23}\sin\varphi\sin 2\psi \end{aligned} \tag{6.5}$$

So a minimum of six measurements in different measurement directions $\varphi$, $\psi$ allow the determination of the triaxial strain tensor.

For reasons of accuracy and for checking the homogeneity of the material state as far as texture and stress gradients are concerned, it is however recommended to measure at least two strain values for each unknown strain tensor component. Furthermore, it has to be noted that in the case of three-dimensional strain/stress states, the determination of the stress-free interplanar lattice distance $d_0$ has to be considered carefully (see Section 6.4).

Once the strain tensor is determined the stress tensor can be calculated by applying Eq. (2.11) (see Chapter 2). However, the macroscopic elastic constants $E$ and $\nu$ should be replaced by the diffraction elastic constants (DEC):

$$s_1(hkl) = \frac{-\nu}{E} \tag{6.6}$$

$$\frac{1}{2}s_2(hkl) = \frac{1+\nu}{E} \tag{6.7}$$

The precise values of $s_1(hkl)$ and $\frac{1}{2}s_2(hkl)$ depend on the reflection $hkl$ measured, so taking into account the anisotropy of the crystalline elasticity (see Section 6.7).

This way the basic equation for stress analysis using diffraction methods [3, 6] is obtained:

$$\begin{aligned} \varepsilon'_{33} &= \varepsilon'_{\varphi,\psi} \\ &= s_1(hkl)(\sigma_{11} + \sigma_{22} + \sigma_{33}) + \frac{1}{2}s_2(hkl)\sigma_{33} \end{aligned}$$

$$+\frac{1}{2}s_2(hkl)[(\sigma_{11}-\sigma_{33})\cos^2\varphi\sin^2\psi+(\sigma_{22}-\sigma_{33})\sin^2\varphi\sin^2\psi]$$

$$+\frac{1}{2}s_2(hkl)[\sigma_{12}\sin 2\varphi\sin^2\psi+\sigma_{13}\cos\varphi\sin 2\psi+\sigma_{23}\sin\varphi\sin 2\psi] \qquad (6.8)$$

This equation can be used for the determination of average stresses in quasi-isotropic crystalline materials.

## 6.4 Determination of the Stress-Free Interplanar Lattice Distance $d_0$

In the case of the presence of triaxial stresses, which are typical for bulk measurements, the $d_0$-value has to be determined for the calculation of absolute strain/stress values. Otherwise the hydrostatic part of the stress tensor cannot be determined. For the determination of the $d_0$-value different approaches are possible:

- The most reliable data are obtained from stress-free powder material. However, this is often not available.
- Sectioning of the material, e.g., fabrication of "combs" or small cubes. This way the macroscopic stresses are relaxed and the $d_0$-value is accessible at the end of the comb teeth or by bathing the cubes in the beam [7].
- Performing measurements in a stress-free region of the component. If the $d$-values obtained in different sample directions are in good agreement, this procedure can be justified.
- Performing measurements at many different locations in the sample. This is called the random walk method and the idea is based on the equilibrium condition, which requires the balancing of stresses over the sample volume.
- Some sample geometries allow us to make use of the equilibrium conditions for the $d_0$ determination. For example, in thin parts of a component with small $x$, $y$ dimensions, it can be possible to scan the complete strain field. Then $\int \sigma_z \, dxdy = 0$ can be used for calculating the $d_0$-value.
- Similarly, in thin plates $\sigma_{33} = 0$ can be assumed. So this condition allows us to determine the $d_0$-value.
- In two-phase materials, the $d_0$-value of one phase must be known. Otherwise sectioning or equilibrium conditions cannot help. So in two-phase materials sometimes etching techniques are used for obtaining the powder of one of the phases.

## 6.5 $\sin^2 \psi$-Technique

In practice the $\sin^2 \psi$-technique [8] can often be used for the determination of residual stresses and it is widely applied in the industrial environment. The requirements for the application of this evaluation method of the experimental strain data are the following:

- The $\sin^2 \psi$-technique can be used in the case of quasi-isotropic materials. This means that the material shows a random or nearly random crystalline orientation distribution (texture) and the grain statistics in the gauge volume measured is sufficient for an isotropic approximation.
- The measured strain values in the different measuring directions $\varphi$, $\psi$ are not affected by stress gradients. This means that the average stress state is the same for each measurement.
- The shear stresses $\sigma_{13}$ and $\sigma_{33}$ can be neglected.

When these conditions are reasonably fulfilled, Eq. (6.8) can be rewritten and simplified:

$$\begin{aligned}\varepsilon'_{33} &= \varepsilon'_{\varphi,\psi} \\ &= s_1(hkl)[(\sigma_{11} - \sigma_{33}) + (\sigma_{22} - \sigma_{33})] + (3s_1(hkl) + \frac{1}{2}s_2(hkl))\sigma_{33} \\ &\quad + \frac{1}{2}s_2(hkl)[(\sigma_{11} - \sigma_{33})\cos^2\varphi + (\sigma_{22} - \sigma_{33})\sin^2\varphi + \sigma_{12}\sin 2\varphi]\sin^2\psi\end{aligned} \tag{6.9}$$

With $\sigma_\phi = \sigma_{11}\cos^2\varphi + \sigma_{22}\sin^2\varphi + \sigma_{12}\sin 2\varphi$ and the differentiation $\dfrac{d\varepsilon'_{\varphi\psi}}{d\sin^2\psi}$ it follows that

$$\frac{d\varepsilon'_{\varphi\psi}}{d\sin^2\psi} = \frac{1}{2}s_2(hkl)(\sigma_\varphi - \sigma_{33}) \tag{6.10}$$

For near surface measurements or in the case of thin samples (plane stress state), it can often be assumed that $\sigma_{33} = 0$. Then the stress value $\sigma_\varphi$ is directly obtained from the slope $m$ of $\dfrac{d\varepsilon'_{\varphi\psi}}{d\sin^2\psi}$

$$\sigma_\varphi = \frac{m}{\frac{1}{2}s_2(hkl)} \tag{6.11}$$

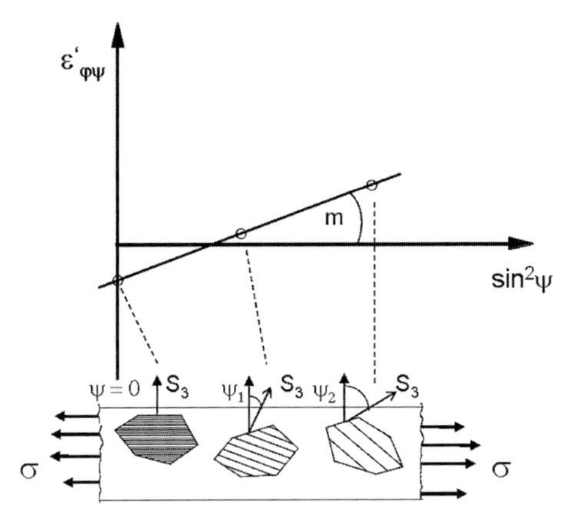


**Figure 6.5** $\sin^2 \psi$ technique.

Figure 6.5 displays the linear relationship between the measured strain values and $\sin^2 \psi$.

So performing measurements in different $\varphi$-directions (e.g., $\varphi = 0°$, $\varphi = 90°$, $\varphi = 45°$) the complete two-dimensional stress tensor is obtained. It has to be noted that the knowledge of the precise $d_0$-value of the stress-free material is not necessary for this evaluation, since the stress value $\sigma_\varphi$ only depends on the slope $m$ in the $\varepsilon'_{\varphi\psi}$ versus $\sin^2 \psi$ diagram.

Moreover, $d_0$ can be obtained form the strain/stress-free direction $\psi^*$, which is given by [6]

$$\sin^2 \psi^* = \frac{-s_1(hkl)}{\frac{1}{2} s_s(hkl)} \left[1 + \frac{\sigma_{22}}{\sigma_{11}}\right] \tag{6.12}$$

## 6.6 Nonlinear Lattice Strain Distributions

If one or more assumptions for the use of the $\sin^2 \psi$- technique are not fulfilled, then nonlinear lattice strain distributions will be obtained. Typical fundamental types of lattice strain distributions are displayed in Figure 6.6.

### 6.6.1 Anisotropy

If experimental nonlinear lattice strain distributions like those in Figure 6.6a are obtained, then an anisotropic material state is documented. Therefore, anisotropy can be due to different reasons, which have to be treated separately:

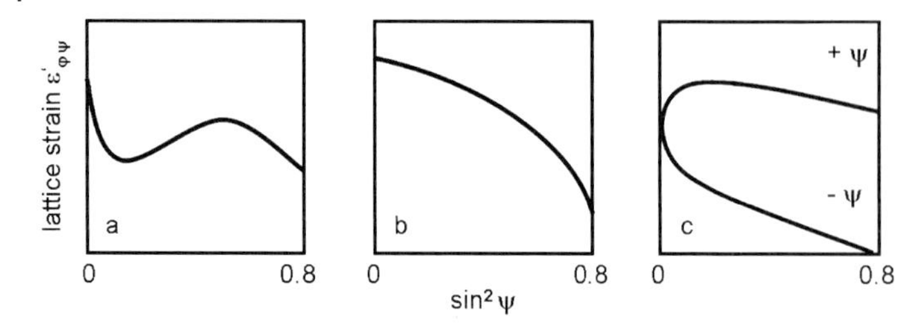


**Figure 6.6** Fundamental types of nonlinear lattice strain distributions.

- *Coarse grain effects*: if the number of crystallites in the gauge volume is too small to fulfill the requirement of quasi-isotropic material state, then the anisotropic strain–stress relationship of the individual crystallites contributing to the diffracted intensity causes oscillations in the $\varepsilon'_{\varphi\psi} - \sin^2\psi$ plot. In this case the number of crystallites illuminated has to be increased either by increasing the gauge volume under study or by oscillating (rotation and/or translation) the sample during the measurement so increasing the number of reflecting crystallites.
- *Texture*: in the case of samples with a nonrandom texture, the elastic and/or plastic anisotropy of the material state leads to the nonlinearity in the $\varepsilon'_{\varphi\psi} - \sin^2\psi$ plot. Here different approaches to obtain reliable stress data are possible. First of all the selection of reflections for the strain measurements has to be taken with care. Therefore, recommendations have been established by the VAMAS working group TWA 20 [7] (see Section 6.8). These recommendations are based on two principles: reflections with high multiplicities are less sensitive to anisotropy, because the diffraction pattern is more randomized. The second principle is connected to the anisotropy of plastic deformation so that a differentiation between more or less sensitive reflections can be made. A similar approach to minimize the nonlinearity of $\varepsilon'_{\varphi\psi} - \sin^2\psi$ plots is the averaging of $\varepsilon'_{\varphi\psi} - \sin^2\psi$ data obtained from measurements of several reflections *hkl*. This is best done by selecting reflection profiles where two or more reflections contribute (e.g., cubic 511 and 333). Another possibility of taking into account the influence of elastic and plastic deformation on the experimental $\varepsilon'_{\varphi\psi}$-values is based on the simulation of the deformation using Taylor [9], Voigt [10], or self-consistent [11] models. It could be shown that this way the oscillation of the experimental data can be reproduced [12, 13] so that reliable macroscopic stress values can be obtained.

### 6.6.2
### Strain/Stress Gradients

Curved lattice strain distributions (Figure 6.6b) are measured, when the gauge volume element under study depends on the measuring direction $\varphi$, $\psi$ and stress or $d_0$ gradients are present. Then each measurement represents another average stress state or mean $d_0$ value. To solve this problem different approaches are possible:

- Especially using high flux sources like synchrotron X-ray sources the gauge volume element can be reduced so that its dimensional variation is small compared to the steepness of the stress or $d_0$ gradient. One limitation, however, is often given by the grain size of the material.
- Another way to handle the case of strain/stress or $d_0$ gradients is to keep a similar size and location of the gauge volume element constant for all measuring directions. This is realized in the case of neutron diffraction [14]. Here the detector is positioned at $2\theta = 90^\circ$, so that a cubic gauge volume element can be realized. In the case of monochromatic neutrons the wavelength of the neutrons has to be adapted to the reflection *hkl* under study so that $2\theta(hkl) = 90^\circ$ is fulfilled. For a polychromatic neutron beam it is sufficient to position the diaphragm for the diffracted beam at $2\theta = 90^\circ$. In the case of high-energy X-radiation reflection intensities can be measured only at small $2\theta$ angles due to the form factor which steeply decreases with increasing $\sin\theta/\lambda$. Small $2\theta$ angles lead to a parallelepiped shape of the gauge volume element. Here a solution has been worked out by movable primary and secondary diaphragrams, which enable to keep the gauge volume element constant for different $\psi$-positions [15].
- Furthermore, evaluation procedures for curved $\varepsilon'_{\varphi\psi} - \sin^2\psi$ distributions have been established. This type of nonlinear lattice strain distribution is mostly observed at the near surface region where steep stress gradients are present, e.g., due to machining processes like grinding. The diffraction methods which have been developed to analyze such nonuniform stress fields are based on the exponential attenuation of the X-rays by matter according to Beer's law. The so-called $1/e$-penetration depth $\tau$ is defined by the condition that the intensity $I$ of the X-rays passing through the material is $1/e$ of the primary intensity $I_0$. Furthermore, $\tau$ depends on the geometrical diffraction conditions, i.e., on the orientation $\phi$, $\psi$ of the scattering vector $\vec{q}(hkl)$, the Bragg angle $\theta$, and the angle $\eta$ that describes the sample rotation around $\vec{q}(hkl)$ (see Chapter 10.1). Consequently, it must be distinguished between the actual

strain/stress depth profiles in the "real-" or $z$-space, $\varepsilon'_{\varphi\psi}(hkl, z)$ and $\sigma_{ij}(z)$ and those obtained from the measurement, $\varepsilon'_{\varphi\psi}(hkl, \tau)$ and $\sigma_{ij}(\tau)$, respectively. They are correlated by

$$\varepsilon'_{\varphi\psi}(hkl, \tau) = \frac{\int \varepsilon'_{\varphi\psi}(hkl, z)\mathrm{e}^{-z/\tau}\, dz}{\int \mathrm{e}^{-z/\tau}\, dz}, \quad \sigma_{ij}(\tau) = \frac{\int \sigma_{ij}(z)\mathrm{e}^{-z/\tau}\, dz}{\int \mathrm{e}^{-z/\tau}\, dz} \tag{6.13}$$

Since the above equations have the form of a Laplace transform with respect to $1/\tau$, the $f(\tau)$-profiles and the methods for their analysis are called "Laplace-space profiles" and "Laplace-space methods," respectively.

To evaluate the real-space stress depth profiles $\sigma_{ij}(z)$, the engineer is mainly interested in, different strategies have been developed. Because it has turned out that the inverse numerical Laplace transform, i.e., the direct conversion of the experimentally obtained $\sigma_{ij}(\tau)$-profiles into $\sigma_{ij}(z)$ leads to ill-conditioned systems of equations, most of the suggested methods go the opposite way and describe the $z$-space profiles of the relevant stress components $\sigma_{11}$ and $\sigma_{22}$ by simple polynomial or exponentially damped functions which can be easily transformed into the Laplace space by Eq. (6.13). Inserting the respective expressions for $\sigma_{ij}(\tau)$ into a "depth-dependent" and "biaxial" form of Eq. (6.8) yields a model function for the strain depth profile $\varepsilon_{\phi\psi}(hkl, \tau)$ which can be fitted to the experimental data.

The so-called section polynomial method [16] is based on fitting spline functions directly to the $d_{\phi\psi}(hkl, \tau)$ versus $\sin^2 \psi$-data. This method allows a rapid overview over the steepness of stress gradients. The use of higher order polynomials, however, may lead to instable solutions. The Universal-plot method [17] allows to combine the measurements on different $hkl$, each with different information depth, in one $\sigma_{ij}(\tau)$ diagram by applying one evaluation procedure. This increases the reliability of the so evaluated stress distributions. A further access to handle curved interplanar strain distributions is given by the scattering vector method [18]. Here lattice strain depth profiles are recorded in fixed orientations $\phi$, $\psi$ of the scattering vector $\vec{q}(hkl)$ by a stepwise rotation of the sample around $\vec{q}(hkl)$ (Figure 6.7). By this rotation different penetration depths are realized (except for $\psi = 0°$). This method is particularly effectively applied in textured samples (e.g., thin coatings), since here only few intensity poles are available for measurements. Furthermore, it allows for an evaluation of pseudomacroscopic $\sigma^{\alpha}_{33}(\tau)$ depth profiles, if compositional gradients $d_0(z)$ may be excluded from the considerations.

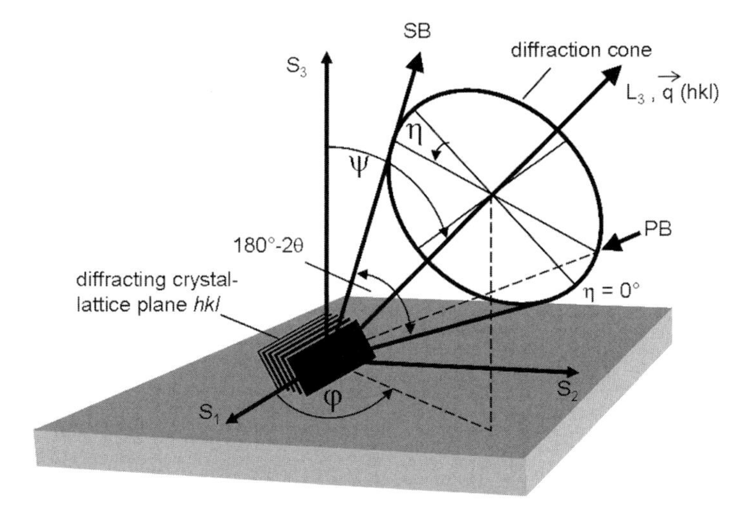


**Figure 6.7** Scattering vector method [18]. PB = primary beam, SB = secondary beam, $\eta$ = sample rotation angle around $\vec{q}(hkl)$.

## 6.6.3 Shear Strains/Stresses

Nonlinear interplanar lattice strain distributions with a splitting between the positive and negative $\psi$-branches are due to the presence of nonvanishing $\sigma_{13}$, $\sigma_{23}$ stress tensor components (Figure 6.6c). This type of nonlinear interplanar lattice strain distribution is characteristic for a stress state after intense surface machining like, e.g., turning. This intense process leads to an inclination of the main tensor components relative to the sample system resulting in different experimental $\varepsilon'_{\varphi\psi}$ values for the positive and negative $\psi$-branches.

For the evaluation of splitted $\varepsilon'_{\varphi\psi} - \sin^2\psi$ diagram, an evaluation procedure has been proposed by Dölle/Hauk [19]. Equation (6.5) can be written as follows:

For $\psi > 0$:

$$\varepsilon'_{\varphi\psi} = \varepsilon_{33} + f_1(\varphi, \varepsilon_{ij}) \sin^2\psi + f_2(\varphi, \varepsilon_{ij}) \sin 2\psi \tag{6.14}$$

For $\psi < 0$:

$$\varepsilon'_{\varphi\psi} = \varepsilon_{33} + f_1(\varphi, \varepsilon_{ij}) \sin^2\psi - f_2(\varphi, \varepsilon_{ij})|\sin 2\psi| \tag{6.15}$$

The factors $f_1$ and $f_2$ are given in Table 6.1.

The experimental values $\varepsilon'_{\varphi+\psi}$ and $\varepsilon'_{\varphi-\psi}$ are used for the calculation of the functions $a_1$ and $a_2$:

$$a_1 = \frac{\varepsilon'_{\varphi+\psi} + \varepsilon'_{\varphi-\psi}}{2} = \varepsilon_{33} + f_1 \sin^2\psi \tag{6.16}$$

**Table 6.1** Factors $f_1$ and $f_2$ for different azimuth angles $\varphi$.

| $\varphi$ (deg) | $f_1$ | $f_2$ |
|---|---|---|
| 0 | $\varepsilon_{11} - \varepsilon_{33}$ | $\varepsilon_{13}$ |
| 45 | $\frac{1}{2}(\varepsilon_{11} - \varepsilon_{33}) + \frac{1}{2}(\varepsilon_{22} - \varepsilon_{33}) + \varepsilon_{12}$ | $\frac{1}{2}\sqrt{2}\,(\varepsilon_{13} + \varepsilon_{23})$ |
| 90 | $\varepsilon_{22} - \varepsilon_{33}$ | $\varepsilon_{23}$ |
| 0 | $\frac{1}{2}s_2(hkl)(\sigma_{11} - \sigma_{33})$ | $\frac{1}{2}s_2(hkl)\sigma_{13}$ |
| 45 | $\frac{1}{2}s_2(hkl)(\frac{1}{2}\sigma_{11} + \frac{1}{2}\sigma_{22} + \sigma_{12} - \sigma_{33})$ | $\frac{1}{2}s_2(hkl)\frac{1}{2}\sqrt{2}(\sigma_{13} + \sigma_{23})$ |
| 90 | $\frac{1}{2}s_2(hkl)(\sigma_{22} - \sigma_{33})$ | $\frac{1}{2}s_2(hkl)\sigma_{23}$ |

$$a_2 = \frac{\varepsilon'_{\varphi+\psi} - \varepsilon'_{\varphi-\psi}}{2} = f_2 \sin|2\psi| \tag{6.17}$$

$\varepsilon_{33}$ is given by $a_1$ for $\sin^2\psi = 0$. The other strain tensor components are then obtained from $\dfrac{\delta a_1}{\delta \sin^2\psi}$ and $\dfrac{\delta a_2}{\delta \sin|2\psi|}$.

## 6.7 Diffraction Elastic Constants

Diffraction elastic constants (DEC) are used for the calculation of stresses from the strains measured by diffraction methods (see Eq. 6.8). The DEC take an intermediate position between the macroscopic strain–stress relationship described by the Youngs modulus $E$ and the Poisson ratio $\upsilon$ and the anisotropic elastic behavior of individual crystallites described by the single crystal elastic constants $c_{ijkl}$. In a macroscopic deformation test, crystallites of all orientations contribute to the measured strains. In a diffraction experiment, only those crystallites contribute to the diffracted intensity which fulfill the Bragg equation for the reflection *hkl* measured and for the chosen measuring direction $\varphi$, $\psi$. So an equivalency between the DEC and the macroscopic elastic constants would require an integration over the diffraction measurements in all measurement directions and for all reflections *hkl*. Comparing the single crystal elastic deformation with the information obtained in a reflection profile *hkl*, then an integration over all crystallites contributing to the diffracted intensity has to be performed. This integration delivers a single crystal elastic constant specific for the reflection *hkl* under study and for the selected measuring direction. However, in a polycrystalline aggregate there are additional crystallite–crystallite interactions, which are due to the different orientations of the crystallites. These crystallite–crystallite interactions cannot be calculated without introducing further assumptions.

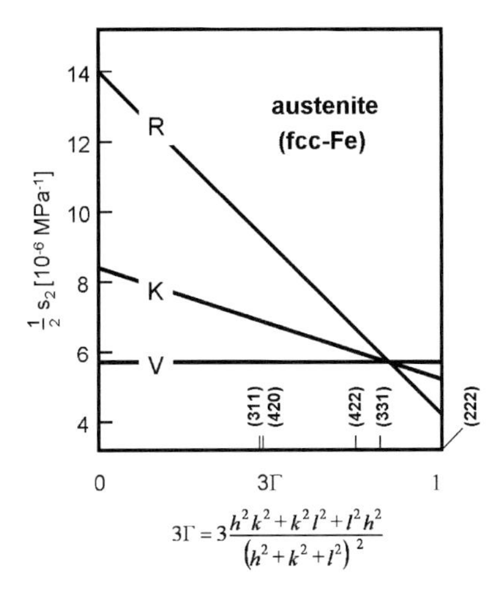


**Figure 6.8** Diffraction elastic constant $\frac{1}{2}s_2$ of austenite (fcc-Fe) according to Voigt, Reuss, Kröner and Eshelby [20–23]. 3Γ-anisotropy factor.

## 6.7.1
## Calculation of DEC

For the calculation of the DEC out of the single crystal elastic constants, three models are in use:

- In the model according to Voigt [20] it is assumed that the strain is uniform in the polycrystalline aggregate so that all crystallites exhibit the same strain.
- In the model according to Reuss [21] it is assumed that the stress is uniform in the polycrystalline aggregate so that all crystallites exhibit the same stress.
- In the model according to Kröner [22] and Eshelby [23] the polycrystalline elastic constants are calculated using an anisotropic spheroidal crystallite embedded in an infinite, elastically isotropic matrix.

The results of these models are displayed for fcc-Fe (austenite) in Figure 6.8.

In practice, the results obtained from the Kröner/Eshelby model are in good agreement with the experimentally determined DEC.

## 6.7.2
## Experimental Determination of the DEC

For checking the results for the DEC obtained from the Kröner/Eshelby method or in the case when single crystal elastic constants are not available, the DEC can

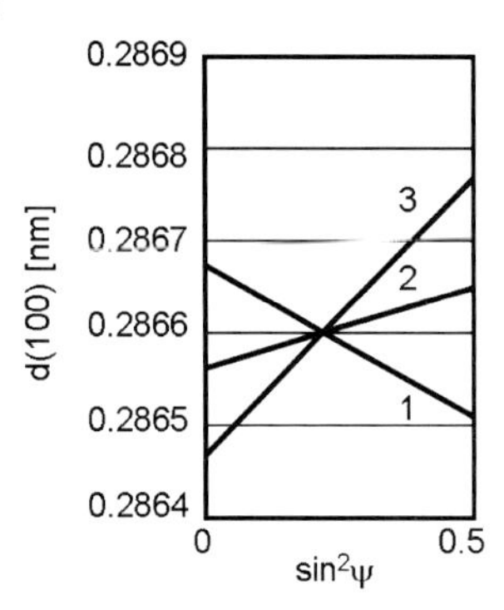


**Figure 6.9** Change of the slope of $d$ versus $\sin^2\psi$-distributions as a function of applied load. 1, residual stress $\sigma_{11} = 100$ MPa; 2, applied stress: 200 MPa; 3, applied stress: 400 MPa.

be experimentally determined by tensile or by bending tests. Figure 6.9 displays the change of the slope of $d$ versus $\sin^2\psi$-distributions as a function of the applied load.

From the data shown in Figure 6.9 the DEC can be calculated according to

$$s_1(hkl) = \frac{1}{d_0}\frac{\partial}{\partial\sigma_{11}} d_{\psi=0}(hkl) \tag{6.18}$$

$$\frac{1}{2}s_2(hkl) = \frac{1}{d_0}\frac{\partial}{\partial\sigma_{11}}\frac{\partial}{\partial\sin^2\psi} d_{\varphi=0,\psi}(hkl) \tag{6.19}$$

It has to be noted that the load measurements have to be performed with decreasing load in order to exclude the possible effects of yielding. In the case of bi- or multiphase materials Eqs. (6.18) and (6.19) give the DEC for the composite. For the evaluation of phase-specific DEC the reader is referred to specialized literature [6].

## 6.8 Experimental Set-up and Measuring Procedures

### 6.8.1 Experimental Set-up

#### 6.8.1.1 Diffractometers

In most cases experimental stress analysis requires strain measurements in different sample directions. Therefore, two basic types of set-ups are possible. Using the set-up of an $\Omega$-diffractometer the inclination angle $\psi$ is realized by a rotation around the vertically oriented $\psi$-axis coinciding with the $2\theta$-axis (Figure 6.10a). This arrangement is often used for neutron diffraction and high-energy X-ray

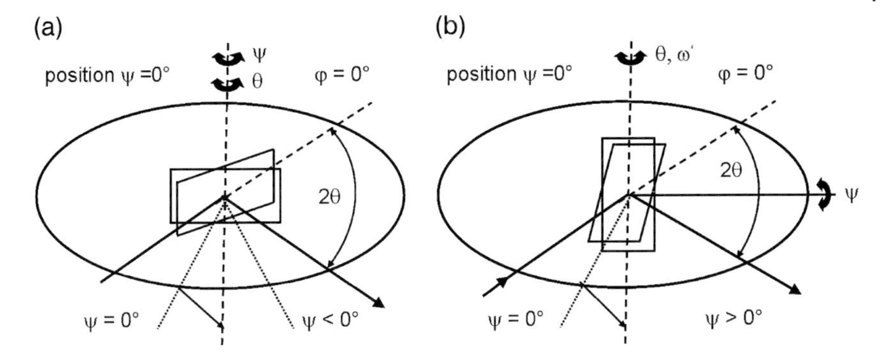


**Figure 6.10** Diffractometer set-ups: (a) Ω-diffractometer, (b) $\psi$-diffractometer.

synchrotron strain analyses since here measurements can be performed at $\psi = 90^\circ$ (transmission mode) due to high penetration depth. Using X-rays with characteristic radiation, however, the accessible $\psi$-range is limited to $|\psi| < \theta$ due to the absorption of either the primary or the secondary beam. Here the set-up of the $\psi$-diffractometer (Figure 6.10b) is favorable. The $\psi$-positions are realized by rotation around the horizontal $\psi$-axis perpendicular to the measuring direction.

For both diffractometer set-ups it is useful to mount a $\varphi$-rotation table and $x$-, $y$-, $z$-translations, e.g., for the measurement of strain/stress fields. Furthermore, Eulerian cradles can be used. Due to the requirements of residual stress analysis, Eulerian cradles with integrated $x$-, $y$-, $z$-translations are available. Usually their $\chi$-segment is opened to avoid beam absorption at high $2\theta$-values (Figure 6.11).

The disadvantage of the Eulerian geometry is the limited space for the samples and for the sample environment, e.g., furnaces or stress rigs. Therefore, other sample movement devices are also now in use, e.g., hexapods, which are able to incline and rotate the complete table with the mounted sample [24].

**Figure 6.11** Open Eulerian cradle with integrated $x$-, $y$-, $z$-translations.

#### 6.8.1.2 Diaphragms
Diaphragms for defining the primary beam size as well as the diffracted beam size are of paramount importance for residual stress analysis. These diaphragms must be easily changeable in their horizontal aperture and vertical height in order to define the gauge volume as required by the measuring task. In the case of divergent radiation, e.g., neutrons, the diaphragms should have translation axes so that they can be positioned as near as possible to the sample for avoiding the smearing out of the sampling gauge volume. For this purpose also radial collimators are in use and it could be shown that reflection shifts due to surface or interface effects can be reduced this way [25]. However, the experimental flexibility suffers from the use of the radial collimators, because they have to be exchanged for adapting the gauge volume size.

#### 6.8.1.3 Detectors
In the case of monochromatic X-rays, mostly point detectors are in use so that the reflection profiles are registered by $\Delta 2\theta$-step measurements. This procedure can be sped up by linear position sensitive detectors which, however, suffer from the higher background due to their larger aperture. The most effective way to exploit the diffracted intensity is the use of two-dimensional position sensitive detectors, available as image plates or as wire nets in gas atmosphere. Using characteristic X-radiation complete Debye-fringes can be simultaneously registered this way. In the case of neutron diffraction, where $2\theta = 90^\circ$ can be realized, the data along the vertical detector axis allow the evaluation of $\Delta \sin^2 \psi$ sections for the $\varepsilon'_{\varphi\psi}$ versus $\sin^2 \psi$ plots. Using monochromatic high-energy X-radiation, special secondary beam diaphragms are necessary for the precise localization of the diffracting gauge volume element (see Chapter 9 for further details).

Using white radiations, time-of-flight detectors are necessary for neutrons (see Chapter 8 for further details) and energy dispersive detectors for white synchrotron X-rays (see Chapters 10.1 and 10.2 for further details).

### 6.8.2 Measuring Procedures

Residual stress analysis applying diffraction methods requires precisely adjusted diffractometers since an accuracy of $\Delta d/d \leq 10^{-4}$ has to be achieved. Therefore, it is recommended to check the adjustment of the diffractometer on a regular basis by performing measurements on a stress-free powder (e.g., Au). The evaluation of the diffractogram allows us to calibrate the wavelength and the $2\theta$-offset used. The evaluation of $\sin^2 \psi$ measurements in the range of $-85^\circ \leq \psi \leq +85^\circ$ should deliver a straight line with $\Delta 2\theta = \pm 0.01^\circ$ and zero slope. A splitting of the positive and negative $\psi$-branches as well as a slope $m \neq 0$ of the $\varepsilon'_{\varphi\psi}$ versus $\sin^2 \psi$ diagram would indicate that the sample is not in the center of the diffractometer, which is defined by the intersection of the rotation axes.

In the case of using monochromatic radiation, the reflection *hkl* for the experimental strain analysis has to be selected. Therefore, different aspects have to be considered:

- As a first step an identification of all reflections in the diffractogram should be performed. This way it can be checked if there are reflections, which are overlapped, e.g., due to the presence of different crystalline phases. For the residual stress analysis of course only reflections belonging to a single crystalline phase are admitted.
- According to Eq. (6.3), the strain analysis is most sensitive at high angles $2\theta$. So preferably a reflection with $2\theta > 90^\circ$ should be selected.
- Evidently measurements can be most time effectively performed on reflections with high intensities.
- For the case of strongly texture affected sample material states, a list of reflections with small and large intergranular strains has been established by the VAMAS TWA 20 group (Table 6.2) [7].

**Table 6.2** Choice of diffracting plane for strongly texture affected sample material states.

| Material | Planes with small intergranular strains | Planes with large intergranular strains |
|---|---|---|
| fcc (Ni, Fe, Cu) | 111, 311, 422 | 200 |
| fcc (Al) | 111, 311, 422, 220 | 200 |
| Bcc (Fe) | 110, 211 | 200 |
| Hcp (Zircalloy, Ti) | $10\,\bar{1}2$, $10\bar{1}3$ (pyramidal) | 0002 (basal)<br>$10\bar{1}0$, $11\bar{2}0$ (prism) |
| hcp (Be) | $20\bar{2}1$, $11\bar{2}2$ (second-order pyramidal) | $10\bar{1}2$, $10\bar{1}3$ (basal, prism and first-order pyramidal) |

Additional hints for measurement and evaluation strategies are given in Section 6.6.1.

For the evaluation of reflection profiles intensity corrections may be necessary depending on the radiation used. If necessary, Lorentz, polarization, and absorption corrections have to be applied [6]. For the reflection position determination $2\theta$ furthermore background corrections are recommended.

## 6.9 Overview on In-depth and Local Residual Stress Analysis

A summary of radiations and information depths is shown in Figure 6.12.

Chapters 7 to 10 outline the limitations and possibilities of the different measuring schemes.

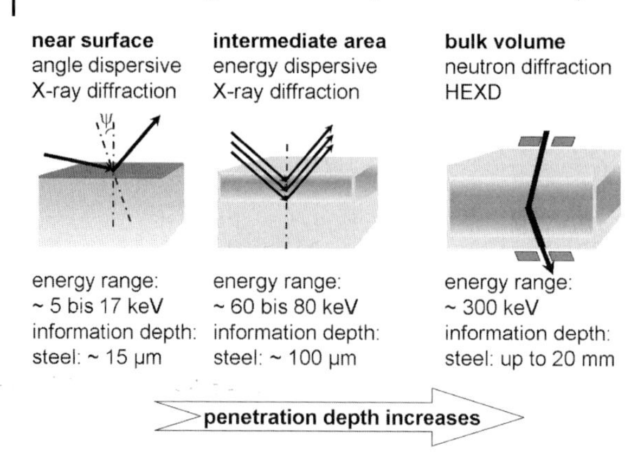


**Figure 6.12** Radiations and corresponding information depths.

Concerning the local resolution of residual stress analysis with diffraction methods two different strategies have been developed on the basis of high X-ray intensities delivered by synchrotron sources:

- Using a large incoming X-ray beam, the local resolution down to approximately $10 \times 10\ \mu m^2$ is achieved by a capillary-array mounted to the front of a 2D-detector (see Chapter 11).
- Using synchrotron X-ray beams with diameters of some µm, the residual stress state of individual crystallites embedded in their polycrystalline environment can be investigated (see Chapters 19 and 20). This way the local resolution of the residual stress state corresponds to the crystallite size or in the case of coarse grained materials to the diameter of the X-ray beam.

## References

1 T.G. Fawcett, J. Faber, F. Needham, S.N. Kabekkodu, C.R. Hubbard, J.A. Kaduk, Developments in Formulation Analysis by Powder Diffraction Analyses, *Powder Diffraction* 2006, **21**, 105–110.

2 G. Faninger, U. Hartmann, Physikalische Grundlagen der quantitativen röntgenographischen Phasenanalyse (RPA), *Härterei. Techn. Mitt.* 1972, **27**, 233–244.

3 I.C. Noyan, J.B. Cohen, *Residual Stress*, Springer, Berlin 1987.

4 L.B. Mc Cusker, R.B. von Dreele, D.E. Cox, D. Louer, P. Scardi, Rietveld refinement guidelines, *J. Appl. Cryst.* 1999, **32**, 36–50.

5 E.J. Mittemeijer, P. Scardi(eds.), *Diffraction Analysis of the Microstructure of Materials*, Springer, Berlin, 2004.

6 V. Hauk, *Structural and Residual Stress Analyses by Nondestructive Methods*, Elsevier, Amsterdam, 1997.

7 Non-destructive testing, Standard test method for determining residual stresses by neutron diffraction (ISO/TS 21432:2005).

8 E. Macherauch, P. Müller, Das $\sin^2 \psi$-Verfahren der röntgenographischen Spannungsmessung, *Z. Angew. Phys.* 1961, **13**, 305–312.

9 G.J. Taylor, The mechanism of plastic deformation of crystals, Part I, *Proc. R. Soc.* 1934, 362–387.

10 G. Sachs, Zur Ableitung einer Fließbedingung, *Zeitschrift des VDI* 1928, **72**, 734–736.

11 C. Tome, P. Maudlin, R. Lebensohn, G. Kaschner, Mechanical response of xirconium: I. Derivation of a polycrystal constitutive law and finite element analysis, *Acta Mat.* 2001, **49**, 3085–3096.

12 S. Aris, R.V. Martins, V. Honkimäki, A. Pyzalla, Simulation of texture and crystallite microstrain development during compression and tension of an Al alloy and comparison to experimental results, *Comput. Mater. Sci.* 2000, **19**, 116–122.

13 B. Reetz, Mikrostruktur und Eigenschaften stranggepresster sowie kaltverformter Messinglegierungen, Cuvillier, 2006.

14 T. Lorentzen, M.T. Hutchings, P.J. Withers, *Introduction to Characterization of Residual Stress by Neutron Diffraction*, Taylor and Francis, London, 2005.

15 I.A. Denks, M. Klaus, Ch. Genzel, Determination of real space residual stress distributions $\sigma_{i,j}(z)$ of surface treated materials with diffraction methods: Part II. Energy dispersive approach, *Mater. Sci. Forum* 2006, 524–525, 37–42.

16 T. Leverenz, B. Eigenmann, E. Macherauch, Das Abschnitt-Polynom-Verfahren zur zerstörungsfreien Ermittlung gradientenbehafteter Eigenspannungszustände in den Randschichten von bearbeiteten Keramiken, *Z. Metallkunde* 1996, **87**, 616–625.

17 H. Ruppersberg, I. Detemple, J. Krier, Evaluation of strongly non-linear surface-stress fields $\sigma_{xx}(z)$ and $\sigma_{yy}(z)$ from diffraction experiments, *Phys. Stat. Sol. (a)* 1989, **116**, 681–687.

18 Ch. Genzel, Evaluation of stress gradients $\sigma_{ij}(z)$ from their discrete Laplace transforms $\sigma_{ij}(\tau_k)$ obtained by X-ray diffraction performed in the scattered vector mode, *Phys. Stat. Sol. (a)* 1996, **156**, 353–363.

19 H. Dölle, V. Hauk, Röntgenographische Spannungsermittlung der Eigenspannungssysteme allgemeiner Orientierung, *Härterei Techn. Mitt.* 1976, **31**, 165–168.

20 W. Voigt, *Lehrbuch der Kristallphysik*, 1. Aufl., Teubner, 1928.

21 A. Reuss, Berechnung der Fließgrenze von Mischkristallen auf Grund der Plastizitätsbedingung für Einkristalle, *Zeitschrift f. Angew. Math. u. Mech.* 1929, **9**, 49–58.

22 E. Kröner, Berechnung der elastischen Konstanten des Vielkristalls aus den Konstanten des Einkristalls, *Z. Phys.* 1958, **159**, 504–518.

23 J.P. Eshelby, The determination of the elastic field of an ellipsoidal inclusion and related problems, *Proc. R. Soc. A* 1957, **241**, 376–396.

24 T. Pirling, G. Bruno, P.J. Withers, SALSA: advances in residual stress measurement at ILL, *Mater. Sci. Forum* 2006, 524–525, 217–222.

25 T. Pirling, Neutron strain scanning at interfaces, *Mat. Sci. Forum* 2000, 347–349, 107–112.

# 7
# Stress Analysis by Angle-Dispersive Neutron Diffraction

*Peter Staron*

## 7.1
## Introduction

Diffraction-based techniques for the analysis of residual stress (RS) states in crystalline solids have been used for a long time [1]. Angle-dispersive X-ray diffraction using laboratory sources are widely employed for determining surface stresses [2]. The great advantage that comes along with neutrons is their large penetration depth, which is about 1000 times larger than for X-rays from a Cu-tube in most materials. Therefore, with neutrons information can be obtained nondestructively from the interior of a specimen. Residual stress analysis (RSA) with neutrons is comprehensively described in [3]. When high-energy X-rays from synchrotron sources became available, they were also used for RSA. The main advantage of synchrotron sources is the high intensity, facilitating high spatial resolutions up to single-grain measurements or fast *in situ* measurements of time-dependent processes. These three techniques – using X-ray tubes, neutrons, or medium and high-energy X-rays from synchrotron sources – should in general be regarded as complementary techniques, each offering its special advantages.

The specialty of neutrons for RSA is the *combination* of large penetration depth and wavelengths that enable a scattering angle of 90°. For example, measuring internal strains throughout the interior of a 2 cm thick steel plate is possible with neutrons within reasonable data acquisition times. In addition to absorption ($\sigma_a$), also scattering power ($\sigma_{coh}$) and incoherent background ($\sigma_{inc}$) determine data acquisition times. Consequently, RSA can be performed in thick aluminum alloys, while it is much more difficult in titanium alloys because of strong absorption and high incoherent background combined with low scattering power (Table 7.1). The maximum penetration thickness also depends on the strength of the source and the spatial resolution that is required. As concerns the diffraction geometry, the 90° scattering angle not only leads to a near-cubic gauge volume, but also makes strain measurements possible even in large parts with a complicated geometry. This can be more difficult with high-energy X-rays because of the much smaller scattering angles (cf. Chapter 9).

*Neutrons and Synchrotron Radiation in Engineering Materials Science: From Fundamentals to Material and Component Characterization*
Edited by Walter Reimers, Anke Rita Pyzalla, Andreas Schreyer, and Helmut Clemens

ISBN: 978-3-527-31533-8

**Table 7.1** Coherent ($\sigma_{coh}$), incoherent ($\sigma_{inc}$), and absorption ($\sigma_{abs}$) cross sections [3] as well as 50% attenuation length ($l_{1/2}$) for neutrons with a wavelength of 1.6 Å in several engineering materials.

| Material | Al | Ti | Fe | Ni | Cu |
|---|---|---|---|---|---|
| $\sigma_{coh}$ (barn) | 1.495 | 1.485 | 11.22 | 13.3 | 7.485 |
| $\sigma_{inc}$ (barn) | 0.008 | 2.87 | 0.4 | 5.2 | 0.55 |
| $\sigma_{abs}$ (barn) | 0.205 | 5.41 | 2.28 | 3.99 | 3.36 |
| $l_{1/2}$ (cm) | 6.74 | 1.24 | 0.59 | 0.34 | 0.72 |

Angle-dispersive diffractometers for RSA are usually limited to a single lattice reflection at a time. In contrast, with energy-dispersive neutron diffraction (Chapter 8) a complete spectrum is recorded with one measurement, giving information about several reflections. The latter can be especially useful for multiphase materials or when microstresses are to be analyzed. However, time-of-flight spectrometers generally work most efficient at pulsed sources, while angle-dispersive spectrometers are well suited for continuous sources. In addition, for many engineering applications the amount of information gained with a single-peak measurement is sufficient.

Fields of application for angle-dispersive neutron strain scanning are, generally speaking, the determination of RS in the interior of components. To be more specific, one broad field of engineering application is welding, because RS and distortion are always connected with welding; they can decrease the load carrying capacity and the lifetime of a component and, thus, information about stress states in welds is often crucial. Another field of application is the verification of models such as finite element models (FEM). Models are used for stress prediction, e.g. in welds, in thermally treated or in deformed material. In Section 7.4 two examples will be given for RS in a weld and FEM predictions of RS in water-quenched components.

## 7.2 Diffractometer for Residual Stress Analysis

### 7.2.1 Set-up of a Diffractometer for Strain Scanning

Today, dedicated instruments are used for RSA, or strain scanning. They are optimized for using a small gauge volume, precisely defined by slits, and achieving a resolution of lattice parameters better than $10^{-4}$, corresponding to a stress resolution of about 20 MPa for steel. For this, thermal neutrons are used with wavelengths between about 1 Å and 3 Å for obtaining the 90°-diffraction geometry for

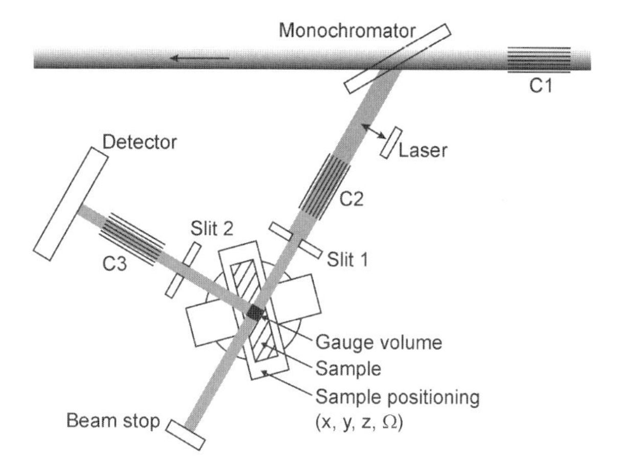


**Figure 7.1** Schematic drawing of a constant-wavelength strain scanner set-up.

most relevant engineering materials. Strain scanners are often found close to the neutron source where they can make use of the large vertical divergence of the neutron beam for increasing the intensity. Another possibility is a place at a thermal neutron guide with the advantage of lower background.

The basis of diffraction strain measurements is Bragg's equation, $n\lambda = 2d \sin \theta$, where $\lambda$ is the neutron wavelength, $d$ is the lattice spacing, $2\theta$ is the scattering angle, and $n$ is a positive integer (cf. Chapter 6). A change $\Delta d$ in the lattice spacing due to internal strains will result in a shift of a Bragg peak position $\Delta\theta$ when a single wavelength is used. The shift of the peak positions is of the order of $0.01°$–$0.1°$; thus, diffraction strain measurements are precision measurements and all components of the diffractometer are required to be sufficiently precise and stable.

The main components of a strain scanner are monochromator, collimating system, slit system, sample stages, and detector (Figure 7.1). The collimating devices C1–C3 define the divergence of the beam at the given positions. For an instrument position at a thermal neutron guide, C1 is defined by the transmission of the guide. Since residual stress distributions are always inhomogeneous a precise and reproducible positioning of a sample in the beam is required. For this, the precision of the positioning devices has to be sufficient and appropriate sample positioning procedures have to be used. In the following, the most important components will be introduced.

### 7.2.2
### Monochromator

Neutron sources offer a broad spectrum of wavelengths; however, neutrons of a single wavelength are needed for an angle-dispersive diffractometer. These can be delivered from a monochromator using either perfect single crystals or mosaic

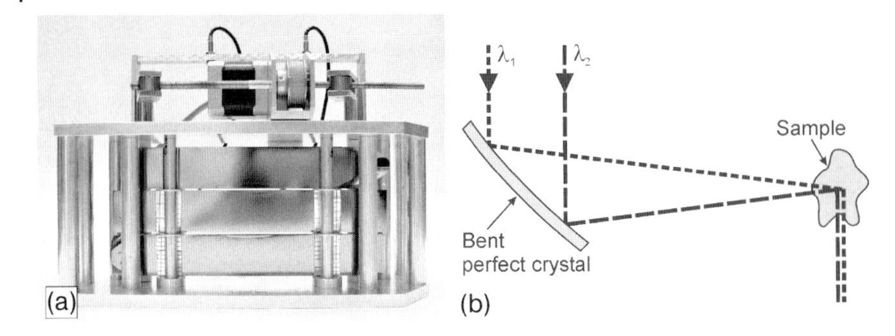


**Figure 7.2** (a) Photograph of a horizontally and vertically focusing monochromator with three elastically bent perfect Si single crystals (produced by the group of P. Mikula, Nuclear Physics Institute, Rez, Czech Republic). (b) Quasiparallel diffracted beam emerging from a sample when perfect crystals are used as monochromator.

crystals. A good choice of a monochromator for a strain scanner is the use of elastically bent perfect Si single crystals [4, 5]. As an example, the monochromator currently used at the strain scanner ARES-2 at the FRG-1 at GKSS is shown (Figure 7.2a). It essentially consists of a four-point bending device for motor-controlled variable elastic bending of the silicon crystals, which is used for horizontal focusing. Three crystals with a size of 30 mm × 210 mm and a thickness of 4 mm are used for vertical focusing.

The advantage of perfect single crystals for strain scanning is that no collimators are required [5]. The diffracted beam emerging from a polycrystalline sample is nearly parallel when the radius of curvature is adapted to the specific diffraction angle because of the angle–wavelength correlation established by the perfect crystals (Figure 7.2b). When the sample does not produce significant broadening of diffraction peaks, the divergence of the diffracted beam is mainly given by the thickness of the monochromator crystals [4]. Thus, a well-defined gauge volume can be achieved.

Conventional monochromators are used with mosaic crystals that contain domains of crystallites with slightly varying orientations, increasing the reflected intensity. Often, pyrolytic graphite (PG) mosaic crystals are used for monochromators because of their good reflectivity. Stacked Ge wafers also have good reflectivity. However, usually collimators have to be used with mosaic crystals to obtain a good spatial resolution.

### 7.2.3 Slit System

The slit system is used to define size and shape of the gauge volume within a sample. It consists of two slits: the first one in the incoming beam and the second one in a diffracted beam. Typical slit widths are between 0.5 mm and

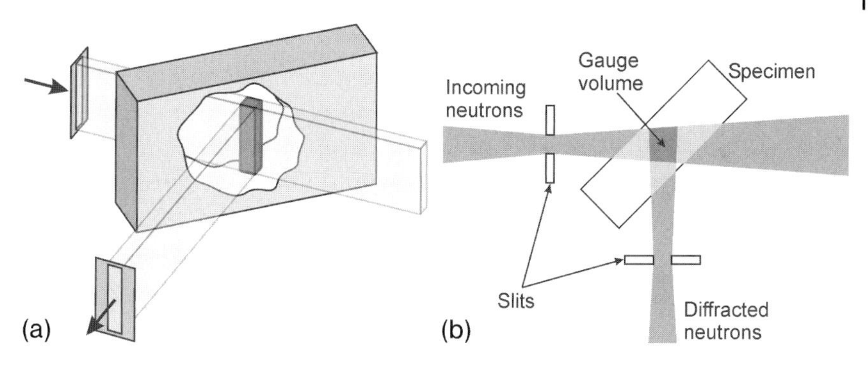


**Figure 7.3** (a) Schematic drawing of a matchstick-like gauge volume within a sample, formed by two slits. (b) Cross section of the gauge volume for divergent beams.

5 mm; slit heights can be from 1 mm up to about 30 mm (Figure 7.3). The true gauge volume is larger than the nominal gauge volume given by the size of the slits due to the divergence of incoming and diffracted beam. The scattering angle should be around 90° to produce a gauge volume with a rectangular cross section. Both slits have to be mounted on linear devices for adapting their distance from the sample.

### 7.2.4 Sample Positioning

"Strain scanning" is actually performed by moving a specimen through the beam and recording the position of a diffraction peak for different specimen positions. A specimen has to be moved in three perpendicular directions $x$, $y$, and $z$ for bringing a specific location within the sample into the beam. Thus, three linear movements are needed. When two linear tables are mounted on top of a goniometer, they can be rotated to align the specimen movement with the scattering vector $q$. A useful accessory is a Eulerian cradle for orienting and positioning of samples with masses up to about 5 kg.

### 7.2.5 Detector

Area detectors with a height of 200–300 mm cover a significant part of a diffraction cone for collecting as much intensity as possible without reducing the resolution of the strain direction too much (Figure 7.4). The readout of the detector electronics gives an intensity matrix (Figure 7.5); therefore, a calibration is needed for the calculation of angles.

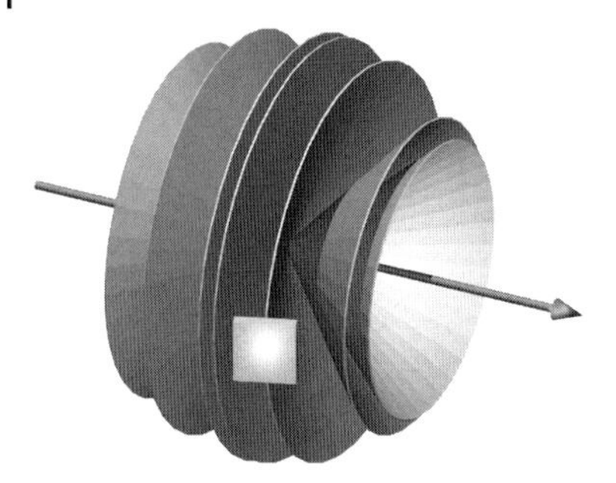

**Figure 7.4** Schematic drawing of the first seven diffraction cones of Al at a wavelength of 1.64 Å. The box is at a detector position of 70° for recording the (2 2 0) reflection.

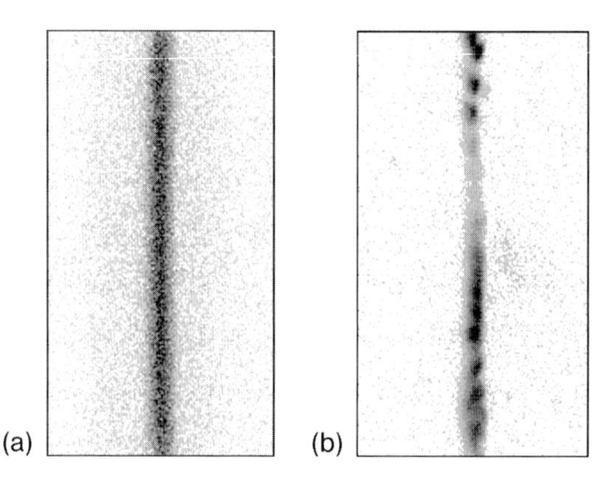


**Figure 7.5** Detector images of the (a) (3 1 1) reflection of fine-grained IN718 (cf. Section 7.4.2) and a nominal gauge volume of 27 mm$^3$ ($2\theta \approx 100°$); (b) (2 2 0) reflection of a coarse-grained Cu alloy and a nominal gauge volume of 90 mm$^3$ ($2\theta \approx 80°$). The width of the images covers a scattering angle interval of approximately 10°.

## 7.3 Measurement and Data Analysis

### 7.3.1 Gauge Volume and Sample Positioning

A matchstick-like gauge volume can be used for scanning strains in directions where the symmetry of the stress state allows integration over one direction (cf. Figure 7.3a) [6]. For example, in welded sheets this would be the longitudinal direction parallel to the weld. This is done to increase the intensity for saving beam time as well as to increase the number of diffracting grains. The size of the gauge volume can range from 1 $mm^3$ up to several 100 $mm^3$, depending on material, specimen thickness, grain size, and available intensity.

The cross section of a gauge volume can be measured by moving a wire through it and recording the diffracted intensity. With a monochromator using elastically bent perfect crystals the divergence of the diffracted beam is small, resulting in a very small increase of the gauge volume in that direction (Figure 7.6).

In the application of diffraction-based RSA it is often tacitly assumed that the assumptions the method is based on are fulfilled. One assumption is that the diffracting grains are representative for the material. However, crystallites are not isotropic, nor are their elastic properties; this fact is accounted for by the concept of *hkl*-dependent diffraction elastic constants (DEC; cf. Chapter 6). But this concept only works when the diffracting subset of grains is truly representative for the material, which is only the case when the number of grains in the gauge volume is large enough. The reason is that diffraction is not sensitive to a grain rotation around an axis parallel to $q$, but the elastic properties of the grain can be. So, only when enough of these grain orientations are included, the experimental strain value will be a representative one.

With a fine-grained specimen like the Ni-alloy discussed in Section 7.4.2 (mean grain size 5 μm) the number of diffracting grains is large and the intensity distribution on the diffraction cone is homogeneous (Figure 7.5a). In contrast, when the grains are large, reflections from single grains will become visible on the area detector and the intensity distribution will become spotty (Figure 7.5b). In such a case measures have to be taken to increase the number of diffracting grains. Otherwise, one should be aware of the fact that the error bar for stress results can be much larger than just that resulting from the statistical error of the measurement. Possible ways to increase the number of diffracting grains are (1) increasing the gauge volume, (2) moving the sample in an appropriate direction during the measurement, or (3) rotating the sample for small angles during the measurement, which also brings new grains into diffraction condition.

Utmost care has to be taken while positioning a sample in the beam. Especially when large stress gradients are present in a sample, positioning errors can change results drastically. Different accessories can be used for supporting sample positioning, like a plumb-line, lasers, telescopes, or video cameras with

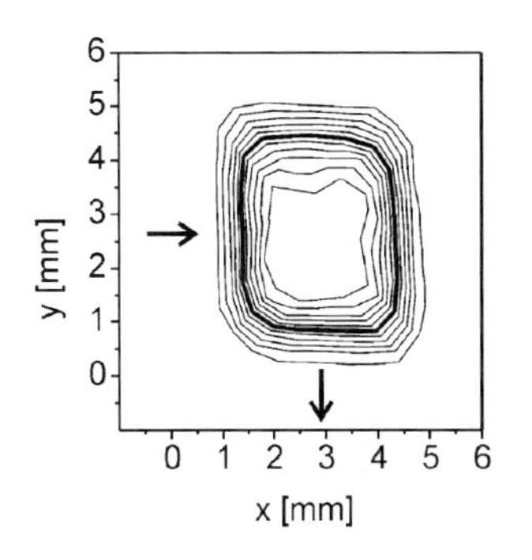


**Figure 7.6** Measured cross section of a gauge volume formed by 3 mm wide slits at a distance of 90 mm from the centre. The thick isointensity line marks half of the maximum intensity giving a size of 3.8 mm × 3.0 mm. The arrows mark the directions of incoming and diffracted beams.

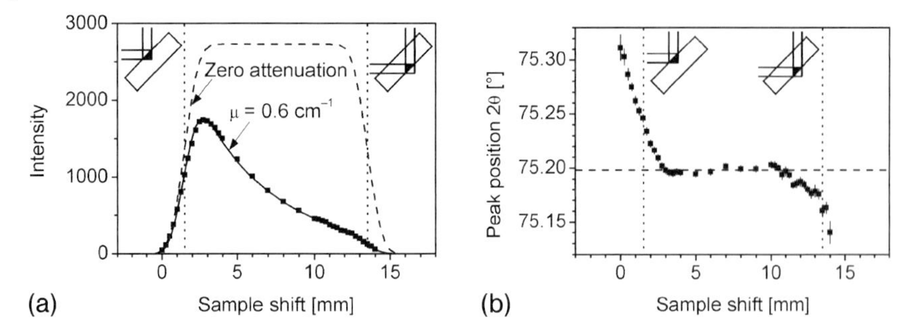


**Figure 7.7** Recorded peak intensity (a) and position (b) for a 12 mm thick container filled with Cu powder that is shifted through the gauge volume. The solid and dashed lines in (a) are calculated functions.

cross-hairs. However, the best results are usually obtained by using the neutron beam for positioning.

For this, a sample is shifted through the beam and the integral intensity of the desired reflection is recorded. Several scans can be required, depending on the geometry of the part. The result of such a scan could be a curve like that shown in Figure 7.7a, although usually measured with less intensity for positioning purposes. The whole curve can be fitted or the position of just one sample surface can be determined from the left part of the curve. Note that for the configuration shown in Figure 7.7a absorption leads to a shift of the position where the intensity is half of the maximum intensity relative to the true sample surface. This shift depends on the material and the size of the gauge volume and has to be corrected for precise sample positioning.

Partial immersion of the gauge volume in the sample leads to shifts in the peak position (Figure 7.7b). The reason is that the centroid of the immersed volume shifts when the gauge volume is only partially filled. Therefore, care has to be taken that the gauge volume is always completely within the sample during normal measurements. Depending on the configuration of the diffractometer, the use of a radial collimator can reduce or eliminate such shifts [7].

## 7.3.2 Data Reduction and Analysis

The basics of diffraction-based stress analysis are given in Chapter 6; therefore, only a few important equations will be reported in the following. Comprehensive descriptions are found in the textbooks [1–3]. In the following it will be assumed that the investigated material is an isotropic single phase material with a close-to-random texture, and microstresses can be neglected.

### 7.3.2.1 Data Reduction and Peak Fitting

The raw data delivered from an area detector is an intensity matrix. At scattering angles different from 90° the curvature of the cone section is visible on the area detector and has to be taken into account for integrating the intensity vertically. Detector channels have to be converted to angles. Note that the wavelength has to be known with a high precision only when absolute lattice parameters shall be determined, which is normally not required for strains measurements. For the latter, only the difference between a peak position and the position of the stress-free reference has to be determined with high precision.

The recorded diffraction peak is a convolution of the diffractometer resolution function and the reflectivity of the sample. In many cases, simple Gaussian peak profiles are observed. When there is significant peak broadening due to the microstructure of the sample, other peak profile functions can be involved. Peak profile analysis is a topic itself and shall not further be discussed in this context; a comprehensive treatment can be found, e.g., in [8].

It is important to fit the background under a peak as well. In many cases it is sufficient to account only for a linear background, however, in some cases a parabolic contribution should be taken into account as well. In these cases, using a linear background can lead to systematic deviations of the calculated peak position from the true position.

When the background is small, a few hundred counts in the peak maximum are sufficient to achieve small enough errors in the peak position. Increasing the data acquisition time only slowly reduces this error; therefore, the required precision should be determined before setting the data acquisition time in order to save beam time. In general, the required counting time will depend on the signal-to-background ratio.

#### 7.3.2.2 Calculation of Stresses

The strain $\varepsilon$ in the direction normal to the reflecting lattice plane (direction of the scattering vector) is obtained from measured peak positions $2\theta$ by

$$\varepsilon = \frac{d - d_0}{d_0} = -\frac{\theta - \theta_0}{\tan \theta_0} \tag{7.1}$$

where $d_0$ is the stress-free lattice parameter. The error in the strain can be calculated from the errors in the lattice parameters by

$$(\Delta\varepsilon)^2 = \left(\frac{\Delta d}{d}\right)^2 + \left(\frac{\Delta d_0}{d_0}\right)^2 \tag{7.2}$$

Since the stress state is described by a symmetric third rank tensor, in general six unknowns have to be determined. Therefore, strain measurements in at least six directions would be required; over-determining the problem by measuring more than six directions reduces the error. However, when the principal stress axes are known, only strain measurements in these three perpendicular directions are required to fully determine the stress tensor, which is diagonal in this case. For specimen geometries such as sheets, plates, tubes, cylinders, or discs the principal stress axes often are assumed to coincide with the principal geometrical axes

**Table 7.2** Diffraction angle of lattice plane families for residual stress analysis for a wavelength of 1.6 Å and corresponding DEC [3] for some engineering materials.

| Material | *hkl* | *d* (Å) | $2\theta$ (1.6 Å) (°) | $E_{hkl}$ (GPa) | $\nu_{hkl}$ |
|---|---|---|---|---|---|
| α-Fe | 211 | 1.170 | 86.3 | 225.5 | 0.28 |
| γ-Fe | 311 | 1.083 | 95.2 | 183.5 | 0.31 |
| Al | 311 | 1.221 | 81.9 | 70.2 | 0.35 |
| Ni | 311 | 1.063 | 97.6 | 185.0 | 0.33 |
| Cu | 311 | 1.090 | 94.4 | 122.0 | 0.35 |

of a specimen. However, one has to bear in mind that this must not always be true, especially in inhomogeneous material regions such as, e.g., welds. But even if the axes chosen for strain measurements are not the true principal stress axes within the specimen, the stresses in the chosen orthogonal directions will be determined.

Triaxial residual stresses in an isotropic polycrystal are calculated from the strains measured in three perpendicular directions by [9]

$$\sigma_i = \frac{E_{hkl}}{1+\nu_{hkl}}\left[\varepsilon_i + \frac{\nu_{hkl}}{1-2\nu_{hkl}}(\varepsilon_1+\varepsilon_2+\varepsilon_3)\right] \tag{7.3}$$

where $i = 1, 2, 3$. Diffraction elastic constants (DEC) have to be used for the modulus of elasticity $E_{hkl}$ and Poisson number $\nu_{hkl}$ (Table 7.2). They can be found in the literature, can often be calculated to a good approximation from single-crystal elastic constants, or can be determined for a specific material in uniaxial tensile tests.

When the error in the stresses $\sigma_i$ is calculated from the errors in the strains $\varepsilon_i$, the correlations via $d_0$ have to be taken into account and the full covariance matrix has to be calculated. Therefore, it is much easier to calculate the error in the stresses from the errors in the uncorrelated lattice parameters (or peak positions) according to

$$(\Delta\sigma_i)^2 = \left(\frac{A}{d_0}\right)^2 [B(\Delta d_i)^2 + C^2(\Delta d_0)^2 + D^2((\Delta d_1)^2 + (\Delta d_2)^2 + (\Delta d_3)^2)]$$

$$A = \frac{E}{1+\nu} \quad B = \frac{1}{1-2\nu} \quad C = \frac{1+\nu}{1-2\nu} \quad D = \frac{\nu}{1-2\nu} \tag{7.4}$$

Stresses can be separated into hydrostatic and deviatory parts; the former is the mean of the diagonal elements of the stress tensor, the latter is the deviation from the mean value in each direction. Note that $d_0$ is not required for calculating deviatoric stresses. Moreover, using Eq. (7.3) it is easy to see that also the difference of any two of the three stress components as well as the effective stress after Von Mises can be calculated without precise knowledge of $d_0$.

Strong textures can have a significant influence on the measured strains [2, 10]. The recommendation is to measure the texture, if possible, and add this information to the documentation of the RS results [11]. Methods have been developed for the treatment of residual stress analysis in textured materials; some of them are introduced in the textbook by Hauk [2].

#### 7.3.2.3 **Macro and Microstresses**

The stresses determined by diffraction methods are the sum of macro and microstresses that lead to peak shifts [12]. Microstresses can result, e.g., from plastic deformation of a material. Microstresses can lead to peak broadening as well as peak shifts. Since the engineer often is interested only in macrostresses, which influence load carrying capacity, or distortion, macrostresses have to be separated from microstresses.

Intergranular microstresses have different influences on different lattice planes due to the elastic anisotropy of the crystallites [13]. Therefore, besides the fact that this effect can be used to study intergranular stresses connected with the plastic history of a material, it is recommended to avoid certain reflections for the determination of macroscopic stresses but to use instead such reflections for which the influence of intergranular strains is small [11]. With these reflections, the best approximation to the macrostresses is obtained. In fcc and bcc materials the use of the (200) reflection is strongly discouraged; instead the use of, e.g., the (111) or (311) reflection (fcc) or the (110) or (211) reflection (bcc) is recommended. In hexagonal materials such as, e.g., Ti the use of the pyramidal planes (10–12) or (10–13) is recommended while the use of the basal (0002) and prism planes (10–10) or (10–13) can be problematic.

#### 7.3.2.4 **Stress-Free Reference**

Determination of strains requires the knowledge of the unstrained (stress-free) lattice parameter $d_0$. This is sometimes problematic because it has to be made sure that the specimen that is chosen as stress-free reference actually has the same chemical composition as the specimen under investigation. In welds, the heat treatment that is introduced by welding can significantly change the chemical composition in the weld and the heat affected zone, depending on the material. One way to obtain reliable reference samples is to cut small pieces out of the specimen under investigation at those positions where stresses are to be determined. The macrostresses are known to relax largely in such small pieces [14, 15].

Other methods like making use of conditions for the balance of forces inside a specimen can also be used to determine $d_0$. In thin specimens with a plane stress state, $d_0$ can be calculated by setting $\sigma_z = 0$ in Eq. (7.3) [9, 16].

Errors in $d_0$ lead to hydrostatic stress components in the results, deviatoric stresses remain unchanged.

## 7.4 Examples

### 7.4.1 Residual Stresses in Friction Stir Welded Aluminum Sheets

A broad field of application of RSA comes from welding. During welding the material is heated with a point-like heat source and cools again while neighboring material is heated. This inhomogeneous heating and cooling induces RS and distortion. RS can reduce the life or load carrying capacity of a weld or influence the service performance of the welded components with respect to fatigue and corrosion properties; therefore, the stress state of a weld has to be assessed before usage. In addition, the effectiveness of stress reducing measures like post-weld heat treatments can be checked.

Friction stir welding (FSW) is a solid state joining process that takes place below the melting point of the materials to be joined [17]. FSW uses a rotating cylindrical tool with a pin to heat the material by friction. The tool pin stirs the plasticized material and joins two pieces together when it is moved along the welding line. Advantages of this technique include, e.g., joining of materials that are difficult to fusion weld, low distortion, and excellent mechanical properties. A potential field of application is mainly Al alloys used in modern aerospace structures where cost and weight are to be reduced by using new joining techniques [18, 19].

Aluminum sheets (AA2024-T351) of size 125 mm × 380 mm and 6.3 mm thickness were butt-joined together by FSW at the BAE SYSTEMS, UK. The welding and rotation speeds of the tool were 95 mm min$^{-1}$ and 350 rpm. Residual strain measurements were performed with the neutron diffractometer ARES at GKSS [20]. Neutrons of 0.164 nm wavelength from an elastically bent perfect silicon (3 1 1) monochromator were used. Strain measurements in the three perpendicular directions $x$ (longitudinal direction, parallel to the weld), $y$ (transverse), $z$ (through-thickness) were performed with the orientations of a sheet as shown in Figure 7.8. A matchstick-like gauge volume with a length of 30 mm was used for scanning $\varepsilon_y$ and $\varepsilon_z$ for improving the intensity [6], assuming that

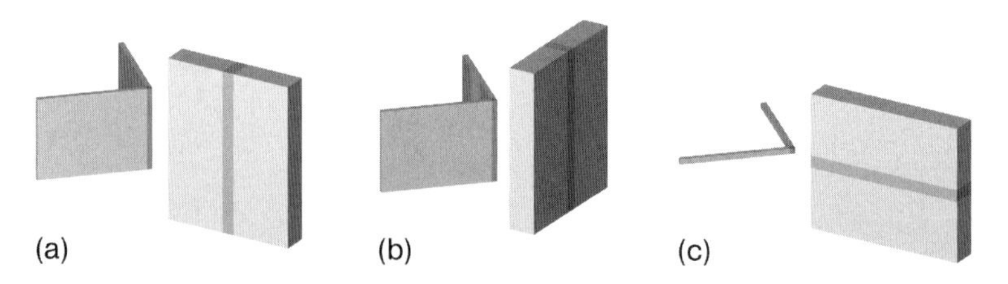


**Figure 7.8** Orientations of a welded sheet relative to the incoming and diffracted beams for strain measurements in the (a) transverse, (b) normal, and (c) longitudinal directions.

the stresses are constant along $x$ over the length of the volume element. A cubic gauge volume was used for scanning $\varepsilon_x$ since larger stress gradients are present along the transverse ($y$) direction. The nominal cross section of the gauge volume was $3 \times 3$ mm$^2$. The Al (3 1 1) diffraction peak was recorded with an area detector at an angle of about 84°. The scan line was 150 mm away from the edge near the weld finishing point. The crystallographic anisotropy of the elastic constants is small for Al and therefore the macroscopic elastic constants were used ($E = 73$ GPa and $\nu = 0.33$).

AA2024 is a heat treatable alloy in which the solute concentration (mainly Cu) in the matrix depends on the heat treatment. But as the lattice parameter can change considerably with the solute content, the $d_0$-value also depends on the heat input the material has experienced during welding [21]. This problem can be solved, e.g., by cutting a "comb" with thin teeth out of the weld region by spark erosion in which most of the macroscopic stresses are known to relax [14]. Such a comb can therefore serve as unstrained reference material. However, this procedure requires quite an effort and unfortunately is a destructive method. Therefore, it was assumed that there is a plane stress state in the relatively thin sheets [9, 16]. From the condition $\sigma_z = 0$, $d_0$ was calculated for each point. The relative change of $d_0$ in the weld determined in this way is up to $-4.6 \times 10^{-4}$ (Figure 7.9a). This is a significant magnitude in terms of strain and thus has to be accounted for.

There are tensile stresses in the longitudinal direction with peak values of about 130 MPa (Figure 7.9b). The double-peak feature of the stress distribution with maximum stress in the heat affected zone (HAZ) and a minimum in the weld nugget has already been observed earlier [22] and seems to be typical for FSW joints in Al sheets of this thickness. As expected, no significant stresses are present in the transverse direction.

Welding a sheet under an external longitudinal tensile load and releasing the load after welding leads to large compressive longitudinal stresses in the weld (Figure 7.9b) [23].

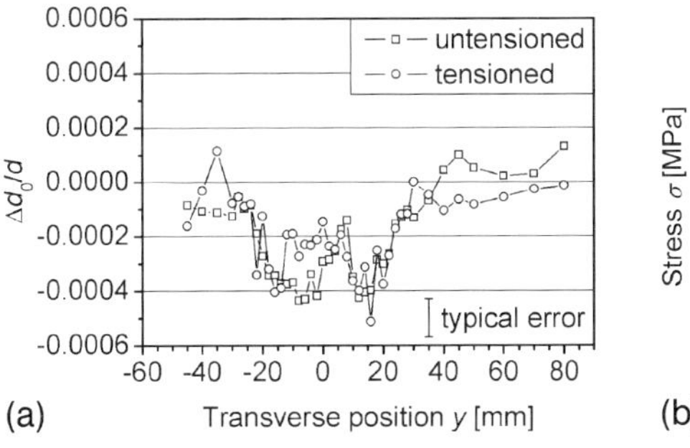


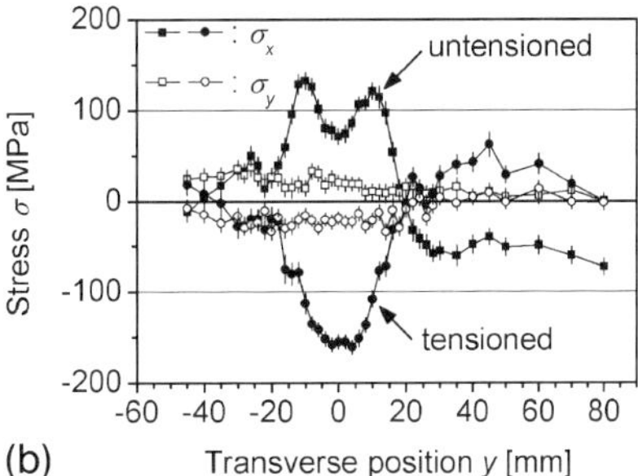


**Figure 7.9** (a) Variations in $d_0$ as calculated from the plane stress assumption for the data of the AA2024 sheets welded with and without tensioning. (b) Residual stresses for sheets welded without and with an external tensile longitudinal load applied during welding.

### 7.4.2
### Residual Stresses in Water-Quenched Turbine Discs

Another field of application of experimental residual stress analysis is the verification and optimization of numerical models. Finite element (FE) models are often used for the simulation of technical processes. These can be processes involving thermal treatments (e.g., welding), or mechanical treatments (e.g., deforming), or a combination of both (e.g., forging). Models are often very good in describing the development of residual stress states; nevertheless they have to be checked against measurements in a real component before their predictions can be fully trusted.

Nickel-based superalloys are being employed for aeroengine components due to their unique combination of high-temperature strength and considerably high fatigue strength [24, 25]. A special thermomechanical processing route has been developed for the Inconel alloy 718 (IN718) in order to obtain enhanced high-temperature strength in turbine discs used for aeroengines. The so-called direct age process requires water quenching directly after forging [26, 27], in contrast to the conventional heat treatment where forging is followed by slow air cooling [28]. A high number of dislocations are frozen in by quenching, acting as potential nucleation sites for the strengthening precipitates that are formed in the subsequent annealing treatment. These precipitates are fine semicoherent $\gamma''$-$Ni_3$(Nb, Ti) and coherent $\gamma'$-$Ni_3$(Al, Ti) particles.

However, the high cooling rate connected with water quenching leads to much higher RS than the standard slow air cooling. Therefore, problems can arise during machining of a disc to its final shape when RS are not completely relieved during the subsequent annealing treatment, which often seems to be the case. If the RS state within the disc is known prior to turning, the process can be adjusted in order to minimize distortion considerably.

Therefore, an FE model was used for the simulation of the water quenching step. In order to verify the residual stress predictions of the FE simulation, RS in a turbine disc that was taken out of the commercial production (Böhler Schmiedetechnik GmbH & Co KG) after forging and water quenching were studied by neutron diffraction. Due to the relatively large dimensions of the part (Ø 320 mm and thickness up to 25 mm), neutron diffraction is the only experimental technique to study bulk residual stresses within the discs. Forging leads to a fine-grained microstructure with a mean grain size of 5 μm [26]. Therefore, the number of grains within the gauge volumes used for neutron measurements is high enough for representing the bulk.

Strain measurements were performed with the neutron diffractometer ARES at the GKSS Research Centre in Geesthacht, Germany, using a constant wavelength of $\lambda = 0.164$ nm [20]. For the measurement of water-quenched IN 718, which is dominated by the face-centered cubic (fcc) Ni-matrix, the (3 1 1) reflection was used, because it is generally recommended for the evaluation of macrostrains in fcc lattices [13, 29].

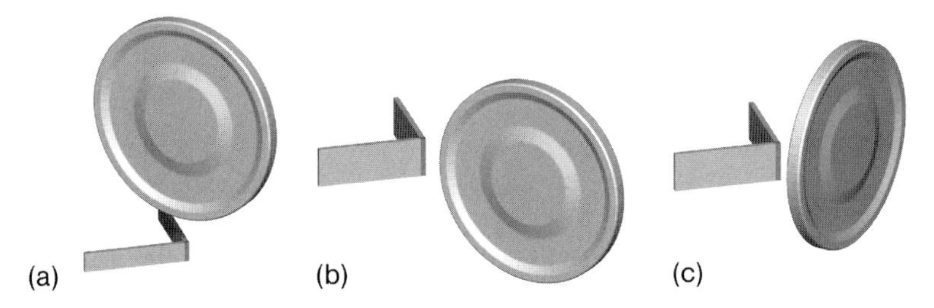


**Figure 7.10** Orientations of a disc relative to the incoming and diffracted beams for strain measurements in the (a) tangential, (b) radial, and (c) axial directions.

In the case of the disc the symmetry axes (radial, tangential, axial) were used as coordinate system and the measurements were performed accordingly (Figure 7.10). A matchstick-like gauge volume with a cross section of $2 \times 2\ mm^2$ was chosen in order to increase intensity; its length was up to 30 mm for measurements of axial and radial strains and 10–15 mm for tangential strains. It is thus assumed that the stresses are constant in the hoop direction over the length of the gauge volume [9]. The diffraction elastic constants (DEC) $E_{hkl} = 195$ GPa and $\nu_{hkl} = 0.31$ for the (3 1 1) reflection were calculated using the Kröner model [30, 31] with the following single crystals values: $C_{11} = 248$ GPa, $C_{22} = 152$ GPa, $C_{44} = 125$ GPa [32].

The unstrained lattice parameter $d_0$ was determined by measuring small cubes with 4–5 mm edge length. The cubes were cut out of the discs by electro-discharge machining with their faces parallel to the axial, radial, and hoop direction. Due to their small size they were considered to be free of macrostresses [15].

The FE model predicts tensile radial and tangential stresses up to about 500 MPa in the interior of the water-quenched disc; these tensile stresses are

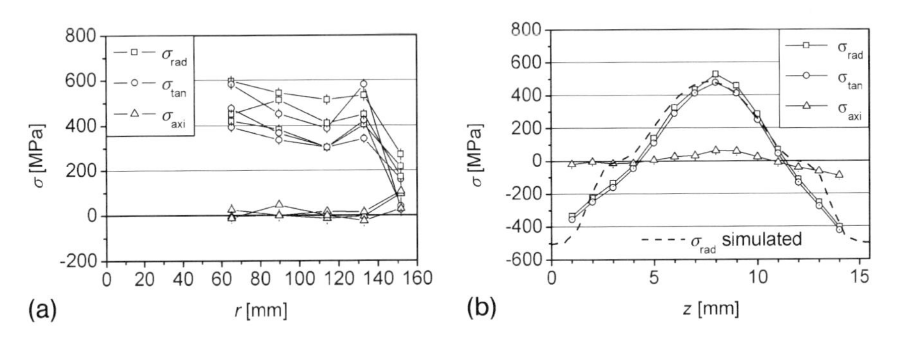


**Figure 7.11** (a) Stress distributions in three different radial directions of a disc, separated by 120° from each other. (b) Stresses through the thickness of a disc (symbols: as measured; dashed line: FE result).

balanced by compressive stresses near the surfaces of the disc. The distributions of radial and tangential stresses are very similar except for the outer rim of the disc. Axial (through-thickness) stresses are zero except for a region near the circumference of the disc. Repeated measurements along three different radii, 120° apart from each other, showed a rotational symmetry of the stress state (Figure 7.11a) [33].

A through-thickness scan was conducted at the thin section of a disc at a radius of 110 mm. A comparison shows that the FE model describes the measured stresses quite accurately (Figure 7.11b) [34].

## 7.5
## Summary and Outlook

Angle-dispersive neutron residual stress analysis is a powerful method for the nondestructive determination of residual stress states in the interior of large specimens. The technique has been developed for about 20 years and by now can be used routinely. This makes the technique more and more interesting for industrial users.

The need for strain measurements in "real-life" components leads to sophisticated designs of new dedicated strain scanners with enough space at the sample position and load carrying capacity for precise positioning of components of several hundred kilograms.

The availability of new neutron sources with more intensity at the sample position will be another step towards measurements in "real-life" components producing not only one-dimensional stress scans, but two- or three-dimensional stress maps that will contribute to a more comprehensive understanding of stress distributions inside components.

## References

1 I.C. Noyan, J.B. Cohen, *Residual Stress, Measurement by Diffraction and Interpretation*, Springer, New York, 1987.
2 V. Hauk, *Structural and Residual Stress Analysis by Non-Destructive Methods*, Elsevier, Amsterdam, 1997.
3 M.T. Hutchings, P.J. Withers, T.M. Holden, T. Lorentzen, *Characterization of Residual Stress by Neutron Diffraction*, Taylor and Francis, Boca Raton, FL, 2005.
4 V. Wagner, P. Mikula, P. Lukas, *Nucl. Instrum. Methods Phys. Res. A* 1993, **338**, 53–59.
5 P. Mikula, V. Wagner, *Mater. Sci. Forum* 2000, **347–349**, 113–118.
6 L. Pintschovius, V. Jung, E. Macherauch, O. Vöhringer, *Mater. Sci. Eng.* 1983, **61**, 43–50.
7 T. Pirling, *Mater. Sci. Forum* 2000, **347**, 107–112.
8 R.L. Snyder, F. Fiala, H.J. Bunge, *Defect and Microstructure Analysis by Diffraction*, Oxford University Press, New York, 1999.
9 A.J. Allen, M.T. Hutchings, C.G. Windsor, C. Andreani, *Adv. Phys.* 1985, **34**, 445–473.
10 P. Van Houtte, *Mat. Sci. Forum* 1993, **133–136**, 97–110.

11 *ISO Polycrystalline Materials – Determination of Residual Stresses by Neutron Diffraction*, ISO/TTA #3:2001, International Organisation for Standardisation (ISO), Geneva, Switzerland, 2001.

12 T.M. Holden, C.N. Tome, R.A. Holt, *Metall. Mater. Trans. A* 1998, **29**, 2967–2973.

13 B. Clausen, T. Lorentzen, T. Leffers, *Acta Mater.* 1998, **46**, 3087–3098.

14 E. O'Brien, in *Proceedings of the Sixth International Conference on Residual Stresses ICRS-6*, IOM Communications Ltd., London, UK, 2000, p. 13.

15 A.D. Krawitz, R.A. Winholtz, *Mater. Sci. Eng. A* 1994, **185**, 123–130.

16 G. Albertini, G. Bruno, B.D. Dunn, F. Fiori, W. Reimers, J.S. Wright, *Mater. Sci. Eng. A* 1997, **224**, 157–165.

17 C.J. Dawes, W.M. Thomas, *Weld. J.* 1996, **75**, 41.

18 S.W. Williams, Air and Space Europe 2001, **3** (3/4) 64–66.

19 X.L. Wang, Z. Feng, S.A. David, S. Spooner, C.R. Hubbard, in *Proceedings of the Sixth International Conference on Residual Stresses ICRS-6*, IOM Communications Ltd., London, UK, 2000, p. 1408.

20 P. Staron, H.-U. Ruhnau, P. Mikula, R. Kampmann, Physica B 2000, **276–278**, 158–159.

21 H.G. Priesmeyer, *Measurement of Residual and Applied Stress Using Neutron Diffraction* (Eds.: M.T. Hutchings and A.D. Krawitz), NATO ASI Series E, **216**, 1992, 277–284.

22 P. Staron, M. Kocak, S. Williams, in *Proceedings of the Sixth International Conference on Trends in Welding Research*, Pine Mountain, GA, USA, ASM International, 2003, pp. 253–256.

23 P. Staron, M. Kocak, S. Williams, A. Wescott, Physica B 2004, **350**, e491–e493.

24 E.A. Loria (Ed.), Superalloys 718 2005, *The Minerals, Metals & Materials Society (TMS)*, Warrendale, PA, USA, 2005.

25 W. Betteridge, *Materials Science and Technology*, VCH, Weinheim, New York, Basel, Cambridge, 1992, Vol. 7, p. 641.

26 W. Horvath, W. Zechner, J. Tockner, M. Berchthaler, G. Weber, E.A. Werner, Superalloys 718, 625, 706 and Derivates, *The Minerals, Metals & Materials Society (TMS)*, Warrendale, PA, USA, 2001, 223–228.

27 G.A. Rao, M. Kumar, M. Srinivas, D.S. Sarma, *Mater. Sci. Eng. A* 2003, **355**, 114.

28 S.J. Hong, W.P. Chen, T.W. Wang, *Mater. Sci. Trans. A* 2001, **32**, 1887–1901.

29 T.M. Holden, C.N. Tomé, R.A. Holt, *Met. Mater. Trans. A* 1998, **29**, 2967–2973.

30 E. Kröner, *Z. Phys.* 1958, **151**, 504–518.

31 F. Bollenrath, V. Hauk, E.H. Müller, *Z. Metallk.* 1967, **58**, 76–82.

32 D. Dye, S.M. Roberts, P.J. Withers, R.C. Reed, *J. Strain Anal.* 2000, **35**, 247–259.

33 U. Cihak, P. Staron, W. Marketz, H. Leitner, J. Tockner, H. Clemens, *Z. Metallk.* 2004, **95**, 663–667.

34 U. Cihak, P. Staron, H. Clemens, J. Homeyer, M. Stockinger, J. Tockner, *Mater. Sci. Eng. A* 2006, **437**, 75–82.

# 8
# Stress Analysis by Energy-Dispersive Neutron Diffraction

*Javier Roberto Santisteban*

## 8.1
## Introduction

An energy-dispersive diffraction experiment is conceptually very simple: a polychromatic neutron beam hits a sample, and the neutrons scattered at a fixed scattering angle are discriminated in energy by a detector. For a polycrystalline sample, the spectrum recorded in the detector consists of several diffraction peaks at the incident wavelengths that fulfill the Bragg condition. The study of internal and applied stresses by the execution and analysis of such experiments is the topic of this chapter.

## 8.2
## Time-of-Flight Neutron Diffraction

Pulsed sources are specially suited for energy-dispersive diffraction experiments because the neutrons are created over a wide energy range, and because their wavelength is easily defined by the time-of-light (TOF) technique. The speed $v$ of a neutron is related to its wavelength $\lambda$ by the de Broglie relation $h\lambda = mv$, or $v$ (m s$^{-1}$) = 3956/$\lambda$ (Å). As all neutrons leave the source at the same time, the wavelength of a neutron traveling a distance $L_1$ from source to sample and $L_2$ from sample to detector, is readily defined from its TOF $t$,

$$\lambda = \frac{h}{mv} = \frac{h}{m}\frac{t}{(L_1 + L_2)} \tag{8.1}$$

Neutrons with a wavelength of 1 Å take around 2.5 ms to travel 10 m; a time easily resolved by modern detection systems. The TOF spectrum from a stainless steel specimen for a detector 50 m away from the source is shown in Figure 8.1. The spectrum is typical of a polycrystalline sample, showing several diffraction peaks corresponding to different $(hkl)$ families of lattice planes, as given by

*Neutrons and Synchrotron Radiation in Engineering Materials Science: From Fundamentals to Material and Component Characterization*
Edited by Walter Reimers, Anke Rita Pyzalla, Andreas Schreyer, and Helmut Clemens

ISBN: 978-3-527-31533-8

Bragg's law $\lambda_{hkl} = 2d_{hkl} \sin \theta_B$. So, the $d$-spacing for each family is directly obtained from the position $t_{hkl}$ of the peak in the TOF spectrum,

$$d_{hkl} = \frac{h}{2 \sin \theta_B m(L_1 + L_2)} t_{hkl} \tag{8.2}$$

where $\theta_B$ is given by the detector position, mostly $2\theta_B = \pm 90°$.

Peak positions are very precisely defined by least-squares refinement of the peaks, with a typical sensitivity of $\delta\varepsilon = \Delta t_{hkl}/t_{hkl} = \Delta d_{hkl}/d_{hkl} \sim 50$ με. A typical fit is displayed in the inset of Figure 8.1. Basically, the atomic planes are used as microscopic strain gauges and the elastic strain is determined from the variation in the interplanar distances $d_{hkl}$, as compared to the value $d^0_{hkl}$ measured in the unstressed condition, $\varepsilon_{hkl} = (d_{hkl} - d^0_{hkl})/d^0_{hkl}$.

Each diffraction peak provides information from only a small fraction of all the crystallites within the sampled volume, i.e., only those favorably oriented to fulfill the Bragg condition. As several crystal families are measured simultaneously, TOF neutron diffraction is specially suited for the study of intergranular or Type II stresses, which develop as a result of the elastic or plastic anisotropy of the individual crystallites. This is, because both internal and applied stresses usually produce different strains $\varepsilon_{hkl}$ for the different crystal families, which can be simultaneously measured by TOF neutron diffraction.

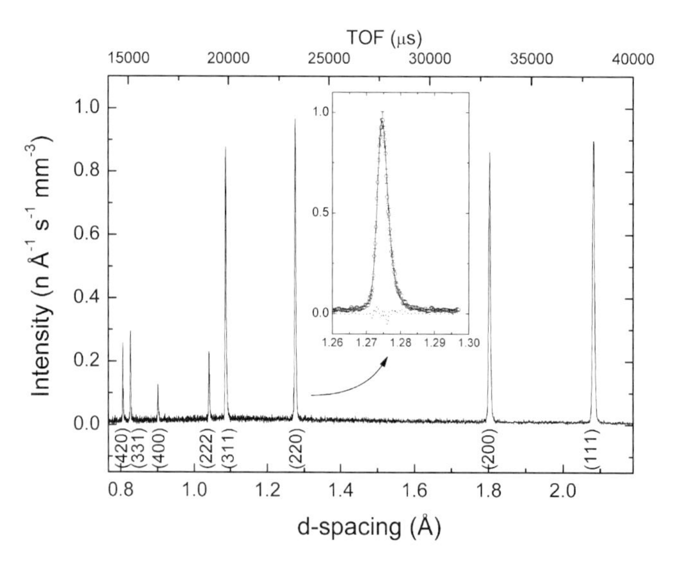


**Figure 8.1** A typical time-of-flight diffraction spectrum, in this case for a stainless steel specimen measured on ENGIN-X. Elastic strains are measured from the shift observed in the position of the diffraction peaks, defined by least-squares refinement as that shown in the inset. Alternatively, average or macroscopic strain can be defined through the change in lattice parameter obtained from a full-pattern refinement.

TOF diffraction can also provide a very good approximation to the macroscopic or Type I elastic strain. This is achieved by averaging several single-peak strains $\varepsilon_{hkl}$, as ultimately the macroscopic strain represents the average response of all crystallites within the sampled volume [1].

### 8.2.1 TOF Peak Shape and Data Analysis Packages

TOF diffraction spectra display asymmetric peak profiles, as a result of experimental uncertainties preventing the precise definition of the travel time of the neutrons. Pulsed sources create high-energy neutrons (~MeV), which need to be slowed down to energies suitable for Bragg diffraction (~meV) before bringing them into the sample. This is achieved within a moderator, a hydrogenous material where neutrons lose their energy through inelastic collisions with protons, broadening the initially sharp pulse. Hence, the total TOF of a neutron is actually divided into two parts:

$$t = t_m + t_L = t_m + \frac{m}{h} L\lambda \tag{8.3}$$

where $t_m$ is the time spent in the moderator, $t_L$ is the flight time from moderator to detector, and $L = L_1 + L_2$ is the total flight path. The sharpness and the energy distribution of the neutron pulse depend on the choice of moderator material and temperature (e.g., water, methane, liquid hydrogen). The emission time from the moderator ($t_\mathrm{m}$) has associated a very asymmetric probability distribution usually described as a truncated exponential decay profile (Figure 8.2a), with a decay constant $\tau$ that depends on the neutron wavelength. On the other hand, the uncertainties in flight path and neutron wavelength produce a rather symmetric probability distribution for $t_L$. Several profiles have been proposed for the actual experimental peak shape, resulting from the convolution of the two distributions. The simplest model is that from Kropff and collaborators [2], consisting of the convolution of a truncated exponential with a Gaussian profile describing the symmetric broadening. A more popular profile is that proposed by Jorgensen et al. consisting of separate rising and falling exponentials for the moderator term, convolved with a Gaussian representing the symmetric broadening [3]. More complex profiles are needed to describe the peaks from high-resolution instruments at small $d$-spacings [4].

Whatever the model, the instrumental peak asymmetry must be properly incorporated in the analysis of the peak positions, as large shifts in peak positions or *pseudostrains* can arise in least-squares fits that use wrongly defined peak shapes. The shape and width of the diffraction peaks vary with neutron wavelength. Figure 8.2b shows such a variation for the ENGIN-X strain scanner, described in the following section. In this case, the peak profile has been defined as the convolution of a truncated exponential, characterized by a decay constant $\tau$; with a Voigt function, which itself is the convolution of a Gauss (of width $\sigma$) and

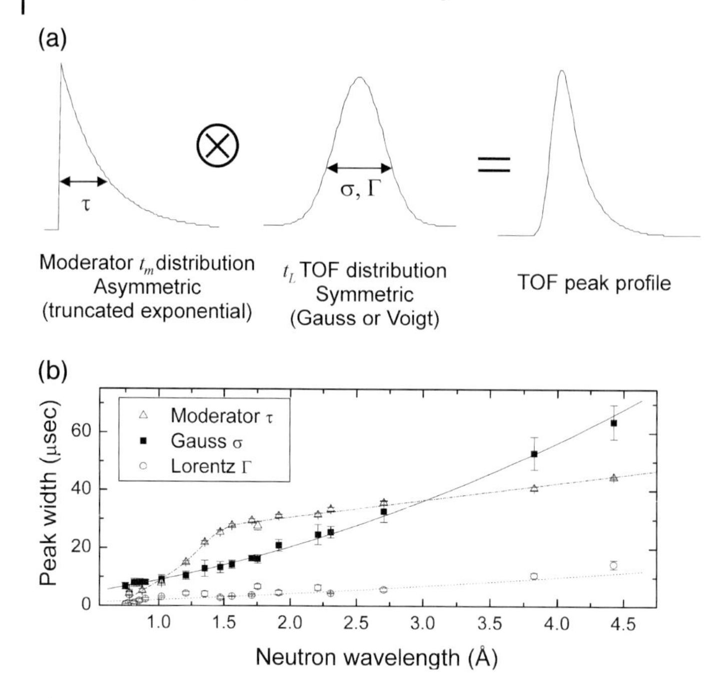


**Figure 8.2** (a) Contributions to the TOF peak shape. In this case, the profile has been represented as the convolution of a truncated exponential (described by the decay time $\tau$) and a Voigt function (described by the widths $\sigma$ and $\Gamma$). (b) Variation of the peak-width parameters for the ENGIN-X diffractometer, as measured on a $CeO_2$ reference.

a Lorentzian (of width $\Gamma$). The figure shows the variation of these three parameters as measured on a reference $CeO_2$ sample, considered to be a good approximation to the instrument resolution. As sample-induced broadenings are typically symmetric, strain analysis of real specimens are usually performed by least-squares fits keeping the asymmetry parameter $\tau$ fixed to its calibrated value.

The main appeal of TOF diffraction lies in exploiting all the information contained in the complete diffraction spectrum. This is achieved through a least-squares refinement of the full diffraction pattern, as originally proposed by Rietveld for angle-dispersive neutron diffraction [5]. The main advantage of the Rietveld method is that it incorporates information from overlapping or idle-defined peaks, which for strain analysis translates in a reduction of counting times over a single-peak-based analysis. There are several computer packages that can deal with TOF diffraction spectra. Maybe the most popular is the General Structure Analysis System (GSAS) code developed by Larson and Von Dreele [6], which is very powerful but not terribly user-friendly, so a graphical user interface (EXPGUI) has been recently developed for it [7]. Other popular programs include Fullprof [8] very strong for magnetic scattering; and MAUD oriented to materials science studies and capable of determining orientation distribution functions from TOF diffraction data [9].

Regarding the use of full pattern refinements for strain analysis, Daymond and collaborators have shown that engineering Type I strains can be well approximated from the change in the average lattice parameters obtained from Rietveld refinements of TOF diffraction spectrum, even neglecting the elastic anisotropy of the material. This method is very popular because Type I or *engineering* stresses can be reasonably calculated using the macroscopic elastic constants of the material. Phenomenological demonstrations of this approach for face centered and hexagonal materials have been given in [10] and [11] respectively, whilst a theoretical discussion has been presented in [12]. On the other hand, a method to obtain the full strain tensor from simultaneous Rietveld refinement of spectra collected at different specimen orientations has been implemented in the GSAS code [13].

## 8.3 TOF Strain Scanners

A TOF neutron strain scanner is essentially a TOF diffractometer, but with some special features to perform measurements at precise locations within the bulk of a specimen. As seen in Chapter 7, this is achieved by defining a small instrumental gauge volume (IGV) with the help of slits and collimators. The strain distribution within a sample is easily explored by moving the sample across the IGV using a translation stage. The neutron strain scanning technique is conceptually very simple but its practical application can be time-consuming. Very long counting times are required when small gauge volumes and large penetration depths are involved, which effectively dictates the range of problems that can be studied with this technique. Withers has defined a *maximum acceptable acquisition time* for neutron and synchrotron strain scanning when measuring strain deep within materials [14]. This is an important concept because neutron strain scanning experiments are performed on large facilities, where the experimenter is granted only a limited amount of beam time, under competitive review. Hence, a minimization of the experimental time required for an accurate determination of peak positions (and hence strain) can be considered the first goal of a neutron strain scanner.

Figure 8.3 shows a schematic diagram of ENGIN-X, an optimized TOF strain scanner installed at the Isis Facility, UK [15]. The instrument is composed of two diffraction banks, plus slits and collimators used to define the IGV as the intersection of the incident and diffracted beams. The measured strain gives the component of the strain tensor along the direction of the scattering vector $\mathbf{q}$, which bisects the incident and diffracted beams. So, in a TOF strain scanner two strain directions can be measured simultaneously. In the following we will explain how the different instrumental parameters (diffraction angle, incident,

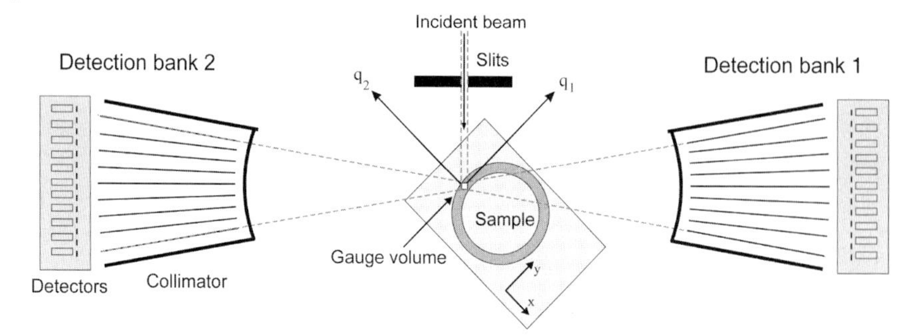


**Figure 8.3** The ENGIN-X strain scanner, located at the ISIS Facility, Rutherford Laboratory, United Kingdom. The instrument gauge volume (IGV) is defined by slits and collimators. The use of two detection banks allows two perpendicular strain directions (identified by $\mathbf{q}_1$ and $\mathbf{q}_2$ in the figure) to be measured simultaneously. The sample is scanned through the IGV using a positioning table.

and diffracted beam paths and collimations, detector size) have been chosen, as the criteria are also applicable to other TOF strain scanners.

The choice of diffraction angle is driven by the geometric shape of the IGV. For most problems, the principal strain axes can be inferred from the geometry of a sample. So in practice, macroscopic strain measurements are usually performed only along three mutually perpendicular directions. The detectors are placed at approximately $2\theta_B = \pm 90^\circ$ from the incident beam because this results in a cubic IGV, ensuring that the same volume of the specimen is measured at each sample orientation by both detectors.

On the other hand, in an optimized instrument the choice of primary and secondary flight paths and the detectors' size are not independent, as it is learnt from a minimization of the experimental counting times.

## 8.3.1 Counting Times and Resolution

Ideally, the uncertainty in the measured strain is directly proportional to the uncertainty in the determination of the position of the Bragg peaks. Withers et al. [16] have shown that the time $T$ required to measure the position of a zero-background peak with a strain precision $\delta\varepsilon$ is

$$T = \frac{1}{I_{hkl}} \left( \frac{\delta t_{hkl}}{t_{hkl} \delta\varepsilon} \right)^2 = \frac{1}{I_{hkl}} \left( \frac{\delta_\varepsilon t_{hkl}}{\delta\varepsilon} \right)^2 \tag{8.4}$$

where $I_{hkl}$ is the integrated peak intensity per unit time, $t_{hkl}$ is the peak position, and $\delta t_{hkl}$ is the peak width. On the right-hand side of Eq. (8.4), $T$ is expressed in terms of the relative peak width $\delta_\varepsilon t_{hkl} = (\delta t_{hkl}/t_{hkl})$, a dimensionless parameter that allows direct comparison between peaks appearing at different TOF, as well as with the required strain precision $\delta\varepsilon$. Through this chapter the notation $\delta_\varepsilon X$ will be used to represent the dimensionless relative width of the probability distribution of the variable $X$. The relative peak width, or *resolution*, of a TOF diffractometer is [17]

$$(\delta_\varepsilon t_{hkl})^2 = (\delta_\varepsilon t)^2_{\mathrm{mod}} + \frac{\cot^2\theta_B}{4}[(\alpha^x_{\mathrm{in}})^2 + (\alpha^z_{\mathrm{det}})^2] + (\delta_\varepsilon d_{hkl})^2 \tag{8.5}$$

where $(\delta_\varepsilon t)_{\mathrm{mod}}$ is the moderator contribution to the peak width, $\alpha^x_{\mathrm{in}}$ and $\alpha^z_{\mathrm{det}}$ are the divergences in the diffraction plane for the incident and detected beams, respectively, and $(\delta_\varepsilon d_{hkl})$ is the sample-induced broadening that appears from microstresses or a local variation in composition. On the other hand, the integrated intensity of a TOF diffraction peak [18] depends on the scattering power and absorption of the sample as well as on instrumental parameters such as the incident neutron flux $\Phi_0$ (expressed in neutrons $\mathrm{s}^{-1}\,\mathrm{cm}^{-2}\,\mathrm{Å}^{-1}\,\mathrm{steradian}^{-1}$), the total divergence in the diffraction plane, the incident beam divergence perpendicular to the diffraction plane $\alpha^y_{\mathrm{in}}$, the size of the illuminated gauge volume $\delta V$, and the detector characteristics

$$I_{hkl} = \Phi_0\alpha^y_{\mathrm{in}}\left(\frac{h_{\mathrm{det}}\%_{\mathrm{det}}}{\pi L_2}\right)\sin\theta_B[(\alpha^x_{\mathrm{in}})^2 + (\alpha^z_{\mathrm{det}})^2]^{1/2}P_{hkl}\mathrm{e}^{-\mu_{hkl}l}\delta V \tag{8.6}$$

where $h_{\mathrm{det}}$ and $\%_{\mathrm{det}}$ are the detector height and efficiency, respectively. The beam attenuation is given by the absorption coefficient $\mu_{hkl}$, and $l$ is the neutron flight path inside the sample. The factor $P_{hkl}$ depends only on the material

$$P_{hkl} = \frac{m_{hkl}|F^2_{hkl}|d^4_{hkl}}{v_0} \tag{8.7}$$

with $v_0$ the volume per atom, $F^2_{hkl}$ the structure factor (including the Debye–Waller factor), and $m_{hkl}$ the multiplicity of the reflection. Equation (8.6) can be used to normalize integrated TOF peak intensities in order to asses the effect of texture as exemplified in [19] for a Zircaloy-4 weld. By replacing Eqs. (8.5) and (8.6) in Eq. (8.4), it can be shown that the minimum in counting time is achieved when

$$\frac{\cot^2\theta_B}{4}((\alpha^x_{\mathrm{in}})^2 + (\alpha^z_{\mathrm{det}})^2) = (\delta_\varepsilon t)^2_{\mathrm{mod}} + (\delta_\varepsilon d_{hkl})^2 \tag{8.8}$$

So in an optimized TOF strain scanner, the total divergence in the diffraction plane must match the combined peak broadenings introduced by the neutron moderation process, and by the sample. The moderator contribution to the peak width is inversely proportional to the total neutron flight path [17],

$$(\delta_\varepsilon t)^2_{\text{mod}} = \left( \frac{\hbar \Delta t_0}{(L_1 + L_2) m \lambda} \right)^2 \tag{8.9}$$

where $\Delta t_0$ is the intrinsic width of the neutron pulse; a measure of the time spent by the neutrons in the moderator. In practical terms, the moderator contribution to the resolution is defined by a careful choice of the moderator and the incident flight path $L_1$. Increasing the secondary flight path $L_2$ is not effective because there is an associated decay in intensity of the form $(1/L_2)^2$. On the other hand, the primary flight path $L_1$ can be increased without a significant loss of intensity by bringing the neutrons to the sample position using a neutron guide. The actual choice of $L_1$ will depend on a variety of parameters including building costs and the materials likely to be studied with the diffractometer. Besides this, Eq. (8.8) calls for an instrument having a variable divergence, which could be adjusted to match the peak broadening introduced by the sample – since typically we are willing to increase the divergence if the incident flux can also be increased.

The ISIS methane moderator was chosen for ENGIN-X due to its combination of narrow pulse width and high flux over the 1 Å to 3 Å wavelength range relevant to most engineering materials. With this choice, Johnson and Daymond showed that the counting times are minimized when the primary flight path of the instrument is ~50 m and the horizontal angular divergence of the diffracted beam is 0.002 radians [20]. For detectors located at a typical secondary flight path of 1.5 m, this gives a detector width $\delta z \sim 3$ mm. The height of the detector $\delta y$ can be increased considerably (~200 mm) as the divergence perpendicular to the diffraction plane does not affect the resolution. In ENGIN-X, the incident beam divergence can be adjusted with two sets of slits located a 1.5 m and 4 m from the IGV. Typical counting times in ENGIN-X for a 50 με resolution are presented in Table 8.1 [15]. A typical 20× reduction in counting times on ENGIN-X over its predecessor ENGIN [21] was achieved through this optimized design.

**Table 8.1** Estimated counting times (in minutes) for strain measurements on ENGIN-X with an uncertainty of 50 με. For each material, the estimated times are those from a Rietveld refinement of the complete diffraction pattern.

| Gauge volume | $2 \times 2 \times 2$ mm³ | | | | $4 \times 4 \times 4$ mm³ | | | |
|---|---|---|---|---|---|---|---|---|
| Depth (mm) | 2 | 10 | 20 | 30 | 4 | 20 | 40 | 50 |
| Al | 5.9 | 6.4 | 7.1 | 7.8 | 0.8 | 0.9 | 1.1 | 1.2 |
| γ-Fe | 0.9 | 2.2 | 6.5 | 19.7 | 0.1 | 0.8 | 7.4 | 22.5 |
| α-Fe | 0.9 | 2.1 | 6.1 | 18.3 | 0.1 | 0.8 | 6.8 | 20.4 |
| Ni | 0.9 | 4.2 | 31.5 | 233.4 | 0.2 | 3.9 | 216.2 | 1602.9 |
| Cu | 1.3 | 2.9 | 7.9 | 21.1 | 0.2 | 1 | 7.1 | 19.3 |
| Zr | 1.5 | 1.6 | 1.9 | 2.2 | 0.2 | 0.2 | 0.3 | 0.4 |
| Ti | 6.5 | 11.4 | 23.1 | 46.7 | 0.9 | 2.9 | 11.8 | 24 |

## 8.3.2
## Neutron Optics and Time Focusing

The measured strain corresponds to an average over the volume $\delta V$ and the solid angle $\delta\Omega$ sampled by the experiment. Figure 8.4a schematically shows the main parameters affecting the shape and size of the IGV. The beam cross section is defined by slits of height $S_y$ and width $S_x$ located at a distance $S_z$ from the geometric center of the diffractometer. The incident beam has a divergence $\alpha^x_{\text{in}}$ in the diffraction plane and $\alpha^y_{\text{in}}$ perpendicular to it. Due to this divergence, the beam cross section is larger at the IGV center than at the slits position and the edges gets blurred. The edge profile is described by an error function, with a width defined by the beam divergence and slit distance. The FWHM of the neutron beam profile at the IGV center is [22]

$$\Delta x, y = S_{x,y}\left[1 + 2.35\left(\frac{S_z}{S_{x,y}}\tan\alpha_{x,y}\right)^3\right]^{1/3} \tag{8.10}$$

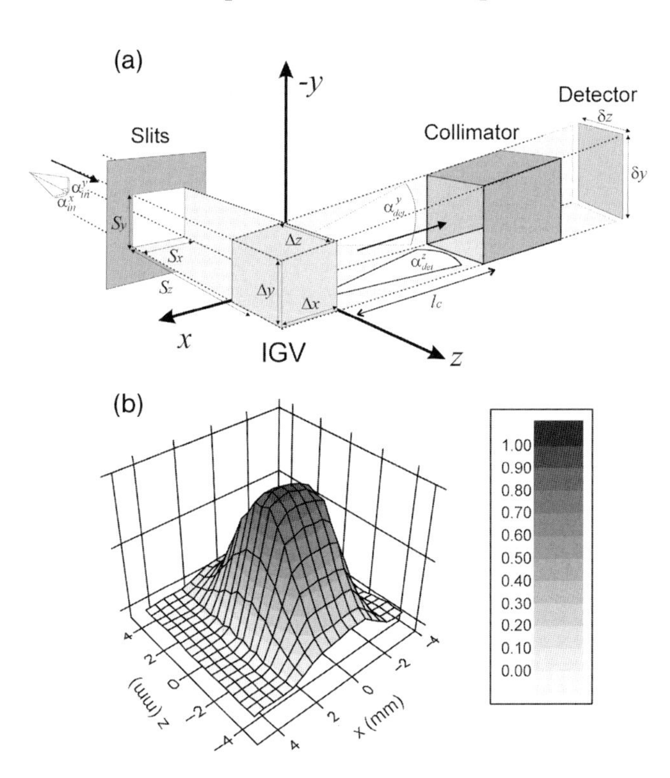


**Figure 8.4** (a) Schematic representation of the instrumental gauge volume (IGV) in a TOF neutron strain scanner. The figure identifies the experimental parameters responsible for the shape and dimensions of the IGV. (b) Experimental intensity map on the horizontal plane of a $2 \times 2 \times 2\ \text{mm}^3$ IGV, measured on ENGIN-X.

Hence, the effective dimensions of the beam – as defined by the FWHM – increase rather slowly with the slit distance $S_z$, even when the actual cross section of the beam increases markedly. The size of the IGV along the beam direction $\Delta z$ is usually defined by a collimator of the specified divergence. But as TOF neutron count rates are relatively low (compared to angle-dispersive diffraction) usually large arrays of detectors covering horizontal and vertical ranges of about 30° need to be used. Radial collimators made of thin absorbing blades are employed for this purpose (Figure 8.2), to collimate the diffracted beam and ensure that all detectors look at exactly the same volume of the sample[22]. Each ENGIN-X detection bank is composed of 1200 TOF detectors, arranged in 240 columns across $2\theta_B$ angles of 75° and 105°. The intensity profile along the beam direction seen from a radial collimator is triangular for the line passing through the focal point, and gradually becomes a Gaussian profile at increasing distances from this point. A horizontal map of the experimental intensity distribution for the ENGIN-X IGV is shown in Figure 8.4b.

From the present discussion, it follows that a precise definition of the IGV is a rather complex matter. For operational purposes, we can define the IGV as a cuboid of sides $\Delta x$, $\Delta y$, and $\Delta z$, so the illuminated gauge volume becomes $\delta V = \Delta x \Delta y \Delta z$. The dimensions of $\Delta x$ and $\Delta y$ are the FWHMs given by Eq. (8.10), and $\Delta z$ is the width of the intensity profile seen from the collimators. For measurements performed near surfaces the IGV may be only partially immersed within the specimen, and the effective centroid of the IGV differs from its geometrical center. This may introduce an instrumental peak shift that needs to be discounted from the total peak shift.

The TOF spectra recorded by each individual detector are transformed into a common $d$-spacing scale and added together in a single diffraction spectrum, in a process commonly known as electronic time focusing [3]. Larger angular arrays are not feasible due to space and access requirements, and due to the increased uncertainty in the direction of the measured strain [23].

## 8.4
## A Virtual Laboratory for Strain Scanning

In parallel to the optimization of the neutron optics, fast and reliable positioning can be achieved by using detailed 3D models of the samples, and a pair of theodolites for precise alignment on the instrument. In ENGIN-X the processes of planning, alignment, and data analysis have been simplified by writing SSCANSS [24]; a computing program providing a "virtual laboratory," i.e., a 3D representation of ENGIN-X that can be easily manipulated through a graphical-user interface. This dedicated software (1) helps in setting up the initial measuring strategy, (2) automates the sample alignment and all the routine aspects of the measurement process, and (3) provides data analysis in almost real time, allowing decisions on changes to the measurement strategy.

(a)

(b)

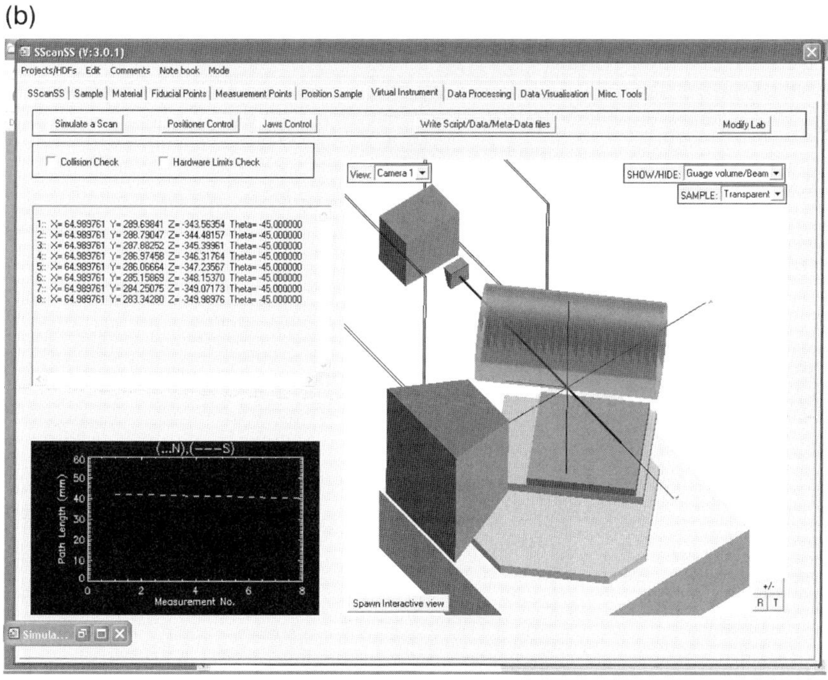


**Figure 8.5** The ENGIN-X virtual laboratory. (a) Measurement of axial strain for a 190 kg pipe containing a repair weld and (b) the corresponding experiment simulation within SSCANSS. (c) Simulation for the radial and hoop strain directions.

(c)

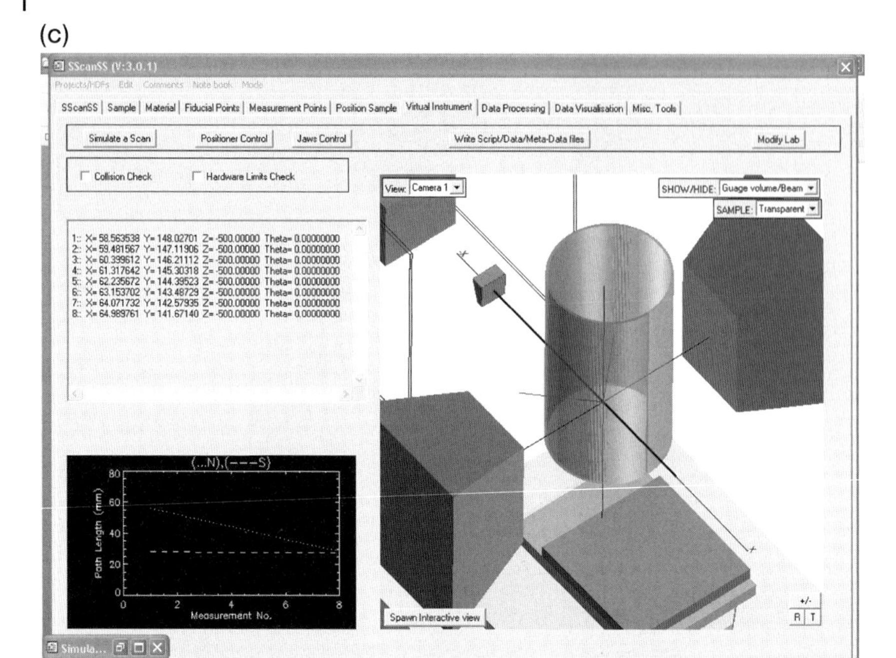


**Figure 8.5** (continued)

The capabilities of SSCANSS are better exemplified by a case study: the analysis of residual stresses introduced by weld repairs in nuclear power plants; a work in collaboration with British Energy and described in [25]. Repair welds are introduced into structures either to remedy initial fabrication defects found by routine inspection, or to rectify in-service degradation of components. A typical example is the repair-welded component shown in position on ENGIN-X in Figure 8.5a. The pipe was fabricated by manual metal arc girth welding of two ex-service stainless steel power station steam headers. In order to simulate a typical repair process, a section of the original weld was removed, and subsequently filled with new material, as schematically shown by Figure 8.6a. Scans of the full strain tensor were performed along several lines in the heat affected zone and in the original weld. A map of the axial strain in the HAZ of the repair is shown in Figure 8.6c, which displays two main concentrations of tensile strain beneath the outer surface of the pipe, at the stop-end and about midway. These short range stress concentrations revealed the importance of stop-end effects for a correct description of the repair process.

Critical to this work has been the use of SSCANSS for the planning and execution of the experiment. The experimental configuration for the measurement of the axial strain is shown in Figure 8.5a, together with the virtual laboratory model of Figure 8.5b. The 3D model of the laboratory includes the collimators,

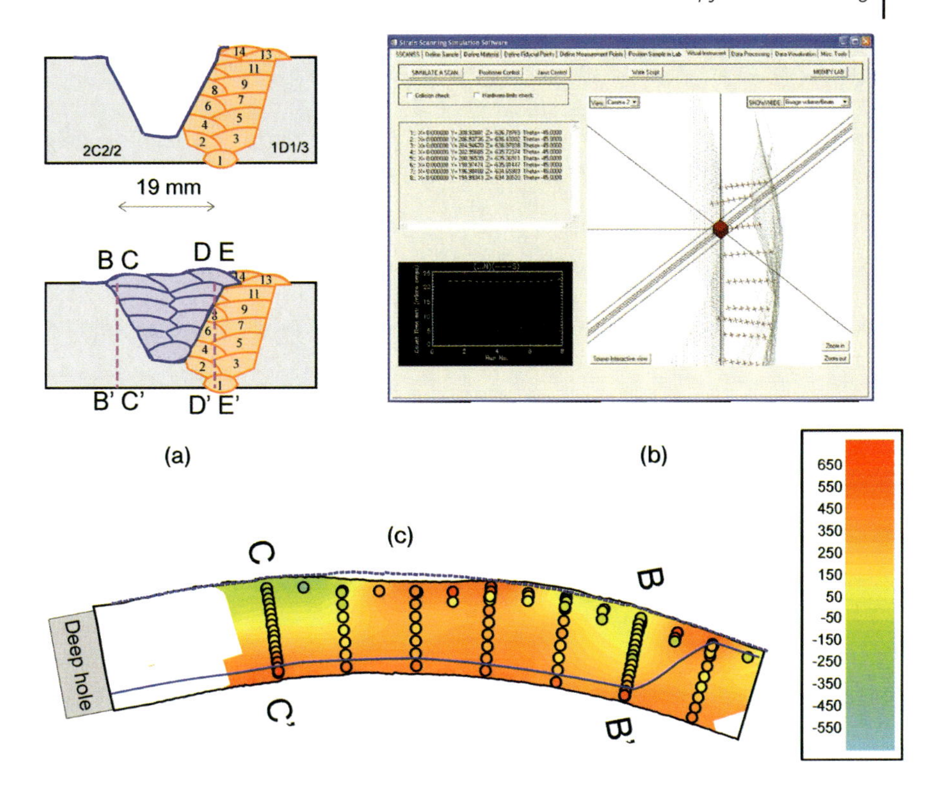


**Figure 8.6** Strain mapping in a repaired weld. (a) Hoop view of repairs introduced in the weld. A section of the original weld was removed and then filled with new material. (b) Detailed 3D model of the specimen surface in the vicinity of the repair. (c) Axial strain map in the heat affected zone of the repair. The profile of the cross section was produced using a coordinate measuring machine.

the slits, and the positioning table. The 3D model of the sample can be produced from basic primitive objects (cuboids, cylinders, pipes, etc.), exemplified in this case by the simple model used to describe the pipe. The program allows exploring alternative experimental arrangements for measuring the different strain components, and alerts in case of collisions. It also provides an estimation of the required counting time using the expressions described in the previous section. For more complex shapes or regions requiring improved spatial resolution, a very precise model can be created using a coordinate measuring machine (CMM). Figure 8.6b shows such a model for the region of interest of the pipe, i.e., in the vicinity of the weld repair. The shape of the IGV (dark cube) and the measurement positions (stars) are also displayed within the 3D model of the sample. The points to be measured are defined interactively in a graphical visualization of the sample, or alternatively their $(x, y, z)$ coordinates in the sample system can be imported from a file.

The positioner movements required to bring the sample to the correct position for measurement may be generated automatically. This facility requires the determination of the transformation relating the sample and laboratory coordinate systems. In order to do this, the positions of a number of fiducial points on the surface of the sample are measured using a pair of precision theodolites [24]. A least-squares procedure is then used to find the transformation matrix that most closely maps the fiducial points on the sample to their measured positions in the laboratory. Since the sample is a rigid body, a minimum of three noncollinear fiducial points are sufficient for this purpose, though using a larger number may improve accuracy. A positional accuracy of better than 0.1 mm is usual for this alignment procedure.

The spectra recorded at each measurement position by the detectors belonging to the same bank are time focused into a single diffraction spectrum, as that shown in Figure 8.1. The program provides automatic single peak and full pattern refinement of the diffraction spectra specially devised for strain determination, through a library of common engineering materials. This enables researchers who are not experts in crystallography to keep pace with the experiment, and be able to modify the experimental plan in the light of experience gained during earlier parts of the measuring process.

## 8.5 Evolution of Intergranular Stresses

Whilst engineering materials are commonly assumed to be mechanically homogeneous and isotropic, in reality this is never the case. For example, composite materials are composed of phases with differing mechanical properties, and even single phase alloys are comprised of many differently oriented anisotropic crystallites with different properties along a particular sample direction. As a consequence of the incompatibility between the constituent parts, interphase and intergranular stresses develop as functions of applied stress or plastic strain. The collection of neutron diffraction profiles *in situ* during mechanical testing provides an effective means of tracking the evolution of these internal stresses. Energy-dispersive diffraction is specially suited in this respect, because a single TOF spectrum is both phase- and orientation-selective, as different phases and different grain orientations contribute to different diffraction peaks. This is because only those grains oriented with an $(hkl)$ plane normal parallel to the scattering vector $\mathbf{q}$ contribute to the $hkl$ diffraction peak.

A typical *in situ* testing experiment is exemplified by Oliver et al. [26]. In that work, intergranular and interphase strains were compared in a single (ferrite) phase low carbon steel and a two phase (ferrite + spheroidized cementite) high carbon steel. A hydraulic stress rig was mounted on the beamline with its loading axis oriented horizontally at an angle of 45° to the incident beam, allowing simultaneous collection of diffraction profiles for lattice planes oriented parallel and perpendicular to the loading axis. Tensile tests were performed in a stepwise

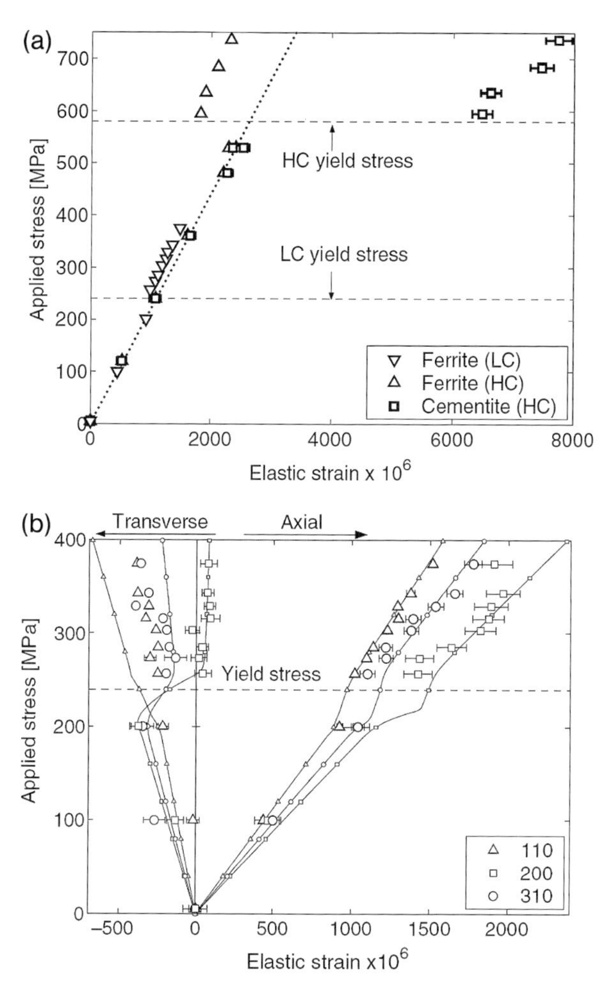


**Figure 8.7** (a) Phase-specific axial strains versus applied uniaxial tensile stress. The dotted line is the linear elastic response of the ferrite phase in HC steel. (b) Modeled and experimental evolution of average grain family elastic strains.

fashion – increasing the applied load and interrupting for neutron data acquisition. Phase-specific elastic strains were determined by Rietveld full profile refinement of the lattice parameters, using the lattice parameters determined prior to loading as the stress-free references. Grain family specific elastic strains were determined by single peak fitting of individual peaks. The elastic phase strains observed along the axial direction are shown as functions of applied stress in Figure 8.7a. As seen, the ferrite elastic strain continues to increase almost linearly with applied stress even after yielding in the single phase low carbon steel. However, in the two phase steel, whilst both phases develop similar elastic strains before yield (indicating that they have similar elastic constants), at the yield point there is a sudden redistribution of strain from the majority ferrite phase to the minority cementite ($Fe_3C$) phase. This can be understood on the basis that the ferrite yields whilst the cementite continues to deform elastically. Thus a strain misfit develops between the phases, introducing compressive residual stress (−170 MPa) into the ferrite and tensile residual stress (+640 MPa) into the cementite. Similar effects can be observed in other composite or multiphase materials such as Al/SiC metal matrix composites or $\gamma/\gamma'$ Ni superalloys.

Figure 8.7b shows the development of grain family specific elastic strains with applied stress in the ferrite phase of the low carbon steel. It is noticeable that differently oriented families of grains behave in significantly different ways. Firstly, the different slopes prior to yield reflect the significant elastic anisotropy of ferrite single crystals, which are stiffest along the ⟨1 1 1⟩ direction and most compliant along ⟨1 0 0⟩. Secondly, the strains diverge during yielding. This reflects a similar effect at the granular level to that causing the interphase stress between ferrite and cementite in the high carbon steel. The experimental intergranular strains can be used to validate models of intergranular stress evolution. The continuous lines in Figure 8.7b represent calculations of an elastoplastic self-consistent model, which is in good agreement with the experimental results.

## 8.6 TOF Transmission Analysis

In a TOF diffractometer, the neutrons that pass through the sample also contain a wealth of crystallographic information about the specimen. For instance, a neutron beam transmitted by a polycrystalline sample displays sudden, well defined increases in intensity as a function of neutron wavelength (Figure 8.8a). These *Bragg edges* occur because for a given ($hkl$) reflection, the Bragg angle increases as the wavelength increases until $2\theta_B$ is equal to 180°. At wavelengths greater than this critical value no scattering by this particular ($hkl$) lattice spacing can occur, and there is a sharp increase in the transmitted intensity. From Bragg's law the wavelength at which this occurs is $\lambda = 2d_{hkl}$, giving a measure of the ($hkl$) $d$-spacing in the direction of the incoming beam.

By placing a detector just behind the sample, the position and magnitude of these edges can be determined by the TOF technique. As the position, height, and width of a Bragg edge depend on essentially the same laws as the position, intensity and width of conventional diffraction peaks, information about the stress state [27], texture [28] or the phases present in the sample [29] is readily available from an analysis of the transmission pattern.

The transmission $Tr$ is simply the ratio between the beam intensities recorded with and without the sample in front of the detector. It is related to the microscopic properties of the sample by $Tr(\lambda) = \exp(-nw\sigma_{\text{tot}}(\lambda))$, where $n$ is the number of scattering centers per unit volume, $w$ is the sample thickness and $\sigma_{\text{tot}}$ is the microscopic total cross-section of the material.

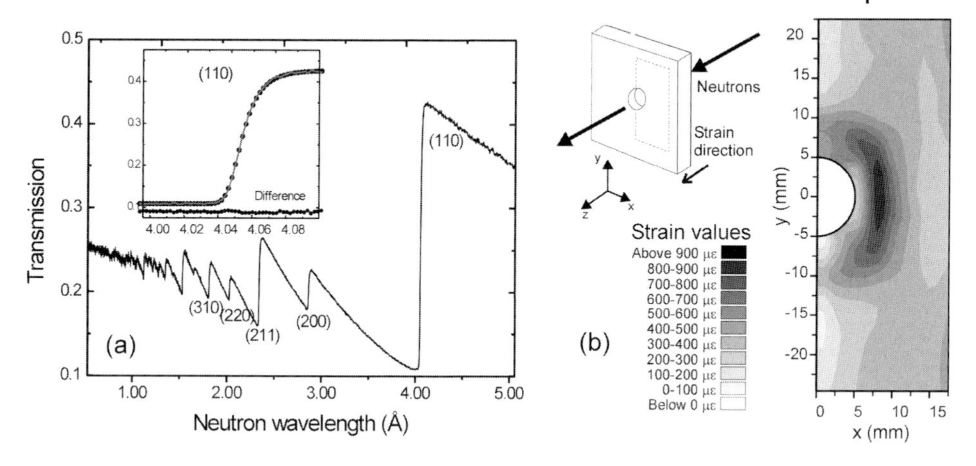


**Figure 8.8** (a) Spectrum of neutrons transmitted by 25 mm of iron powder, displaying characteristic Bragg edges. A least-squares fit of the (1 1 0) edge is shown in the inset. (b) Variation of the elastic strain around a cold expanded hole in a 12 mm thick iron plate. The measured strains correspond to the through-thickness average of the out-of-plane residual elastic strain around the hole.

## 8.6.1 Bragg Edges

In practice, the Bragg edges are broadened by the finite resolution of the instrument and by imperfections within the sample, and so the transmission near a Bragg edge is described by a characteristic edge profile [30]

$$Tr(t) = \exp(-nw\sigma_0)\left\{\left[\exp(-nw\sigma_{hkl}) + (1-\exp(-nw\sigma_{hkl})) \times \frac{1}{2}\right.\right.$$
$$\left.\left. \times \left[\operatorname{erfc}\left(-\frac{t-t_{hkl}}{\sqrt{2}\sigma}\right) - \exp\left(-\frac{t-t_{hkl}}{\tau}+\frac{\sigma^2}{2\tau^2}\right)\operatorname{erfc}\left(-\frac{t-t_{hkl}}{\sqrt{2}\sigma}+\frac{\sigma}{\tau}\right)\right]\right]\right\} \tag{8.11}$$

with $n$ and $w$ are defined above. The edge position is $t_{hkl}$, and the edge width is defined by $\tau$ and $\sigma$, the asymmetric and symmetric broadenings discussed in Section 8.2.1. The edge height is defined by the jump $\sigma_{hkl} = P_{hkl}/d_{hkl}$ in the total cross section of the material, and the edge baseline is given by $\sigma_0$; the value at the long wavelength side of the edge. A fit for the (1 1 0) edge of iron using this profile is shown in the inset of Figure 8.8a. A more complex edge profile including two truncated exponentials has been proposed by Vogel [36].

### 8.6.2
### Strain Mapping

The edge position $t_{hkl}$ can be precisely defined by a least-squares fit using the edge profile of Eq. (8.11), and converted into *d*-spacing after a simple linear calibration (with $\Delta d_{hkl}/d_{hkl} \sim 50$ με). In analogy to the diffraction case, the macroscopic strain can also be defined from the variation in the lattice parameter *a*, as defined from simultaneous least-squares fit of several Bragg edges [32]. This opens up the possibility of using an array of detectors to produce a radiographic "image" of the strain variation in the sample. However, it also means that the measurement is an average over the complete transmission path through the sample, making the method useful for examining plates and other essentially two-dimensional objects. With this in mind, ENGIN-X has a pixellated transmission detector consisting of a $10 \times 10$ array of $2 \times 2$ mm$^2$ scintillator detectors on a 2.5 mm pitch [31]. An example of the use of this detector is shown in Figure 8.8b, corresponding to the strain around a cold expanded hole in a 12 mm thick ferritic steel plate. The fatigue lives of fastener holes may be increased if the hole is cold expanded prior to insertion of the fastener. The most common method presently used in the aircraft industry involves expansion of a lubricated split sleeve by an oversize mandrel using commercially available equipment. Fatigue life predictions of structures containing such expanded holes rely critically on estimates of the residual stress distribution surrounding the hole.

### 8.6.3
### Quantitative Phase Analysis

For multiphase materials, the transmitted spectrum displays Bragg edges from all crystallographic phases. Hence, the volume fractions of the constituent phases can be determined from a Rietveld-type analysis of the transmitted spectrum. Under certain circumstances, the transmission geometry provides some advantages over conventional neutron diffraction: for relatively thick samples transmission patterns of sound statistics can be collected within seconds, as the transmitted count rate is high compared to the intensity diffracted into a certain solid angle. This allows studying kinetics of structural phase transitions in solids lasting for much less than 1 h. Moreover, the technique is not sensitive to the sample location along the beam, giving both high flexibility and relatively simple sample environments. Meggers et al. used this technique to study phase transformations in steel [33], and Vogel et al. the reduction of nickel oxide to nickel [34]. The changes occurring in the transmission of EN24 steel during the isothermal decomposition of Austenite (fcc or $\gamma$-Fe) to Bainite is exemplified in Figure 8.9 (Bainite is essentially a mixture of ferrite bcc $\alpha$-Fe and carbides). EN24 is a high-strength engineering steel whose properties are critically dependant on heat treatment. Creation of a given property profile using differential heat treatment depends on reliable time-temperature-transformation (TTT) data describing the

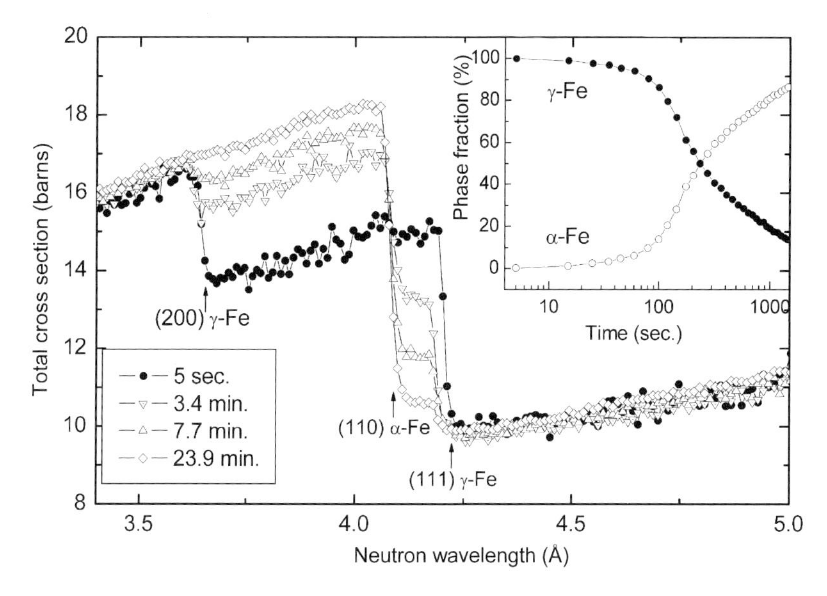


**Figure 8.9** Detail of the total cross section of EN24 steel at 380 °C after austenization at 830 °C [35]. The times are measured from the time of insertion in the furnace. The inset shows the evolution of the phases, obtained after a Rietveld analysis of 32 spectra.

temporal evolution of its constituent phases. The evolution of Bainite in EN24 was studied using two tubular furnaces aligned with the neutron beam. The sample was austenized in the 830 °C furnace, and subsequently moved into a 380 °C furnace after cooling with an air blast. The processes were performed *in situ*, and transmission spectra were saved with counting times as short as 10 s. The disappearance of the $\gamma$-(1 1 1) and $\gamma$-(2 0 0) edges and the appearance of the $\alpha$-(1 1 0) are clearly seen in the figure. The inset shows the time evolution of the $\gamma$ and $\alpha$ phases, obtained after a Rietveld analysis of the transmission spectra using the code BetMan developed by Vogel [36].

For other systems, a full-pattern refinement of transmission data can become much more complicated than the corresponding diffraction case. This is because the presence of texture affects not only the edge intensities, but also the shape of the transmission between edges. Besides this, noncoherent scattering contributions are usually more important and have more complex dependences on neutron wavelength that for transmission than for diffraction. In view of this, several analysis strategies have been proposed which avoid these complications by restricting the refinements to "well-behaved" wavelength windows [29].

Besides kinetic studies, phase distributions across a specimen can be mapped with a pixellated transmission detector, as it has been demonstrated in [29]. The possibility of TOF-enhanced phase-sensitive neutron radiography has been recently reported [41].

### 8.6.4
### Other Applications

The $\sin^2 \psi$ technique, traditionally used in X-ray diffraction, can also be successfully implemented in TOF neutron transmission. In [27] Steuwer and collaborators performed uniaxial tensile tests on both ferritic and austenitic steel specimen; recording the transmitted spectra at different sample inclinations. Their results have demonstrated that TOF transmission can be used to determine anisotropic strains, diffraction elastic constants, the residual and applied stress state, and the unstrained lattice parameter. An actual application of the transmission technique to map the unstressed lattice parameter on a thin slice from a weld has been reported in [37].

The energy-dispersive neutron transmission of single crystals is rather different to the polycrystalline case. Instead of Bragg edges, the transmitted beam presents a series of dips in intensity as a result of reflection in the crystal planes [38]. The position of these dips can be exploited for the definition of the crystal orientation with a resolution of 1 min of arc, whilst the width gives a measure of the crystal mosaicity. In [38] a Cu monochromator was plastically deformed by uniaxial tension along the [1 1 0] direction *in situ*, and the evolution of crystal orientation and mosaicity was tracked by analyzing the transmitted TOF spectra. Several crystal reflections at different locations of the sample were simultaneously studied during the experiments, revealing the anisotropic increase in mosaicity.

Finally, the transmission of thermal neutrons through textured materials is strongly dependent on the neutron wavelength as well as on the specimen orientation. So, large variations exist in the neutron total cross section of naturally occurring and manmade objects, as opposed to the simple Bragg edges expected for randomly oriented crystallites. Such differences have been exploited for the identification of the manufacturing process of ancient bronze artifacts [39, 40]. The possibilities of TOF transmission for the nondestructive characterization of the spatial variation of texture across macroscopic specimens is discussed in [28], in particular the relation between the orientation-dependent total cross section and the pole figures measured on traditional diffraction experiments. The transmission of materials presenting a fiber texture has been separately discussed by Vogel [36].

## 8.7
## Conclusions

Energy-dispersive neutron diffraction presents some real advantages for the study of internal and applied stresses. This is because the elastic strain along different sample directions and for different phases or crystal families can be measured simultaneously, which represents an ideal scenario for the study of Type-II stresses. Thus, *in situ* loading experiments have provided unique experimental information about deformation mechanisms in polycrystalline materials. Besides this, the technique can also provide a very good approximation to the macroscopic or engineering elastic strain. As several crystal families (and hence several crystal orientations) are measured on a single diffraction spectrum, the average of the experimental single-peak strains is usually very close to the macroscopic response of the material. This allows the use of macroscopic elastic constants for the calculation of stresses, hence providing a rather reliable and simple technique for nondestructive stress analysis.

Pulsed neutron sources are specially suited for energy dispersive experiments because the TOF technique offers a simple energy analysis of the diffracted beam. Important reductions in counting times in TOF neutron strain scanning can be achieved by making the right choice of neutron flight paths and beam collimation.

Besides conventional neutron diffraction, energy-dispersive neutron transmission at pulsed neutron sources can also provide information about the spatial variation of strain across plate specimens. We must remember however that in this case the information corresponds to the through-thickness average of the out-of-plane elastic strain.

## Acknowledgments

I am indebted to Ed Oliver, Lyndon Edwards, Mark Daymond, Axel Steuwer, John Bouchard, Sven Vogel, Phil Withers, Jon James, Mike Fitzpatrick, and Hans Priesmeyer, for sharing with me most of the knowledge contained within this chapter.

## References

1 J.D. Kamminga, T.H. De Keijser, E.J. Mittemeijer, R. Delhez, *J. Appl. Cryst.* 2000, **33**, 1059–1066.
2 F. Kropff, J.R. Granada, R.E. Mayer, *Nucl. Instrum. Methods* 1982, **198**, 515–521.
3 J.D. Jorgensen, J. Faber Jr., J.M. Carpenter, R.K. Crawford, J.R. Haumann, R.L. Hitterman, R. Kleb, G.E. Ostrowski, F.J. Rotella, T.G. Worlton, *J. Appl. Cryst.* 1989, **22**, 321–333.
4 S. Ikeda, J.M. Carpenter, *Nucl. Instrum. Methods A* 1985, **239**, 536–544.
5 H.M. Rietveld, *J. Appl. Cryst.* 1969, **2**, 65–71.
6 A.C. Larson and R.B. Von Dreele, *General Structure Analysis System* 1994 (GSAS), Los Alamos National Laboratory Report LAUR 86-748 (GSAS is available from http://www.ccp14.ac.uk/ccp/ccp14/ftp-mirror/gsas/public/gsas/).
7 B.H. Toby, *J. Appl. Cryst.* 2001, **34**, 210–213 (EXPGUI and GSAS are available at http://www.ncnr.nist.gov/programs/crystallography/software/expgui/expgui.html).
8 J. Rodriguez-Carvajal, *Physica B* 1993, **192**, 55–69 (FullProf is available at http://www-llb.cea.fr/fullweb/fp2k/fp2k.htm).
9 S. Matthies, J. Pehl, H.-R. Wenk, L. Lutterotti, S.C. Vogel, *J. Appl. Cryst.* 2005, **38**, 462–465 (MAUD is available at http://www.ing.unitn.it/~maud).

10 M.R. Daymond, M.A.M. Bourke, R.B. Von Dreele, B. Clausen, T. Lorentzen, *J. Appl. Phys.* 1997, **82**, 1554–1562.
11 M.R. Daymond, M.A.M. Bourke, *J. Appl. Phys.* 1999, **85**, 739–747.
12 M.R. Daymond, *J. Appl. Phys.* 2004, **96**, 4263–4272.
13 D. Balzar, R.B. Von Dreele, K. Bennet, H. Ledbetter, *J. Appl. Phys.* 1998, **84**, 4822–4833.
14 P.J. Withers, *J. Appl. Cryst.* 2004, **37**, 596–606.
15 J.R. Santisteban, M.R. Daymond, J.A. James, L. Edwards, *J. Appl. Cryst.* 2006, **39**, 812–825.
16 P.J. Withers, M.R. Daymond, M.W. Johnson, *J. Appl. Cryst.* 2001, **34**, 737–743.
17 C.G. Windsor, *Pulsed Neutron Scattering*, Taylor & Francis, London, 1981.
18 B. Buras, *Nukleonika* 1963, **8**, 259–260.
19 D.G. Carr, M.I. Ripley, T.M. Holden, D.W. Brown, S.C. Vogel, *Acta Mater.* 2004, **52**, 4083–4091.
20 M.W. Johnson, M.R. Daymond, *J. Appl. Cryst.* 2002, **35**, 49–57.
21 M.W. Johnson, L. Edwards, P.J. Withers, *Physica B* 1997, **234**, 1141–1143.
22 P.J. Withers, M.W. Johnson, J.S. Wright, *Physica B* 2000, **292**, 273–285.
23 M.R. Daymond, Physica B 2001, **301**(3–4), 221–226.
24 J.A. James, J.R. Santisteban, M.R. Daymond, L. Edwards, *Proc. NOBUGS2002, NIST*, Gaithersburg, USA, 2002 (http://atarXiv.org/abs/cond-mat/0210432).
25 P.J. Bouchard, D. George, J.R. Santisteban, G. Bruno, M. Dutta, L. Edwards, E. Kingston, D.J. Smith, *Int. J. Pressure Vessels Piping* 2005, **82**, 299–310.
26 E.C. Oliver, T. Mori, M.R. Daymond, P.J. Withers, *Acta Mater.* 2003, **51**, 6453–6464.
27 A. Steuwer, J.R. Santisteban, P.J. Withers, L. Edwards, M.E. Fitzpatrick, *J. Appl. Crystall.* 2003, **36**, 1159–1168.
28 J.R. Santisteban, L. Edwards, V. Stelmukh, *Physica B* 2006, **385–386**, 636–638.
29 A. Steuwer, J.R. Santisteban, P.J. Withers, L. Edwards, *J. Appl. Phys.* 2005, **97**, 074903.
30 J.R. Santisteban, L. Edwards, A. Steuwer, P.J. Withers, *J. Appl. Crystall.* 2001, **34**, 289–297.
31 J.R. Santisteban, L. Edwards, M.E. Fitzpatrick, A. Steuwer, P.J. Withers, M.R. Daymond, M.W. Johnson, N. Rhodes, E.M. Schooneveld, *Nucl. Instrum. Methods A* 2002, **481**, 255–258.
32 A. Steuwer, P.J. Withers, J.R. Santisteban, L. Edwards, G. Bruno, M.E. Fitzpatrick, M.R. Daymond, M.E. Johnson, D. Wang, *Phys. Stat. Sol. (a)* 2001, **185**, 221–230.
33 K. Meggers, H.G. Priesmeyer, J. Trela, M. Dahms, *Mater. Sci. Eng. A* 1994, **188**, 301–304.
34 S. Vogel, E. Ustundag, J.-C. Hanan, V.W. Yuan, M.A.M. Bourke, *Mater. Sci. Eng. A* 2002, **333**, 1–9.
35 J.R. Santisteban, L. Edwards, A. Steuwer, P.J. Withers, M.E. Fitzpatrick, *Appl. Phys.* 2002, **A75**, 1433–1436.
36 S. Vogel, Ph.D. Thesis, Kiel University, 2000.
37 J.-R. Santisteban, A. Steuwer, L. Edwards, P.J. Withers, M.E. Fitzpatrick, *J. Appl. Crystall.* 2002, **35**, 497–504.
38 J.R. Santisteban, *J. Appl. Crystall.* 2005, **38**, 934–944.
39 S. Siano, L. Bartoli, J.R. Santisteban, W. Kockelmann, M.R. Daymond, M. Miccio, G. De Marinis, *Archaeometry* 2006, **48**, 77–96.
40 J.R. Santisteban, S. Siano, M.R. Daymond, *Mater. Sci. Forums* 2006, **524–525**, 421–434.
41 W. Kockelmann, G. Frei, E.H. Lehmann, P. Voutobel, J.R. Santisteban, *Nucl. Instr. Meth. A* 2007, **578**, 421–434.

# 9
# Residual Stress Analysis by Monochromatic High-Energy X-rays

*René Valéry Martins*

In Chapters 7 and 8, the use of the angle dispersive and time-of-flight method for residual stress analysis using neutron diffraction was explained. This chapter and Chapter 10 similarly will deal with residual stress analysis using monochromatic high-energy synchrotron radiation (angle dispersive set-up) respectively white high-energy synchrotron radiation (energy dispersive set-up).

## 9.1
## Basic Set-ups

The aim of all the diffraction experiments described here is the determination of the (residual) lattice strain from diffraction peak positions with an accuracy of at least $100 \times 10^{-6}$ (100 μstrain). Figure 9.1 shows schematically four basic experimental configurations for residual strain measurements using high-energy synchrotron radiation. The set-ups aim at different length scales, ranging from the macroscopic scale to the size of individual grains. They are designed for transmission geometry, meaning that the incoming beam penetrates the sample at the front side and exits it at the backside delivering diffraction cones. The incoming beam is usually blocked by a beam stop behind the sample (not always shown in Figure 9.1) while the diffracted rays are detected with a detector system such as a point, line, or area detector. The latter usually speeds up data acquisition considerably and allows, depending on its size and the experimental technique employed, the monitoring of complete diffraction rings.

The simplest set-up (Figure 9.1a) consists of a sample mounted on a positioning device, a pair of slits or a pinhole to confine the size of the incoming beam, and a detector. The detector can either be a point, line, or area detector. The gauge volume size can only be horizontally and vertically defined by the beam-size. Therefore, only a two-dimensional (2D) spatial resolution can be achieved. The data obtained represents information integrated along the beam path and no depth-resolved information is contained. Because of the small Bragg angle of high-energy synchrotron radiation the scattering vectors for the diffraction cones

*Neutrons and Synchrotron Radiation in Engineering Materials Science: From Fundamentals to Material and Component Characterization*
Edited by Walter Reimers, Anke Rita Pyzalla, Andreas Schreyer, and Helmut Clemens

ISBN: 978-3-527-31533-8

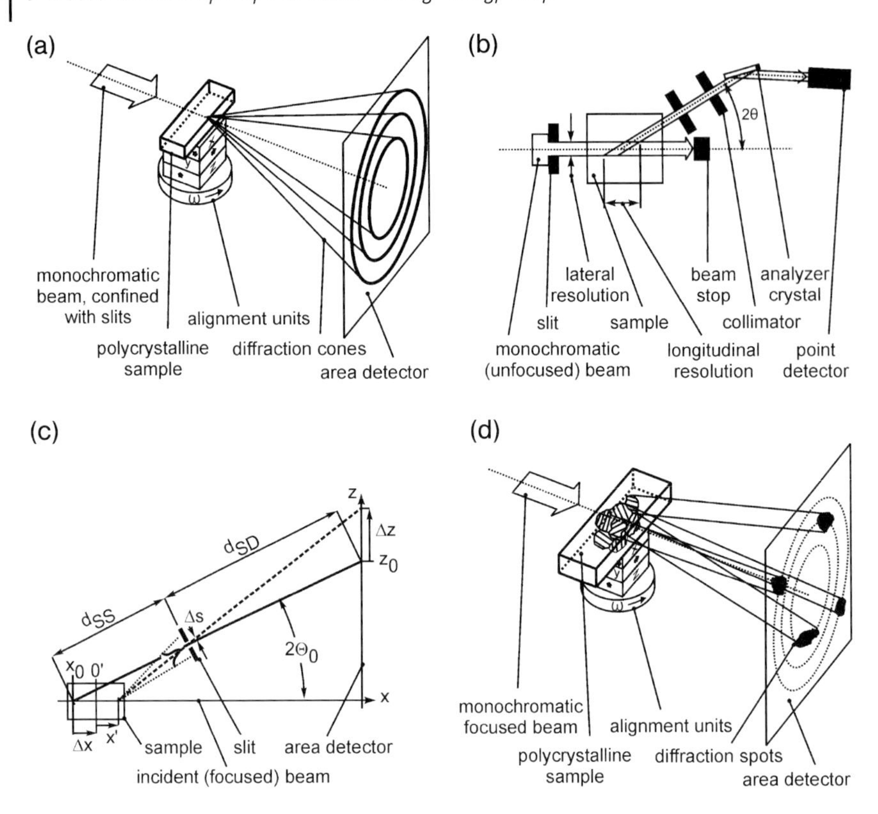


**Figure 9.1** Basic experimental set-ups in transmission mode for strain measurements: (a) integrating over the sample thickness; (b) crossed beam technique with collimator; (c) crossed beam principle with slit imaging showing ray diagram and coordinate system describing the relations between sample position, gauge volume position, and beam positions on the area detector placed perpendicularly to the *x*-axis at the position of the *z*-axis; (d) bulk grain studies applying 3DXRD technique.

lie almost in a plane (e.g., for the Al 311 reflection the diffraction angle $2\theta$ is about 7.3° at an energy of 80 keV; the scattering vector is then only by the Bragg angle of 3.65° out of plane). Therefore, one data image with at least a quarter of a diffraction ring containing the principal axes allows already the calculation of an in-plane biaxial stress if it can be assumed that the stress component parallel to the incoming beam is zero, which is e.g. often valid for thin samples. If two additional images are recorded with the sample rotated around one of the in-plane principal axes by 45° and 90° the complete strain tensor information is available. To avoid systematic errors due to changes of the sample-to-detector distance it is very useful to attach directly to the sample a strain-free reference sample made from a different material than the investigated one to avoid peak overlap between the different materials (e.g., Al/Cu is a useful combination). The measurement of line shifts relatively to the reference peaks reduces the influence of errors originating from wavelength instabilities and uncontrolled sample movements. For the determination of the strain-free lattice parameter $d_0$ please consult Chapter 6.4. Provided an image collection for a sufficient number of sample orientations is available the same data can be used for texture analyses (cf. Chapter 3). Set-ups of this kind are typically found at second-generation synchrotrons (e.g., the HARWI II beamline at the HASYLAB/DESY) or bending magnet beamlines. The beams available there can have sizes in the cm range. This can be of advantage when a coarse grain structure makes macroscopic measurements difficult (see Section 9.6). Narrowing down such a beam in the direction where steep stress or texture gradients are expected in the sample, but leaving it large in the other direction allows then to increase the number of illuminated grains without sacrificing spatial resolution in the direction of interest.

The set-ups in Figure 9.1b and c can be grouped under the term crossed-beam techniques. They are characterized by an adjustable 3D spatial resolution. The lateral resolution (i.e., perpendicular to the incoming beam) is given by the beamsize and the longitudinal (i.e., parallel to the beam) is defined by an optical element placed behind the sample. Figure 9.1b shows a set-up with a collimator behind the sample, thus defining an elongated diamond shaped gauge volume whose geometrical size depends on the dimensions of the incoming beam, the diffraction angle, and the collimator opening behind the sample. A collimator can be made, for example, from two distant slits that meet the condition that the opening angle is smaller than the divergence of the radiation to be collimated. The first slit should be as close to the sample as possible. The geometrical elongation of the gauge volume can be calculated according to Eq. 9.1:

$$l = \frac{b + a \cdot \cos(2\theta)}{\sin(2\theta)} \tag{9.1}$$

$l$ gauge volume length
$a$ width of the incoming beam
$b$ width of the collimator
$2\theta$ diffraction angle

The obviously lower spatial resolution in longitudinal direction as compared to the horizontal and vertical one can sometimes be compensated by a clever sample alignment. In the case of strain investigations one would try to align the sample such that the longitudinal gauge volume axis lies in direction of a shallow strain gradient. Equation (9.1) is also useful for a rough estimation in the case of the other experimental configurations (Figure 9.1c). However, it has to be noted that the effective resolution can be worse by a factor of three or more, depending on the beam divergence and the energy-dependent spatial resolution of the detector used (cf. Section 9.3.2).

In the case of the collimator set-up, the diffraction peak position for the gauge volume defined is measured by scanning the collimator and the point detector across the peak. It is clear that a simultaneous monitoring of a reference signal is not possible and that the system is very sensitive to geometrical errors in the collimator orientation. Furthermore the technique is rather time consuming since only a small fraction of on diffraction ring can be measured at a time. The data acquisition time can be sped up considerably when using several collimators simultaneously. This was realized in the so-called MAXIM set-up (cf. Chapter 11) [1]. In this set-up, a capillary array is matched to a CCD chip, ideally such that each CCD pixel captures the information from only one capillary of the array. For a chip with $1024 \times 1024$ pixels this means that about $10^6$ small gauge volumes can be scanned simultaneously. At present this technique is extended to the use with high-energy X-rays [2].

To circumvent the influence of geometrical changes due to sample surface movements or collimator movements an analyzer crystal can be inserted between collimator and detector. This is shown in Figure 9.1b. Again, each peak has to be scanned individually. This experimental set-up is one of the standard configurations available, e.g., at beamline ID31 of the ESRF. The data quality obtained there allows also peak profile analyses for the investigation of microstrains and microstructural phenomena. An example for depth-resolved studies using this technique can be found in [3].

Figure 9.1c shows a so-called slit imaging set-up. In contrast to the beforementioned set-ups with a collimator the slit as the gauge volume-defining element stays fixed and the sample is scanned along the beam. At each sample position an image of the diffracted signal is recorded. The implications for the data analysis are addressed in Section 9.2.

Figure 9.1d shows a set-up for the simultaneous investigation of several individual bulk grains, applying the 3DXRD technique (cf. Chapter 19). The set-up resembles, in principle, very much the one in Figure 9.1a. However, investigations on this length scale require special arrangements and approaches in the set-up and the data analysis respectively. An example is presented in Section 9.5.

## 9.2 Principle of Slit Imaging and Data Reconstruction

The aim of depth-resolved strain measurements by slit imaging is, as a function of the sample position, the determination of the difference between the peak position of a diffracted beam, emanating from a local volume within a sample, and the peak position from an ideally unstrained reference sample. In Figure 9.1c, this reference beam position is represented by the solid line at an angle of $2\Theta_0$ to the incoming beam. The origin of this line coincides here with the origin of the laboratory coordinate system $x_0$ and the point where it impinges on the area detector, placed at the $z$-axis, is the detector reference point $z_0$. The beam passes the slit that defines the local gauge volume within the strained sample. The slit opening is $\Delta s$. The respective distances between sample, slit, and detector are called $d_{ss}$ and $d_{sd}$. The strained sample is positioned with its center $0'$ at the position $\Delta x$. Due to the finite width of the reflection, only part of the rays emanating at the position $x'$ are recorded on the detector at the distance $\Delta z$, relative to the reference position $z_0$. It is clear that beams impinging on the detector can have their origin at different positions $x'$. Therefore, it is not possible to obtain the peak position for a specific $x'$ from a single recorded intensity distribution. A series of images has to be taken in a sufficient step width and range along $x$ in order to assure that the signal originating from $x'$ can be completely detected, though split up on several images. The diffraction peaks for each position $x'$ can now be reconstructed using Eq. (9.2) [4]. The equation permits the calculation of the source point $x'$ of each beam detected at a position $\Delta z$ on the detector.

$$x' = b \cdot \Delta z \left[ 1 + \frac{\Delta z}{d_{SD} \cdot \sin(2\theta_0)} \right]^{-1} - \Delta x \quad \text{with } b = \frac{d_{SS}}{d_{SD} \cdot \tan(2\theta_0)} \tag{9.2}$$

$d_{SS}$ sample-to-slit distance
$d_{SD}$ sample-to-detector distance
$2\theta_0$ diffraction angle of unstrained sample
$\Delta x$ sample displacement along incoming beam
$\Delta z$ displacement on detector
$\Delta s$ slit opening

For the practical application of the formula an approximate value can be used for $2\theta_0$ instead of its exact value. Furthermore, the data reconstruction is only necessary, when $d_{SD}$ is less than ten times $d_{SS}$. At larger $d_{SD}$ values the coupling between sample position and peak position becomes negligible and the data correction can be omitted.

## 9.3 The Conical Slit

### 9.3.1 Working Principle

The conical slit set-up (Figure 9.2) is a special case of the slit imaging set-up (cf. Sections 9.1 and 9.2). The idea behind the conical slit is to use a set of concentric conical apertures with a common apex instead of a slit made from a pair of straight blades that are aligned for a specific reflection at a specific azimuthal angle on the diffraction ring. The advantage of the conical slit is obvious, instead of observing just a small fraction of a diffraction cone emanating from a gauge volume it becomes possible to monitor simultaneously different complete diffraction rings emanating from the same gauge volume. The conical apertures have to match the diffraction cones of the material to be investigated. The distance of their common apex to the slit surface can be considered as the focal length of the system. Note that usually a series of images has to be taken to achieve, after data reconstruction, the highest spatial resolution (cf. Section 9.2).

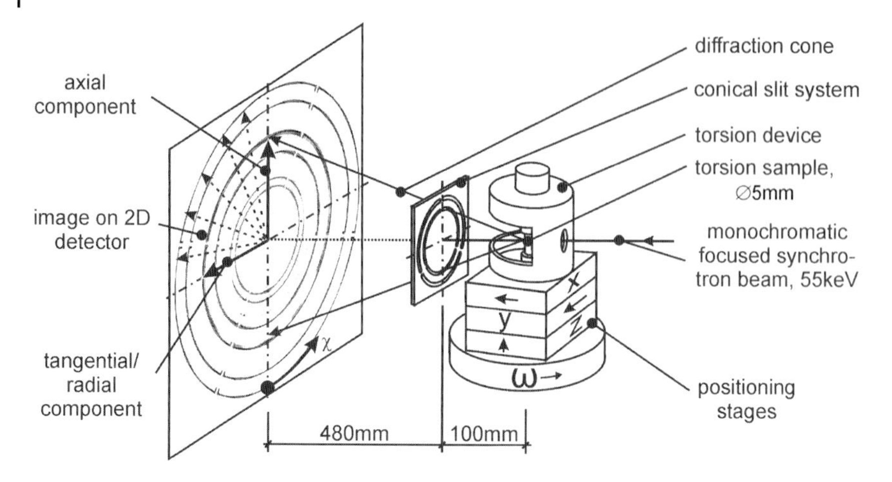


**Figure 9.2** Experimental set-up for depth-resolved measurements, using conical slit system and 2D detector.

### 9.3.2
### Capabilities

At present three conical slits have been manufactured, corresponding to the first, second, and third-generation type, respectively. The first generation was made for fcc materials with a focal distance of 10 mm, however, it is not in use anymore. A description, which is in principle also valid for the other conical slits, can be found in [5]. The second-generation slit has a focal length of 100 mm, thus enabling the easy accommodation of bulky sample environment. It was designed for fcc and bcc materials. An overview of the reflections accepted, including the experimentally determined gauge length and geometrical parameters of the slit system are given in Table 9.1.

Each of the seven apertures has a gap of 25 μm. Due to the fixed accepted diffraction angles the energy of the incoming beam has to be adjusted so that the diffraction cones match the aperture pattern. For Al this is 55 keV and for $\alpha$-Fe it is 77.7 keV. In case of other crystal symmetries it is still possible to adjust the energy in a way so that at least one diffraction ring is passing the slit.

The second-generation slit is available at the beamline ID11 at the European Synchrotron Radiation Facility (ESRF).

The third-generation slit is identical to the second one except for its focal length of 50 mm and some mechanical improvements. It is available at the beamline 1-ID at the Advanced Photon Source (APS).

To achieve a high-spatial resolution parallel to the incoming beam the dimensions of the incoming beam need to be below 10 μm × 10 μm. A major requirement is, therefore, the focusing of the incoming beam. Just cutting down the beam size with slits would result in a very low photon flux and unacceptably long exposure times. Looking at Eq. (9.1) and the gauge lengths given in Table 9.1 it can be seen that the gauge lengths, determined experimentally for the second-

**Table 9.1** Overview of the reflections selected by the second-generation conical slit, geometrical slit parameters, and the corresponding measured longitudinal gauge lengths.

| *h k l* | Material | $2\theta$ (°) | Radius (mm) | Gauge length (μm) |
|---|---|---|---|---|
| 1 1 1 | Al[a] | 5.529 | 9.6755 | 750 |
| 2 0 0 | Al/α-Fe | 6.383 | 11.1854 | 640 |
| 2 1 1 | α-Fe | 7.821 | 13.7314 | 560[b] |
| 2 2 0 | Al/α-Fe | 9.029 | 15.8931 | 500 |
| 2 2 2 | Al/α-Fe | 11.064 | 19.5573 | 420 |
| 3 3 1 | Al | 13.934 | 24.8159 | 370 |
| 4 2 2 | Al/α-Fe | 15.676 | 28.0596 | 320 |

a) For Al, a photon energy of 55 keV has to be used and for α-Fe 77.7 keV to match the conical slits.
b) Interpolated value.

generation slit, are a factor of about three bigger than the geometrical ones. The main reasons for that are the $2\theta$ and optics-dependent beam divergence behind the sample and the energy-dependent spatial resolution of the detector. In the present case the highest divergences behind the sample ranged between 0.696 mrad for the 1 1 1 and 1.41 mrad for the 4 2 2 reflection at the beamline ID11 at the ESRF with a bent Laue crystal and an elliptically bent multilayer [6] for vertical and horizontal focusing respectively. The spatial resolution was found to be 335 μm at 55 keV for a MAR345 image plate detector (with a nominal resolution set to 100 μm). Improving the energy-dependent detector resolution by a factor of two would lead to an improvement of the gauge length by the same factor.

The great advantage of the conical slit is the possibility to acquire almost complete depth-resolved orientation space data (texture), which is, in addition, suitable for strain analyses. The data acquisition is orders of magnitude faster than with conventional crossed beam techniques.

A drawback is that each set-up has to be adjusted via the wavelength to the unit cell of the material investigated. Therefore, in most cases, phase transformations cannot be observed.

### 9.3.3 Example

A typical set-up with a conical slit is shown in Figure 9.2. It consists of devices for sample translation and rotation, the conical slit, and a large area detector. The conical slit itself is mounted on a mechanically stable high-precision positioning unit to allow the alignment within 1 μm.

The incoming beam is focused below a spot size of 10 μm × 10 μm. Presently available detectors for high-energy X-rays have a maximum size of 350 mm × 350 mm. This allows a maximum sample to detector distance of about 600 mm

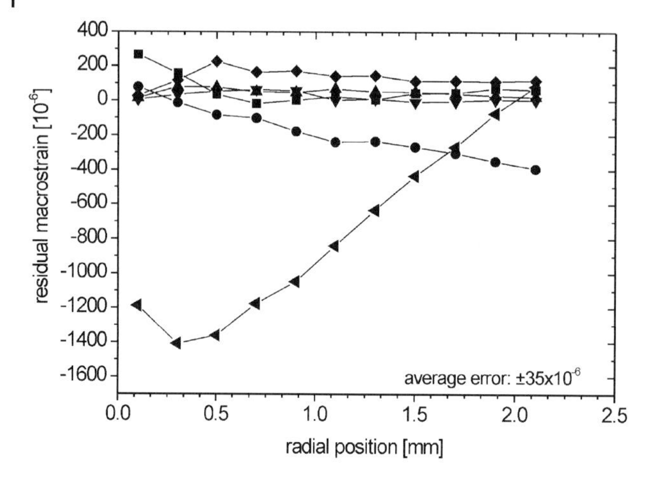


**Figure 9.3** Macrostrain tensor components $\varepsilon_{11}$ (● tangential), $\varepsilon_{12}$ (◄), $\varepsilon_{13}$ (▼), $\varepsilon_{22}$ (■ axial), $\varepsilon_{23}$ (▲), and $\varepsilon_{33}$ (♦ radial) as a function of the radius within a solid torsion sample made from Al alloy AlMg3 with a diameter of 5 mm and deformed to $\gamma = 1.5$.

for the monitoring of all available complete Debye–Scherrer rings. In the present example, torsion samples were investigated in an *in-situ* sample deformation device. The samples were made from aluminum alloy AlMg3. Aspects such as radial-dependent texture, macrostrain, and microstrain evolution were investigated [7]. The set-up also allowed investigations of the dynamic texture and strain tensor evolution during deformation [8]. Figure 9.3 shows the radial dependence of the six macrostrain tensor components as a function of the radius. Apart of the expected maximum compressive shear strain close to the sample center a tensile axial strain component close to the center could be observed. The axial strains can be explained by the texture related, radially dependent development of axial forces.

## 9.4 The Spiral Slit

The spiral slit [9] was developed at the beamline ID15 at the ESRF and is available there for user experiments. As the conical slit, the spiral slit is a special case of the slit imaging technique too (cf. Sections 9.1 and 9.2).

### 9.4.1 Functional Principle

The spiral slit consists of twelve equidistant concentric spiral apertures cut into a Tungsten plate. All paths through the 2 mm thick tungsten plate lie on cones, which have one common apex (i.e., the focal point of the slit system; here at 50 mm). The slit gap is adjustable from 0 to 100 μm because the spirals were actually cut into two 1 mm thick plates, which can be rotated against each other. Instead of having just one slit selecting a gauge volume within the sample for a given (small) azimuthal range and one $2\theta$ angle, the spiral slit accepts all rays emerging from the sample within the angular range of $2\theta = 5^\circ$ to $35^\circ$. However, in contrast to the conical slit system discussed in Section 9.3, the spiral slit does not allow the observation of complete rings. Each spiral cuts out a small fraction of the diffraction cone. The great advantage of the spiral slit is the fact that one single image of the diffraction pattern behind the slit contains all the depth- and phase-resolved information for the sample volume illuminated by the X-ray beam. This can be explained in the following way: Considering Figure 9.1a, no slit at all would give rise to information integrated over the sample thickness. By inserting a single slit in the beam path behind the sample as shown in Figure 9.1c, depth-resolved information can be obtained. But, to collect the complete depth-resolved information either the sample needs to be translated along the incoming beam or the slit has to be moved to different positions and the diffraction data has to be recorded at each position of either the sample or the slit. Now, the spiral slit makes this movement unnecessary, because each point on the spiral accepts a different scattering angle and azimuth (angle around the incoming beam, cf. Figure 9.4). In a manner of speaking, the spiral form of the aperture itself leads to a movement. In addition, the depth acceptance of several mm to cm of the spiral slit (grossly comparable to the depth of focus of an optical lens system) allows that the rays from different depths along the beam path in the sample can pass the slit (see inset in Figure 9.4). As a result the depth-resolved signal is azimuthally spread over the image.

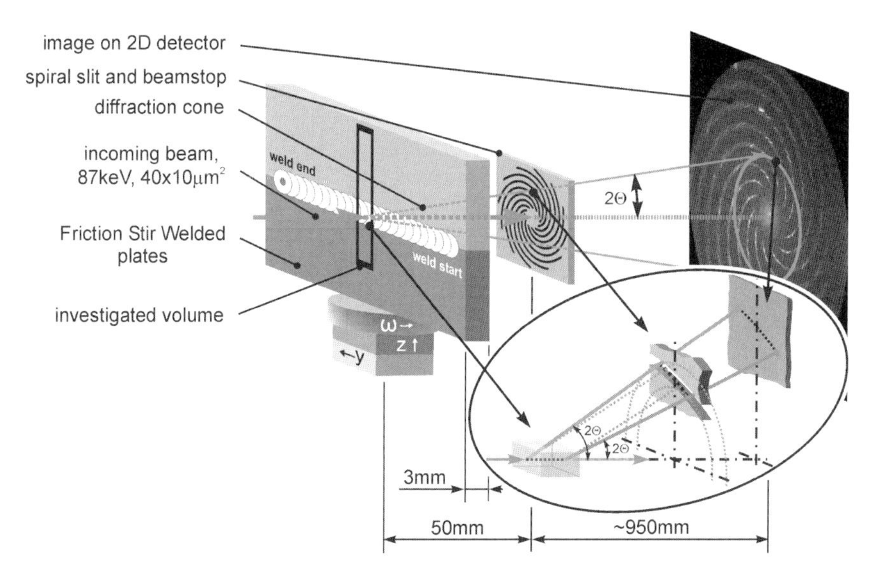


**Figure 9.4** Schematic of the experimental set-up with spiral slit and area detector for depth-resolved investigations. The inset illustrates the functional principle of the spiral slit technique.

### 9.4.2 Capabilities

The great advantage of the spiral slit is its capability to monitor in only one image the entire depth-resolved $2\theta$ range from 5° to 35°. This makes it very interesting for the fast depth-resolved investigation of phase transformations [10] and residual strain. The time resolution for the observation of dynamic processes can be in the range of a second, depending on the detector used.

An inherent drawback of the spiral slit technique is that the diffraction rings recorded are not complete. Therefore, depth-resolved texture investigations, strain tensor measurements on single grains or investigations of very coarse-grained materials (cf. Section 9.6) could in many cases better be made by other means, for example, with a conical slit. However, if time is not an issue a solution to obtain quasicontinuous rings is either the rotation of the sample around the beam in several steps or a scanning of the sample within the depth acceptance of the slit. Then, for each angular or longitudinal position an image has to be recorded.

### 9.4.3 Example

Figure 9.5 shows experimental results from the investigation of a dissimilar friction stir welded sample [11] made from AA2024 in T3 condition and AA6082 in T6 condition (single phase alloys) [9]. A sketch of the experimental set-up is shown in Figure 9.4 (cf. Section 9.4.1).

Due to the depth acceptance parallel to the X-ray beam a spinning capillary (not shown in the figure), containing standard reference Si powder and Al powder, could be placed before the sample providing the reference information for each recorded diffraction pattern. The patterns were monitored with a large area detector system, in the present case a MAR345 online image plate scanner. Each recorded image contains the depth-resolved information for a 2D strain distribution and the material composition. In order to determine a triaxial stress state, an additional image was recorded at each position with the sample rotated by 60° around the $\omega$-axis. This obviously degrades the spatial resolution parallel to the weld. However, the effect is rather small taking into consideration the plate thickness of 3 mm and the fact that the gradients parallel to the weld are not very steep in the region investigated. The strain distribution was measured along a line crossing the weld at 80 mm from the weld start. The step size along this line was of 1 mm in the weld region and of 4 mm for the outer regions. In order to improve the grain averaging, the sample was oscillated during the data collection over a range of 10 mm parallel to the weld.

The residual macrostrain tensor components were determined by fitting the stress-induced distortion of the recorded fractioned Debye–Scherrer rings. The depth-resolved phase distribution was derived from the differences in the recorded diffraction patterns of the two materials. A part of the results is shown in Figure 9.5. For better comprehension the tool is schematically drawn in Figure 9.5a. The

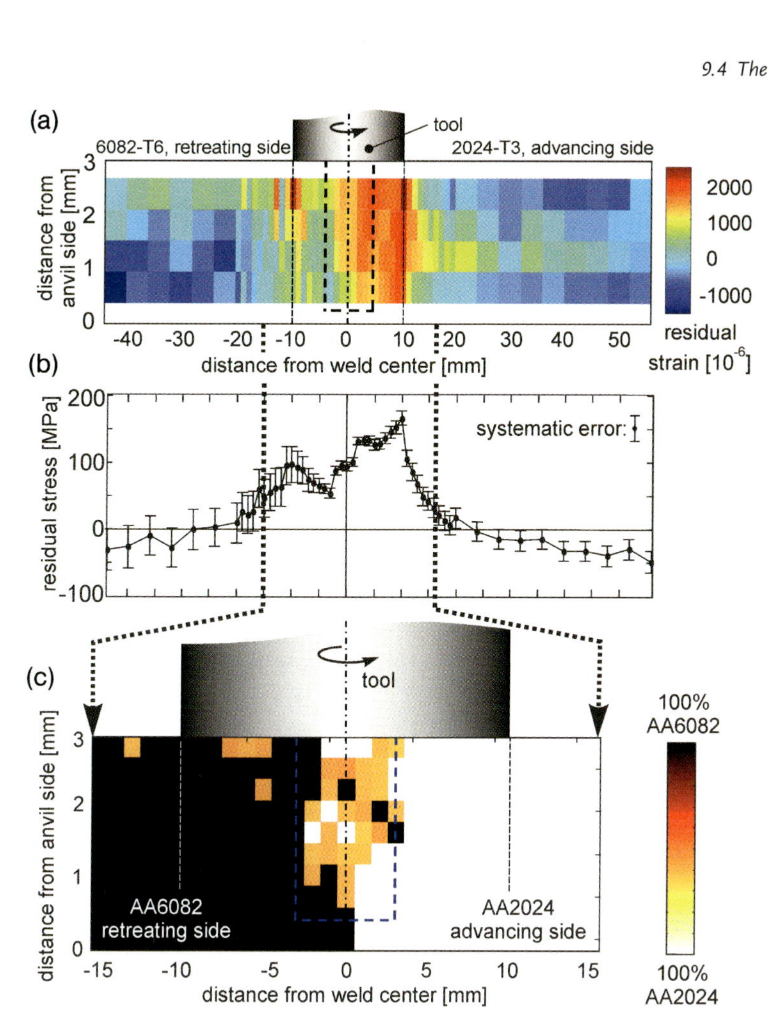


**Figure 9.5** (a) Depth-resolved residual macrostrains in the longitudinal direction (parallel to weld); (b) through thickness longitudinal stress, calculated from strain measurements in three dimensions; (c) enlarged view of the plate material distribution (Al alloys: AA6082 and AA2024) in the stirring zone.

depth-resolved strain map for the direction parallel to the weld is presented in Figure 9.5a. It is particularly inhomogeneous on the retreating side with high tensile strains up to $+2000 \times 10^{-6}$ under the tool shoulder (about $-10$ mm from the center) and values around zero close to the anvil side. The distribution of the longitudinal residual stress across the weld and integrated over the sample thickness can be seen in Figure 9.5b. The stress rises up to $+160$ MPa under the tool shoulder at $+10$ mm. The strong asymmetries of the strain and stress distributions arise from the mixing of the two materials and the different response of the material to the tool force. The material distribution is shown in Figure 9.5c and a close correlation to the depth-resolved strain distribution can be observed.

## 9.5 Simultaneous Strain Measurements in Individual Bulk Grains

An introduction to the principles of three-dimensional X-ray diffraction (3DXRD) and its capabilities can be found in Chapter 19. Here, an example for strain measurements in individual bulk grains (i.e., strain of kind II) of a tensile test sample is presented [12]. The set-up resembles much the one shown in Figure 9.1a and can be seen as a special case with much higher requirements concerning the accuracy of sample positioning. Furthermore, the difference to experiments where the information is integrated over the thickness is that the number of illuminated grains is sufficiently small so that distinct diffraction spots arise rather than continuous diffraction rings. In fact, the technique only works well, as long as the spots can be separated from each other, and, as a consequence, be analyzed individually. To reduce the number of grains observed, one can also use the conical slit described in Section 9.3.

The experiment was carried out on the 3DXRD microscope [6] at beamline ID11 at the ESRF. The synchrotron X-ray beam with an energy of 54.9 keV ($\lambda = 0.2258$ Å) was focused vertically to 100 μm with a bent Laue monochromator and confined in horizontal direction by slits to 150 μm.

The sample was mounted in a 24 kN Instron stress rig, which in turn was mounted on a sample alignment tower, to allow lateral and angular movements. The tensile axis of the sample was aligned perpendicularly to the X-ray beam and the normal direction of the flat side of the sample was parallel to the beam direction at an angular position of $\omega = 0°$. The volume illuminated by the X-ray beam is given by the horizontal and vertical dimensions of the beam and by the path length of the beam through the sample. The diffraction spots originating from grains fulfilling the diffraction condition are recorded with an image intensifier with an active circular area of 180 mm in diameter, coupled to a CCD with an effective pixel size of 87.5 μm. The detector system was placed at 442 mm from the sample center. The 3DXRD method was applied. Space constraints limited the rotation range to $-44° \leq \omega \leq 46°$.

The sample was loaded in steps of 12 MPa to a maximum of 48 MPa and then unloaded to 0 MPa. The total plastic strain was 2.12%. The data was recorded at each load level at five lateral positions along the tensile axis of the sample in order to increase the grain statistics without increasing the problem of diffraction spot overlap. The spot overlap would give rise to larger uncertainties in the attribution of diffraction spots to specific grains. One particular diffraction spot was used to realign the sample after each deformation step in order to allow for the sample extension.

The data presented in Figure 9.6 is a subset of the complete set collected at the five lateral sample positions. It shows the strain evolution in a deeply embedded grain as a function of load. Upon unloading the measured strain components are close to zero, indicating no or very little plastic deformation for this grain. In total the data for ten individual grains were evaluated. The results were grouped according to slip patterns as expected for the single crystals after the Taylor

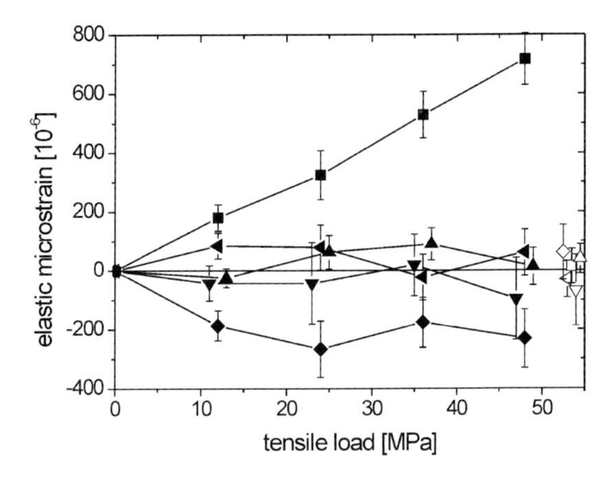


**Figure 9.6** Strain evolution measured for one of the investigated grains as a function of load for the components $\varepsilon_{12}$ (◄), $\varepsilon_{22}$ (■ axial), $\varepsilon_{13}$ (▼), $\varepsilon_{23}$ (▲), and $\varepsilon_{33}$ (♦ transverse). Unfilled symbols relate to the strain state after the final unloading. A slight scatter along the abscissa is imposed in the plot to enhance the visibility of the error bars.

model. The different groups exhibited significant differences in the strain evolution (data not shown here).

An example for an investigation of intragranular strains, i.e., strain of kind III, can be found in [13].

## 9.6
## Coarse Grain Effects

One of the general problems when performing strain or texture measurements with X-rays using a small gauge volume is the average grain size of the material investigated. Conclusions on the macroscopic level can be false due to the fact that too few grains have contributed to the measured interference line. Estimations of the influence of the grain size and the gauge volume have been proposed [14, 15]. Here, an approach is presented which is optimized for small gauge volumes. To diminish the influence of the coarse grain effects, the sample can be oscillated, either in rotation or in translation but it has to be noted that a combination of several oscillations requires an increased speed for each additional movement in order to irradiate the resulting gauge volume homogeneously. More than one oscillation direction becomes impractical when detector exposure times are of the order of a few seconds. Furthermore, the oscillation range should not significantly deteriorate the spatial resolution. In the experiments discussed here, tests have shown that the linear oscillation was by far the most effective compared to angular oscillation. For the experiments with the torsion samples (cf. Section 9.3.3) an angular oscillation would have been limited to a range of

$\pm 5^\circ$ due to the deformation device and the requirement not to degrade significantly the spatial resolution.

To assess how many grains within the irradiated effective volume are contributing to the signal detected at a specific position, first the number of illuminated grains has to be calculated. In fact, it should be considered how many grains are intersected by the effective gauge volume. In the experiments discussed here, the horizontal size of the beam is much smaller than the length of the illuminated effective volume and its height, given by the linear oscillation range. The division of the gauge volume by the average grain volume would give misleading results. This is easily understandable when considering a gauge volume of say 10 μm × 10 μm × 500 μm and a grain size of 60 μm. The result would be that less than one grain contributes to the diffraction pattern. Instead, based on the intersections, the number of (partially) irradiated grains can be estimated according to:

$$N_g = \text{ceil}\left(\frac{l}{d_{gl}}\right) \cdot \text{ceil}\left(\frac{v}{d_{gv}}\right) \cdot \text{ceil}\left(\frac{h}{d_{gh}}\right) \tag{9.3}$$

$N_g$ number of (partially) illuminated grains
$l$ gauge volume length
$v$ effective vertical gauge volume size, e.g., the linear oscillation range
$h$ effective horizontal gauge volume size
$d_{gx}$ grain size in longitudinal, vertical, and horizontal direction

The expression *ceil* means that the result of the division is rounded toward the greater integer; therefore, it will be always at least 1.

In the next step, it has to be estimated how many of these grains are actually diffracting, when considering only one reflection. This depends on the mosaic spread of the material, the multiplicity of the reflection of interest, and, if applied, the range of the angular oscillation. The fraction of diffracting grains can then be estimated based on Eq. (9.4) [14], assuming a uniform texture.

$$N_{dg} = N_g \cdot \frac{J(\Delta\omega_g + \Delta\omega_o)}{2} \tag{9.4}$$

$N_{dg}$ number of diffracting grains, assuming uniform texture
$N_g$ number of (partially) illuminated grains
$J$ multiplicity of the reflection
$\Delta\omega_g$ mosaic spread of the grains in radians
$\Delta\omega_o$ angular oscillation range in radians

This number relates to the complete diffraction ring. Therefore, in the last step it has to be estimated how many grains are contributing to the observed azimuthal section of the diffraction ring. This can be, for example, the section of the diffraction ring accepted by a point detector or in case of an area detector the azimuthal range over which the diffraction rings are averaged for the later evaluation [15].

$$N_{odg} = N_{dg} \cdot \frac{\chi_a}{2\pi} \quad (9.5)$$

$N_{odg}$ number of observed diffracting grains
$N_{dg}$ number of diffracting grains
$\chi_a$ azimuthal averaging range in radians

Values for the mosaic spread of Al grains of up to 0.12 rad were reported by [16, 17].

According to a common rule of thumb 1000 illuminated grains are considered to be sufficient for macroscopic investigations when using a reflection with high index and low symmetry (e.g., the Al 311 reflection). However, recent modeling efforts by [18] seem to indicate that the number should be considerably higher and that in some cases a low-index and high-symmetry reflection might yield better macroscopic results. Their investigations related to single phase materials with no initial texture. For that case 10,000 sampled grains started to give reliable results.

## 9.7 Analysis of Diffraction Data from Area Detectors

Each image contains usually several complete Debye–Scherrer rings. After background subtraction and determination of the beam center, i.e., the common center of all the rings (note that this must be done only once for one data set), the images are transformed to polar coordinates using, for example, the software Fit2D [19]. A fine azimuthal slicing (in the order of 1000 slices), as proposed by [20], allows the analysis of diffraction rings that have no smooth azimuthal intensity distribution. Before the least square fitting of a linear background and Pseudo Voigt profiles to the peak pattern of single pixel rows, a reconstruction procedure might need to be applied to the data [4] (cf. Section 9.2). Afterward the results of the fit are rebinned to azimuthal sections of 5°.

Applying the appropriate coordinate transformations and absorption corrections to the integrated intensities, obtained from the least square fitting, the data can be used to calculate the orientation distribution function (ODF) (cf. Chapter 3).

The strain tensor can be calculated for each measured reflection from the distortion of the diffraction rings. This requires the knowledge of the unstrained lattice parameter $d_0$ (cf. Chapter 6.4), and the correction of the data for systematic distortions, caused by, for example, the detector system itself and the geometry of the experimental set-up. The distortions are corrected by calculating the strains relatively to the supposedly strain-free diffraction rings of a powder reference. The strain tensor components are then determined, for example, by fitting the parameters of the general Eq. (9.6) [21] to the corrected peak positions.

$$f_{11}\varepsilon_{11} + f_{12}\varepsilon_{12} + f_{22}\varepsilon_{22} + f_{13}\varepsilon_{13} + f_{23}\varepsilon_{23} + f_{33}\varepsilon_{33} = -\frac{\Delta\theta}{\tan\theta_0} \tag{9.6}$$

$\varepsilon_{ij}$ are the components of the strain tensor, $f_{ij}$ are scattering vector and sample orientation dependent factors and $\theta$ is the Bragg angle of the observed $h\ k\ l$.

**Table 9.2** Overview of different experimental techniques and their capabilities. Note that exceptions are always possible and that the table is just meant as a rough guideline.

| Technique | Application | Spatial resolution | Acquisition time for one data set | Remarks |
|---|---|---|---|---|
| Transmission with integrated signal | Strain, texture, and phase mapping | Lateral: beamsize | Seconds to min/image/sample position | 2D information, no depth resolution |
| Crossed beam with collimator | Strain scanning | Longitudinal: ≥50 μm | ≥60 s/peak/depth position/sample orientation | Focused beam, 1D-strain info., scanning of diffraction peak |
| Slit imaging | Strain scanning | Longitudinal: ≥50 μm | ≥1 s/image/peak/depth position/sample orientation | Focused beam, 1D-strain info., imaging of peak |
| Polycapillary | Strain and phase mapping | ≥6 μm | ≥15 min/60 images/$h\ k\ l$/sample orientation | Large beam, 1D-strain info., simultaneous scanning of thousands of peaks |
| Conical slit system | Texture and strain mapping, single grain studies | Longitudinal: ≥260 μm | ≥1 s/image/depth position/sample orientation | Focused beam, 2D-strain info./image, recording of complete rings |
| Spiral slit system | Strain and phase mapping | Longitudinal: ≥80 μm | ≥1 s/sample orientation | Focused beam, 2D-strain info. with complete depth info. in one image |
| 3DXRD | Simultaneous strain measurements in individual grains | Grain size | ≤2.5 s/image/$\omega$-pos.<br>≤600 s for data set with 240 $\omega$ positions and a $\Delta\omega$ of 0.5° | Focused beam, grain indexing provides orientation of each grain, grain position can be calculated. |

## 9.8 Matrix for Comparison and Decision Taking Which Technique to Use for a Specific Problem

When planning an experiment one has to decide which length scale the experiment should aim at and on which time scale. Macroscopic investigations with no time resolution will require a different set-up than single grain investigations with a time resolution of minutes or seconds. Furthermore, the application of some techniques described before also depends on the light source. Focusing of high-energy X-rays, for example, is usually not available at second-generation synchrotron sources, though exceptions exist [22]. Special techniques involving, for example, a spiral or conical slit are until now only available at specific beamlines. Table 9.2 is meant to provide an informative basis concerning the capabilities of the techniques and their performance in terms of time and spatial resolution. The numbers given in the table can vary considerably, for example, because of the material investigated or the detector system used. In general, one can say that the numbers change for the better because of continuous improvements in set-up and detector systems.

## References


1 T. Wroblewski, A. Bjeomikhov, *Nucl. Instrum. Meth. A* 2005, **538**, 771–777.

2 T. Wroblewski, A. Bjeomikhov, B. Hasse, *Mater. Sci. Forum* 2006, **524–525**, 273–278.

3 P.J. Webster, L. Djapic Osterkamp, P.A. Browne, D.J. Hughes, W.P. Kang, P.J. Withers, G.B.M. Vaughan, *J. Strain Anal. Eng.* 2001, **36** (1), 61–70.

4 U. Lienert, R. Martins, S. Grigull, M. Pinkerton, H.F. Poulsen, Å. Kvick, *Mat. Res. Soc. Symp. Proc.* 2000, **590**, 241–246.

5 S. F. Nielsen, A. Wolf, H. F. Poulsen, M. Ohler, U. Lienert, R. A. Owen, *J. Synchrotron Rad.* 2000, **7**, 103–109.

6 U. Lienert, H.F. Poulsen, Å. Kvick, *Proceedings of the 40th Conference of AIAA on Structures, Structural Dynamics and Materials, St. Louis, 1999*, pp. 2067–2075.

7 R.V. Martins, U. Lienert, L. Margulies, A. Pyzalla, *Mat. Sci. Eng. A-Struct.* 2005, **402**, 278–287.

8 R.V. Martins, U. Lienert, L. Margulies, A. Pyzalla, *Mater. Sci. Forum* 2002, **404–407**, 115–120.

9 R.V. Martins, V. Honkimäki, *Texture Microstruct.* 2003, **35**(3/4), 145–152.

10 A. Hagen, H.F. Poulsen, T. Klemmsø, R.V. Martins, V. Honkimäki, T. Buslaps, *Fuel Cells* 2006, **5**, 361–366.

11 W.M. Thomas, E.D. Nicholas, J.C. Needham, M.G. Murch, P. Temple-Smith, C.J. Dawes, International Patent No. PCT/GB92/02203, 1991, GB Patent No. 9125978.8, 1991, US Patent No. 5,460,317, 1995.

12 R.V. Martins, L. Margulies, S. Schmidt, H.F. Poulsen, T. Leffers, *Mat. Science Eng. A- Struct.* 2004, **387–389**, 84–88.

13 W. Pantleon, H.F. Poulsen, J. Almer, U. Lienert, *Mat. Sci. Eng. A- Struct.* 2004, **387–389**, 339–342.

14 U. Lienert, S. Grigull, Å. Kvick, R.V. Martins, H.F. Poulsen, *Proc. ICRS-6*, Oxford, United Kingdom, 2000, 2, pp. 1050–1057.

15 A. Pyzalla, *J. Nondestruct. Eval.* 2000, **19** (1), 21–31.

16 L. Margulies, G. Winther, H.F. Poulsen, *Science* 2001, **291**, 2392–2394.

17 H.F. Poulsen, D. Juul Jensen, T. Tschentscher, E.M. Wcislak, *Texture Microstruct.* 2001, **35** (1), 39–54.

18 B. Clausen, T. Leffers, T. Lorentzen, *Acta Mater.* 2003, **51**, 6181–6188.

19 A.P. Hammersley, S.O. Svensson, M. Hanfland, A.N. Fitch, D. Häusermann, *High Pressure Res.* 1996, **14**, 235–248.

20 A. Wanner, D.C. Dunand, *Mat. Res. Soc. Proc.* 2000, 590.

21 BB. He, K.L. Smith, *Proc. SEM Spring Conf. on Experimental and Applied Mechanics, Houston, Texas, June 1–3*, 1998, pp. 217–220.

22 Z. Zhong, C.C. Kao, D.P. Siddons, J.B. Hastings, *J. Appl. Crystallogr.* 2001, **34**, 646–653.

# 10
# Residual Stress Analysis by White High Energy X-Rays

## 10.1
## Reflection Mode

*Christoph Genzel*

### 10.1.1
### Motivation

Mechanical surface processing such as grinding or shot-peening is used extensively in industry to generate beneficial near-surface residual stress fields in critical components, which may extend up to depths of some hundred microns into the material [1]. A phase specific and structural analysis of these stresses is usually performed by angle-dispersive (AD) diffraction methods using the characteristic radiation provided by conventional X-rays tubes [2, 3]. However, due to the low photon energies, which lie in a range between about 5 and 17 keV, only small penetration depths $\tau$ of some tens of microns can be achieved in this way. Locally resolved neutron diffraction, on the other hand, requires rather large gauge volumes of at least $1 \times 1 \times 1\ mm^3$ and, therefore, is used mostly to analyze residual stress fields in the bulk, where the gauge is dipped completely into the material.

Consequently, there remains a information gap which concerns a depth range between some tens and some hundreds of microns below the surface, where the conventional X-ray methods are no longer present and the neutron methods are not yet sensitive. This range, however, which may be called the "intermediate zone" between surface and volume, is of particular interest from both the scientific and the technological point of views, because it is the transition zone between the biaxial surface stress state and the triaxial volume stresses in the interior of the material.

It will be shown in this section that energy-dispersive (ED) diffraction performed in reflection geometry using white high-energy synchrotron radiation up to about 150 keV offers new possibilities and prospects to close this gap and to obtain valuable information on the residual stress state in the intermediate zone.

*Neutrons and Synchrotron Radiation in Engineering Materials Science: From Fundamentals to Material and Component Characterization*
Edited by Walter Reimers, Anke Rita Pyzalla, Andreas Schreyer, and Helmut Clemens

ISBN: 978-3-527-31533-8

### 10.1.2
### Basic Relations in-Depth-Resolved Energy-Dispersive X-Ray Stress Analysis (XSA) in Reflection Geometry

Bragg's equation which reads in its AD form $n\lambda = 2d_{hkl}\sin\theta_{hkl}$ relates for a fixed wavelength $\lambda$ a set of lattice planes $hkl$, with distance $d_{hkl}$ to a certain angle $\theta_{hkl}$ (the Bragg angle). Recording diffraction patterns in the $\theta$–$\theta$-mode, interference lines are observed for scattering angles $2\theta_{hkl}$. In ED diffraction using a white beam with a continuous photon energy spectrum, on the other hand, the diffraction angle $\theta$ and the scattering angle $2\theta$ under which the diffracted energy spectrum is observed, can be chosen freely and remain fixed during the measurement. The correlation between the lattice spacing $d(hkl)$ and the corresponding diffraction line $E(hkl)$ on the energy scale follows immediately by inserting the energy relation $E = h\nu = hc/\lambda$ into the Bragg equation

$$d(hkl) = \frac{hc}{2\sin\theta}\frac{1}{E(hkl)} = \text{const.}\,\frac{1}{E(hkl)} \tag{10.1.1}$$

($h$: Planck's constant, $c$: velocity of light). With Eq. (10.1.1), the lattice strain $\varepsilon_{\phi\psi}(hkl)$ determined at some orientation $(\phi, \psi)$ with respect to the sample system becomes

$$\varepsilon_{\phi\psi}(hkl) = \frac{d_{\phi\psi}(hkl)}{d_0(hkl)} - 1 = \frac{E_0(hkl)}{E_{\phi\psi}(hkl)} - 1 \tag{10.1.2}$$

where $E_0(hkl)$ denotes the energy that corresponds to the strain-free lattice spacing $d_0(hkl)$. With respect to a depth-resolved analysis of the strain/stress fields below the surface it must be further regarded that each reflection $E(hkl)$ on the energy scale comes from another (average) information depth $\tau$. A general formulation of the $1/e$ penetration depth $\tau$ which is defined by the condition that the intensity $I$ of the X-rays passing through the material is $1/e$ of the primary intensity $I_0$ is given by [4]

$$\tau = \frac{\sin^2\theta - \sin^2\psi + \cos^2\theta\,\sin^2\psi\,\sin^2\eta}{2\mu(E)\sin\theta\cos\psi} \tag{10.1.3}$$

where $\mu$ denotes the linear absorption coefficient which depends on the energy $E$ (i.e., the wavelength $\lambda$) of the radiation, and $\eta$ describes the rotation of the sample around the diffraction vector $\mathbf{q}(hkl)$ (Figure 10.1.1).

In AD-XSA experiments a continuous variation of $\tau$ can be achieved only by the geometry parameters $\psi$ and $\eta$, respectively. In ED diffraction, on the other hand, $\tau$ can also be varied continuously by the diffraction angle $\theta$. From Eq. (10.1.1) it follows directly that an increase of $\theta$ leads to a "compression" of the diffracted energy spectrum, i.e., a shift of the diffraction lines $E(hkl)$ toward smaller energies and vice versa (Figure 10.1.2).

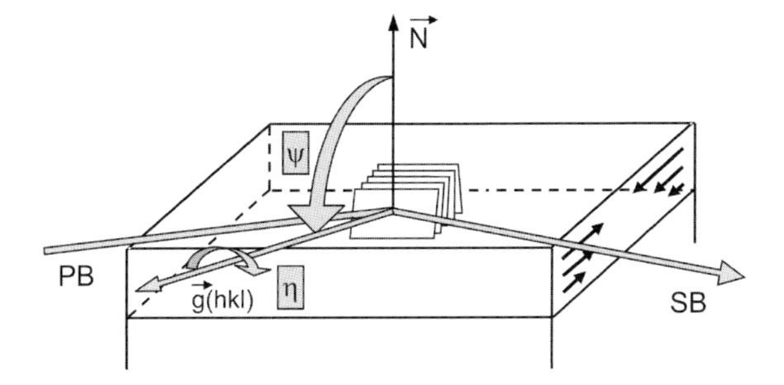


**Figure 10.1.1** Diffraction geometry in X-ray stress analysis. PB and SB are the incident (primary) and the diffracted (secondary) beam, respectively. $\psi$ denotes the inclination between the surface normal **N** and the diffraction vector **q**($hkl$), $\eta$ describes the sample rotation around **q**($hkl$) [5]. The azimuth angle $\phi$ which defines the inplane orientation of the measuring direction was omitted in the figure, because it is of no relevance for the variation of the X-ray penetration depth.

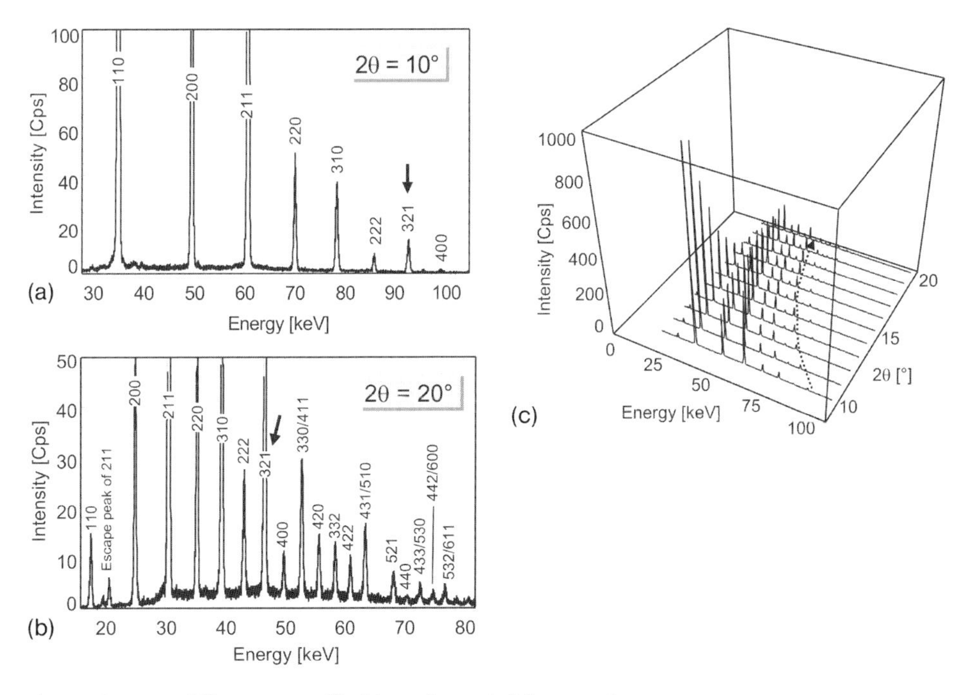


**Figure 10.1.2** ED diffractograms of ferritic steel recorded for scattering angles $2\theta$ of 10° (a) and 20° (b), respectively (HMI materials science beamline EDDI at BESSY). The total counting time for each ED spectrum was 60 s. (c) Stack plot of diffracted spectra which were measured in a $2\theta$-range between 10° and 20° with $\Delta 2\theta = 1°$. The dotted line marks the shift of the 321 reflection line on the energy scale.

The dependence of $\tau$ on the scattering angle $2\theta$ and on the inclination angle $\psi$ according to Eq. (10.1.3) is demonstrated in Figure 10.1.3 by the example of ferritic steel. The maxima in Figure (10.1.3a) result from two effects which counteract each other. Setting $\eta = 90°$ ($\Psi$-mode of the XSA) and $\psi = 0°$ in Eq. (10.1.3) yields $\tau = \sin\theta/[2\mu(E)]$. An increase of $\theta$ in the numerator of this expression increases $\tau$ due to the steeper incidence (exit) of the primary (diffracted) beam with respect to the sample surface. This "geometrically induced" shift of $\tau$ is superimposed by some "energy induced" shift due to the increase of $\mu$ with decreasing energy (i.e., increasing $\theta$, cf. Figure 10.1.2).

Besides the high penetration depths achieved in ED diffraction experiments, the multitude of reflections recorded in one energy spectrum is an important additional parameter that can be used in the depth-resolved XSA. So Figure (10.1.3b) shows that depending on its energy for each reflection *hkl* another $\tau$-range is covered in a $\sin^2\psi$ experiment.

If the residual stress/strain state depends on the investigated depth $z$ below the surface, the basic equation of XSA takes the form

$$\varepsilon_{\phi\psi}(hkl, z) = F_{ij}(hkl, \phi, \psi)\sigma_{ij}(z) \quad (10.1.4)$$

with the stress factors $F_{ij}$ acting as proportional factors between the measured strain and the individual stress components defined in the sample reference sys-

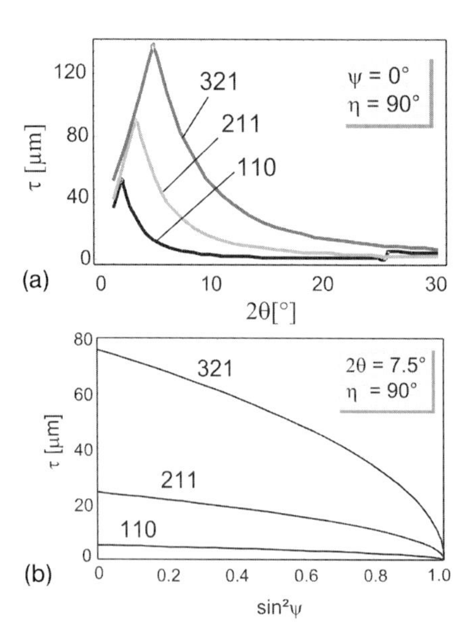


**Figure 10.1.3** The dependence of the penetration depth $\tau$ on the scattering angle $2\theta$ (a) and the inclination angle $\psi$ (b), respectively, calculated for a ferritic steel according to Eq. (10.1.3) for $\eta = 90°$, which corresponds to the $\Psi$ -mode of the XSA [5].

tem [2]. Taking into account Beer's exponential attenuation law, one has to distinguish between the actual ("real space") $z$-depth profiles on the one hand and the exponentially weighted ("Laplace space") $\tau$-depth profiles, which are obtained experimentally, on the other hand. They are correlated by the equations

$$\varepsilon_{\phi\psi}(hkl,\tau) = \frac{\int \varepsilon_{\phi\psi}(hkl,z)\mathrm{e}^{-z/\tau}\,\mathrm{d}z}{\int \mathrm{e}^{-z/\tau}\,\mathrm{d}z} \qquad \sigma_{ij}(\tau) = \frac{\int \sigma_{ij}(z)\mathrm{e}^{-z/\tau}\,\mathrm{d}z}{\int \mathrm{e}^{-z/\tau}\,\mathrm{d}z} \tag{10.1.5}$$

which have the form of a Laplace transform with respect to $1/\tau$ [2]. In Section 10.1.4 it will be shown by practical examples how the residual stress depth profiles obtained for the individual reflections $hkl$ can be joined in a so-called Universal or Master plot [6].

## 10.1.3 Experimental Set-up

Figure 10.1.4 shows the set-up of a white beam synchrotron experiment for ED diffraction. The high-energy X-ray beam is provided by a multipole wiggler, which emits photons in a horizontal fan of some degrees but highly collimated in the vertical plane. The emitted energy spectrum mainly depends on the electron storage ring parameters and on the magnetic field of the wiggler. The optical elements of the beamline consist of an absorber mask to predefine the primary beam cross-section, a filter system to attenuate low energy photons giving rise to sample heating due to absorption and a cross-slit system to define the final beam shape. The primary beam meets the sample, which is mounted on a four-circle diffractometer that enables us to realize any diffraction geometry applied in stress and texture analysis.

Because synchrotron radiation is linearly polarized in the horizontal plane (i.e., in the storage ring plane), synchrotron diffraction experiments in most cases are performed in the vertical plane. In the ED mode using high-energy photons rather small scattering angles $2\theta$ between about 4° and 20° are chosen,

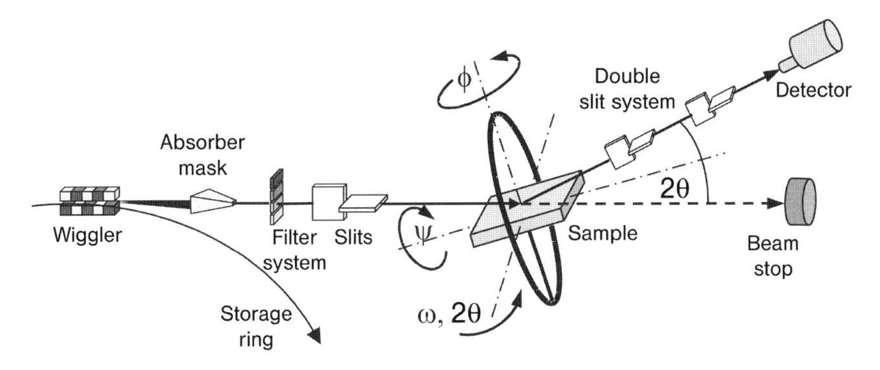


**Figure 10.1.4** Set-up of a white beam synchrotron beamline for ED diffraction.

because higher angles $2\theta$ would shift and compress the diffracted energy spectrum toward smaller energies (see Eq. (10.1.1) and Figure 10.1.2) and therefore reduce the penetration depth $\tau$ in the case of reflection mode experiments (cf. Figure 10.1.3). Furthermore, the rather low experimental resolution $\Delta E/E$ in ED diffraction would lead to diffraction line overlapping and thus prevent a separate evaluation of individual reflections $E(hkl)$. Energy broadening $\Gamma_G$ in ED diffraction depends on two factors [7]:

$$\Gamma_G(E) = [(\Gamma_0^2 + 5.55F\varepsilon E) + (E \cot \theta \Delta\theta)^2]^{1/2} \tag{10.1.6}$$

The first term in Eq. (10.1.6) denotes the intrinsic detector resolution – $\Gamma_0$: noise (about 100 eV), $F$: Fano factor (0.1–0.13), $\varepsilon$: energy for creating an $n^-/n^+$ pair (3 eV for germanium). For Ge-solid state detectors used in high-energy ED diffraction the intrinsic resolution is about 160 eV for $E = 10$ keV and about 420 eV for $E = 100$ keV. Consequently, the resolution $\Delta E/E$ resulting from the detector electronics lies between about $5 \times 10^{-3}$ and $1 \times 10^{-2}$, which is rather poor on the absolute scale. Relative shifts $\Delta E(hkl)$ of the diffraction lines which are the basis of strain analysis (cf. Eq. 10.1.2), however, do not depend on this absolute intrinsic energy broadening and can be determined with high accuracy.

The second term in Eq. (10.1.6) describes the energy broadening resulting from the divergence $\Delta 2\theta$ in the diffracted beam, which is due to the finite aperture of the double slit system (Figure 10.1.5a). Because a sample tilt $\psi$ toward larger inclination angles (i.e., grazing diffraction conditions) leads to a significant increase of the irradiated sample surface area, the geometrical situation concerning the divergence in the diffracted beam also varies with $\psi$. Consequently, a systematic shift $\Delta E(hkl)$ of the diffraction lines as a function of $\psi$ is observed even for stress-free material. The effect can be limited to maximum shifts of $\Delta E(hkl) < 20$ eV if the equatorial (i.e., the vertical) slit aperture is kept small (Figure 10.1.5b).

In any case, however, careful calibration of each residual stress analysis by means of measurements performed at stress-free reference samples under exactly the same experimental conditions is necessary to avoid systematic errors. Finally, it should be noted that the stability of the energy line positions, which means their insensitivity with respect to a variation of the incoming counting rate, strongly depends on the correct set-up of the electronic detector parameters.

### 10.1.4
### Example for Depth-Resolved Residual Stress Analysis by ED Diffraction

The multitude of reflections *hkl* recorded simultaneously in one ED diffractogram offers new and exciting possibilities for a depth-resolved analysis of the macroresidual stress fields in the near surface region of polycrystalline materials and technical parts. A practical example is shown in Figures 10.1.6 to 10.1.8. $\sin^2\psi$ measurements were performed along and perpendicular to the grinding direction of a uniaxial ground plate made of the nickel-based superalloy IN 718.

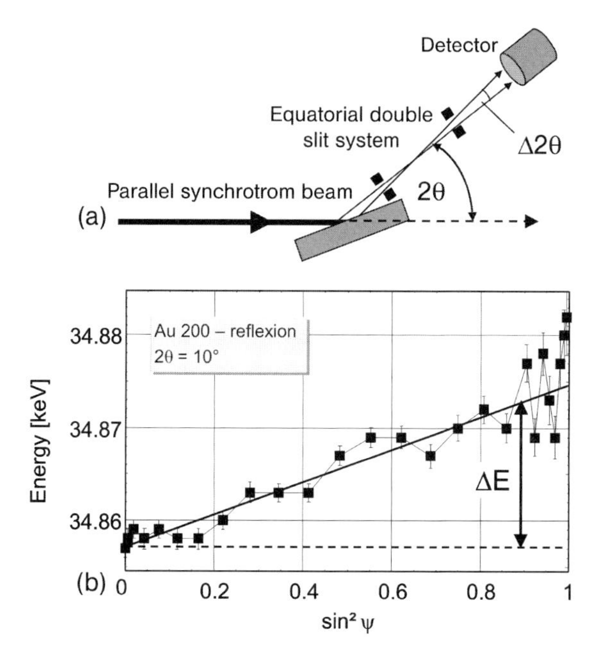


**Figure 10.1.5** (a) The equatorial divergence in the diffracted beam as the reason for geometrical energy broadening in ED diffraction performed in reflection geometry. (b) Geometrically induced energy shift of the Au-200 diffraction line in a $\sin^2 \psi$ experiment. The slope would correspond to a "ghost" stress of about 50 MPa. The angular divergence $\Delta 2\theta$ in the diffracted beam was 0.006° (equatorial aperture of the secondary slit system 20 μm, distance between the slits 350 mm).

The $2\theta$ angle of 10° ensures that about 10 evaluable reflections *hkl* lie within the energy range between about 30 keV and 100 keV (Figure 10.1.6a).

Because each diffraction line $E(hkl)$ is related to another linear absorption coefficient $\mu(E)$, the residual stress values obtained from the individual $d(hkl)$ versus $\sin^2 \psi$ distributions by means of the $\sin^2 \psi$ method have to be assigned to different average penetration depths $\langle \tau(hkl) \rangle$ which may be defined by

$$\langle \tau(hkl) \rangle = \tfrac{1}{2} [\tau(hkl)_{\min} + \tau(hkl)_{\max}] \tag{10.1.7}$$

where $\tau(hkl)_{\min}$ and $\tau(hkl)_{\max}$ correspond to $\psi_{\max}$ and $\psi_{\min} = 0°$, respectively (cf. Eq. 10.1.3). Plotting the residual stresses obtained this way for the individual reflections *hkl* against the average penetration depth $\langle \tau(hkl) \rangle$ gives a first idea of the residual stress depth distribution (Figure 10.1.7). Therefore, this method may be regarded as a modification of the so-called multiwavelength method [8] used in AD diffraction to the ED mode.

However, if steep residual stress gradients are present within the penetration depth of the X-ray beam, the $d(hkl)$ versus $\sin^2 \psi$ distributions are not linear but

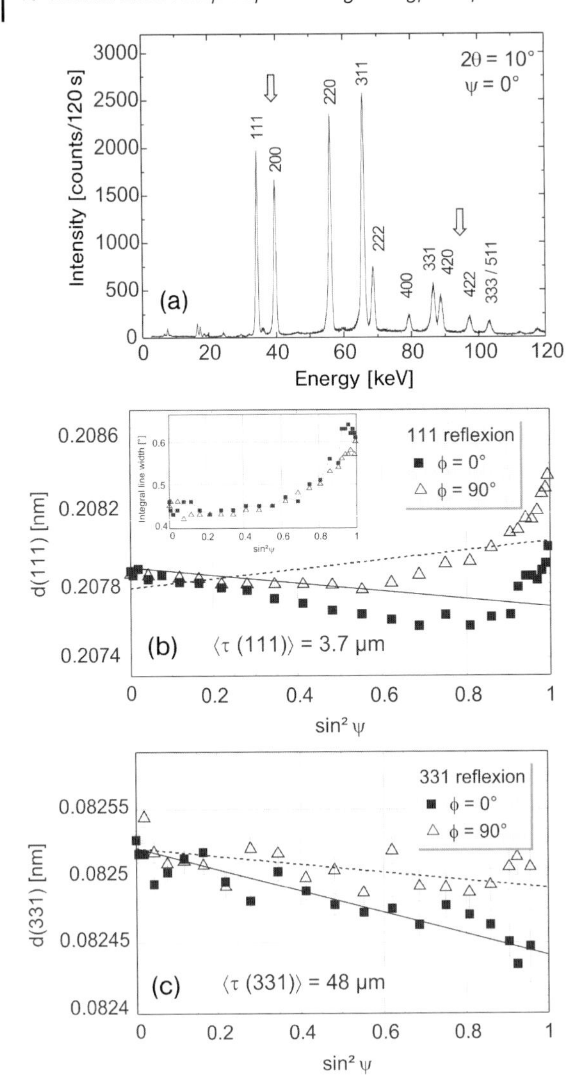


**Figure 10.1.6** (a) ED diffractogram of a uniaxial ground Inconel IN 718 plate obtained for $2\theta = 10^\circ$ (HMI materials science beamline EDDI at BESSY). (b, c) $d(hkl)$ versus $\sin^2 \psi$ plots for the 1 1 1 and 3 3 1 reflections (see marks in the diffractogram above), $\phi = 0^\circ$ and $90^\circ$ correspond to the grinding and the transverse direction, respectively. $\langle \tau(hkl) \rangle$ is the average penetration depth for the reflection *hkl* according to Eq. (10.1.7). The inset in (b) shows the distribution of the integral line width.

show a more or less strong curvature (see Figure 10.1.6b). If (as in the present case) other possible reasons giving rise to nonlinear $d(hkl)$ versus $\sin^2 \psi$ distributions, like texture and/or effects related to the plastic material anisotropy, may be excluded from the considerations, use can be made of more sophisticated

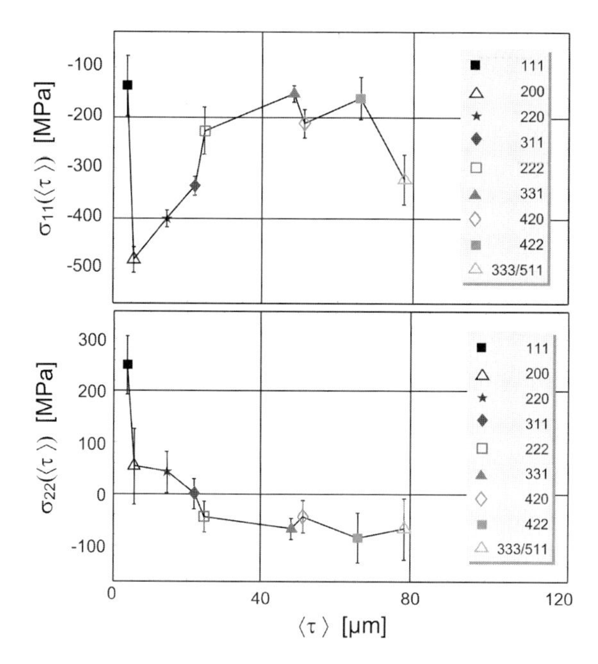


**Figure 10.1.7** Residual stress depth profiles $\sigma_{11}(\langle\tau\rangle)$ (grinding direction) and $\sigma_{22}(\langle\tau\rangle)$ (transverse direction) in the near surface region of a uniaxial ground Inconel IN 718 plate. The residual stresses were calculated for the individual reflections *hkl* from the slopes of *d*(*hkl*) versus $\sin^2\psi$ plots (cf. Figure 10.1.6) using the diffraction elastic constants (DEC) evaluated by the model of Eshelby/Kröner (see, for example, [2]).

methods to evaluate the near surface residual stress gradients. For a biaxial in-plane residual stress state discrete depth distributions $\sigma_{11}(\tau)$ and $\sigma_{22}(\tau)$ may be obtained directly from the $\sin^2\psi$ data by means of the following formalism:

$$\sigma_{11}(\tau) = \tfrac{1}{2}[f^+(\tau) + f^-(\tau)] \tag{10.1.8a}$$

$$\sigma_{22}(\tau) = \tfrac{1}{2}[f^+(\tau) - f^-(\tau)] \tag{10.1.8b}$$

$$f^+(\tau) = \frac{\varepsilon_{0\psi}(hkl,\tau) + \varepsilon_{90\psi}(hkl,\tau)}{\tfrac{1}{2}s_2(hkl)\sin^2\psi + 2s_1(hkl)} \tag{10.1.8c}$$

$$f^-(\tau) = \frac{\varepsilon_{0\psi}(hkl,\tau) + \varepsilon_{90\psi}(hkl,\tau)}{\tfrac{1}{2}s_2(hkl)\sin^2\psi} \tag{10.1.8d}$$

In (10.1.8c) and (10.1.8d) $s_1(hkl)$ and $1/2s_2(hkl)$ are the diffraction elastic constants (DEC) used in X-ray stress analysis of materials with random texture. Furthermore, $\varepsilon_{0\psi}(hkl,\tau)$ and $\varepsilon_{90\psi}(hkl,\tau)$ denote the lattice strains depth profiles obtained in the azimuths $\phi = 0^\circ$ and $90^\circ$, respectively. For a biaxial residual stress state with $\sigma_{i3} \equiv 0$ $(i = 1, 2, 3)$, the right-hand side of Eqs. (10.1.8a) and (10.1.8b) contain the pure experimental information, whereas the unknown inplane stresses stand alone on the left-hand side. Therefore, Eqs. (10.1.8a) and (10.1.8b) are of "universal" nature, and all experimental data, independently of the radiation and/or reflections $E(hkl)$ used, can be plotted versus the corresponding penetration depth $\tau$ into one "Universal" or "Master" plot [6]. For a detailed discussion of the "Universal-plot" method and its application to the ED case of diffraction the reader is referred to [5, 9].

A comparison of the results obtained by applying the "Universal-plot method" (Figure 10.1.8) with those achieved by the modified "multiwavelength method" shown in Figure 10.1.7 reveals a good qualitative agreement of the corresponding depth profiles for $\sigma_{11}$ and $\sigma_{22}$. The better the agreement becomes the smaller the gradients are, which in the present example applies for rather large penetration depths $\tau$. In this case the $d(hkl)$ versus $\sin^2\psi$ distributions remain nearly linear (see the 3 1 1 reflection in Figure 10.1.6c) and, thus, allow for the application of the $\sin^2\psi$ method. The steep residual stress gradients close to the surface, on the other hand, lead to strongly nonlinear $\sin^2\psi$ plots (see the 1 1 1 reflection in Figure 10.1.6b). Consequently, the assumptions underlying the $\sin^2\psi$ method are here not fulfilled, which leads to significant uncertainties in stress analysis. In those cases, the "Universal-plot" method yields more reliable results.

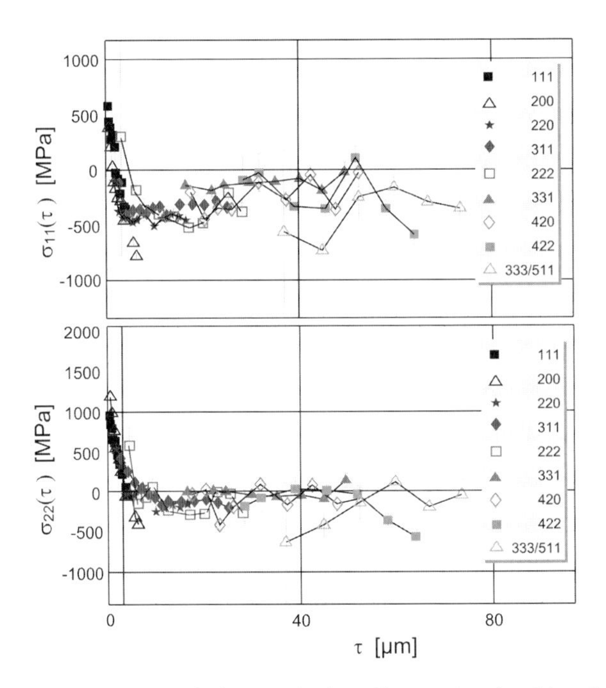


**Figure 10.1.8** Residual stress depth profiles $\sigma_{11}(\tau)$ and $\sigma_{22}(\tau)$ evaluated from the $\sin^2\psi$ measurements (cf. Figure 10.1.6) by means of the Universal-plot method introduced in [6]. See the text for further details.

The example shown in Figures 10.1.6 to 10.1.8 illustrates that the uniaxial grinding process obviously results in very steep residual stress gradients in both, the grinding and the transverse direction. The high tensile stresses at the direct surface decrease rapidly within a very thin surface layer and are compensated by compressive residual stresses in deeper material zones. Furthermore, it is clearly to be seen that the in-plane residual stress state induced by the unidirectional mechanical surface treatment is not of rotational symmetry. Within the first 20–30 μm below the surface the complete residual stress state $\sigma_{22}(\tau)$ in the transverse direction is shifted by an amount of about 300–500 MPa toward higher tensile stresses, whereas the gradient concerning its form and steepness is similar in both directions. The larger scattering of the "Universal"-plot data in Figure 10.1.8 with increasing depth is due to the rather weak intensities of the high-energy diffraction lines (cf. Figure 10.1.6a), which leads to increasing uncertainties in stress evaluation using the formalism in Eqs. (10.1.8a) to (10.1.8d).

Finally, it should be mentioned that valuable additional information on the microstructure is available from the integral width $\beta^{\text{int}}$ of the diffraction lines. From line profile analysis it is well known that microstrain caused by lattice defects like dislocations and stacking faults leads to line profile broadening [10]. Taking $\beta^{\text{int}}$ as a characteristic measure for describing the lattice imperfections due to the plastic deformation induced by the grinding process, one realizes from the inset in Figure 10.1.6b that the line width increases significantly with $\sin^2\psi$, i.e., decreasing depth. Thus, the conclusion may be drawn that the highest defect incorporation occurs at the direct surface whereas deeper zones are less influenced by the surface treatment process. It should be noted, however, that a full quantitative line profile analysis requires additional measurements at reference samples being free of structural line broadening effects in order to separate instrumental broadening (in the case of ED diffraction experiments especially due to the energy resolution of the detector, see Eq. 10.1.6) from the pure physical contributions coming from the investigated sample.

### 10.1.5 Concluding Remarks and Prospects

Compared with the long tradition of AD methods in XSA, the application of ED diffraction methods in XSA is strongly associated with the availability of powerful synchrotron sources of the third generation and therefore, came up no more than some years ago. The most important difference between AD and ED diffraction consists in mode of data acquisition. In the latter case the positions of the sample and the detector remain fixed during the measurement and complete diffraction spectra with a multitude of diffraction lines are recorded simultaneously within very short times of some seconds up to a few minutes. Apart from this difference the same experimental methods as used in AD-XSA can be applied to ED diffraction mode. However, the higher information content in the form of complete diffraction spectra instead of single diffraction lines allows for decisive advancements of the stress evaluation procedures. With respect to XSA performed in the reflection mode this especially concerns new possibilities for strain/stress depth profiling. In this section it was shown by the example of a "conventional" $\sin^2\psi$ measurement, how the plenty of information included in the ED diffraction spectra can be used to evaluate near surface residual stress depth profiles on different levels of approximation.

There are still other possible methods that can be applied for strain depth profiling in ED diffraction with white synchrotron radiation. From Figures 10.1.2 and 10.1.3, for example, it follows immediately that a variation of the diffraction angle $\theta$, which here is a free parameter, can be used to adjust the penetration depth $\tau$ and, therefore, to keep it constant even for different inclination angles $\psi$. That way it becomes possible to perform residual stress analysis in predefined depths below the sample surface [5]. But, as also shown in [5], even more sophisticated methods for a triaxial analysis of near surface residual stress fields can be adapted and expanded to the ED case of diffraction, such as the "scattering vector method," where depth profiling is done by a stepwise sample rotation $\eta$ around the diffraction vector $\mathbf{q}(hkl)$ (see Figure 10.1.1).

The XSA methods outlined in this section yield the near surface residual stress fields as a function of the penetration depth $\tau$, which according to Eq. (10.1.5) are called the "Laplace stresses" $\sigma(\tau)$. The calculation of the actual "real space stresses" $\sigma(z)$ by means of the inverse numerical Laplace transform, however, is an extremely ill-conditioned mathematical problem, which has not been solved satisfactorily so far. An "experimental solution" of this problem using ED diffraction in the reflection mode is topic of current research work [10]. Defining a very small strain gauge having a height of $<10\ \mu m$ by slits in the primary and the diffracted beam, respectively, high resolution strain/stress scanning becomes possible even in the real or $z$-space and, therefore, allows for a comparison of the $\sigma(\tau)$ and $\sigma(z)$ residual stress profiles [11]. Furthermore, use can be made of the high information content contained in the ED diffraction spectra in order to study effects related to the elastic as well as plastic material anisotropy.

## References

1 B. Scholtes, Eigenspannungen in mechanisch randschichtverformten Werkstoffen. Ursachen, Ermittlung und Bewertung, DGM Informationsgesellschaft, Oberursel, 1991.
2 V. Hauk, *Structural and Residual Stress Analysis by Nondestructive Methods*, Elsevier, Amsterdam, 1997.
3 L. Spieß, R. Schwarzer, H. Behnken, G. Teichert, *Moderne Röntgenbeugung*, Teubner, Wiesbaden, 2005.
4 Ch. Genzel, Formalism for the evaluation of strongly non-linear surface stress fields by X-ray diffraction in the scattering vector mode. *Phys. Stat. Sol. (a)* 1994, **146**, 629–637.
5 Ch. Genzel, C. Stock, W. Reimers, Application of energy-dispersive diffraction to the analysis of multiaxial residual stress fields in the intermediate zone between surface and volume. *Mater. Sci. Eng. A* 2004, **372**, 28–43.
6 H. Ruppersberg, I. Detemple, J. Krier, Evaluation of strongly non-linear surface-stress fields $\sigma_{xx}(z)$ and $\sigma_{yy}(z)$ from diffraction experiments. *Phys. Stat. Sol. (a)* 1989, **116**, 681–687.

7 D. E. Cox, in *Handbook on Synchrotron Radiation*, Vol. 3, Chap. 5 (Eds.: G. Brown, D. E. Moncton), Elsevier, Amsterdam, 1991.
8 B. Eigenmann, B. Scholtes, E. Macherauch, Eine Mehrwellenlängenmethode zur röntgenographischen Analyse oberflächennaher Eigenspannungszustände in Keramiken. *Mat.-wiss. u. Werkstofftech.* 1990, **21**, 257–265.
9 C. Stock, Analyse mehrachsiger Eigenspannungsverteilungen im intermediären Werkstoffbereich zwischen Oberfläche und Volumen mittels energiedispersiver Röntgenbeugung. Dissertation, TU Berlin, 2003.
10 E. J. Mittemeijer, P. Scardi (Eds.), *Diffraction Analysis of the Microstructure of Materials*, Springer Series in Materials Science, Vol. 68, Springer, New York, 2004.
11 I. Denks, M. Klaus, Ch. Genzel, Determination of real space residual stress distributions $\sigma_{ij}(z)$ of surface treated materials with diffraction methods: Part II. Energy dispersive approach. *Mater. Sci. Forum*, 2006, **524–525**, 37–42.

## 10.2
## Transmission Mode

*Anke Rita Pyzalla*

### 10.2.1
### Motivation

White high-energy synchrotron radiation in transmission mode – like experiments in reflection mode (see Section 10.1) – gives access to the whole energy range and, thus, can be advantageous for a number of experiments [1, 2]:

- The high energies available in a white beam allow for large penetration depths, thus, enabling the determination of phase volume fractions, texture, and/or residual stresses in the bulk of materials and small components. In addition, the white high-energy synchrotron radiation beam passes through sample environments, thus, enabling measurements in samples mounted, e.g., in vacuum furnaces. Even in steel samples penetration depths in the order of several mm can be reached using a white high-energy beam (medium energies only give access to the intermediate zone between surface and volume, see Section 10.1).
- Simultaneous information about phases, texture, and (residual) stresses can be obtained from each diffractogram.
- The availability of a multitude of reflections for residual stress analyses enables the determination of intergranular strains/stresses in samples subjected to plastic deformation, respectively, a more precise estimation of the uncertainty of the strain values in samples without marked intergranular strains.

- A white beam produces a photon flux at the sample position (under equal geometrical conditions), which is several times higher than when using a monochromatic beam. Therefore, a white beam is often advantageous for the observation of time-dependent processes, such as stress relaxation.

## 10.2.2
## Experiment Set-up and Experimental Details

### 10.2.2.1 Penetration Depth

One measure characterizing the penetration depth is the 50% transmission thickness, which is specified in Table 10.2.1 [3] for thermal neutrons in the case of the four elements Fe, Ni, Ti, and Al, which represent the majority components of commonly used engineering alloys.

In contrast to neutrons, the penetration depth of X-rays depends on the atomic number as well as on the radiation energy (see Figure 10.2.1 and Chapter 4).

**Table 10.2.1** 50% transmission thickness of Fe, Ni, Ti, and Al for thermal neutrons.

| | Fe | Ni | Ti | Al |
|---|---|---|---|---|
| 50% transmission thickness (cm) | 0.58 | 0.39 | 1.39 | 6.93 |

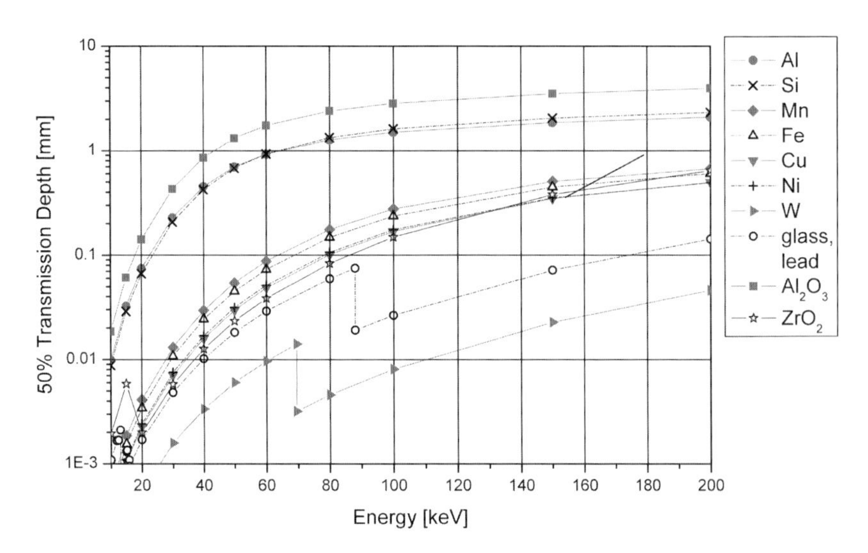


**Figure 10.2.1** Dependence of the 50% transmission depth on the synchrotron radiation energy.

The absorption edges of metals used as matrices or constituents in structural materials and the most engineering ceramics are at energies below 10 keV (Figure 10.2.1). At low energies of several keV, corresponding to low-energy X-ray tubes, the 50% transmission thickness is in the order of only a few micrometers. Thus, most X-ray residual stress analyses are performed in the near-surface zone. Due to the decrease of the attenuation coefficient with increasing radiation energy [4], the 50% transmission thickness for high energies increases up to about 2 mm at 100 keV in Fe and Ni and even to about 15 mm in Al. In addition, the possible sample thickness exceeds the 50% transmission thickness significantly due to the high photon flux available. Thus, Pyzalla et al. [5] determined, e.g., strains in 35-mm-thick Al-MMC hot extruded profiles and, recently, Croft et al. [6] and Steuwer et al. [7] analyzed the strain state inside several mm thick steel specimens.

For W, however, and similar heavy elements the penetration depth at high energies remains less than about 1 mm. Thus, some materials, such as cermets (WC-Co- and WC-Ni-composites), even in the case of a high energy and flux of the beam, are not very suitable for residual stress analysis in the bulk using high-energy synchrotron radiation, but these materials are used, e.g., as beam stoppers and slit edges.

#### 10.2.2.2 **Gauge Volume and Spatial Resolution**

Early attempts at using high-energy X-rays in transmission geometry were performed in the laboratory [8] using high-voltage X-ray tubes commercially employed in industrial radiography, e.g., in the nondestructive testing of welds. Due to the low photon flux emitted by these X-ray tubes, gauge volume sizes had to be in the order of several 10 $mm^3$ and often exceeded the size of the samples. The transfer of the method to synchrotron radiation sources due to the high photon flux available today enables measurements with a local resolution several orders of magnitudes higher.

Even using synchrotron radiation reflection intensities is only sufficient for small $2\theta$ angles (usually $2\theta < 15°$) [9],

$$I_{hkl} = CN^2 \lfloor Li_o(E) E^{-2} J |F|^2 A(E, \theta_o, \alpha, \beta) \, p(\alpha, \beta) \rfloor_{hkl} \tag{10.2.1}$$

$C$ is a constant that depends on the scattering geometry:

$$C = (hc)^3 r_e^2 \frac{D}{r} V \frac{1 + \cos^2 2\theta_o}{32\pi \sin^3 \theta_o} \Delta\theta_o \tag{10.2.2}$$

| | |
|---|---|
| $I_{hkl}$ | integrated intensity of a reflection *hkl* |
| $r_e = 2.82 \times 10^{-13}$ cm | classical electron radius |
| $N$ | number of crystallites per unit volume |
| $J$ | multiplicity factor |
| $i_o(\lambda)$ | intensity per unit wavelength range of the incident beam |
| $\lvert F \rvert$ | denotes the structure amplitude |

| | |
|---|---|
| $\Delta\theta_o$ | divergence of the X-ray beam |
| $A(E, \theta_o, \alpha, \beta)$ | absorption factor |
| $p(\alpha, \beta)$ | orientation distribution function, $\alpha$, $\beta$: azimuth res. rotation angle |
| $D$ | height of detector window |
| $r$ | distance sample–detector |

Due to the small $2\theta$ angles necessary strains can only be measured in those directions (i.e., directions of the scattering vector $\vec{q}$), which are almost normal to the incident beam.

As a further consequence of the small $2\theta$ angles the gauge volume shape (Figure 10.2.2) is a parallelepiped also called lozenge [10]. The width-to-length ratio of the gauge volume depends strongly on the $2\theta$ angle chosen. It can be very roughly estimated by

$$\frac{L_V}{B_V} \cong \frac{1}{\tan\theta} \tag{10.2.3}$$

For more sophisticated approaches at determining the gauge volume size refer to Brusch [8]. Equation (10.2.3) reveals that gauge volume length strongly increases with decreasing $2\theta$ and increasing the width of the incoming respectively the diffracted beam. As a consequence of the elongated gauge volume shape the spatial resolution of strain measurements in the direction near the incident beam is degraded strongly (for a $2\theta$ angle of 10.7° about 10-fold [10]) in comparison to the resolution in the perpendicular plane.

In order to overcome the limitation in spatial resolution in the direction of the synchrotron beam, mathematical models and experimental procedures were developed by Brusch [8] and recently by Korsunsky and co-workers [10]. On the other hand, the lozenge shape of the gauge volume is of advantage when determining 2D residual stress states, e.g., in multilayer systems or interfaces (if plane stress conditions can be assumed).

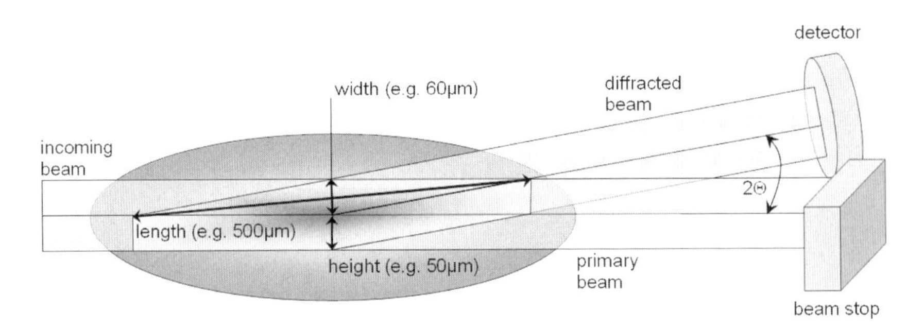


**Figure 10.2.2** Scheme of the geometry of the incident and the diffracted beams and the gauge volume.

At synchrotron sources the spatial resolution of strain/stress analyses in technical materials does not appear to be limited by the decrease of photon flux introduced by small slits, but, rather from coarse grain effects, which mean that there are not enough crystallites within the gauge volume to permit obtaining statistically relevant reflection line profiles.

As a consequence of the necessity to perform the diffraction experiment at low $2\theta$ angles the beam path in the case of measurements in circular or plate-like components can become very large in at least one direction. A solution might be found, e.g., by a slight inclination of the sample perpendicular to the scattering plane.

#### 10.2.2.3 **Example for an Experiment Set-up**

Energy-dispersive diffraction at synchrotron sources usually is performed at beamlines with a Wiggler insertion device, such as, e.g., the beamline ID15A at ESRF, Grenoble, France [11] or the HARWI-beamline of GKSS at DESY, Hamburg, Germany [12].

The main components of a typical experiment set-up for a white beam stress measurement in transmission geometry consist of slits in the incoming beam, the sample mounted on a translation/rotation state, slits in the diffracted beam and the detector (Figure 10.2.3).

Slits, often made of WC blocks, are used to decrease the radiation background and define the width of the incoming and the diffracted beam and, thus, the gauge volume. In those cases where a sharp definition of the volume element shape is required, the first slit in the diffracted beam should be positioned as near as possible to the sample although due to the small divergence of the synchrotron beam this is not as crucial as it has been found for neutron diffraction. A second slit in the diffracted beam is positioned close to the detector entrance window.

Since strains and stresses are not uniform, but present gradients across samples and components, sample positioning is crucial in strain/stress measure-

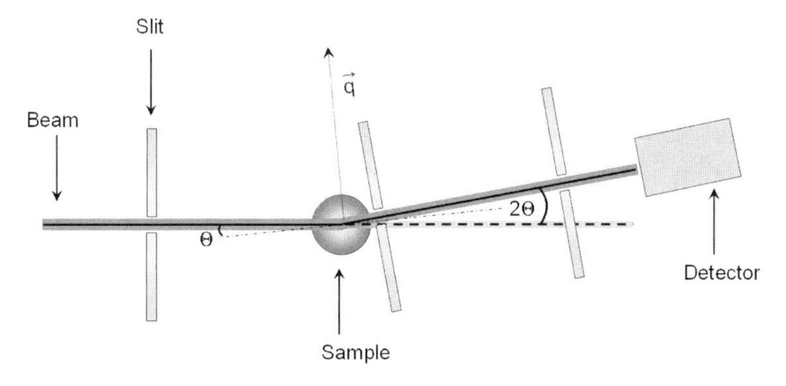


**Figure 10.2.3** Typical experiment set-up for a white beam strain/stress measurement in transmission geometry.

ments. The positions at the surfaces of a sample or component with respect to the incident beam can be determined by moving the sample perpendicular to the incident beam and recording the total diffracted intensity (Figure 10.2.4) [13].

Depending on sample shape, size, and necessary sample movements during the experiment it may be convenient to use an $x$–$y$–$z$-sample stage or a Eulerian cradle. A Eulerian cradle further allows turning the sample in the diffraction plane as well as perpendicular to the diffraction plane and, thus, is in particular useful for strain measurements in the $\sin^2 \psi$-mode (which yield the difference between the principal stresses in the case of a 3D stress state) and for strain measurements on strongly textured samples.

In order to reduce the heat load on the sample as well as the background noise aluminum absorbers often are used, which are placed near the beam entrance into the experimental hutch. Further improvement of the signal-to-noise ratio can be achieved by placing a beam stop in the primary beam directly after the sample and by guiding both the incoming and the diffracted beam, e.g., through brass tubes.

The detectors presently available for energy-dispersive strain/stress measurements at high energies are Ge-detectors. Multichannel analyzers (MCA) are used to bin the pulses from the amplifier. The detectors usually are calibrated by radioactive sources, which emit signals at well-defined energies. In a second step the detector calibration can be refined by using the diffraction pattern collected from standard powders [14, 15], further methods are outlined, e.g., in [16]. Constant cooling of the detectors using liquid nitrogen is necessary in order to avoid detector drifts (changes in the relation between energy and detector channel). In order

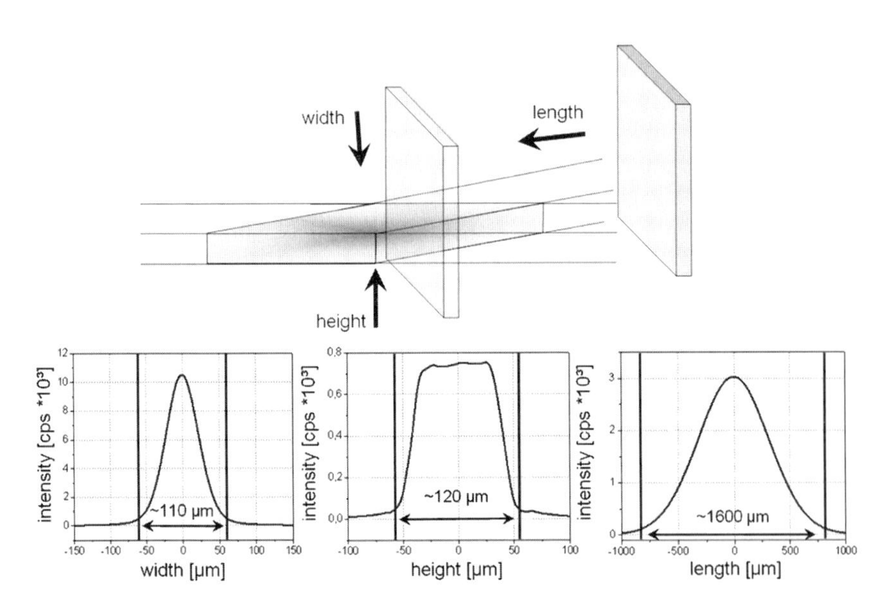


**Figure 10.2.4** Translation scan for determining the position of the sample.

to improve $d$-space coverage as well as for measuring strains in different directions simultaneously set-ups with several detectors are used increasingly often [17, 18].

Focusing optics so far have not been used in energy-dispersive strain/stress analyses, since it is not usually possible to focus a wide band of high-energy X-rays [16]. In the future periodic multilayers may offer some prospect [16, 19].

### 10.2.3
### Data Evaluation

Data evaluation in the case of energy-dispersive diffraction follows in general the same essential strategies available in angle-dispersive diffraction. These are either fitting of a suitable mathematical function to individual reflections or using a Rietveld-type approach [20] to fit the whole diffraction pattern at the same time, e.g., [7, 16, 21]. Due to the comparatively low energy resolution (650 eV at 122 keV [22]) of commercial Ge-semiconductor detectors the reflection profiles (Figure 10.2.5) usually are best described using a Gauss profile [16, 23–25]. Steuwer et al. [26] pointed out that in some cases asymmetries in the reflection profiles resulting from the detection process, beam divergence and slit settings may have to be taken into account.

Stress analyses in the case of materials with a low symmetry crystal structure or multiphase materials might suffer from the comparatively low resolution of commercial Ge-detectors, which does not allow the separation of reflections at high energies, where the diffractogram appears "compressed." While resolution in terms of being able to distinguish between neighboring reflections is worse in

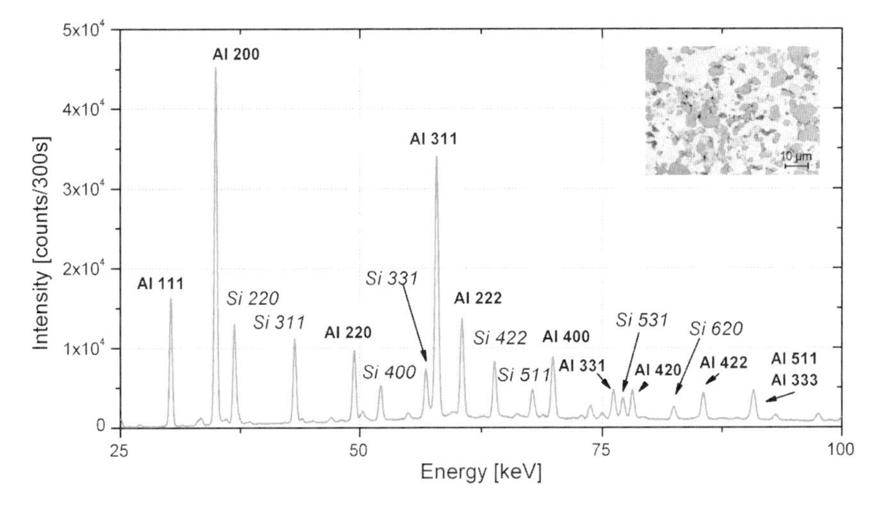


**Figure 10.2.5** Energy dispersive diffratogram of a PM Al–Si-alloy (AlSi25Cu4Mg1).

the case of energy-dispersive analyses than in the case of angle-dispersive stress measurements, the strain resolution of energy-dispersive stress analyses does not appear to be significantly inferior to angle-dispersive stress analyses [26]. Due to the strong effect of the energy-dispersive detection process on the reflections, reflection line profile analyses appear not to be feasible so far.

Due to the large number of reflections available simultaneously (in a similar way to time-of-flight neutron diffraction) energy-dispersive X-ray diffraction appears well suited for texture analyses [27–29], however, only very limited use has been made of these possibilities so far.

### 10.2.4 Examples

The most common applications for energy-dispersive X-ray diffraction in transmission are studies of material behavior and in particular of phase transformations under high pressure and under combined high pressure and temperature [16, 30]. Similar to these applications energy-dispersive strain/stress measurements are usually preferred to angle-dispersive diffraction in experiments needing fast data acquisition, that is, e.g., kinetic studies or strain scanning over large sample areas and in experiments, where the high penetration depth at high energies or the high photon flux in the white beam is crucial, e.g., [31–33].

*Strains in friction stir welds.* Friction stir welding (FSW) is a solid state joining process introduced in the early 1990s by TWI [34, 35]. This welding technique is increasingly commercially used [36–38] for joining of light metals and dissimilar materials. Welding of metal plates or profiles in friction stir welding is achieved by a rotating steel pin, which is rotated at speeds of several hundred rpm and advanced down the contact line between the plates respectively profiles. The rotating tool generates friction heat and shear strains, which provide a superplastic flow mechanism to stir the contiguous metals together into a weld zone on the trailing side of the advancing headpin (see Chapter 7.4) [39].

Compared to conventional fusion welds friction stir welds in general show a lower level of distortion, shrinkage, and defects (such as, e.g., cracks, porosity). Because of the low temperature levels compared to fusion welding techniques, it is generally believed that the residual stress level is lower in friction stir welds. However, the very rigid clamping arrangement exerts a much higher restraint on the deformation of the welded plates than the more compliant clamps used for fixing the parts during conventional welding processes. These restraints impede the contraction due to cooling of the weld seam and heat affected zone in longitudinal as well as in transverse direction, introducing significant transverse and longitudinal residual stresses [40].

While extreme values of residual stresses in welds usually appear on the top or the bottom side of butt welds, either in the weld seam center or the heat affected zone, it was believed that the complex material flow in friction stir welding might introduce a residual stress maximum in the bulk material. Thus, besides using laboratory X-rays residual stress distributions in friction stir welds have also been

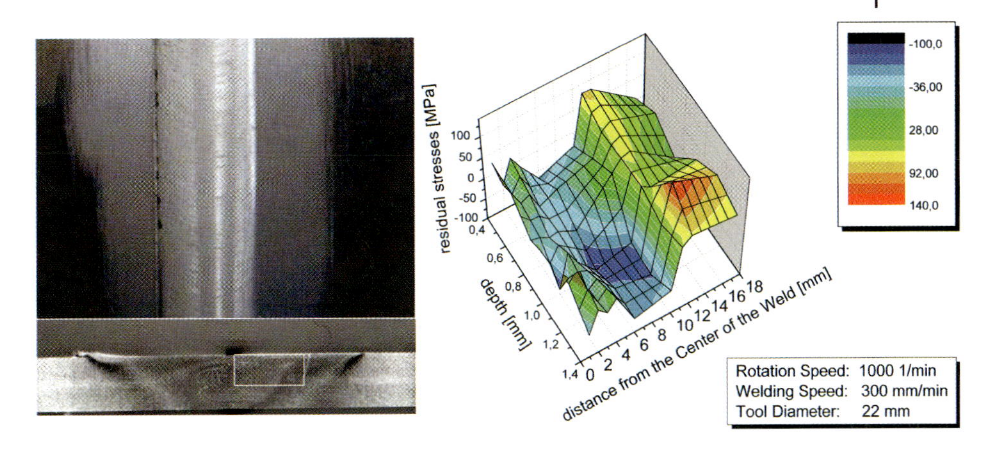


**Figure 10.2.6** Map of the residual stress values in the longitudinal direction of a friction stir weld (residual stress distribution corresponds to the area indicated by the white box in the right part of the figure).

extensively studied using synchrotron X-rays and neutrons, e.g., [41–44]. Using white high-energy synchrotron radiation a strain map of a part of the weld seam and the heat affected zone of an AA6013 T4 (Al alloy subjected to a special heat treatment) but weld could be determined (Figure 10.2.6). The strain map reveals that the longitudinal residual stress values determined in the bulk differ in some areas of the weld from those obtained in the sample near-surface zone.

From the strain map strain profiles at different distances to the weld surface could be extracted. As an example Figure 10.2.7 shows the longitudinal residual stress values obtained 100 μm below the surface of the friction stir weld. For comparison also the distribution of the longitudinal residual stress obtained by using a laboratory X-ray source and neutrons are shown. The comparison reveals that the residual stress distributions obtained by all three methods are even quantitatively very similar. The average values of the longitudinal residual stresses obtained by neutron diffraction indicate well the distance to the weld center of the maximum tensile residual stresses. However, due to the large gauge volume in neutron residual stress analyses the maximum residual stress values are underestimated. The residual stress curve obtained using laboratory X-rays and the curve obtained from the map produced by strain scanning with white high-energy synchrotron radiation about 100 μm below the sample surface agree well, both reveal tensile residual stress maxima of about 90 MPa in about 10 mm distance to the weld center line.

*Strain/stress relaxation processes in MMCs.* Metal matrix composites (MMCs) combine the properties of the ceramic reinforcement with the properties of the metal matrix and the interface between them. MMCs have a wide range of applications [45, 46], due to wear resistance, e.g., [47], due to high specific stiffness or due to an adjustable overall low thermal expansion coefficient, e.g., [48] and im-

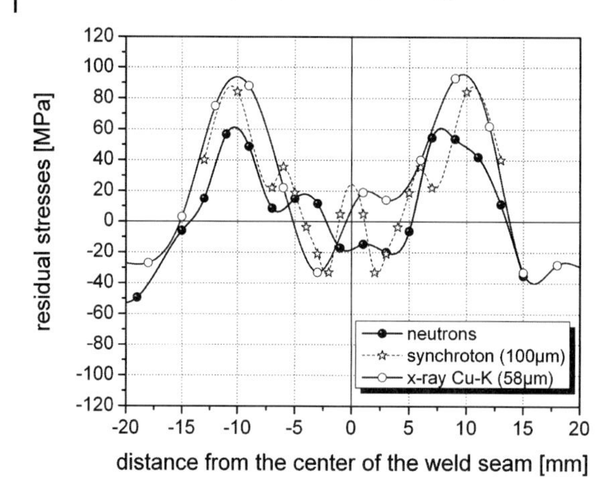


**Figure 10.2.7** Profile of the longitudinal residual stresses in the friction stir weld (see Figure 10.2.6), error about $\pm 20$ MPa.

proved resistance to thermal cycling [49–51]. During manufacturing and during loading of the MMC phase specific strains/stresses develop because of the misfit between the thermal and mechanical properties of matrix and reinforcement. At elevated temperatures the misfit induced phase specific strains/stresses experience significant relaxation, which affects the load bearing capacity and the damage behavior of the MMCs.

In order to quantify thermal stress relaxation of MMCs *in situ* tensile tests and relaxation tests at constant strain were performed using white high-energy synchrotron radiation. The energy-dispersive set-up provided a multitude of reflections simultaneously, thus, in AlSi25Cu4Mg1 (Figure 10.2.7) the elastic strain of the Al-alloy matrix [52] and the Si-particles (Figure 10.2.5) could be determined for several lattice planes in dependence on the total strain of the sample at temperatures up to 400 °C (Figure 10.2.8). The curves obtained reveal at small total strain of the sample that the Si-particles participate in load bearing, whereas at higher total strains of the sample the elastic strain of the Si-particles reaches a limit and does no longer increase with increasing total strain. This indicates strong plastic deformation of the Al-matrix, while the Si particles drift in the Al material flow. The slope of the curves does not show significant differences for the different lattice planes of the Si particles. This can be attributed to the low elastic anisotropy of Si (the anisotropy factor for Si is $a(\text{Si}) = -4$, for comparison $a(\gamma\text{-Fe}) = -17$, $a = c_{11} - c_{12} - 2c_{44}$) [53].

The relaxation of the elastic strain of the Al metal matrix and the Si particles was determined by deforming the sample at a given temperature to a defined total elongation. The macroscopic stress evolution versus time was measured using a load cell, the evolution of the elastic strain in different lattice directions versus time was determined by energy-dispersive diffraction. The results of the diffraction experiment revealed that the relaxation of the elastic strain in the Al metal

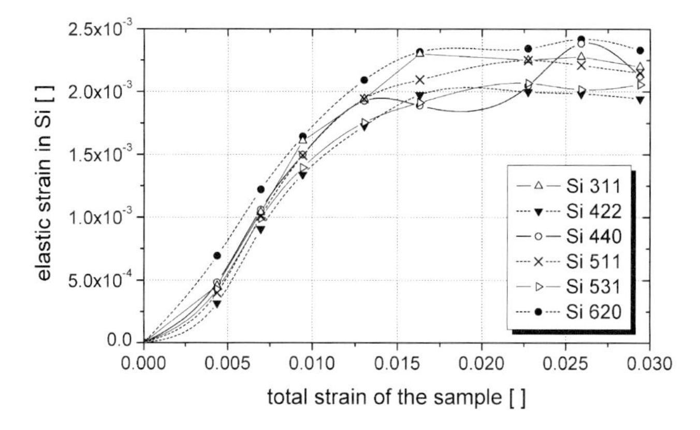


**Figure 10.2.8** Elastic strain of the Si particles versus sample total strain, AlSi25Cu4Mg1, $T = 400$ °C.

matrix does not depend significantly on the lattice plane but only on temperature. A time-dependent characteristic relaxation time and the activation energy for the relaxation of the phase specific strains/stresses could be determined. The data obtained at 400 °C for the 200 and the 400 reflection reveals impressively the speed and accuracy of the elastic strain measurements (Figure 10.2.9). Analyzing the data obtained for strain relaxation at 150 °C, 200 °C, 300 °C, and 400 °C characteristic relaxation times were obtained [52]. The activation energy for stress relaxation in the aluminum matrix of the Si particle containing AlSi25Cu4Mg1 alloy determined by using the approach outlined in [54] is $Q = 0.8 \pm 0.2$ eV.

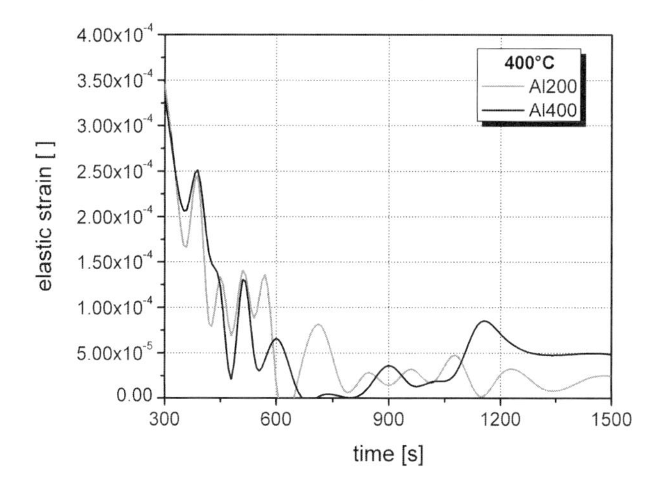


**Figure 10.2.9** Elastic strain relaxation in the Al-matrix of AlSi25Cu4Mg1, total strain = 1%, $T = 400$ °C.

## 10.2.5 Conclusions

The use of white high-energy X-rays for residual stress analyses enables simultaneous access to phase volume fractions, orientation distribution of the crystallites, and strains in different phases. The decrease of the absorption coefficient with radiation energy allows for large penetration depths and, thus, bulk measurements of the three-dimensional residual stress state.

Measurements need to be performed at small $2\theta$ angles, which results in a lozenge-shaped gauge volume. The significant elongation of the gauge volume in the direction of the beam limits the spatial resolution achievable.

Due to the high photon flux, the simultaneous recording of a multitude of reflections and the large penetration depth white high-energy synchrotron radiation is particularly advantageous for strain/stress measurements, e.g., in the case of strain mapping of samples and small regular shaped components, phase specific strain/stress determination in multiphase materials, determination of intergranular strains/stresses and kinetic processes, such as strain relaxation.

### Acknowledgments

Discussions with Professor Dr. W. Reimers and financial support by the DFG, Py9/1-2 is gratefully acknowledged.

### References

1 W. Reimers, M. Broda, G. Brusch, D. Dantz, K.-D. Liss, A. Pyzalla, T. Schmackers, T. Tschentscher, Evaluation of residual stresses in the bulk of materials by high energy synchrotron diffraction, *J. Nondestr. Eval.* 1998, **17** (3), 129–140.

2 W. Reimers, M. Broda, G. Brusch, D. Dantz, K.-D. Liss, A. Pyzalla, T. Schmackers, T. Tschentscher, Use of high-energy synchrotron diffraction for residual stress analyses, *J. Mater. Sci. Lett.* 1999, **18** (7), 581–583.

3 P. J. Webster, X. D. Wang, G. Mills, in *Proc. ECRS 4 (June 4–6, 1996, Cluny, France)*, S. Denis, J.-L. Lebrun, N. Bourniquel, M. Barral, J.-F. Flavenot, eds., pp. 127–134.

4 B. Dziunikowski, in *Energy Dispersive X-ray Fluorescence Analysis, Comprehensive Analytical Chemistry*, G. Svehla, ed., Elsevier, Amsterdam, 1989.

5 A. Pyzalla, K.-B. Müller, J. Wegener, Hot extrusion of a PM AlSi-alloy–influence of extrusion parameters on microstructure and properties, *Z. Metallkde* 2000, **91**, 831–837.

6 M. Croft, I. Zakharchenko, Z. Zhong, T. Tsakalakos, Y. Gulak, Z. Kalman, J. Hastings, J. Hu, R. Holtz, K. Sadananda, *Stress distribution and tomographic profiling with energy dispersive x-ray scattering, Materials Research Society Symposium – Proceedings, vol. 678, 2001*, pp. EE2.1.1–EE2.1.6.

7 A. Steuwer, J.R. Santisteban, M. Turski, P.J. Withers, T. Buslaps, High-resolution strain mapping in bulk samples using full-profile analysis of energy dispersive synchrotron X-ray diffraction data, *Nucl. Instrum. Methods Phys. Res., Sect. B: Beam Interact. Mater. Atoms* 2005, **238** (1–4), 200–204.

8 G. Brusch, Entwicklung eines Mess- und Auswerteverfahrens zur Analyse von Eigenspannungsverteilungen mit Hochenergie-Röntgenbeugung, Berichte aus dem Zentrum für Eigenspannungsanalyse des Hahn-Meitner-Institut Berlin GmbH, HMI-B 550, 1998.

9 L. Gerward, S. Lehn, G. Christiansen, *Texture Crystall. Solids* 1976, **2**, 95–111.

10 A.M. Korsunsky, W.J.J. Vorster, S.Y. Zhang, D. Dini, D. Latham, M. Golshan, J. Liu, Y. Kyriakoglou, M.J. Walsh, The principle of strain reconstruction tomography: determination of quench strain distribution from diffraction measurements, *Acta Materialia* 2006, **54** (8), 2101–2108.

11 http://www.esrf.fr/UsersAndScience/Experiments/MaterialsScience/ID15/ (accessed October 23, 2006).

12 http://www.gkss.de/pages.php?page=w_abt_genesys_harwi.htmllanguage=d&version=g (accessed October 23, 2006).

13 D. Dantz, Dirk, Eigenspannungen in mikrowellengesinterten Ni/8Y-ZrO2 und NiCr8020/8Y-ZrO2 Gradientenwerkstoffen Dissertation: Fakultät III (Prozesswissenschaften), Technische Universität Berlin, 2000.

14 M. Broda, A. Pyzalla, W. Reimers, X-ray determination of load and residual stress distribution in C/SiC – composites, *Appl. Compos. Mater.* 1999, **6**, 51–66.

15 J. Liu, K. Kim, M. Golshan, D. Laundy, A.M. Korsunsky, Energy calibration and full-pattern refinement for strain analysis using energy-dispersive and monochromatic X-ray diffraction, *J. Appl. Crystallogr.* 2005, **38** (4), 661–667.

16 S.M. Clark, Thirty years of energy-dispersive powder diffraction, *Crystallogr. Rev.* 2002, **8** (2–4), 57–92.

17 T. Wroblewski, U. Ponkratz, F. Porsch, *Nucl. Instrum. Methods Phys. Res. A* 2004, **532** (3), 639–643.

18 A. Steuwer, J.R. Santisteban, M. Turski, P.J. Withers, T. Buslaps, High-resolution strain mapping in bulk samples using full-profile analysis of energy-dispersive synchrotron X-ray diffraction data, *J. Appl. Crystallogr.* 2004, **37** (6), 883–889.

19 H. Duval, J.C. Malaurent, P. Dhez, P. Duval, *Opt. Commun.* 1997, **136**, 44–51.

20 R.A. Young(ed.), *The Rietveld Method, International Union of Crystallography*, Oxford University Press, Oxford, 1993.

21 A. M. Glazer, M. Hidaka, J. Bordas, *J. Appl. Cryst.* 1978, **11**, 165–177.

22 Canberra, http://www.canberra.com/Products/485.asp.

23 R.G. Helmer, R.L. Heath, M. Putnam, D.H. Gipson, *Nucl. Instrum. Methods* 1967, **57**, 46–57.

24 A. Pyzalla, Methods and feasibility of residual stress analysis by high energy synchrotron diffraction in transmission geometry using a white beam, *J. Nondestr. Eval.* 2000, **19**, 21–31.

25 J.W. Otto, On the peak profiles in energy-dispersive powder X-ray diffraction with synchrotron radiation, *J. Appl. Crystallogr.* 1997, **30** (6), 1008–1015.

26 A. Steuwer, M. Peel, T. Buslaps, Aspects of residual stress determination using energy-dispersive synchrotron X-ray diffraction, *Mater. Sci. Forum* 2006, 524–525, 267–272.

27 J. Szpunar, L. Gerward, *J. Mater. Sci.* 1980, **15**, 469–476.

28 M. A. Player, Z. Shi, S. M. Clark, M. C. Miller, C. C. Tang, *Physica B* 1998, **248**, 405–410.

29 R. A. Schwarzer, *Steel Res.* 1993, **64**, 570–574.

30 J. Staun Olsen, B. Buras, L. Gerward, S. Steenstrup, *J. Phys. E: Sci. Instrum.* 1981, **14** (10), 1154–1158.

31 K. Suzuki, K. Tanaka, Y. Akiniwa, M. Kawamura, K. Nishio, H. Okado, In-situ stress measurement of bond coatings at high temperature by high-energy synchrotron X-rays, *Zairyo/J. Soc. Mater. Sci.*, Japan 2003, **52** (7), 756–763.

32 A.M. Korsunsky, S.P. Collins, R. Alexander Owen, M.R. Daymond, S. Achtioui, K.E. James, Fast residual stress mapping using energy-dispersive synchrotron X-ray diffraction on station 16.3 at the SRS, *J. Synchrotron Radiat.* 2002, **9** (2), 77–81.

33 M.L. Martinez-Perez, F.J. Mompean, J. Ruiz-Hervias, C.R. Borlado, J.M. Atienza, M. Garcia-Hernandez, M. Elices, J. Gil-Sevillano, R.L. Peng, T. Buslaps, Residual stress profiling in the ferrite and cementite phases of cold-drawn steel rods by synchrotron X-ray and neutron

diffraction, *Acta Materialia* 2004, **52** (18), 5303–5313.

34 C.J. Dawes, Introduction to friction stir welding and its development, *Weld. Metal Fabrication* 1995, **63** (1), 1–3.

35 W.M. Thomas, E.D. Nicholas, J.C. Needham, M.G. Murch, P. Temple-Smith, C.J. Dawes, PCT World Patent Application WO 93/10935 (UK 9125978.8, 6 Dec.) (Publ: 10 June 1991).

36 R.S. Mishra, Z.Y. Ma, Friction stir welding and processing, *Mater. Sci. Eng.* R: Rep. 2005, **50** (1–2), 1–78.

37 E.D. Nicholas, Friction processing technologies, *Welding in the World* 2003, **47** (11–12), 2–9.

38 T.W. Nelson, *Friction stir welding – a brief review and perspective for the future, Friction Stir Welding and Processing III: Proceedings of a Symposium Sponsored by the Shaping and Forming Committee of (MPMD) of the Minerals, Metals and Materials Society, TMS*, 2005, pp. 149–159.

39 Y. Li, L.E. Murr, J.C. McClure, Flow visualization and residual microstructures associated with the friction-stir welding of 2024 aluminum to 6061 aluminum, *Mater. Sci. Eng.* 1999, **A 271** (1–2), 213–223.

40 C. Dalle Donne, E. Lima, J. Wegener, A. Pyzalla, T. Buslaps, *Investigations on residual stresses in friction stir welds, Proc. 3rd International Symposium on Friction Stir Welding (Kobe, Japan, September 27–28, 2001)*, TWI, UK (CD-ROM).

41 E.B.F. Lima, J. Wegener, C. Dalle Donne, G. Goerigk, T. Wroblewski, T. Buslaps, A.R. Pyzalla, Dependence of the microstructure, residual stresses and texture of AA 6013 friction stir welds on the welding process, *Z. Metallk.* 2003, **94**, 908–915.

42 M.B. Prime, T. Gnäupel-Herold, J.-A. Baumann, R.J. Lederich, D.M. Bowden, R.J. Sebring, Residual stress measurements in a thick, dissimilar aluminum alloy friction stir weld, *Acta Materialia* 2006, **54**, 4013–4021.

43 P. Staron, M. Koçak, S. Williams, A. Wescott, Residual stress in friction stir-welded A1 sheets, *Physica B: Condens. Matter* 2004, **350** (1–3, Suppl. 1), e491–e493.

44 J.M.N. James, D.J. Hughes, D.G. Hattingh, G.R. Bradley, G. Mills, P.J. Webster, Synchrotron diffraction measurement of residual stresses in friction stir welded 5383-H321 aluminium butt joints and their modification by fatigue cycling, *Fatigue Fracture Eng. Mater. Struct.* 2004, **27** (3), 187–202.

45 T.W. Clyne, P.J. Withers, *An Introduction to Metal Matrix Composites*, Cambridge University Press, UK, 1993.

46 S. Das, *J. Mater. Sci.* 1997, **16**, 1757.

47 A. Pyzalla, H. Berns, *Mater.-wiss. u. Werkstofftech.* 1997, **28**, 180.

48 U. Göbel, G. Lefranc, H.P. Degischer, *Materialwissenschaften und Werkstofftechnik* 2003, **34**, 375.

49 P. Prader, H.P. Degischer, *Mater. Sci. Tech.* 2000, **16**, 893.

50 G. Requena, A. Schnabl, H.P. Degischer, in *Verbundwerkstoffe und Werkstoffverbunde*, Hrg. H.P. Degischer, Wiley-VCH, Weinheim, 2003, 139.

51 A. Schnabl, H.P. Degischer, *Z. Metallk.* 2003, **94**, 743.

52 A.R. Pyzalla, B. Reetz, A. Jacques, J.P. Feiereisen, O. Ferry, T. Buslaps, W. Reimers, In-situ investigation of stress relaxation in Al/Si – MMCs using high energy synchrotron radiation, *Z. Metallkunde* 2004, **95**, 624–630.

53 L. Spiess, R. Schwarzer, H. Behnken, G. Teichert, *Moderne Röntgenbeugung*, Teubner-Verlag, Wiesbaden, Germany 2005.

54 B. Scholtes, *Eigenspannungen in mechanisch randschicht-verformten Werkstoffzuständen*, DGM-Informationsgesellschaft Verlag, Oberursel, Germany 1992.

# 11
# Diffraction Imaging for Microstructure Analysis

*Thomas Wroblewski*

## 11.1
## Introduction, the Principle of Diffraction Imaging

Microdiffraction is often considered to be a synonym for microbeam diffraction in which a microbeam illuminates a small spot on a specimen. For an efficient realization of such a microbeam not only sophisticated optics are needed but also a source of high brilliance. Thus microbeam diffraction is nowadays mostly performed on third generation, low emittance synchrotron radiation sources. Low emittance sources like first generation storage rings, X-ray tubes, and least of all neutron sources are not well suited for microbeam experiments because most of their intensity is removed if a small beam is selected.

An alternative approach is to use an extended beam illuminating a large sample area and to perform the spatial discrimination behind the sample as it is done in optical imaging. For X-rays and neutrons, however, the deviation of the index of refraction from unity is rather small and optical elements with apertures large enough for efficient imaging are not available. Therefore, imaging has been restricted to radiography, which profits from the straight propagation of the radiation, and techniques where the sample itself acts as optical element, for example in topography, where crystal faults in an otherwise perfectly reflecting (single) crystal are visualized.

A given reflection from a single crystal is collimated by the lattice planes of the crystal into only one direction. Radiation diffracted by the randomly oriented crystallites of a polycrystalline specimen may, however, fall in any direction on a Debye–Scherrer cone. The diffraction cones from different locations of the specimen may overlap, thus preventing spatially resolved investigations with a standard imaging detector. This crossfire can be suppressed by a collimator tube between the sample and the detector. In this case the probed location is selected from the diffracted beam. Applying a large array of parallel tubes in front of a position sensitive detector yields simultaneous information of the radiation scattered from a large sample area into the direction parallel to the tubes axes. In

*Neutrons and Synchrotron Radiation in Engineering Materials Science: From Fundamentals to Material and Component Characterization*
Edited by Walter Reimers, Anke Rita Pyzalla, Andreas Schreyer, and Helmut Clemens

ISBN: 978-3-527-31533-8

contrast to microbeam techniques which require scanning to get two-dimensional information, the collimator array in front of an area detector yields this information in a single shot and is therefore a real imaging technique.

## 11.2
## The MAXIM Experiment at HASYLAB Beamline G3

At HASYLAB beamline G3 a diffractometer for X-ray diffraction imaging has been realized [1]. It resembles a four-circle diffractometer with an $x$–$y$–$z$-table on its $\varphi$-axis (Figure 11.1). The $2\theta$-circle carries a CCD detector with a microchannel plate (MCP) as collimator array. The diameter of the individual collimator tubes in the MCP of 10 μm and their length of 4 mm results in an angular acceptance of 2.5 mrad. At a sample to detector distance of 5 mm the spatial resolution given by acceptance times distance is 12.5 μm. This matches the pixel size of the CCD which is 13 μm. The CCD has $1024^2$ pixels allowing the imaging of a 13 mm* [13 mm/$\sin(2\theta-\omega)$] large sample area (the factor $1/\sin(2\theta-\omega)$ at $\chi=0°$ results from the projection of the sample on the CCD in the diffraction plane). The MCP/CCD combination can be moved along the $2\theta$-arm to realize the minimal sample to detector distance given by the sample shape or its environment (furnace, etc.). An additional detection system consisting of a scintilla-

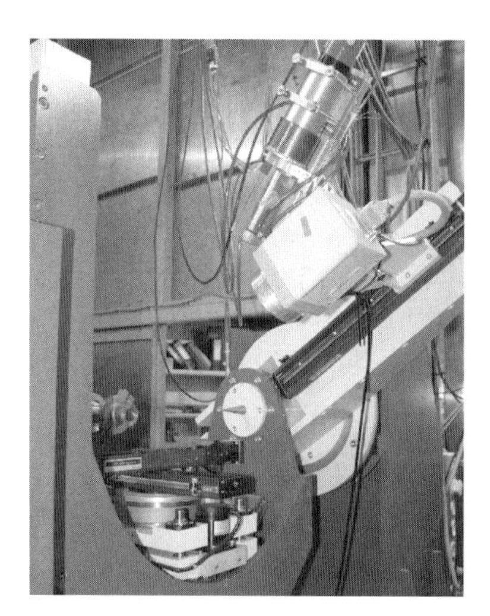

**Figure 11.1** The four-circle MAXIM diffractometer at HASYLAB beamline G3. The MCP/CCD combination can be moved along the $2\theta$-arm to allow optimization of the sample to detector distance. An additional Soller/scintilator arrangement is inclined about 20° with respect to the MCP/CCD combination. The sample can be mounted on the $x$–$y$–$z$ table on top of the $\varphi$-circle. A needle in the $2\theta$-axis can be used for adjustment. The tube for the incident beam is visible on the left.

tion counter behind a Soller collimator on the $2\theta$-arm is inclined about 20° with respect to the MCP/CCD combination. It points to the sample position if the entire ensemble is moved out about 300 mm and can be used for adjustment or measurements integrating over the specimen. While the latter operation mode yields one intensity value per scan step, the measurements with the MCP/CCD yield a series of images during the scan.

## 11.3
## Data Structure

The intensity of each pixel as function of the scanning parameter corresponds to a scan with a point detector at a certain location at the sample. Thus, in principle, up to one million data sets can be collected simultaneously. Figure 11.2 shows a single exposure taken in the aluminum 3 1 1 reflection of a friction stir weld. In the stirred zone the coarse grained structure, visible in the parent material, has been destroyed by the process. The data obtained from series of such exposures taken during an angular scan across the reflection may be rearranged into peak profiles, one for each pixel. In this case, in which no structural variation is expected parallel to the weld, the amount of data can be reduced by integrating along this direction. The right part of Figure 11.2 shows the intensity as a func-

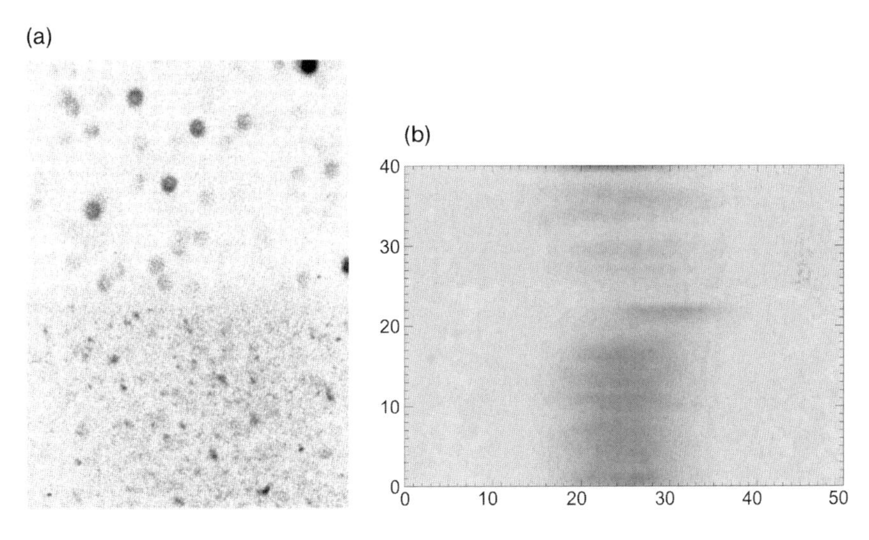


**Figure 11.2** Exposure taken with the MCP/CCD combination adjusted to the aluminum (3 1 1) reflection from a friction stir weld. The coarse grained structure (a) of the parent material is destroyed in the stirred zone (b). The contour plot on the right shows the intensity as a function of $2\theta$ (horizontal) and the direction perpendicular to the weld (vertical). The influence of the coarse grained structure on the reflection profile is clearly visible although the intensity has been integrated in the direction along the weld. Furthermore, the peak shift due to strain in the intermediate zone is evident.

tion of the position perpendicular to the weld and the diffraction angle. The peak shift due to strain in the thermal affected zone between the weld and parent material is clearly visible.

## 11.4
## Strategies for Data Reduction and Visualization

The simultaneous measurement of one million data sets does not imply that the sample consists of one million different constituents. There may be spatial (between adjacent pixels) and/or "spectral" correlations (arising from similar structural properties). The aim is, therefore, to reduce the data by grouping the data according to these properties (classification). In the above example data reduction has been achieved using prior knowledge of the spatial arrangement.

If such prior knowledge does not exist, more sophisticated methods have to be applied. Manifold classification techniques for such series of images from the same scene have been developed in another field of research, namely remote sensing, where images are collected as a function of the optical or infrared wavelength (for an overview of remote sensing strategies see for example [2]). The identical structure of MAXIM and remote sensing data allows easy transfer of the strategies for visualization and evaluation.

A well-known visualization tool is the false color composite using three different images for the image channels red, green, and blue (for example, infrared, red, and green bands for a color infrared (CIR) image). Classification can be done according to the colors (corresponding to intensity ranges in the three exposures) in such a composite. Such simple methods as thresholding typically use only few images.

Other, more sophisticated techniques make use of many or even all images of a series. The intensities of one pixel in $n$ images can be represented as a vector in $n$-dimensional space and being analyzed by classification. Vectors corresponding to similar diffractograms will form clusters in the $n$-dimensional feature space which can be regarded as individual classes. The aim of classification is to determine sample properties for the entire image scene and to obtain a class map with corresponding class spectra. Depending on the *a priori* knowledge about the sample one can apply supervised or unsupervised classification techniques. Whereas one provides *a priori* knowledge in supervised classification (i.e., mean spectra from selected sample regions), unsupervised classification might be used in the case of lack of knowledge about the investigated sample or the expected effects. An iterative statistical approach is used to find clusters/classes in feature space where the number of classes has to be provided. Typically one starts from a random distribution of class centers/seeds in feature space (or class positions along the diagonal) and determines the similarity, e.g., the distance, of all vectors to all class centers. Based on this information the vectors are connected to the classes, the class centers are recalculated with respect to the vectors bound to each class and the obtained seeds (one spectrum for each class) are then used as starting point for the next iteration.

Another approach is the use of so-called spectral transforms, which involves the calculation of new spectra from the measured ones. The ratio of two spectra is a simple example for such a transform. A more sophisticated transform is the principal component rotation in which so-called eigenimages are determined. The principal component analysis (PCA) can be regarded as a rotation in $n$-dimensional space to principal axes. For example, the coordinate system given by the intensities in each image is being rotated such that the PCA directs along the strongest variation inside the data and the obtained first eigenimage shows the highest variance compared to the eigenimages with higher index. While the projections of the vectors corresponding to different classes onto one axis corresponding to an original exposure usually show severe overlap (similar intensities in the exposure) they will be well separated in the projection to the first principal components (different intensities in the first eigenimages). The PCA decorrelates the information stored inside the data set.

A variant of the principal component rotation, the minimum noise fraction rotation was applied to data obtained from a rail [3]. Due to the load of fast and heavy trains the ferrite of the rails transforms into a so-called white etching layer. The reflections from this layer show a severe overlap with those from the ferrite. The sample showed a poor reflectivity due to the small size of the crystallites. Furthermore, the low energy X-rays of 6.5 keV, used to suppress the Fe-fluorescence were attenuated by several beamline components. The exposure taken in the ferrite (1 1 0) reflection is, therefore, rather noisy, resulting in noisy diffractograms for the individual pixels (Figure 11.3).

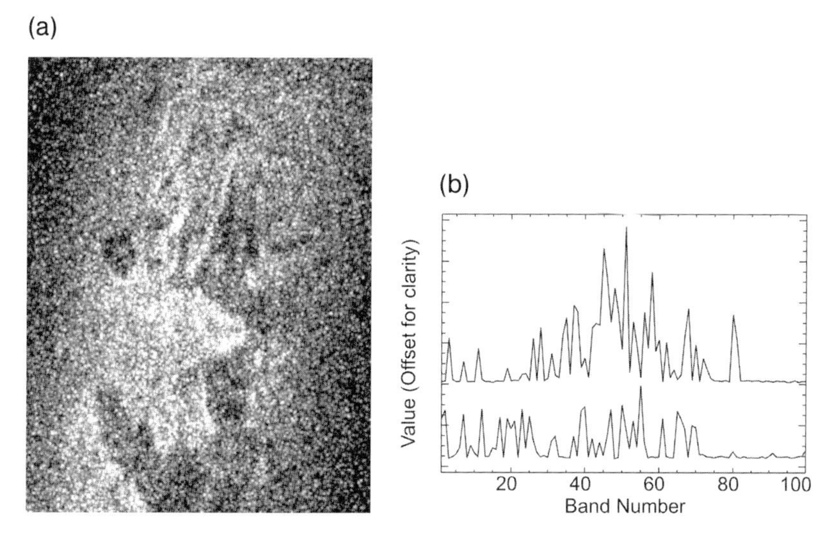


**Figure 11.3** Exposure taken in the Fe (1 1 0) reflection from a perlitic steel rail. The peaks from ferrite and the white etching layer show a severe overlap. Due to several experimental parameters the exposures have been very noisy leading to noisy diffractograms for the individual pixels.

(a)

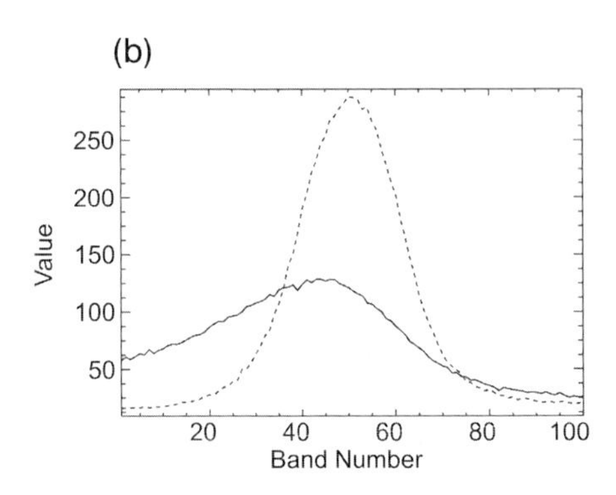


**Figure 11.4** Second eigenimage from a principal component transform using 1 0 1 exposures as in Figure 11.3 showing the distribution of the white etching layer (dark areas) on the ferrite. Averaging over dark areas (solid line) or bright areas (dashed line) yields the peak profiles for the white etching layer and the ferrite, respectively.

Figure 11.4 shows the second eigenimage obtained from the spectral transform. Here the distribution of ferrite and white etching layer is clearly visible. Based on this information the corresponding regions containing the white etching layer or ferrite, respectively, can be defined in the original exposures. The diffractograms extracted from these areas are displayed in Figure 11.4b.

## 11.5 Outlook, Bulk Imaging

A paramount advantage of the MAXIM technique compared to $\mu$-beam methods is that the information from a large area is obtained in a single shot and does not require time-consuming spatial scanning. This enables the investigation of dynamical processes like recrystallization without prior knowledge of where they occur (which indeed is not available) using the information from the images taken at the end of the process which show the recrystallized grains [4]. Figure 11.5 shows growth curves of crystallites in cold rolled copper during annealing at 120°. This experiment demonstrated that the recrystallization cannot be described by the simple models which for example assume homogenous nucleation and growth (a review of recrystallization models can for example be found in [5]). A possible explanation could be the model of self-organized criticality [6]. This rather universal concept describes processes in systems far from equilibrium. The energy deposited in such a system (in this case by cold rolling) is re-

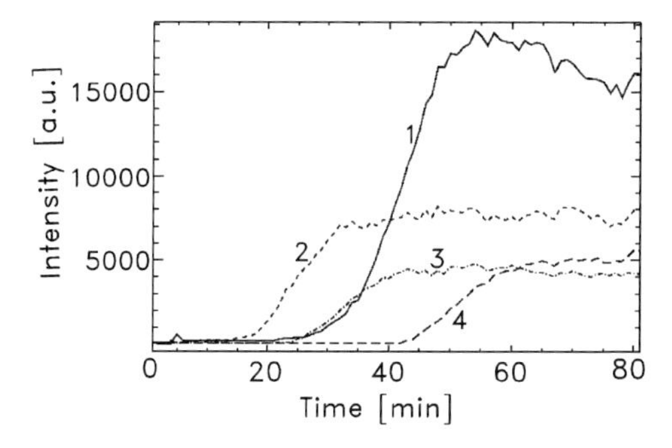


**Figure 11.5** Growth curves for four medium-sized Cu-crystallites during annealing (the size of crystallites extends over several orders of magnitude). The process can clearly be divided into the three phases of nucleation, growth, and saturation. There is no correlation between the time of nucleation and the final size of the crystallites.

leased in avalanches which modify their environment giving rise to further energy releases. Prominent examples of such systems are $1/f$ noise, earthquakes, or avalanches in an evolving sandpile.

Even more surprising results were obtained on aluminum [7]. Here the crystallites could be divided into two categories. One showed growth curves similar to those in Figure 11.5. The crystallites of the other group, however, first grew but than shrank. An analysis correlating the time at which the crystallites obtained half of their maximum size (during growth) with the maximum size revealed that the crystallites showing normal growth grew at a later time and obtained a smaller maximum size than those showing shrinkage. This behavior, that the small crystallites seem to swallow the big ones, is in contradiction to the model of Ostwald ripening.

A possible explanation for this finding might be that only grains near the surface have been observed due to the low penetration depth of the 8 keV X-rays used. The behavior in the bulk could be different. Bulk diffraction imaging, however, requires highly penetrating radiation such as neutrons or high-energy X-rays. Unfortunately, the set-up at beamline G3 is not suited for such experiments because the MCP is transparent and the directly illuminated CCD is insensitive for both kinds of radiation.

With neutrons capillaries of boron silicate glass can be used as collimator in combination with a neutron sensitive area detector [8]. The low intensity of neutron sources, however, does not allow high spatial and/or temporal resolution.

For high-energy X-rays recently long polycapillaries have become available. Initial tests with a collimator bundle of 30 μm capillaries of 300 mm length at energies of 26 keV and 78 keV (third harmonic) demonstrated that these capillaries are straight within their aspect ratio of $10^{-4}$. In combination with a detector sensitive to high-energy X-rays ($Gd_2O_2S$ coupled to a CCD via a magnifying taper) bulk diffraction imaging could successfully be performed [9]. In this feasibility study a slice through an Al-tube was illuminated by a flat beam (in contrast to the studies at low X-ray energies where the sample surface defines the investigated plane). By scanning the sample perpendicular to this flat beam three-dimensional information could be obtained. The long readout time (20 s) of the detector used, however, prevented kinetic studies. Presently a dedicated instrument is under construction at the GKSS high-energy beamline W2 (HARWI) at DORIS/HASYLAB. This instrument is scheduled to be in operation in 2007 and will enable bulk diffraction imaging including dynamic studies of bulk and (simultaneously) surface in combination with tomography [10].

## References


1 T. Wroblewski, O. Clauss, H.-A. Crostack, A. Ertel, F. Fandrich, Ch. Genzel, K. Hradil, W. Ternes, E. Woldt, *Nucl. Instrum. Methods A* 1999, **428**, 570–582.
2 R.A. Schowengerdt, *Remote Sensing*, Academic Press, New York, USA, 1997.
3 T. Wroblewski, E. Wild, T. Poeste, A. Pyzalla, *J. Mater. Sci. Lett.* 2000, **19**, 975–978.
4 T. Wroblewski, *Z. Metallk.* 2002, **93**, 1228–1232.
5 F.J. Humphreys, M. Hatherly, *Recrystallization and Realted Annealing Phenomena*, Pergamon, UK, 1995.
6 P. Bak, C. Tang, K. Wiesenfeld, *Phys. Rev. Lett.* 1987, **59**, 381–384.
7 T. Wroblewski, A. Buffet, Proceedings Fundamentals of Deformation and Annealing, *Mat. Sci. Forum*, 2007, **550**, 631–636.
8 T. Wroblewski, E. Jansen, W. Schäfer, R. Skowronek, *Nucl. Instrum. Methods A* 1999, **423**, 428–434.
9 T. Wroblewski, A. Bjeoumikhov, *Nucl. Instrum. Methods A* 2005, **538**, 771–777.
10 BMBF-Project OD18/9 within SPP1104 (Prof. Dr. W. Reimers, TU Berlin) and DFG-Project 05 KS4KT2/0 (Prof. Dr. B. Odenbach, TU Dresden).

# 12
# Basics of Small-Angle Scattering Methods

*Philipp Klaus Pranzas*

## 12.1
## Common Features of a SAS Instrument

In a SAS experiment a monochromatic, focused beam is scattered by inhomogeneities in the sample. In Figure 12.1, a schematic set-up of a typical SANS instrument is shown exemplary. As a monochromator a velocity selector (neutrons) or focusing mirrors (X-rays) are used. A velocity selector consists of a rotor with absorbing lamellae. Only neutrons in a small wavelength (speed) distribution (e.g., $\Delta\lambda/\lambda = 10\%$) can pass according to the rotation speed of the rotor. The focused beam is obtained by an alignment of apertures, the collimation line (neutrons), or by a slit system (X-rays). The scattering pattern is recorded by a position-sensitive area or linear gas detector (neutrons and X-rays) or by CCD/image plate detectors (X-rays).

In this chapter the fundamental scattering parameters are introduced which are essential to derive information about structures in the sample from the scattering signal.

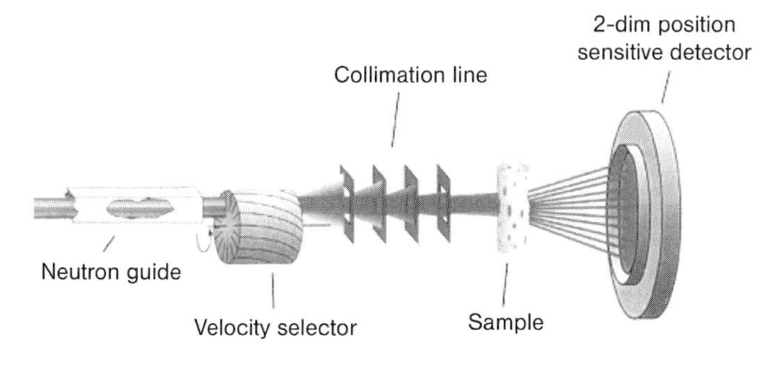


**Figure 12.1** Main components of a SANS instrument.

*Neutrons and Synchrotron Radiation in Engineering Materials Science: From Fundamentals to Material and Component Characterization*
Edited by Walter Reimers, Anke Rita Pyzalla, Andreas Schreyer, and Helmut Clemens

ISBN: 978-3-527-31533-8

## 12.2
## Contrast

The key parameter which determines if neutrons or X-rays can see a certain structure is the contrast between the scattering body and the surrounding matrix. The different interactions of neutrons and X-rays with the scattering atom were already introduced in Chapter 4 as well as the X-ray atomic scattering factors and the neutron scattering lengths/cross sections.

The scattering length density $\eta_s$ of a scattering body consisting of $i$ different elements is calculated by

$$\eta_s = \sum_i b_i \cdot \frac{\delta \cdot N_A}{M_W} \quad (10^{10}\ \mathrm{cm}^{-2}) \tag{12.1}$$

with

$\delta$ bulk density ($\mathrm{g\ cm^{-3}}$)
$N_A$ Avogadro constant ($6.022 \times 10^{23}\ \mathrm{mol^{-1}}$)
$M_W$ molecular weight or atomic mass ($\mathrm{g\ mol^{-1}}$)
$b$ coherent scattering length (neutrons) or atomic scattering factor $f$ (X-rays) ($10^{-12}$ cm)

The contrast is the scattering length density difference (SLDD) $\Delta\eta$ between matrix and scattering body. In neutron scattering of biological samples it is common to substitute hydrogen atoms in a macromolecule by deuterium or to use the so-called contrast variation by changing the $H_2O/D_2O$ ratio in the solution. At a certain $D_2O$ concentration – the matching point – a part of or the whole molecule is "matched out"; it is "invisible" for the neutron because $\eta$ is the same as in the matrix.

## 12.3
## Scattering Curve

The neutron is scattered at structures which have a certain size and contrast to the matrix. Larger bodies scatter to smaller angles and vice versa. The scattering vector or momentum transfer $\vec{q}$ is the vector between the wave vector $\vec{k}_0$ of the incoming neutron and the wave vector $\vec{k}$ of the neutron scattered under the angle $2\theta$ (Figure 12.2).

From Figure 12.2 the modulus of the scattering vector can be calculated

$$|\vec{q}| = \frac{4\pi}{\lambda} \sin\theta \quad (\mathrm{nm}^{-1} \text{ or } \mathrm{Å}^{-1}) \tag{12.2}$$

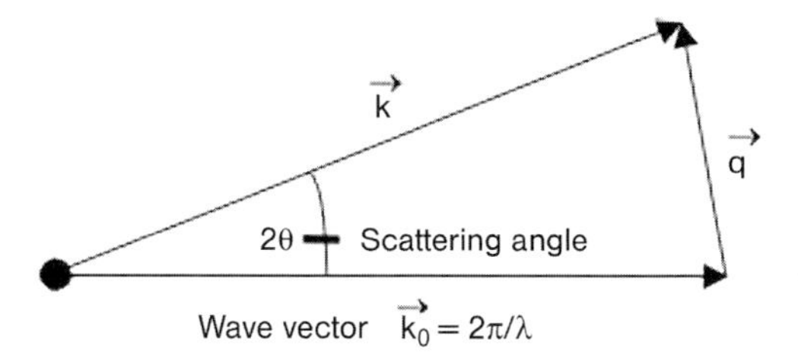


**Figure 12.2** Definition of the scattering vector $\vec{q}$.

$q$ is commonly used because it combines the scattering angle with the wavelength $\lambda$, so that results obtained at different instruments measured at varying wavelengths can directly be compared. In the X-ray community it is also common to use the symbols $s$ or $h$, sometimes differing from $q$ by a factor of $2\pi$. The dimension $d$ of the structures scattering at a certain scattering vector can be estimated by

$$d = \frac{2\pi}{q} \tag{12.3}$$

The resolution of a SAS experiment – the smallest dimension $d_{\min}$ – is therefore defined by $q_{\max}$.

The three-dimensional information of the scattering pattern measured with a position sensitive area detector (Figure 12.3a) is reduced by radial integration to two dimensions. A plot of the scattering intensity $I(q)$ over $q$ – the scattering curve – is obtained (Figure 12.3b).

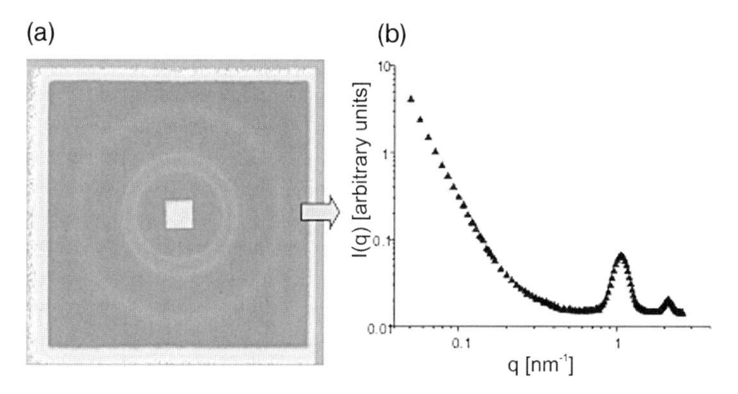


**Figure 12.3** (a) Scattering pattern of silver behenate measured with SANS using a position sensitive area gas detector with 256 × 256 pixels (pixel size 2.2 mm) at a distance of 1 m from the sample. (b) Scattering curve calculated by radial integration from the scattering patterns measured at 1, 3, and 7 m from the sample.

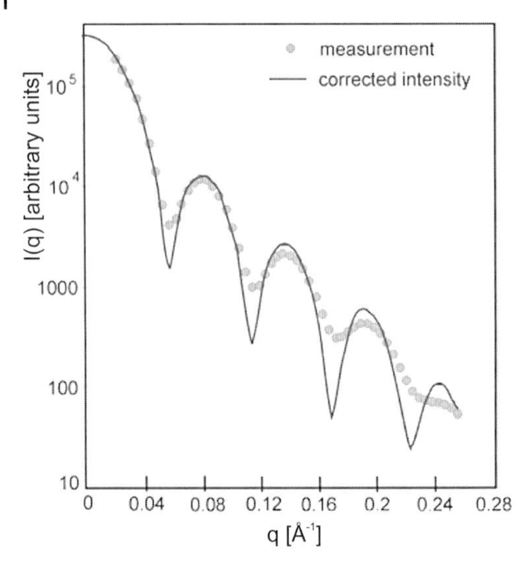


**Figure 12.4** Measured and corrected scattering curve of the proteine Apoferritin [8].

In the scattering pattern the intensity is smeared out, which results, e.g., in a peak broadening, due to smearing effects such as the limited detector resolution, the width of the wavelength distribution of the monochromator, and the beam divergence depending on the slit/aperture (collimation) alignment and size. In Figure 12.4 this is demonstrated showing the measured and theoretically corrected ("desmeared") scattering curve of an Apoferritin solution, an almost ideal monodisperse system of spheres [8]. For the comparison between measurement and theory and for the model calculation of size distributions the measured curve has to be desmeared or the theoretical curve has to be smeared out. Specific examples for scattering curves will be presented in Chapter 13.

Before analyzing the coherent scattering of a SANS curve, the background from the incoherent scattering, which is constant (isotropic) in the whole room angle ($4\pi$), has to be subtracted first. Some elements such as hydrogen have a high incoherent scattering length in comparison to the coherent scattering. A high content of such an element may lead to a bad signal/noise ratio. In SAXS the isotropic background (e.g., fluorescence) is much lower.

## 12.4
## Power Law/Scattering by Fractal Systems

A first characterization of complex systems is possible analyzing the slope(s) of the scattering curve, known as power law behavior or scattering by fractal systems [9]. If the decay of the scattering intensity follows a $q^{-D}$ behavior, with the fractal dimension $D$, an elongated (rod-like) structure in the sample can be

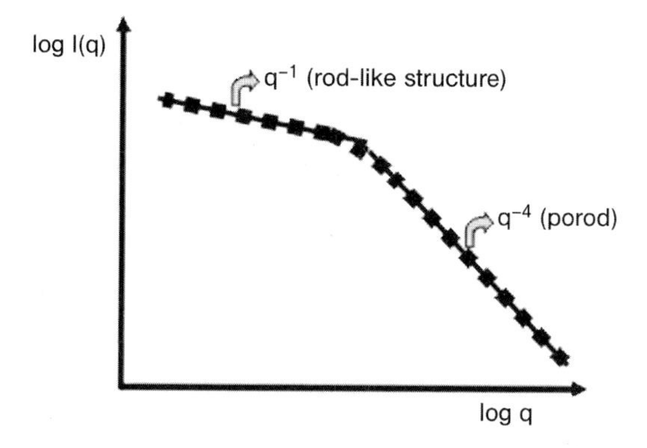


**Figure 12.5** An ideal scattering curve of a rod-like structure.

assumed for $D = 1$ and a flattened (disc-like) structure for $D = 2$. The Porod scattering of sharp interfaces leads to $D = 4$ ("$q^{-4}$ decay"). A schematic scattering curve of an elongated structure is shown in Figure 12.5.

In general, a smaller fractal dimension points to a more open structure. A value of $D$ between 1 and 3 indicates the existence of a mass fractal – a fractal system of small units building larger aggregates with a self-similar structure over a large size range. Surface fractals – fractal systems with a dense core and with fractal structures on the surface – are usually showing a $D$ value between 3 and 4.

## 12.5 Guinier and Porod Approximations

The Guinier equation (12.4) describes the intensity as a function of $q$ [10]. It is valid only in the region of small $q$ values ($q \cdot R_g < 1$).

$$I(q) \approx I(0)e^{-(1/3)q^2 R_g^2} \tag{12.4}$$

$I(0)$ is the intensity extrapolated to $q = 0$. It is used, e.g., for the determination of the molecular weight of particles in diluted solutions. The radius of gyration $R_g$ is in analogy to mechanics the "weight average" of all radii present in the sample. Depending on the model used for describing the structures in the sample, the average radius of the model structures can be calculated from $R_g$. The formulas are listed in the literature [3, 6], e.g., the radius of spheres: $R^2 = 5/3R_g^2$. In a linear fit of the Guinier region in the so-called Guinier plot in which the logarithm of $I(q)$ is plotted versus $q^2$ (Figure 12.6a) the two parameters $R_g$ and $I(0)$ can be determined from slope and ordinate. A not existing linear range indicates the presence of very large structures which scatter at low $q$ values, perhaps outside the accessible $q$ range, and the Guinier approximation should not be applied.

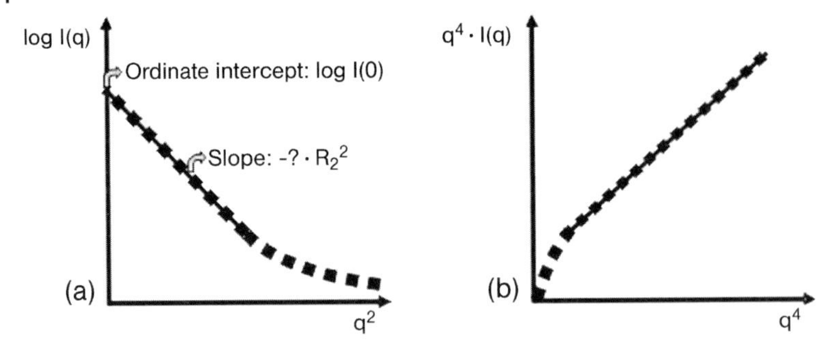


**Figure 12.6** Schematic scattering curves: (a) Guinier plot, (b) Porod plot.

In the Porod plot $q^4 \cdot I(q)$ is plotted versus $q^4$ (Figure 12.6b) [11]. The linear fit $I(q) = C + A/q^4$ at large $q$ values is convenient to determine the incoherent background (constant $C$).

## 12.6
## Macroscopic Differential Scattering Cross Section

For the calibration from the relative intensity to "absolute units", the macroscopic differential scattering cross section $\frac{d\Sigma}{d\Omega}$ is calculated as follows [4]:

$$\frac{d\Sigma}{d\Omega}(q) = \frac{I(q)}{M \cdot \left(\frac{I_0}{M}\right) \cdot D \cdot T \cdot \varepsilon \cdot \Delta\Omega} \quad (\text{cm}^{-1}), \tag{12.5}$$

with

$I(q)$ measured intensity
$M$ monitor counts
$I_0$ intensity of primary beam
$I_0/M$ primary beam/monitor ratio (calibrated by a standard with known scattering cross section, e.g., vanadium or $H_2O$)
$D$ sample thickness
$T$ transmission of sample ($=e^{-\Sigma D}$)
$\varepsilon$ detector efficiency/response
$\Delta\Omega$ solid angle element ($= dx \cdot dy/L^2$, with sample–detector distance $L$)

From the macroscopic differential scattering cross section the form factor (size and shape) and the structure factor (degree of local order) of the scattering structures can be calculated using Eq. (12.6).

$$\frac{\partial \Sigma}{\partial \Omega}(q) = N_P \cdot V_P^2 \cdot (\Delta\eta)^2 \cdot P(q) \cdot S(q) + B_{inc} \tag{12.6}$$

with

$N_P$ number density of scattering bodies
$V_P$ volume of one scattering body
$\Delta\eta$ contrast or scattering length density difference (SLDD)
$P(q)$ form (shape) factor (also $F(q)$)
$S(q)$ interparticle structure factor, dependent on degree of local order, interaction between scattering bodies, for $N_P$ approaching 0, $S(Q)$ becomes 1 (diluted system)
$B_{inc}$ incoherent background (isotropic)

## 12.7 Model Calculation of Size Distributions

The information about both size and shape of the scattering structures cannot be derived at the same time from the scattering curve. A first impression about sizes present in the sample can be derived by Fourier transformation of the scattering curve. The distance distribution function is obtained, showing the frequency distribution of distances between scattering centers. This distribution can be illustrated by all possible cuts through the scattering structures. The distance distribution is model independent and does not include shape information; therefore, a careful interpretation is obligatory.

For the calculation of the size distribution including the shape of the scattering structures a model is chosen using preliminary information and additional results from other methods, e.g., electron microscopy. With this model the theoretical scattering curve is calculated and (least-squares) fitted to the measured curve taking into account polydispersity and instrumental smearing. Commonly used model shapes of the scattering structures are spheres, ellipsoids, shells (hollow spheres), elliptic/hollow cylinders, or flattened/disc-like structures. The form factors of these simple bodies are listed in the literature [4, 6]. The form factor of spheres with the radius $R$ is shown in Eq. (12.7):

$$P(q, R) = \left(3\frac{\sin(qR) - qR\cos(qR)}{(qR)^3}\right)^2 \tag{12.7}$$

In real systems even a "monodisperse" distribution of structures has a certain width. For a model calculation including the determination of the size distribution Eq. (12.6) changes to

$$\frac{d\Sigma}{d\Omega}(q) = (\Delta\eta)^2 \cdot \int_0^\infty n(R) \cdot V(R)^2 \cdot P(q, R) \cdot S(q, R_{HS}, f_{HS})\, dR \tag{12.8}$$

with

| | |
|---|---|
| $n(R)dR$ | number density of particles with sizes between $R$ and $R+dR$ |
| $S(q, R_{HS}, f_{HS})$ | structure factor describing the interparticle interference effect |
| $R_{HS}$ | hard-sphere radius of a particle with radius $R$ |
| $f_{HS}$ | volume fraction of hard spheres |

Normal or lognormal distributions or a certain number of B-splines [11] are used as distribution functions. A lognormal distribution with the width $\sigma$ of a number of spheres $n_0$, with the position of the maximum $R_0$, is expressed by

$$n(R) = \frac{n_0}{\sqrt{2\pi}\sigma R_0} \exp\left(-\frac{(\ln(R/R_0))^2}{2\sigma^2}\right) \tag{12.9}$$

## 12.8 Magnetic Structures

For the characterization of magnetic structures SANS measurements with applied magnetic field can be performed. Besides the nuclear scattering additional magnetic scattering of the magnetic scattering lengths in the sample arises which follows a $\sin^2 \alpha$ behavior, where $\alpha$ is the angle between the scattering vector $\vec{q}$ and the direction of the magnetic field. The resulting scattering pattern is anisotropic with only nuclear scattering parallel ($\alpha = 0°$) and nuclear plus magnetic scattering perpendicular ($\alpha = 90°$) to the magnetic field (Figure 12.7). To obtain information about nuclear and magnetic scattering horizontal and vertical sectors with a width of about 10° are used for the calculation of the scattering curves.

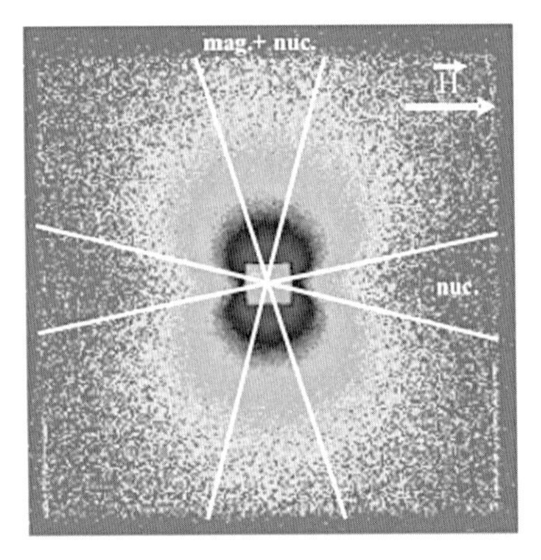


**Figure 12.7** Magnetic scattering pattern of a Fe–Cu (1 mass%) model alloy.

Polarized neutrons are the ideal probe for the investigation of magnetic structures, because the magnetic moment due to the spin of the neutron is used. Only neutrons with a spin direction parallel to the magnetic field are chosen by a polarization device, e.g., a super mirror with polarization ≥95%. These so-called polarized neutrons interact with the magnetic structures in the sample, while interference between nuclear and magnetic scattering takes place. With a spin flipper, e.g., a radio frequency flipper, the inverse (antiparallel) polarization state can also be realized. Using the scattering difference of these two states the magnetic scattering is determined with high precision and without any background.

Synchrotron radiation can also be used to study magnetism in solids, utilizing the large resonant enhancement of the magnetic cross section at the absorption edges of the metals in the sample [12]. The advantage of this method is its element specificity which enables one to distinguish between the magnetic contributions of different elements. However, for many of the magnetic elements the specific absorption edges giving rise to a large enhancement in the X-ray magnetic cross section are in the low energy regime and therefore limit the applicable sample thickness.

## References


1 A. Gunier, G. Fournet, *Small Angle Scattering of X-rays*, Wiley, New York, 1955.
2 O. Glatter, O. Kratky (eds.), *Small Angle X-ray Scattering*, Academic Press, New York, 1982.
3 L.A. Feigin, D.I. Svergun, *Structure Analysis by Small-Angle X-Ray and Neutron Scattering*, G.W. Taylor (ed.), Plenum, New York, 1987.
4 G. Kostorz, Small-angle scattering and its applications to materials science, in *Treatise on Materials Science and Technology*, vol. 15, G. Kostorz (ed.), Academic Press, New York, 1982, p. 227.
5 P. Lindner, Th. Zemb (eds.), *Neutron, X-ray and Light Scattering: Introduction to an Investigative Tool for Colloidal and Polymeric Systems*, North-Holland, Amsterdam, 1991.
6 R.P. May, Small-angle scattering, in *Neutron Data Booklet*, A.-J. Dianoux and G. Lander (eds.), ILL, 2002, pp. 2.1-1–2.1-8.
7 S.M. King, Small angle neutron scattering, in *Modern Techniques for Polymer Characterisation*, R.A. Pethrick and J.V. Dawkins (eds.), Wiley, New York, 1999 (http://www.isis.rl.ac.uk/largescale/loq/documents/sans.htm).
8 J. Zhao, *Dissertation*, Geesthacht, 1995, p. 31.
9 J. Teixeira, *J. Appl. Cryst.* 1988, **21**, 781.
10 A. Guinier, *Ann. Physique* 1939, **12**, 161.
11 G. Porod, *Kolloid. Z.* 1951, **124**, 83.
12 J.P. Hannon, G.T. Trammell, M. Blume, D. Gibbs, *Phys. Rev. Lett.* 1988, **61**, 1245.

# 13
# Small-Angle Neutron Scattering

*Philipp Klaus Pranzas*

A typical SANS instrument is shown in Figure 13.1. As a monochromator a velocity selector is used. This is a fast turning rotor with absorbing lamellae which only neutrons can pass within a certain width of the wavelength distribution, typically 10%. The divergence of the beam is adjusted in the collimation line. Collimators are an alignment of apertures with a certain distance from each other. As a rule of thumb the same collimation length as the distance between the sample and the detector should be chosen. The position sensitive area detector is moving computer controlled in the evacuated scattering tube which typically has a length between about 10 and 30 m. A second detector is useful for parallel measurements at large angles.

For the use of polarized neutrons a polarizer (e.g., super mirror with polarization $\geq$95%) and a spin flipper (e.g., radio frequency flipper) are necessary to obtain polarized neutrons with a spin parallel to the direction of the magnetic field and to flip the neutron spin to the direction antiparallel to the field. A large variety of instruments, such as magnets, refrigerator cryostats, and furnaces, are commonly used at the sample position as well as linear translation, rotary, tilting, and lift stages for the adjustment of the sample. The typical range of the wave

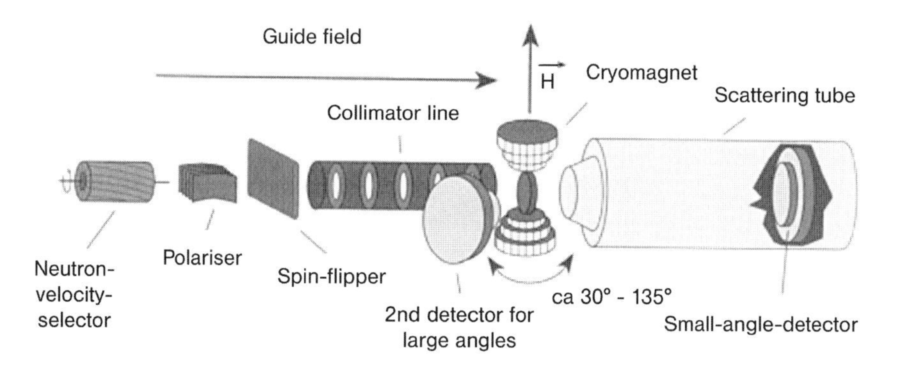


**Figure 13.1** Special features of a SANS instrument (SANS-2@GeNF) [1].

*Neutrons and Synchrotron Radiation in Engineering Materials Science: From Fundamentals to Material and Component Characterization*
Edited by Walter Reimers, Anke Rita Pyzalla, Andreas Schreyer, and Helmut Clemens

ISBN: 978-3-527-31533-8

vector $q$ is between 0.01 and about 5 $nm^{-1}$, corresponding to an accessible size range of 1–100 nm. In the following four different examples of SANS applications are stated to demonstrate the various possibilities of this method in the fields of engineering materials science and soft matter.

## 13.1 Nanocrystalline Magnesium Hydride for the Reversible Storage of Hydrogen

Hydrogen is one of the most interesting renewable energy sources regarding storage, transport, and conversion of energy in the future [2]. Light metal hydrides offer a safe alternative to the storage of hydrogen in compressed or liquid form. Prototypes of cars, motor bikes, and boats using metal hydrides already exist. The usage of metal hydrides in notebooks and video cameras is considered as well. Magnesium hydride is interesting for industrial use because it has a high hydrogen storage capacity of up to 7.6 wt%. Absorption and desorption kinetics of hydrogen on $MgH_2$ are well known as well as cycling behavior and thermal stability. High-energy ball milling is used to obtain a nanocrystalline material – nanoparticles consisting of crystallites separated by grain boundaries – with fast absorption/desorption kinetics. SANS in combination with USANS on a double crystal diffractometer (DCD) is the ideal tool to investigate the influence of microstructure, e.g., crystallite and particle sizes, on the sorption properties and to characterize structural changes due to absorption and desorption of hydrogen over a very wide range of particle sizes [3].

Model $MgH_x$ samples with varying hydrogen content were prepared at 300 °C in a hydrogen titration apparatus after 20 h of high-energy ball milling in several sorption and desorption cycles [3]. The samples were measured with SANS and DCD in quartz cuvettes with a thickness of 1 mm. SANS measurements were performed at the instrument SANS-2 at the Geesthacht Neutron Facility (GeNF) using distances between the sample and the detector of 1, 3, 9 m ($\lambda = 0.58$ nm) and 21 m ($\lambda = 0.58$ nm and 1.16 nm, $\Delta\lambda/\lambda = 0.1$). These five measurements covered the range of scattering vector $q$ between $10^{-2}$ $nm^{-1}$ and 3 $nm^{-1}$, offering sufficient large overlaps between the particular angular ranges. The SANS curves were obtained by radial integration of the two-dimensional scattering pattern (see also Chapter 12). Scattering data were normalized by monitor counts, corrected for sample transmission and detector response. The calibration to the absolute scale (macroscopic differential scattering cross section over $q$) was obtained at this time by an additional vanadium measurement. The scattering cross section of vanadium – predominantly incoherent scattering in the whole room angle – is known. USANS experiments were carried out at the double crystal diffractometer (DCD) at GeNF [4] with a wavelength of 0.443 nm resulting in an accessible $q$-range from $10^{-4}$ to $10^{-2}$ $nm^{-1}$. The scattering curves in absolute values were obtained after desmearing and correction of multiple scattering according to the model calculation procedure described in [5]. In Figure 13.2 the combined USANS-SANS scattering curves of several $MgH_x$ samples are shown. The error bars of the scattering cross section in DCD and SANS curves are in the size

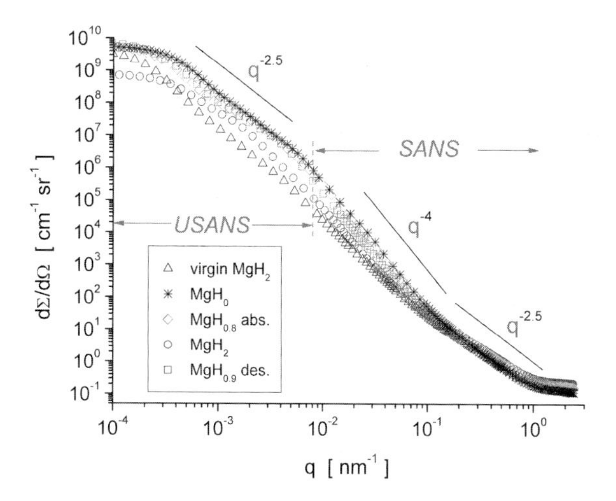


**Figure 13.2** Combined USANS-SANS scattering curves of nanocrystalline $MgH_x$ samples with different hydrogen content and cycling history.

range of the used symbols with a mean error of about 5–10%. In DCD evaluation, the error is obtained by varying the fit parameters of the model calculation [5]. The error bars of the SANS curves are obtained from the intensity variation of the 2D scattering pattern during radial integration.

A first characterization of the scattering curves starts at large $q$ values: the hydrogen content of the samples can be calculated from the different heights of the curves caused by the incoherent scattering of hydrogen. As next step the power law behavior of the scattering curves is analyzed. Two regions with a $q^{-2.5}$ decay are marked in Figure 13.1 at small and large $q$, separated by a Porod $q^{-4}$ region. This indicates the presence of a bimodal particle size distribution. The two $q^{-2.5}$ regimes show the mass fractal behavior of the nanocrystalline sample.

As a first step to a quantitative structural characterization, size distributions were calculated from the SANS/USANS curves using as first approximation a model of hard spheres in a homogenous matrix (two-phase model) and a constant scattering length density differences $\Delta\eta_S$ between $MgH_2$ and Mg of $3 \times 10^{10}$ $cm^{-2}$. The obtained volume fraction distributions are shown in Figure 13.3.

The small differences between the curves up to a radius of about 100 nm correspond to structural changes in the size range of the crystallites inside the $MgH_x$ particles. The peak at 2–3 nm broadens and shifts to larger radii with proceeding cycling, which indicates a coarsening of the crystallites during cycling at 300 °C, which is in agreement with literature. The equal heights of these peaks indicate that there is a constant contrast of the crystallite structures within the particles independent from the hydrogen content. The volume fractions with diameters larger than about 400 nm characterize the particle size of the large $MgH_x$ particles. In this size range the contrast between particle and the surrounding atmosphere is of course expected to change with the hydrogen content. The com-

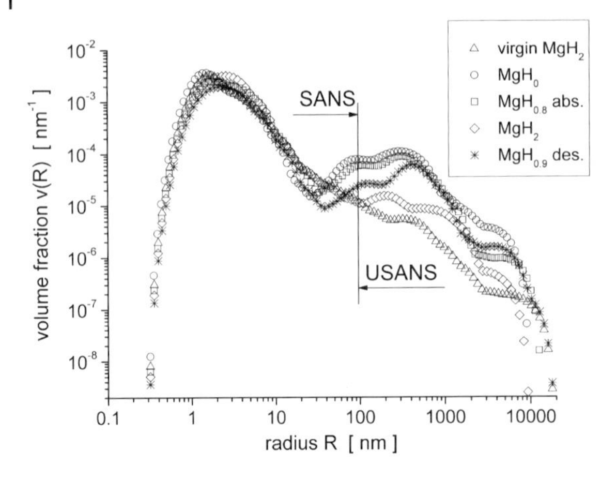


**Figure 13.3** Volume fraction distributions of $MgH_x$ samples calculated from the scattering curves shown in Figure 13.2 [3].

parison of the volume fractions of the samples with similar hydrogen concentrations like virgin $MgH_2$ and $MgH_2$ after one hydrogenation cycle indicates that particles of radii larger than 10 μm break-up and smaller particles form during the cycling process.

The results demonstrate the potential of SANS/USANS to characterize structures in metal hydrides over a large size range. For a further evaluation additional information from other methods, such as electron microscopy or X-ray diffraction, is necessary to use more complex models, e.g., fractal models. This is a good basis to study the influence of milling parameters, such as milling time and milling tool materials (steel or ceramic), on the microstructure of the hydrides. This structural characterization leads to a better understanding of sorption mechanisms in nanocrystalline light metal hydrides and gives useful information for the development of this new type of hydrogen storage material.

## 13.2
## Precipitates in Steel

An excellent example for using SANS in engineering materials science is the investigation of the precipitation behavior in maraging steels [6]. Ni-containing maraging steels belong to a class of high-strength steels with very low carbon content. Due to their strength and toughness maraging steels are used in many technically ambitious areas, e.g., for tools, medical instruments, light weight constructions, or military applications. The mechanical properties are mainly controlled by intermetallic precipitates in the nanometer size range dispersed in a rather soft martensitic matrix. For a further understanding of the influence of composition and heat treatment on the microstructures TEM and SANS studies

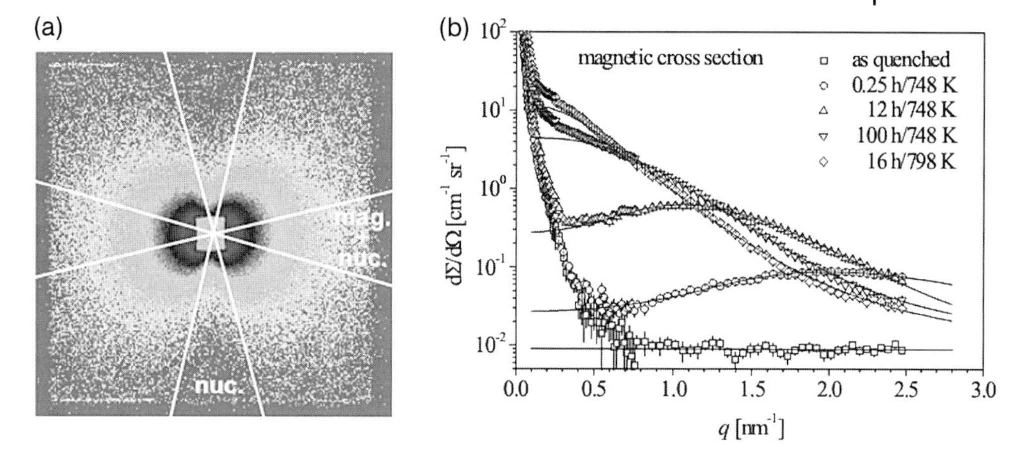


**Figure 13.4** (a) Two-dimensional scattering pattern of a steel sample saturated in a vertical magnetic field of about 2 T and (b) magnetic scattering cross sections calculated from sectors (horizontal–vertical) of the scattering patterns. The solid lines are fits obtained for the calculation of the size distributions shown in Figure 13.5 [6].

were carried out. The samples were solution heat treated for 1 h at 1273 K, air cooled, and subsequently aged for 0.25 h, 12 h, and 100 h at 748 K and for 16 h at 798 K, respectively [6].

A typical two-dimensional scattering pattern of a steel sample saturated in a vertical magnetic field (like, e.g., in the set-up in Figure 13.1) of about 2 T is shown in Figure 13.4a. The magnetic scattering cross sections calculated from sectors of the scattering patterns (the scattering cross section of the vertical sector is subtracted from the cross section of the horizontal sector, see Chapter 12.8) are presented in Figure 13.4b together with the fits (solid lines) obtained during the calculation of the size distributions (Figure 13.5). The SANS measurements were performed using a wavelength of 0.57 nm ($\Delta\lambda/\lambda = 0.1$) and again four detector distances. The error in the absolute scale of the calculated volume fractions obtained by the calibration with vanadium was approximately 5–10% [6].

The magnetic cross section (Figure 13.4b) originates from precipitates which were formed during the aging process. The precipitates are assumed to be nonmagnetic and can therefore be treated as "magnetic holes" in the ferromagnetic matrix with saturated magnetization in the magnetic field of 2 T. As the magnetic scattering length density (S.L.D.) of the precipitates is assumed to be zero, the magnetic scattering length density difference (S.L.D.D. or "contrast") of the precipitates in the matrix is equal to the S.L.D. of the matrix. Using this assumption the volume fraction of the precipitates can be calculated from the scattering curves.

A maximum is observed for the two samples aged at 748 K for 0.25 h and 12 h. This peak originates from the interference of a large number of precipitates

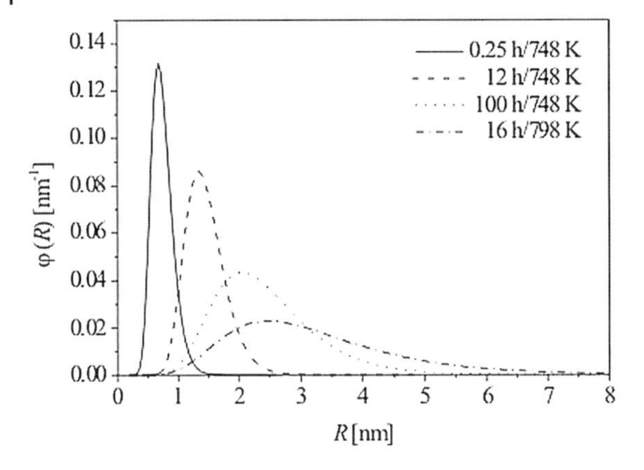


**Figure 13.5** Size distributions $\phi(R) = n(R)V(R)$ of spherical precipitates with diameter $D$ obtained from the magnetic SANS cross sections shown in Figure 13.4 [6].

distributed in the matrix with short distances between each other. This interparticle interference effect is accounted for by introducing the hard sphere factor $C_{HS}$. This factor is a constant used to calculate the hard-sphere radius of a particle $R_{HS} = C_{HS} \cdot R$ with a radius $R$ including a depleted zone around the particle [7]. For the aging time of 100 h at 748 K larger but fewer precipitates were found with longer distances, that means less interference (smaller $C_{HS}$ value).

In this work a special evaluation procedure was used: as a direct calculation of the size distribution from the scattering curve could lead to several mathematical solutions providing fits with a similar quality, shape and sizes of the precipitates were first determined by an analysis of TEM images. Afterwards, the size distribution was calculated using these parameters obtaining the number of the precipitates with a certain size. Following this evaluation procedure the local information of the TEM image was combined with the statistics of the complete SANS sample volume. In Figure 13.5 the obtained volume fraction distributions of spherical precipitates with diameter $D$ are shown.

The size distributions show the growth of the precipitates with increasing aging time. Precipitates with mean radii between 0.74 and 2.43 nm were formed at a temperature of 748 K. Number densities and volume fractions were determined which indicate fast precipitation kinetics. The heat treatment that produces the maximum hardness in this alloy is 12 h/748 K.

The ratio $A$ of magnetic to nuclear scattering intensity, which depends on the chemical composition of the precipitates, can be calculated from the measuring curves and compared with theoretical values [8]. Following this procedure an intermetallic G-phase ($Ti_6Si_7Ni_{16}$) phase was concluded for the precipitates.

As SANS is providing information about size distribution and composition of precipitates, it is the ideal method for the characterization of steels. Besides maraging steels othcr steel classes were analyzed by SANS as well [9–11]. For the investigation of magnetic structures the use of polarized neutrons is often advantageous because the determination of the magnetic cross section can be performed with high precision and without any background [12].

## 13.3 $SiO_2$ Nanoparticles in a Polymer Matrix – an Industrial Application

SANS is often used for the determination of size and volume fraction of components in composite materials. A typical example is the characterization of $SiO_2$ nanoparticles in aqueous or organic solution. SANS is an outstanding method for such investigations as large sample volumes can be measured which leads to a good statistics, the contrast between the $SiO_2$ particles and the matrix is high and – last but not least – no sample preparation is necessary and therefore no artifacts are formed.

Using an innovative, modified sol–gel process, hanse chemie AG, Geesthacht, Germany manufactures spherical, monodisperse and nonagglomerated silica nanoparticles which are used for coatings or varnishes and as additives for a wide variety of other applications, e.g., in adhesives. The small particle size and the narrow width of the size distribution and therewith the degree of agglomeration have significant influence on the properties of this new class of materials. The important step in the production process is the transfer of the monodisperse $SiO_2$ nanoparticles prepared from aqueous sodium silicate into a polymer or epoxy resin matrix.

Measurements at the SANS-2 instrument at GeNF were performed using again a wavelength of 0.58 nm and four detector distances. The evaluation of the resulting SANS curves showed that a very narrow particle size distribution of silica nanoparticles with sizes between 5 and 30 nm was obtained by the manufacturing process. It could easily be verified that the size distribution of the $SiO_2$ particles does not change significantly during the transfer from aqueous to polymer matrix as shown in Figure 13.6.

As model system interacting hard spheres were chosen. The relative sharp maximum in the SANS curve of the $SiO_2$ nanoparticle sample in aqueous solution originates from interparticle interference of the large number of particles in the concentrated solution of 40%. Therefore, the hard sphere constant factor $C_{HS}$ [7] was used again for the calculation of the size distributions. The obtained distributions of the two samples in Figure 13.6 are almost identical. This is a proof that the nanoparticles do not agglomerate during the transfer from aqueous to polymer matrix. The results demonstrate that with SANS size distributions of $SiO_2$ nanoparticles in concentrated and diluted solutions as well as in aqueous and polymer media can easily be determined. On the basis of these results hanse chemie AG applied for a patent for this special production process.

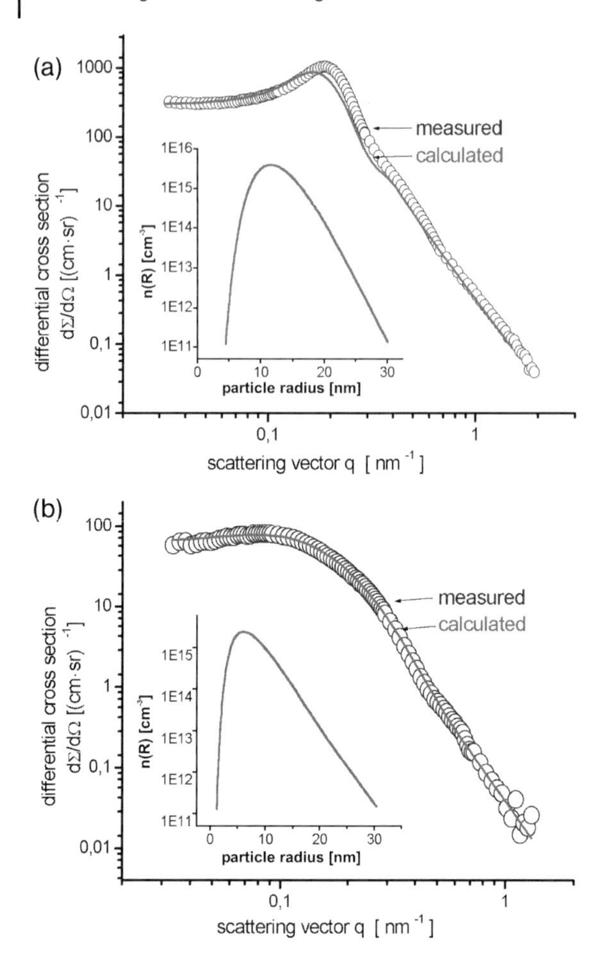


**Figure 13.6** SANS curves and corresponding particle size distributions of (a) 40% $SiO_2$ nanoparticles in water and (b) 5% $SiO_2$ nanoparticles in cycloaliphatic bisepoxide. The solid lines are the obtained fits during the calculation of the size distributions [1].

## 13.4 Green Surfactants

Glycolipids are amphiphilic molecules consisting of saccharide residues linked to an alkyl chain. They are applied in biology as biosurfactants and in environmental technology in waste water treatment [13]. They are less environmentally damaging than synthetic surfactants and can be produced from renewable sources; therefore, they are usually referred to as natural or "green" surfactants. To study the phase and aggregation behavior of glycolipids, especially the formation of micelles, SANS and SAXS investigations were performed [14]. The SANS

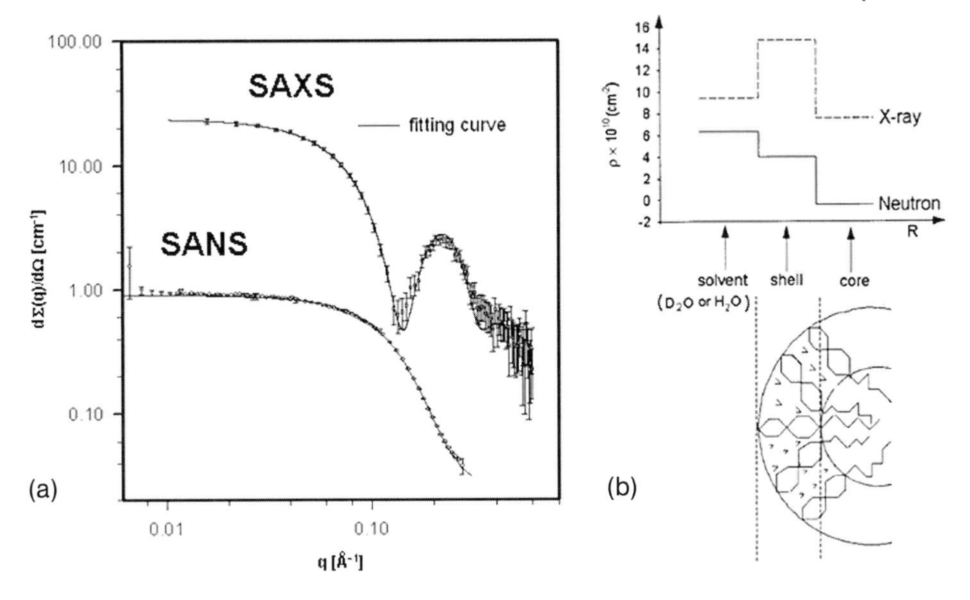


**Figure 13.7** (a) SANS and SAXS curves of OM measured in $D_2O$. (b) Scattering length densities for neutrons and X-rays of bulk solvent, hydrophilic shell and micelle core, and shell model [14].

experiments were carried out at the SANS-1 instrument at GeNF. A wavelength of 0.85 nm ($\Delta\lambda/\lambda = 0.1$) and four sample-to-detector distances (from 0.7 to 7 m) were used to cover the $q$-range from 0.07 to 2.5 $nm^{-1}$. As an example surfactant *N*-octyl-b-maltopyranoside (OM) was used. SANS and SAXS curves of OM are shown in Figure 13.7a, and the shell model and scattering length density profiles for neutrons and X-rays of bulk solvent, hydrophilic shell, and micelle core in Figure 13.7b.

The SANS and SAXS curves are obviously different due to the different scattering length density differences. The scattering-length density $\rho$ (or $\eta$) is calculated by $\rho = \sum_i^n b_i/v$, where $v$ is the volume of the head or tail group and $b_i$ is the scattering length of the atoms. For neutron scattering in $D_2O$ seven H atoms of the OH groups in the head group were assumed to exchange with D atoms in the solvent.

The monodisperse spherical shell model shown in Figure 13.7b (hydration water in the shell of the micelle is not considered in this figure), which describes SANS and SAXS data, was applied to determine the radii of core and whole micelle as well as the contrast ratio between the core and hydrophilic shell. The resulting fits are the solid lines in Figure 13.7a. Using this model it was shown that the micelle size is constant over a large range of OM concentrations and it is not sensitive to temperatures from 10 °C to 50 °C [14].

This example from the field of soft matter science shows the complementary information which can be obtained by SANS and SAXS. The micellar structure was solved in this case using the combination of both methods. Additional results from other methods are often useful for completing the information obtained by SANS.


## Acknowledgments

Valuable contributions of my colleagues Regine Willumeit, Helmut Eckerlebe, Peter Staron and Vasyl Haramus are gratefully acknowledged.


## References


1 A. Schreyer, J. Vollbrandt, *Experimental facilities at the Geesthacht Neutron Facility GENF 2006.*

2 L. Schlapbach, A. Zuettel, Hydrogen-storage materials for mobile applications, *Nature* 2001, **414**, 353–358.

3 P.K. Pranzas, M. Dornheim, D. Bellmann, K.-F. Aguey-Zinsou, T. Klassen, A. Schreyer, SANS/USANS investigations of nanocrystalline $MgH_2$ for reversible storage of hydrogen, *Physica B*, in press (online available, 2006). Experimental Facilities at the Geesthacht Neutron Facility GeNF, GKSS, Geesthacht 2006, *Physica B* 2006, **385–386**, 630–632.

4 D. Bellmann, P. Staron, P. Becker, Performance of a double crystal diffractometer with different channel-cut perfect Si crystals, *Physica B* 2000, **276**, 124–125.

5 P. Staron, D. Bellmann, Analysis of neutron double-crystal diffractometer scattering curves including multiple scattering, *J. Appl. Cryst.* 2002, **35**, 75–81.

6 P. Staron, B. Jamnig, H. Leitner, R. Ebner, H. Clemens, Small-angle neutron scattering analysis of the precipitation behaviour in a maraging steel, *J. Appl. Cryst.* 2003, **36**, 415–419.

7 D.J. Kinning, E.L. Thomas, Hard-sphere interactions between spherical domains in diblock copolymers, *Macromolecules* 1984, **17**, 1712–1718.

8 M. Große, A. Gokhman, J. Böhmert, Dependence of the ratio between magnetic and nuclear small angle neutron scattering on the size of the heterogeneities, *Nucl. Instrum. Methods B* 2000, **160**, 515–520.

9 H. Leitner, P. Staron, M. Bischof, H. Clemens, E. Kozeschnik, J. Svoboda, F.D. Fischer, *Coarsening of Secondary Hardening Carbides in a Hot-Work Tool Steel: Experiments and Simulation, MS&T 2004 Conference Proceedings 2004*, pp. 623–632.

10 H. Leitner, P. Staron, H. Clemens, S. Marsoner, P. Warbichler, Analysis of the precipitation behaviour in a high-speed steel by means of small-angle neutron scattering, *Mater. Sci. Eng. A* 2005, **398**, 323–331.

11 M. Bischof, S. Erlach, P. Staron, H. Leitner, C. Scheu, H. Clemens, Combining complementary techniques to study precipitates in steels, *Z. Metallk.* 2005, **96**, 1074–1080.

12 R. Coppola, R. Kampmann, M. Magnani, P. Staron, Microstructural investigation using polarized neutron scattering of a martensitic steel for fusion reactors, *Acta Mater.* 1998, **46**, 5447–5456.

13 S.C. Lin, Biosurfactants: recent advances, *J. Chem. Technol. Biotechnol.* 1996, **66**, 109–120.

14 L. He, V.M. Garamus, S.S. Funari, M. Malfois, R. Willumeit, B. Niemeyer, Comparison of small-angle scattering methods for the structural analysis of octyl-, maltopyranoside micelles, *J. Phys. Chem. B* 2002, **106**, 7596–7604.

# 14
# Decomposition Kinetics in Copper–Cobalt Alloy Systems: Applications of Small-Angle X-ray Scattering

*Günter Goerigk*

## 14.1
## Introduction

The macroscopic properties (mechanical, electrical, optical, catalytic ...) of numerous materials – examples are alloys, glasses, semiconductors, metal nanoparticles prepared on microporous support structures, biomembranes, or proton-conducting technical membranes – are strongly influenced by the materials nanostructure. These nanostructures are built up by chemical and/or topological inhomogeneities on a mesoscopic length scale between 1 and 1000 nm, which is difficult to access by microscopic methods. In this chapter a scientific objective from metallurgy is introduced as an example for applying synchrotron radiation techniques for the analysis of scientific problems from basic research, which are of strong interest for applied physics and materials sciences.

## 14.2
## ASAXS Fundamentals

The structural properties of materials on the nanometer length scale can be analyzed by small-angle scattering experiments mainly small-angle neutron scattering (SANS) or small-angle X-ray scattering (SAXS) experiments. In the foregoing decades SAXS experiments with synchrotron radiation offered big advantages in comparison to classical SAXS experiments carried out with X-ray tubes. The high photon flux of the synchrotron radiation offers the possibility of following, for example, phase transitions and the related changes of the nanostructure in alloys or glasses by *in-situ* time-resolved scattering experiments and the time dependence of the structural and the quantitative parameters can be obtained. The latter are strongly related to important thermodynamic parameters of the phase transition under investigation, which govern the alloys macroscopic properties. Important structural parameters are, for instance, the size and the shape of small precipitates of an alloy or a glass and quantitative parameters are volume

*Neutrons and Synchrotron Radiation in Engineering Materials Science: From Fundamentals to Material and Component Characterization*
Edited by Walter Reimers, Anke Rita Pyzalla, Andreas Schreyer, and Helmut Clemens

ISBN: 978-3-527-31533-8

fractions or chemical compositions of such nanophases and the related thermodynamic parameters are diffusion coefficients, equilibrium solubility, and interfacial energy. The scattering of X-ray photons on atoms shows for all elements a strong variation in the vicinity of the so-called X-ray absorption edges and due to this, the small-angle scattering, which origins specifically from one nonhomogeneously distributed component of an alloy, changes with the X-ray energy in the neighborhood of the related X-ray absorption edges. For this technique – called anomalous small-angle X-ray scattering (ASAXS) – synchrotron radiation is needed, because synchrotron radiation covers a wide range of X-ray energies. Meanwhile, numerous storage rings provide synchrotron radiation with energy (X-ray) spectra from bending magnets, wigglers, or undulators, which cover the energy range of the X-ray absorption edges of most of the elements and the ASAXS technique became an important tool for the element-specific nanostructural characterization of numerous classes of composite materials in solid state physics, chemistry, and materials science. In the following chapter we will demonstrate how the nanostructure specifically related to the Co component in a crystalline Cu–Co alloy can be analyzed by ASAXS and how from the obtained structural and quantitative parameters important decomposition parameters of this alloy have been deduced.

In Figure 14.1 the phase diagram of the alloy copper–cobalt is shown partially. The solid lines depict the borders between different phases of the alloy depending on the temperature and the concentration of Co. The solidus–liquidus line in the top is the border between the solid and the liquid phase of the alloy. Below this line in the solid state at high Cu concentrations (>90%) the so-called solution

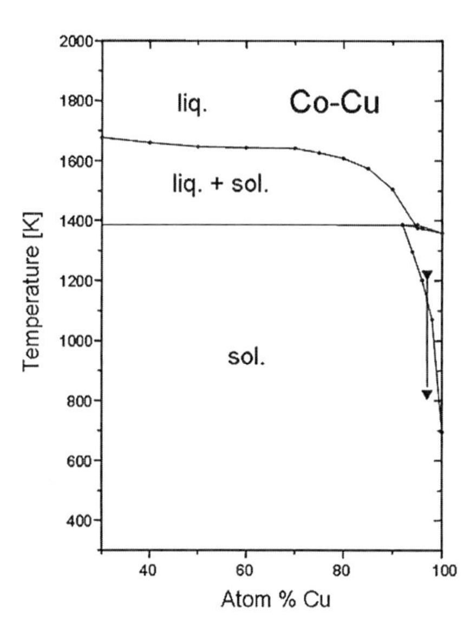


**Figure 14.1** Phase diagram of the binary alloy Cu–Co [18].

line with nearly vertical direction is drawn. When passing this line from higher to lower temperature with Co concentrations of several percent one enters the so-called two-phase area. The vertical arrow in Figure 14.1 connects the point in the phase diagram, where the solution treatment of the alloy was applied (4 h at 1223 K) with the point, where the *in-situ* study with ASAXS was performed. Beyond the phase border both components of the alloy can co-exist in a homogeneous solid solution. Crossing the phase border to lower temperatures by cooling causes the origin of a second phase built up by small precipitates, which consists nearly of pure cobalt. Details of the thermodynamics of the decomposition kinetics can be found in [1, 2].

The phase diagram provides no information about the distribution and the morphology of the precipitated phase and of its development in time. These can be analyzed by SAXS. The photons are scattered by the electrons of the atoms. Because the number of electrons of Cu (Z = 29) and Co (Z = 27) are different, a scattering contrast originates from the Co precipitates with respect to the surrounding Cu matrix. X-ray photons, which penetrate the alloy, are scattered at these inhomogeneities and a characteristic scattering pattern due to the geometrical shape and the size of the precipitates originates, when measured with a suitable detector system. The scattered intensity contains the information about the number of the precipitates in the alloy. The items are summarized in Eq. (14.1):

$$\begin{aligned} d\sigma/d\Omega(q) &= N_p \cdot \Delta\rho^2 \cdot V_p^2 \cdot S(q) \\ \Delta\rho &= |n_p \cdot f_p - n_m \cdot f_m|^2 \end{aligned} \tag{14.1}$$

$q = 4\pi/\lambda \sin(\Theta)$ is the amount of the so-called scattering vector, which is related to the scattering angle $2\Theta$. $\lambda$ is the wavelength of the X-ray photons. $d\sigma/d\Omega$ denotes the scattering cross section (in electron units), $N_p$ and $V_p$ are the number and the volume of the scattering precipitates, respectively, and $\Delta\rho$ is the contrast between the electron density in the precipitates and the surrounding matrix. $S(q)$ is the form factor of the scattering precipitates and is related to its geometrical shape and size via the Fourier transform. The electron density contrast, $\Delta\rho$, is calculated from the difference of the averaged atomic number densities $n_p$, $n_m$ in the precipitates and the surrounding matrix and the atomic scattering factors $f_p$, $f_m$ of the related atoms. A detailed description of SAXS can be found in [3].

Additionally to the scattering of the precipitates the scattering of other nanostructures like dislocations in the crystal, voids, and the surface roughness of the analyzed sample contribute to the SAXS. The scattering pattern represents the superposition and the scattering contribution specifically related to the Co precipitates cannot be distinguished. This is the classical problem of scattering experiments, for instance, performed with an X-ray tube.

The separation of the scattering contribution specifically related to the Co precipitates can be obtained when synchrotron radiation is employed, because the synchrotron radiation provides an X-ray spectrum over a wide energy range. As can be seen from Eq. (14.1), the atomic scattering factors of Eq. (14.2) enter the contrast. The atomic scattering factors contain the number of electrons, $Z$, and the so-called anomalous dispersion corrections $f'(E)$, $f''(E)$, which are generally energy dependent:

$$f(E) = Z + f'(E) + if''(E) \tag{14.2}$$

The atomic scattering factor $f$ in Eq. (14.2) is written in units of the classical electron radius, $r_e = 2.8 \times 10^{-13}$ cm. In the vicinity of the X-ray absorption edges (K, L, M...) the anomalous dispersion corrections and the atomic scattering factors show a strong variation with the energy. The scattering of the Co atoms changes strongly in the neighborhood of the K-absorption edge at 7.709 keV, while the scattering of the Cu atoms remains nearly unchanged. The nanostructures composed from Co (the Co precipitates) show an energy dependent small-angle scattering due to their energy dependent contrast (contrast variation), while the scattering originating from nanostructures composed mainly from Cu (dislocations, voids, surface roughness) remain nearly unchanged. The small-angle scattering of the Co precipitates can be separated from the superimposed contributions by measurements at two different energies in the vicinity of the K-absorption edge of Co. This technique generally can be applied to a large amount of alloys (binary, ternary ...) and other materials like glasses, semiconductors, catalysts, macromolecular solutions, etc. The specific structure information of one of the materials constituents can be accessed by tuning the scattering contrast in the vicinity of the X-ray absorption edge of the related element. Details of the ASAXS technique can be found under [4] and [5].

## 14.3 Results of ASAXS Experiments Characterizing the Decomposition in Copper–Cobalt Alloys

In Figure 14.2 results from a solution treated alloy are shown. The analyzed Cu–Co alloy with a Co concentration of 0.5at% was solution treated under reducing Ar/$H_2$-atmosphere over 4 h at a temperature of 1223 K. As can be seen from the phase diagram (Figure 14.1) in the thermodynamic equilibrium the Co component is homogeneously distributed in the Cu matrix. The solution treatment was followed by a rapid quench into water. The small-angle scattering curves measured at two energies in the vicinity of the K-absorption edge of Co at 7.709 keV show no difference. Thus the homogeneous distribution of the Co atoms within the Cu matrix was preserved by quenching.

The solution treated alloy was reheated to the aging temperature of 813 K. As can be seen from Figure 14.1, at this temperature an alloy with a Co concentration of 0.5at% is in the miscibility gap and thus decomposition of small Co precipitates takes place. The origin and the evolution of the Co precipitates was followed *in-situ* by time-resolved ASAXS measurements. Figure 14.3 summarizes the results obtained after heating the sample for 1/2 hour. Figure 14.3a shows the

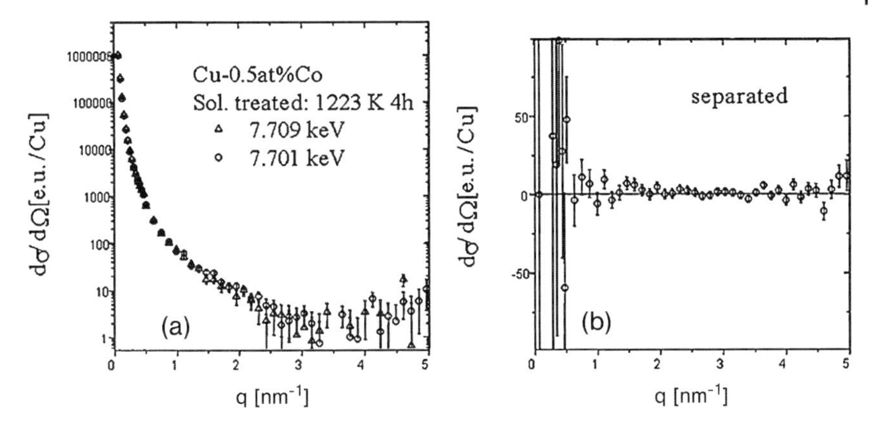


**Figure 14.2** The two scattering curves of the solution treated alloy Cu-0.5at%Co measured at two energies: (a) The points of the difference curve scatter around zero; (b) the precipitates are not yet generated.

two scattering curves obtained from SAXS measurements at 7.709 and 7.998 keV, respectively. The significant difference of the two scattering curves can be attributed to the origin of the small Co precipitates. From the difference scattering curve in Figure 14.3b, the averaged radius of the Co precipitates can be obtained. ASAXS measurements at 15 time increments between 0.5 and 6 h have been performed and the small-angle scattering of the Co precipitates was separated. The results are summarized in Figure 14.4. In Figure 14.4b the radii of the Co precipitates obtained from the separated scattering curves are plotted versus the time,

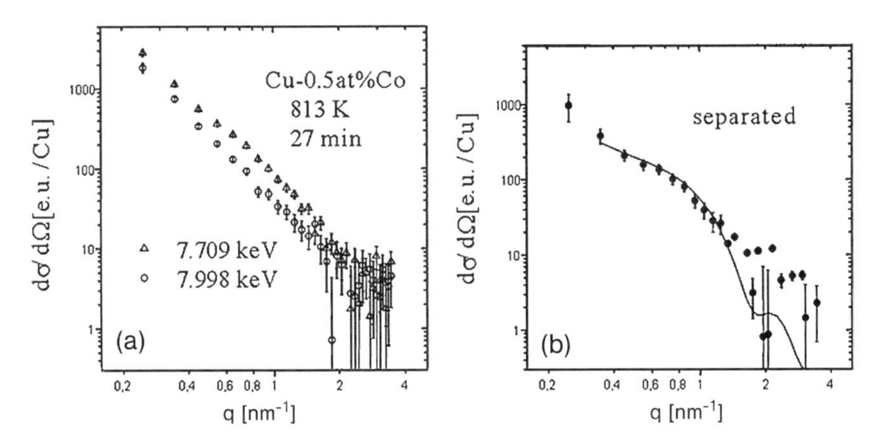


**Figure 14.3** After heating the alloy to a temperature of 813 K significant differences are observed after 1/2 hour in the scattering curves measured at different energies. The differences can be attributed to small Co precipitates. From the two scattering curves (a) the scattering curve related to the Co precipitates can be separated (b).

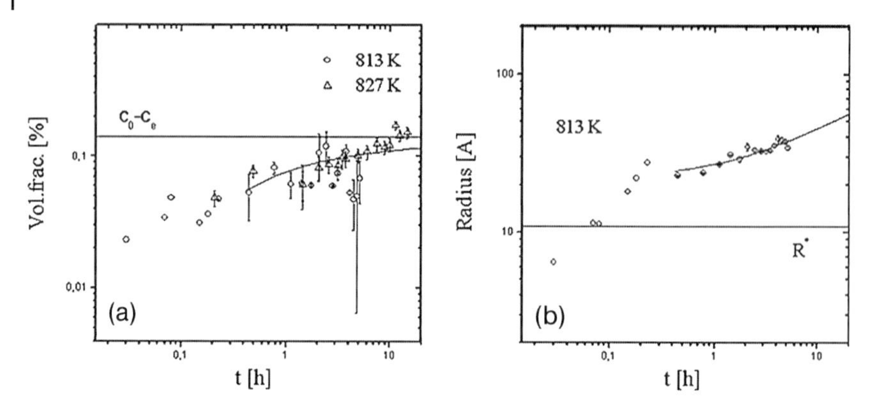


**Figure 14.4** Summarized results obtained from 15 ASAXS measurements between $\frac{1}{2}$ and 6 h. The development in time of the volume fraction (a). The time dependence of the averaged radius of the precipitates (b). The points in the left of the solid lines represent time stages within the first 15 min and have been measured at only one energy. These points were not used for the further analysis of the decomposition parameters.

while in Figure 14.4a the volume fraction of the precipitates is shown. The solid lines are fitted potential laws (Eq. 14.3), which give the time dependence of the averaged radius $R_0$ and the volume fraction $v_f$ of the precipitates as deduced from the theory of Lifshitz, Slyozov, and Wagner on the decomposition kinetics in supersaturated solutions [6, 7].

$$R_0^3(t) - R^3(0) = \alpha_{LSW} \cdot (t - t_0)$$
$$v_f(t) = c_0 - c_e - \Delta(t) = \Lambda \cdot t^{-1/3} \qquad (14.3)$$

$c_0$ is the Co concentration of the alloy and $c_e$ represents the solubility, which the alloy finally reaches in thermodynamic equilibrium. The supersaturation $\Delta(t)$ defines the excess concentration of the alloy at time $t$. From the constants $\alpha_{LSW}$ and $\Lambda$ the values of the interfacial energy, $\sigma = 180$ mJ/cm$^2$, and the critical radius, $R^* = 1.1$ nm have been calculated [8]. The interfacial energy is the amount of energy, which is necessary when an interface between the second phase and the surrounding copper matrix is formed, while the critical radius gives the minimum size of the precipitates. Beyond this size the precipitates are stable and can enter the growth regime. The time dependence of the decomposition kinetics and the calculated decomposition parameters are in good agreement with the predictions of the decomposition theories given in [1, 6, 7]. The measurements represent a time-resolved analysis of the decomposition kinetics in Cu–Co alloys performed with synchrotron radiation for the first time. With the technique of ASAXS the scattering contributions, which are specifically related to the Co precipitates could be separated and the increase of the averaged radius and the volume fraction with time, respectively, were calculated. The time dependence and the calculated value of the interfacial energy give evidence of a classical decomposition process as predicted by the theory of Lifshitz, Slyozov, and Wagner [6, 7].

## 14.4 Outlook

Precipitates play an important role in the context of hardening of alloys, i.e., in order to strengthen an alloy against deformation. They serve as obstacles for dislocations, which are responsible for plastic deformation and influence the materials mechanical and thermal stability [2]. Other examples for materials inhomogeneities, which influence the macroscopic properties, are metal-containing amorphous hydrocarbon films [9], which are used for coating of materials that possess high hardness, excellent wear behavior, and very low friction coefficients. Additionally, due to their metal content, the electrical conductivity may be varied over several orders of magnitude, making them interesting, e.g., for electrical sliding contacts and a large number of industrial protective applications. Further examples are the embrittlement of reactor steel caused by radiation-induced metal carbides [10], glass ceramics with magnetic nanocrystals for magnetic storage devices [11] or amorphous Si–Ge alloys used in solar cell techniques [12–15]. The optoelectronic properties of these semiconductor alloys are strongly correlated with the inhomogeneous distribution of the Ge component in the amorphous matrix. All these materials have been analyzed by ASAXS in the last years.

Catalysts systems, which contain metal nanoparticles, for example platinum or palladium, prepared on porous support structures of silica or porous carbon, are used in fuel cell technologies. The catalytic properties of these materials are strongly influenced by the size and the size distribution of the nanoparticles, but probably also from the support structure. Significant contributions to the structural analysis of those systems came from ASAXS, because a relatively weak scattering originating from the nanoparticles must be separated from an overwhelming superimposed scattering of the porous support structure (up to 99% of the total scattering intensity) [16, 17].

## 14.5 Summary

Anomalous small-angle x-ray scattering (ASAXS) using synchrotron radiation has become an important tool for the element-specific nanostructural characterization of numerous classes of composite materials in solid state physics, chemistry, and materials science. As an example nanostructure characterization of a crystalline Cu–Co alloys, in particular related to the Co component is treated here. The structural and quantitative parameters obtained by ASAXS revealed that the decomposition of the Cu–Co alloys investigated can be described by the Lifshitz, Slyozov, and Wagner theory.

## References


1 J.W. Gibbs, *The Scientific Papers of J.W. Gibbs*, Vol. 1, New York: Longmans Green (1906)

2 P. Haasen, *Physikalische Metallkunde*, Berlin: Springer (1974)

3 A. Guinier, G. Fournet, *Small-Angle Scattering of X-rays*, New York: Wiley (1955)

4 H.B. Stuhrmann, G. Goerigk, B. Munk, *Anomalous X-Ray Scattering, Handbook of Synchrotron Radiation* (Ed. by S. Ebashi, M. Koch, E. Rubenstein), 1991, Vol. 4, 555–580, Amsterdam: Elsevier.

5 H.-G. Haubold, R. Gebhardt, G. Buth, G. Goerigk, *Structural characterization of compositional and density inhomogeneities by ASAXS, Resonant Anomalous X-ray Scattering* (Edited by G. Materlik, C.J. Sparks and K. Fischer), 1994, 295–304, Amsterdam: Elsevier.

6 I.M. Lifshitz, V.V. Slyozov, The kinetics of precipitation from supersaturated solid solutions, *J. Phys. Chem. Solids* 1961, **19**, 35–50.

7 C. Wagner, Theorie der Alterung von Niederschlägen durch Umlösen, *Z. Elektrochem.* 1961, **65**, 581–591.

8 G. Goerigk, H.-G. Haubold, W. Schilling, Kinetics of Decomposition in Copper–Cobalt: a Time-Resolved ASAXS Study, *J. Appl. Cryst.* 1997, **30**, 1041–1047.

9 K.I. Schiffmann, M. Fryda, G. Goerigk, R. Lauer, P. Hinze, A. Bulack, Sizes and distances of metal clusters in Au-, Pt-, W- and Fe-containing diamond-like carbon hard coatings: a comparative study by small angle X-ray scattering, wide angle X-ray diffraction, transmission electron microscopy and scanning tunnelling microscopy, *Thin Solid Films* 1999, **347**, 60–71.

10 M. Große, F. Eichhorn, J. Böhmert, G. Brauer, H.-G. Haubold, G. Goerigk, ASAXS and SANS investigations of the chemical composition of irradiation-induced precipitates in nuclear pressure vessel steels, *Nucl. Instrum. Methods Phys. Res. B* 1995, **97**, 487–490.

11 U. Lembke, A. Hoell, R. Müller, W. Schüppel, G. Goerigk, R. Gilles, R.A. Wiedenmann, Formation of magnetic nanocrystals in a glass ceramic studied by small-angle scattering, *J. Appl. Phys.* 1999, **85**, No. 4, 2279–2286.

12 G. Goerigk, D.L. Williamson, Nanostructured Ge distribution in a-SiGe:H alloys from anomalous small-angle X-ray scattering studies, *Sol. Stat. Comm.* 1998, **108**, No. 7, 419–424.

13 G. Goerigk, D.L. Williamson, Quantitative ASAXS of Germanium Inhomogeneities in Amorphous Silicon–Germanium Alloys, *J. Non-Cryst. Solids* 2001, **281**, 181.

14 G. Goerigk, G.D.L. Williamson, Comparative ASAXS study of hotwire and PECVD grown amorphous silicon–germanium alloys, *J. Appl. Phys.* 2001, **90**, 5808.

15 G. Goerigk, D.L. Williamson, Temperature induced differences in the nanostructure of hot-wire deposited silicon–germanium alloys analyzed by small-angle X-ray scattering, *J. Appl. Phys.* 2006, **99**, 084309.

16 H.-G. Haubold, X.H. Wang, H. Jungbluth, G. Goerigk, W. Schilling, In situ anomalous small-angle X-ray scattering and X-ray absorption near-edge structure investigation of catalyst structures and reactions, *J. Mol. Struct.* 1996, **383**, 283.

17 A. Benedetti, L. Bertoldo, P. Canton, G. Goerigk, F. Pinna, P. Riello, S. Polizzi, ASAXS study of Au, Pd, and Pd–Au catalysts supported on active carbon, *Catalysis Today* 1999, **49**, 485–489.

18 *Smithells Metalls Reference Book, 6th ed.*, Brandes, E.A. (Ed.), London: Butterworth & Co. Ltd. London (1983).

# 15
# B3 Imaging

*Wolfgang Treimer*

## 15.1
## Radiography

### 15.1.1
### Fundamentals

A radiograph is the two-dimensional attenuation distribution of a ray-path-integrated projection from a three-dimensional object. Historically established are films in the case of X-rays and gas counters in the case of neutrons. For neutron radiography also film-converter foil systems were developed and used. In the last decades more and more electronically based systems (image foils) and electronic recording systems (CCD cameras) became available with improved image properties and qualities.

Today only in very special cases film imaging is used; much more common are detectors, based on CCD cameras that supply a hardware resolution (pixel size) down to some micrometers which is of the same order of films. The advantages of electronic recording are the possibilities of easy image processing, high bit depth (optical density), real-time imaging, and information handling (delivery, storage). The next chapter will give all interesting details about the recording cameras; here we will summarize only the some necessary facts which are important for our considerations.

The method radiography seems to be a simple technique, which needs only a few parameters to work with. First some definitions must be established which are also important for the discussion of tomography. The step from radiography = projection to tomography = many projections is then rather small. Only some mathematical tools have to be applied to get a 3D image from a series of radiographs (projections).

For a neutron or X-ray radiography one must have in general a beam, an object, and a detector system for imaging. The beam is specified by the

*Neutrons and Synchrotron Radiation in Engineering Materials Science: From Fundamentals to Material and Component Characterization*
Edited by Walter Reimers, Anke Rita Pyzalla, Andreas Schreyer, and Helmut Clemens

ISBN: 978-3-527-31533-8

- flux $\Phi = \dfrac{\text{number of photons (neutrons)}}{\text{unit time} \times \text{unit area}}$
- wavelength distribution $(\Delta\lambda/\lambda)$
- divergence $\phi$
- polarization $P$ (or spin state in the case of neutrons, here not discussed)

For radiography and tomography $\Phi$, $\Delta\lambda/\lambda$, $\phi$ are of principal interest; in the case of polarized neutron radiography one has to use polarized neutrons in front of the sample, a spin analyzer in front of the sample and/or a spin flipper and a spin analyzer in front of the detector (film). The flux used for a radiograph is a quantity, which strongly depends on the monochromaticity of the beam, i.e., its wavelength distribution $(\Delta\lambda/\lambda)$ and divergence $\phi$. In order to illustrate the dependence of the available intensity consider a source $S$ emitting radiation with a wavelength spectrum as shown in Figure 15.1.

The same holds for X-rays. The Bremsstrahlung spectrum has a characteristic shape that is nearly independent of the target material (the kinetic energy of the accelerated electrons is much larger than the binding energy of the bound electrons). In the case of X-rays the maximum photon energy is determined by the maximum acceleration voltage, yielding the shortest wavelength in the spectrum. It is given by ($h$ = Planck constant, $c$ = speed of light, $\nu_{max}$ = maximum frequency)

$$\lambda_{\min} = \frac{hc}{h\upsilon_{\max}} \tag{15.1}$$

One can see from Eq. (15.1) that the wavelength is of the order of 0.2 nm and less and the energy $h\nu$ ranged from approximately 10 keV until 250 keV. In the case of neutrons the kinetic energy is given by

$$E_{\text{kin}} = \frac{m \cdot v^2}{2} \tag{15.2}$$

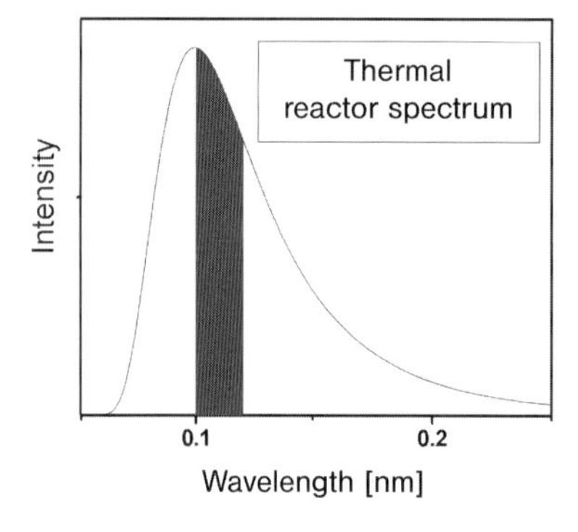


**Figure 15.1** Cold reactor spectrum: black bar denotes a certain $\Delta\lambda/\lambda$-band in the spectrum, the area of the band is proportional to the intensity of monochromatic radiation.

with the corresponding neutron velocity calculated by $v = \frac{h}{m\lambda} = 3.95603 \times 10^3\ \text{ms}^{-1}$ ($\lambda$ in 0.1 nm or Å). The neutron energy $E_{kin}$ is then given by $E_{kin} = \frac{m \cdot v^2}{2} \rightarrow E(\lambda) = 0.081804 \left[\frac{\text{eV}}{\lambda}\right]$. That yields for $\lambda = 0.1$ nm $\sim 0.081\,804$ eV (with $m$ = neutron mass = $1.674\,9286(10)10^{-27}$ kg, 1 (eV) = $6.241\,495\,96 \times 10^{18}$ (J)).

The neutron flux $\Phi$ from a research reactor has usually the shape of a Maxwell spectrum due to the thermal equilibrium of neutrons with the moderator. The spectrum is different in the case of neutron guides, because of their different curvatures and coatings. Neutron radiography (and tomography) uses mainly the whole spectrum, accepting both a very intense neutron beam and an integrated absorption behavior of the sample.

### 15.1.2
### Interactions of Neutrons with Matter

There are three possibilities that neutrons can interact with matter, they can be scattered, absorbed or transmitted (Figure 15.2).

Commonly all interactions are described by so-called cross sections, referring to the particular interaction the neutrons undergo. The interactions of neutrons with matter can be distinguished into macroscopic $\Sigma$ and microscopic cross sections $\sigma$. One can define $\sigma$ and $\Sigma$ as follows (definition after [1, 2]). The cross section $\sigma$ is called microscopic if it refers to a single nucleus. The dimension $[\sigma] = (\text{cm}^2)$; unit = 1 barn = $10^{-24}$ cm$^2$. The differential cross section

$$\frac{d\sigma}{d\Omega} = \frac{\text{number of interacting particles/unit time} \times \text{unit cone}\, d\Omega}{\text{number of incident particles/unit time} \times \text{unit area} \times \text{unit cone}\, d\Omega}$$

$$= (\text{area})$$

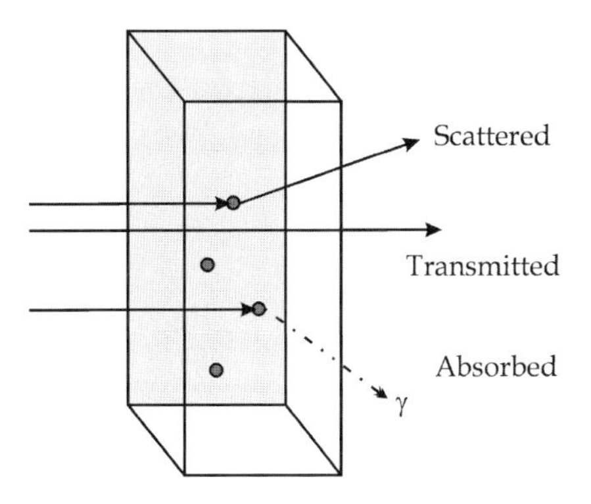


**Figure 15.2** Possible interactions of a neutron with matter.

The cross section $\Sigma$ is called macroscopic if $\Sigma = N \times \sigma$, and $N$ denotes the number of nuclei per $cm^3$. The dimension of $\Sigma$ is ($cm^{-1}$). From the change of the intensity $I$ (or neutron flux $\Phi$)

$$-\frac{dI}{I} = \Sigma \times dz \tag{15.3}$$

one gets the probability $\Sigma \times dz$ that neutrons will interact with matter along the path $dz$. With $A_0$ (Loschmidt or Avogadro number $= 6.02 \times 10^{23}$) the number of atoms $cm^{-3}$ $N$ is given by

$$N = \frac{\rho}{A} \times A_0 \tag{15.4}$$

if $A$ is the atomic weight and $\rho$ the density. The macroscopic cross section then becomes

$$\Sigma = N \times \sigma = \frac{\rho}{A} \times A_0 \times \sigma \tag{15.5}$$

*Example*: Determine the macroscopic cross section of iron.

*Solution*: Iron has a density of 7.86 g $cm^{-3}$, the microscopic cross section for absorption of iron is 2.56 barns, and the gram atomic weight is 55.847 g. Thus $\Sigma = 0.217\ cm^{-1}$.

For a material consisting of $j$-different atoms, $\Sigma$ becomes the weighted sum of the cross sections of the constituents

$$\Sigma = \sum_j N_j \sigma_j \tag{15.6}$$

where $N_i$ is the density of the nuclei of kind $i$ and $\sigma_i$ the corresponding microscopic cross section of the nuclei. From this equation the attenuation law can be expressed as

$$I(z) = I_0 \cdot e^{-\int_{path} \Sigma(z)\,dz} \tag{15.7}$$

if $\Sigma(z)$ means the $z$-dependent cross section along the path of the neutron through the sample. Note that the reciprocal value, $1/\Sigma = \lambda_{free}$, is the mean free path of neutrons in the material. Referring to the absorption law one can define a "half-thickness" $z_{1/2}$ as

$$z_{1/2} = \frac{\ln(2)}{\int_{path} \Sigma(z)\,dz} \tag{15.8}$$

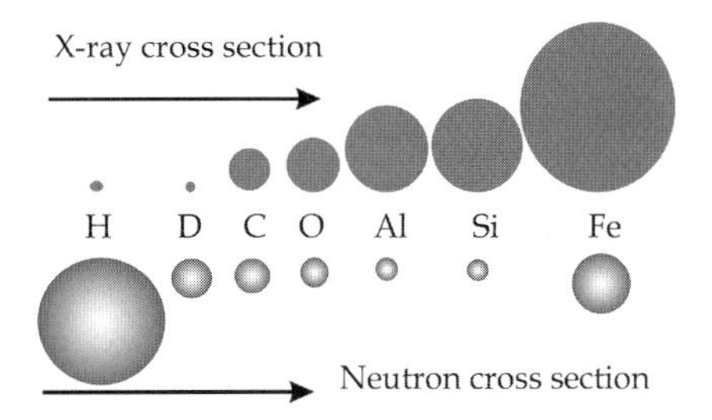


**Figure 15.3** For X-rays the "amount or strength" of interaction increases with the atomic number (and number of electrons) that does not hold for neutrons.

In the case of scattering the macroscopic scattering cross section is

$$\Sigma_s(z) = \sum_j N_j(z) \times \sigma_{s,j} \tag{15.9}$$

i.e., the weighted sum of all contributing nuclei $(\sigma_{s,j})$ along the $z$-direction. For neutron radiography the absorption cross section is of major importance, i.e., $\Sigma = N_j\sigma$ is replaced by the absorption cross section $\Sigma_a = N_{j,a}\sigma_{j,a}$ if the dominant interaction is absorption:

$$\Sigma_a(z) = \sum_j N_j(z) \times \sigma_{a,j} \tag{15.10}$$

The total macroscopic cross section $\Sigma_t(z)$ is then defined as

$$\Sigma_t(z) = \Sigma_a(z) + \Sigma_s(z) \tag{15.11}$$

The scattering process plays an important role in radiography and tomography because scattering contributions usually result in a decrease of sharpness in the picture. (However, scattering processes can also be used as imaging signals.) Figure 15.3 shows approximately the relations between X-ray and neutron cross sections and points out the complementary nature of both radiations.

### 15.1.3 Geometries

For radiography the radiation is usually collimated by slits or a collimator system that reduces the propagation direction of the radiation and therefore the intensity. To estimate the intensity in front of a sample which is situated at a certain distance from the source one can deduce the flux as follows. If a source emits radiation into $4\pi$, the source flux $I_0$ decreases at a certain distance $R$. Within a solid angle $\Delta\Omega = (\phi_h \times \phi_v)$ $(\phi_{h,v}$ is the horizontal and vertical divergence) the intensity

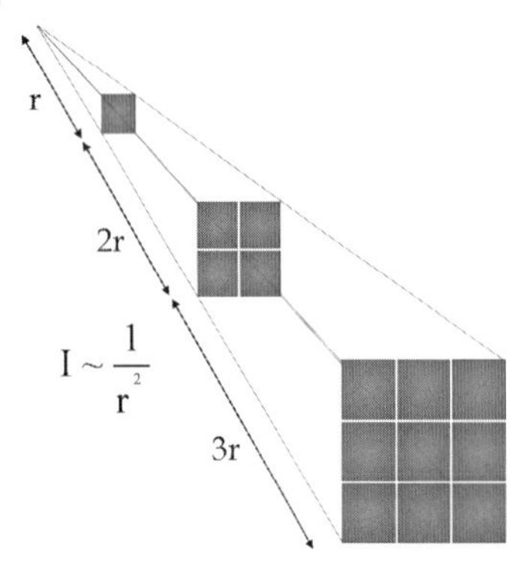


**Figure 15.4** The number of photons or neutrons/unit area decreases as $1/R^2$.

$I$ is proportional to the area of the surface of the sphere $\Delta A = ds \times ds$. The element $ds$ can be determined as $(\phi \times R)$, where $\phi$ is the one-dimensional divergence. The intensity will decrease as $1/R^2$. Figure 15.4 demonstrates the $1/R^2$-law for a given distance $r$, $2r$, and $3r$ from a point source. A given intensity $I$ at a distance $r$ is distributed over a four times larger area at $2r$ and therefore the density is decreased by a factor 4.

For a known divergence $\phi$ (which can be measured) the available intensity $I$ in front of a sample can be calculated from a given source flux $I_0$. The intensity (= neutrons/area × time) at a distance $d$ decreases as shown in Figure 15.4. Moreover, an area in an object $A_{\text{Object}}$ is enlarged and blurred to an area $A_{\text{Image}}$ depending on the following parameter:

- $L/D$ ratio ~ divergence,
- distance object–detector (film, CCD camera),
- blurring due to secondary detection.

### ad (i) and ad (ii) *L/D* Ratio

The (effective) source size $D$, the distance $L$ of the object from the source, and the distance $l$ of the object from the detector play important roles in radiography and determine – together with other parameters – the quality of a radiograph. Every point in a sample is scaled up to an area with the diameter $d$ on the detector (film, CCD camera) as

$$d = \frac{l}{L/D} \tag{15.12}$$

$L/D$ defines the divergence $\phi$, as easily can be seen in Figure 15.5.

*Note*: A large $L/D$ number means a low divergence and neutron flux and vice versa. Depending on the sample problem to be imaged, $L/D$ must be adapted to the boundary conditions (neutron flux, spatial resolution of the detector, etc.) of the experimental set-up.

**Table 15.1** Geometric parameter for a given distance source–object $L = 10$ m and distance object–point detector $l = 50$ mm, the pinhole diameter is varied from 10 mm to 50 mm.

| **Pinhole diameter (mm)** | **10** | **20** | **30** | **40** | **50** |
|---|---|---|---|---|---|
| $L/D$ | 1000 | 500 | 333 | 250 | 200 |
| Divergence $\phi$ (°) | 0.057 | 0.114 | 0.171 | 0.228 | 0.285 |
| l = 50 mm: point spread (mm) | 0.05 | 0.1 | 0.15 | 0.2 | 0.25 |

In radiography and tomography one uses both a pinhole or a collimator. In the case of a pinhole, $D$ is the diameter of the pinhole. In Table 15.1, $L/D$ and divergence $\phi$ are displayed for given pinholes in mm if $L = 10$ m.

*Example*: For an $L/D = 250$ a point in a sample, 50 mm apart from the detector, is imaged as a circle with a diameter of 200 μm; if $l = 100$ mm, the circle becomes 400 μm.

*Note*: The larger $L/D$ and the shorter the distance object–detector (film) $l$ (Figure 15.5) the sharper is the attenuation radiograph. In the case of phase radiography the second condition (the shorter the distance object–detector) does not hold.

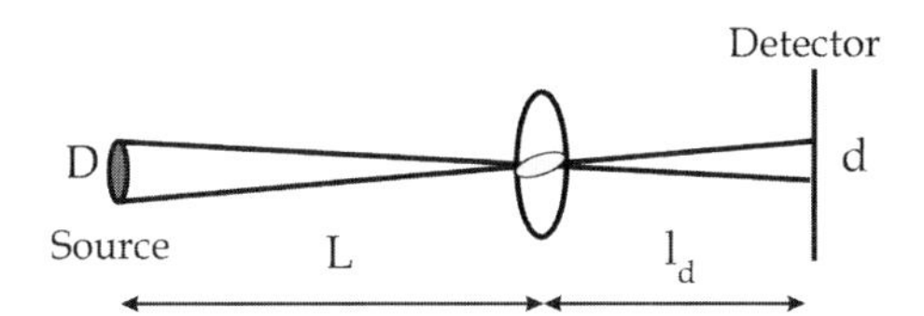


**Figure 15.5** Definition of the $L/D$-ratio: $D$ = size of the (effective) source, $L$ = distance source–object (point), $l$ = distance object–detector.

**ad (iii) Blurring Due to Secondary Detection**
Neutrons can be detected only by means of $(n, \alpha)$, $(n, \gamma)$, or $(n, \beta)$ – reactions to measure an electronic signal, i.e., one needs nuclear reactions to convert neutrons into photons. Due to the finite thickness of the converter and the emission cone of produced photons, an additional uncertainty of localization of neutrons is produced (Figure 15.6).

The capture of a neutron leads to isotropic emission of secondary radiation. A film or CCD camera focused on the converter screen will therefore measure an area spot instead of a point spot. A point or a line is enlarged simply by the converter process, well-defined incident direction – on the converter different spots on the image screen (which forms an despite the fact of the assumption of a perfect parallel beam (which does not exist!). One recognizes the role of the geome-

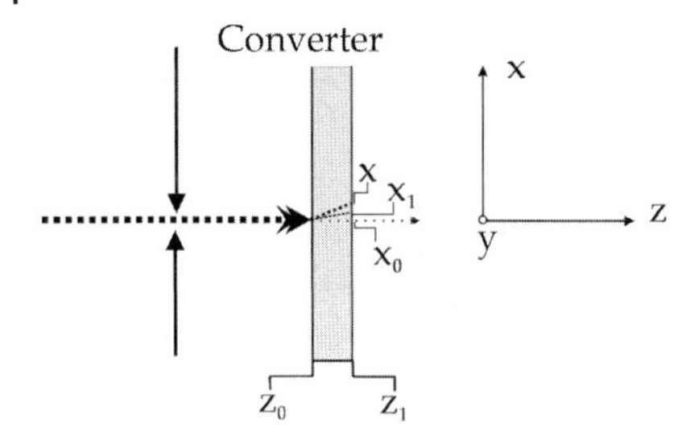


**Figure 15.6** A neutron propagating into the $z$-direction and spatially defined by a fine slit impinges on a converter foil with a thickness $\Delta z = z_1 - z_0$ and produces photons scattered into the directions $(x_0, y_0)$, $(x_1, y_0)$, ....

tries and the converter foil thickness that determine sharpness and resolution of an image.

## 15.1.4
## Resolution Functions

To take this blurring (mentioned above) into account one can define a resulting secondary radiation probability density function represented by an edge or line spread function LSF as [2]:

$$\mathrm{LSF}(x, x_0) = \frac{\delta/\pi}{1 + \delta^2 (x - x_0)^2}, \qquad \delta = \frac{1}{d_{\mathrm{conv}}} \tag{15.13}$$

where $d_{\mathrm{conv}}$ (in the direction of $z$) is the distance of the converter foil to the image plane, and $x_0$ is the center of the incident beam. Usually $d_{\mathrm{conv}}$ can be taken as the thickness of the converter if the film is attached or the CCD camera is focused on it. $\mathrm{LSF}(x, x_0)$ is also called the line spread function for this system. Figure 15.7 shows three different LSF for three different converter thicknesses.

The point spread function (PSF) can be derived from LSF by rotating around the $z$-axis which was kept constant in the case of the LSF. For $x_0 = 0$, $y_0 = 0$ the PSF can be written as [2]

$$\mathrm{PSF}(x, y) = \frac{\delta^2/2\pi}{[1 + \delta^2\{x^2 + y^2\}]^{3/2}} \tag{15.14}$$

In Figure 15.8 three PSF images are calculated for different converter thicknesses ((a) = 50 μm, (b) = 100 μm, and (c) = 300 μm).

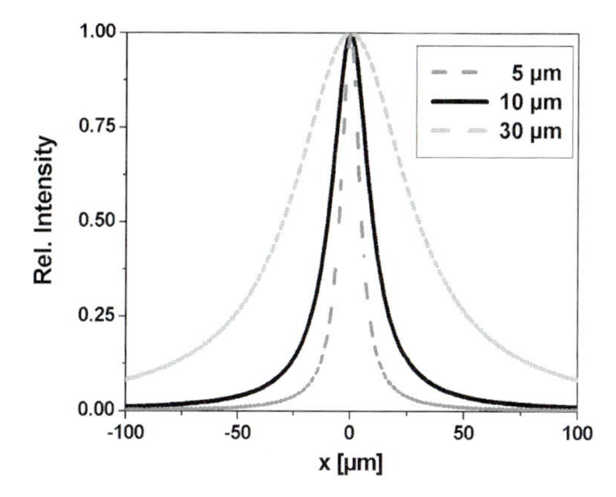


**Figure 15.7** Line spread function (LSF) as a function of converter thickness. 10 mm$^{-1}$ means $d_{\mathrm{conv}} = 0.1$ mm ($x_0$ was set zero).

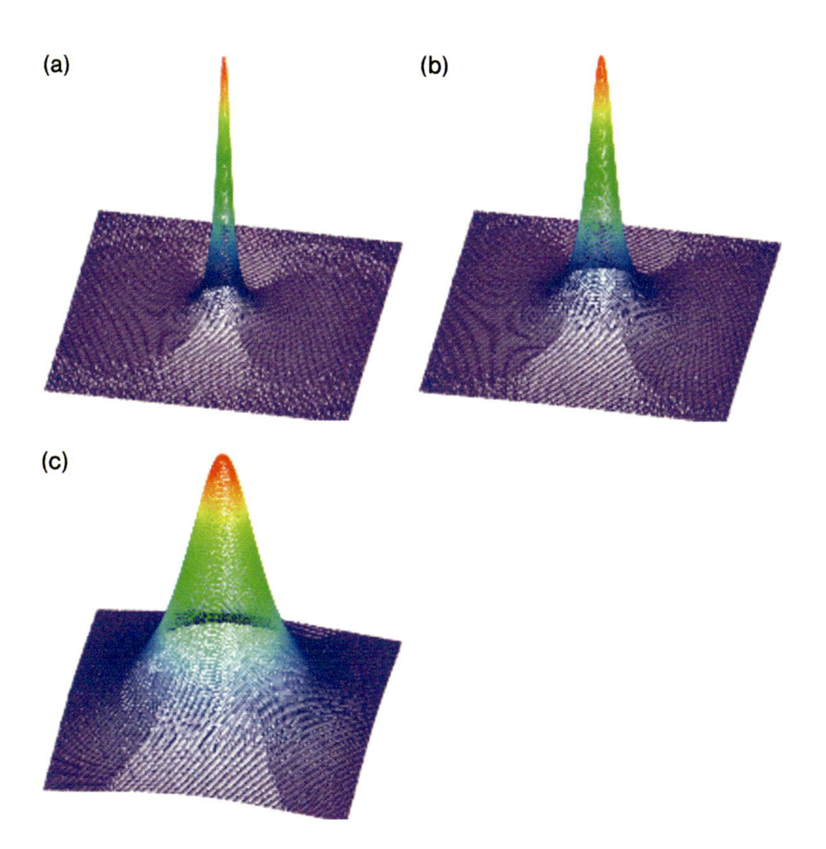


**Figure 15.8** The point spread functions for different line pairs (*l*p) per mm was calculated after Eq. (15.14).

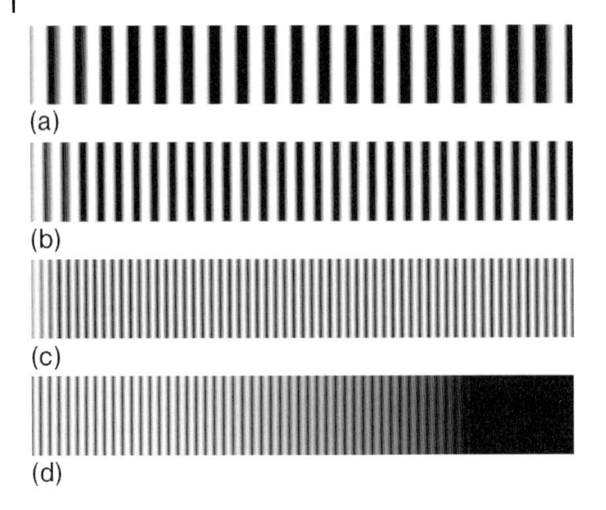


**Figure 15.9** (a–c) Line gratings with different spacing; (d) shows for a constant line spacing from left to right with decreasing contrast.

*Note*: Every action on the picture (measurement, restrictions, filtering, etc.) decreases the information. To determine the blurring in an image one must calculate the 2D-convolution of the (instrumental) PSF with the original object function.

The modulation transfer function (MTF) is known as the modulus of the optical transfer function (OTF) which is the Fourier transform (FT) of the PSF. It characterizes a transfer system such as a radiography or tomography set-up. The question to be answered is how does the contrast in an image depend on the spatial resolution of the transfer system. The MTF can simply be determined by measuring the contrast $C$ given by $(I_{max} - I_{min})/(I_{max} + I_{min})$ for different line pairs $mm^{-1}$ and plotting $C$ as a function of the spatial frequency. For this procedure one uses imaging gratings having different line pairs $mm^{-1}$ (see Figure 15.9). Another grating looks like the one shown in Figure 15.10.

The MTF can also be derived from the FT of the first derivative of the PSF as $|\mathrm{MTF}| = \mathrm{FT}\{\mathrm{PSF}\}$ [3].

**Figure 15.10** From left to right: decreasing line spacing decreases the contrast. The function which describes the contrast as a function of line spacing (line pairs/mm) is called the modulation transfer function (MTF).

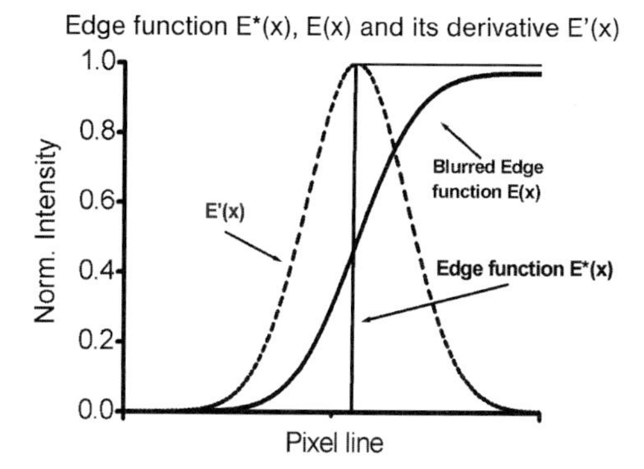


**Figure 15.11** Edge functions: a perfect edge $E^*(x)$ is measured as $E(x)$; its first derivative has the shape of a Gaussian function.

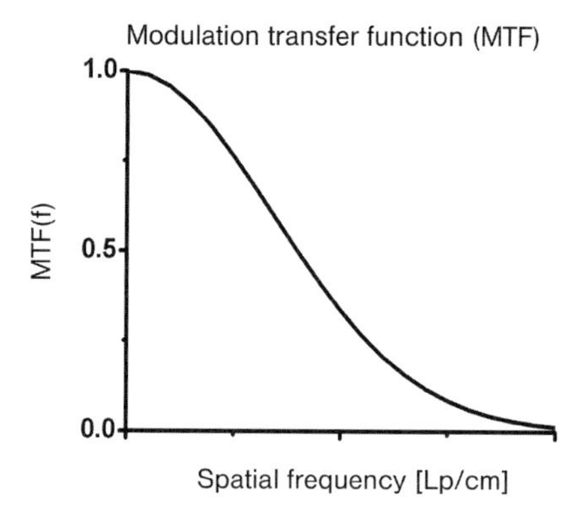


**Figure 15.12** MTF of the edge function of $E'(x)$ (see Figure 15.11).

In Figure 15.11 a perfect edge $E^*(x)$ and a measured one $E(x)$ are shown. Due to divergence and other effects (blurring due to converter foil, finite pixel element of a CCD camera, etc.) a point is spread out and decreases the sharpness of the image. The measured edge $E(x)$ can be approximated either by a function

$$E(x) = \frac{1}{2} + \frac{1}{\pi} \tan^{-1}(\delta \times x) \tag{15.15}$$

or by the error function $\mathrm{erf}(x)$ (Figure 15.12).

### 15.1.5 Image Degradation

The image degradation is based on a number of experimental components that can be summarized as follows:

- beam divergence $\phi$,
- large distance object–detector,
- motion of the sample during an exposure,
- converter unsharpness (roughness of the surface + finite thickness of the converter), and
- object scattering (neutrons are scattered in the object, lose their incident direction, which leads to a reduction of image contrast and artifacts).

The contrast of an image (neighboring pixels or edges in a picture) is given by

$$C = \left| \frac{gr_1 - gr_2}{gr_1 + gr_2} \right| \tag{15.16}$$

$gr_i$ = gray value or pixel value of a pixel. The final contrast $C$ of an image is given by its smallest value.

*Note*: The difference between two gray levels does not determine $C$ uniquely; $C$ changes within the range of gray levels within an image. For the same difference $gr_2 - gr_1 = 10$, one gets different contrasts: if $gr_1 = 10$ and $gr_2 = 20$ one gets a contrast $C = 33\%$; for $gr_1 = 100$ and $gr_2 = 110$ one yields a contrast $C \sim 4.8\%$!

The image contrast is also influenced by the following variables:

- absorption differences in the specimen,
- wavelength (band, spectrum) of the primary radiation,
- beam hardening (the spectrum alters to shorter wavelengths along the path through the sample), and
- scattering in the converter.

For a film-converter system the X-ray film itself must fulfill some conditions, i.e., the film contrast itself is determined by the following items:

- grain size or type of film,
- film processing chemicals,
- concentrations of film processing chemicals,
- time of development,
- temperature of development, and
- degree of mechanical agitation (physical motion).

Modern CCD cameras have nearly superseded the film-converter technique, because the digital image allows a number of image processing actions. All image processing techniques known from medical imaging can also be applied to neutron radiography and tomography. There is a large number of applications of neutron radiography, but also some "new" developments which must be shortly mentioned. Due to better experimental conditions and the need of industrial applications neutron radiography has experienced new attention, and developments such as energy-dependent radiography, (rapid) real-time radiography and phase contrast radiography have increased the interest in this method.

### 15.1.6
### Other Imaging Techniques

In this chapter we summarize some techniques which become more and more important due to better detectors (higher efficiency) and dedicated radiography stations which are optimized for these techniques.

#### Energy Dispersive Radiography

A neutron spectrum from $\lambda = 0.05$ nm to $\lambda = 1$ nm covers a large range of lattice constants of crystals which are parts of materials. Therefore, it can be used for a special imaging technique that will be described as "energy dispersive radiography" (EDR). The transmission of neutrons through a material depends on many factors such as on the composition of the material itself, the particular

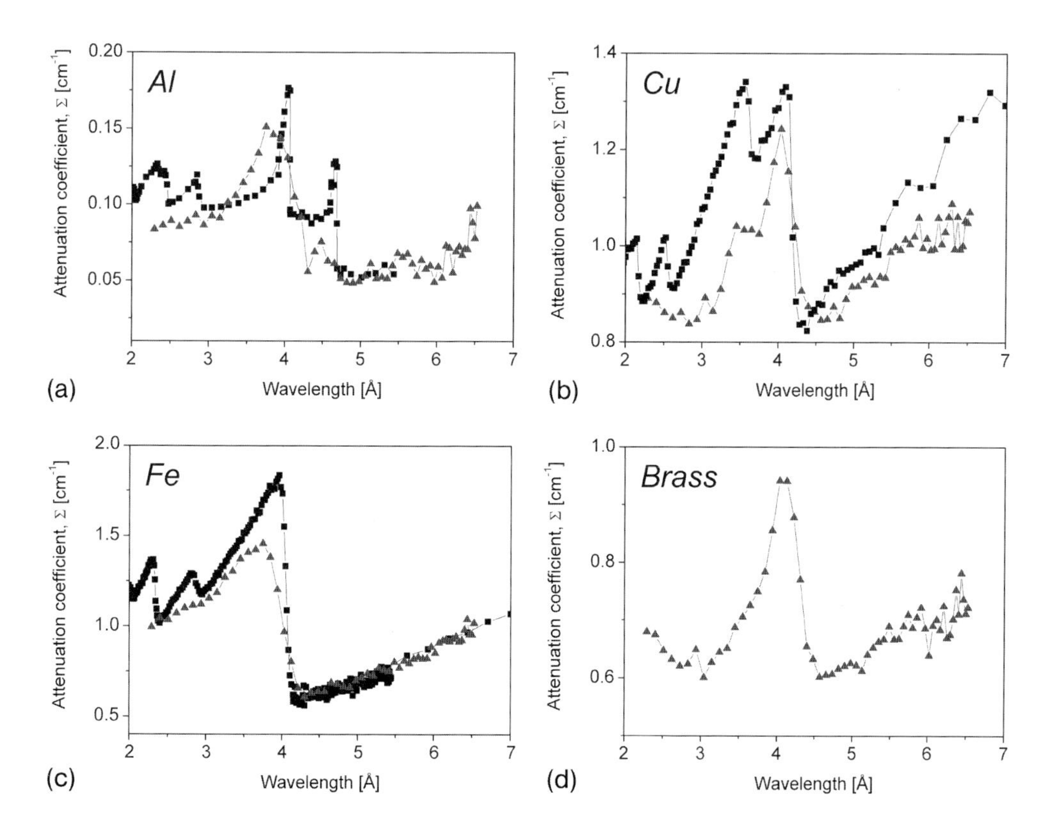


**Figure 15.13** Attenuation coefficients of Al, Cu, Fe, and brass as a function of the neutron wavelength. Data points represented by triangles are own measurements at the HMI in 2005/2006.

absorption (and scattering) cross sections, the wavelength (energy) of the neutrons, temperature, isotopes and the most important factor, the crystalline structure and its orientation. If certain regions or parts in a material contain, e.g., aluminum (Al), copper (Cu), or iron (Fe) and brass, then one can observe the following effect in a radiography (Figure 15.13).

Consider the behavior of the attenuation cross section of, e.g., Al, Cu, brass, and Fe in the close neighborhood of 4 Å or 0.4 nm (see Figure 15.13). These measurements were performed at CONRAD at the BER II reactor (HMI), a radiography and tomography instrument for cold neutrons. Using a double monochromator system of two graphite crystals the energy and therefore the wavelength could be tuned within a range from 0.28 nm up to 0.62 nm. The wavelength dependence enables sophisticated radiographies as shown below because the attenuation can be the same at different energies (different wavelengths).

To show this, an object (a set of step wedges) was imaged by means of neutron radiography at different neutron energies. Theses radiographies require the possibility of using different neutron energies $E_1$ and $E_2$, selected out of the spectrum (e.g., by a chopper or by monochromator crystals). The incident intensities $I_{01}$ and $I_{02}$ in front of the sample are different (due to the spectrum) as well as

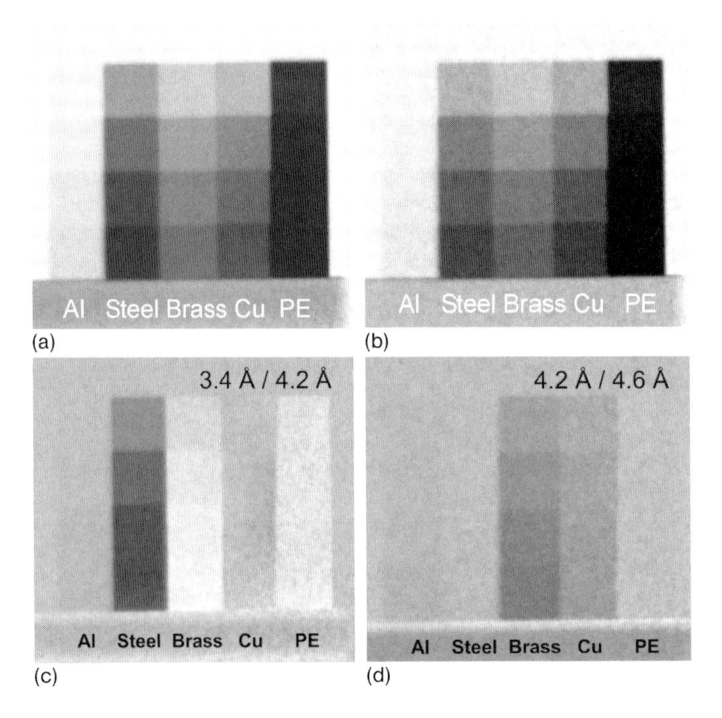


**Figure 15.14** Radiographs of wedges of different materials (Al = aluminum, Cu = copper, PE = polyethylene) at (a) $\lambda = 0.34$ nm and (b) $\lambda = 0.643$ nm; (c, d) show normalized radiographs. Note the different visibilities of the materials for those energies.

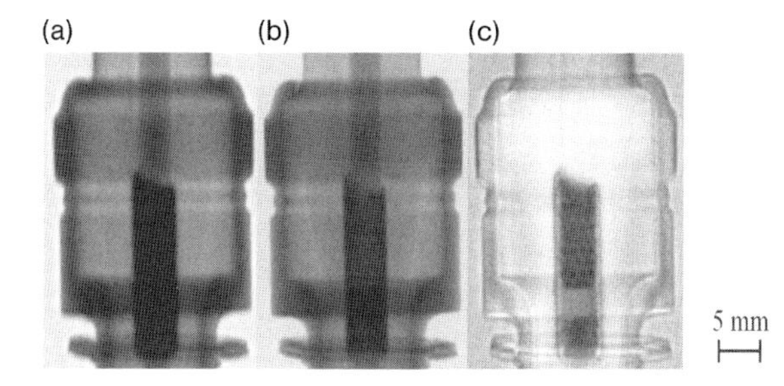


**Figure 15.15** Radiographs of a spark plug at 6.9 Å (a) and 3.2 Å (b). The division of both images yields image (c) (after Kardjilov et al. [4]). Due to application of Eq. (15.17) the surrounding part becomes transparent and the inner part becomes sharper visible.

the transmitted intensities $I_1$ and $I_2$. Figure 15.14 shows five different materials imaged at different energies. The step wedges had thicknesses of 5 mm, 10 mm, 15 mm, and 20 mm.

To use these features for radiography one must normalize the radiographs according to Eq. (15.17). All parts having the same attenuation coefficient will be suppressed. The ratio of the normalized intensities yields an exponential expression,

$$\frac{\frac{I_1}{I_{01}}}{\frac{I_2}{I_{02}}} = \frac{I_1 \cdot I_{02}}{I_2 \cdot I_{01}} = \frac{e^{-\Sigma(\lambda_1)d}}{e^{-\Sigma(\lambda_2)d}} = e^{[\Sigma(\lambda_2)-\Sigma(\lambda_1)]d} \tag{15.17}$$

which decreases or increases. The contribution to attenuation of some materials is also shown in Figure 15.15.

### Real-time Radiography

To visualize rapid processes one has to distinguish between oscillating and simple emerging (nonoscillating) processes. In the latter case each shot (or frame) must image (by absorption) within a certain time interval $\Delta t$ an instantaneous progress of the process. The process within $\Delta t$ itself can be visualized by registering as many frames as possible. In the case of an oscillating piston of a model aircraft engine – as it was done in Figure 15.16 – the motion was triggered, so that different positions of the piston could be imaged at different times. The sharpness of the pictures and therefore of the produced film depends on the number of neutrons which can be collected within one pixel ($\Delta x$) within $\Delta t$ (per frame) and how well defined the triggering can be managed. Therefore, the exposure time $\Delta t$ should be (much) smaller than the velocity of the process within a certain $\Delta x$ ($x + \Delta x$), i.e., any detail of the oscillation process should move less than half of a pixel within the exposure time $\Delta t$ [5, 6].

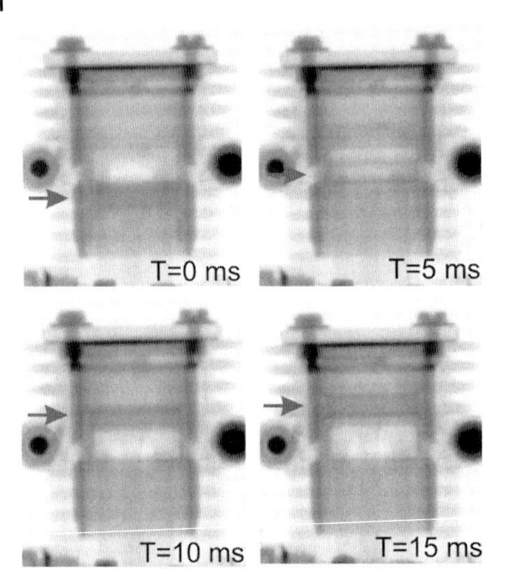


**Figure 15.16** Visualization of the movement of a piston of a small model aircraft machine. The motor ran with approximately 1100 rpm, the exposure time was 1 ms, and 200 frames were added for each picture. Neutron flux $\sim 2 \times 10^8$ neutrons $cm^{-2}$ s. A short film could be produced which shows the movement of the piston in slow motion (HMI 2004, Treimer et al. 2005 [5, 6]).

**Phase Contrast Radiography**

The idea of phase contrast imaging is the following: consider a (neutron) wave which is coherent in a region given by the first Fresnel zone. The size of the first Fresnel zone (FZ) $A_{FZ}$ is

$$A_{FZ} = \frac{R \cdot r}{R + r} \cdot \pi \cdot \lambda \cdot \frac{1}{\sin(\theta)} \qquad (15.18)$$

with $R$ = distance point source–object, $r$ = distance object–detector, and $\theta$ the angle between the ray propagation vector and the plane of observation (usually $\theta = 90^\circ$). This size $A_{FZ}$ determines the coherent source for any interference effects. For standard geometries (e.g., $R \sim 500$ cm, $r \sim 1$ cm, $\lambda \sim 0.5$ nm) $A_{FZ}$ is approximately $1.6 \times 10^{-7}$ cm$^2$ and the diameter of the coherent source $\sim 4$ μm. So one has to create a "point-like" source by a small pinhole that determines, together with the divergence of the beam the size, the coherent area $d_{coh}$ of a real neutron beam. The coherent size of a source $d_{coh}$ can also be estimated by simple geometric considerations as given by [5]

$$d_{coh} = \lambda \cdot \frac{1}{2} \frac{R}{D} \qquad (15.19)$$

where $d_{coh}$ is the size of the monochromatic source, $R/D$ the common known "$L/D$ ratio," and $L$ denotes the distance object source. A short calculation demonstrates that $D$ (size of source) must be kept very small, of the order some hundred microns or even less, to achieve a reasonable size of the coherent "source area." In the case of phase contrast the coherent part of the wave propagates through an object and experiences position-dependent phase shifts, i.e., the propagating wave front is modulated by different indices of refractions in the sample. The modulation occurs quite rapidly in the so-called near field region ($z < d^2/\lambda$), $z$ = distance object–detector, $d$ = diameter of the coherent illuminated part of the object and it broadens with increasing $z$ [6, 7]. The fringe modulation caused, e.g., by an edge is washed out in the case of a too large divergence and/or source size. To observe the strongest phase contrast, one has to operate in the near field region, where $z < d^2/\lambda$, where the modulation of the amplitude is best. It can be shown that the optimal contrast occurs at $z = d^2/2\lambda$. This was (partially) proven by the investigations of contrast to realize experimentally phase images of steel needles at different detector distances and ratios $L/D$ [8]. One approach to realize phase contrast was performed by neutron radiography some years ago. The achieved contrast, however, was not as brilliant as it should be. This was due to a number of experimental conditions, mainly due to the small intensity (several hours exposure time). First experiments based on phase effects (refraction) were already performed in 1989 [9], and the first neutron radiography based on phase contrast was performed in 2000 [10]. However, phase contrast imaging is much more used by synchrotron radiation due to the higher brilliance of the source.

One can calculate the contrast, knowing that the Fourier transform of the structure is measured. The phase development close behind the sample yields strong intensity oscillations which were measured at different distances from the object. However, one can also use simple ray-tracing, adding up all ampli-

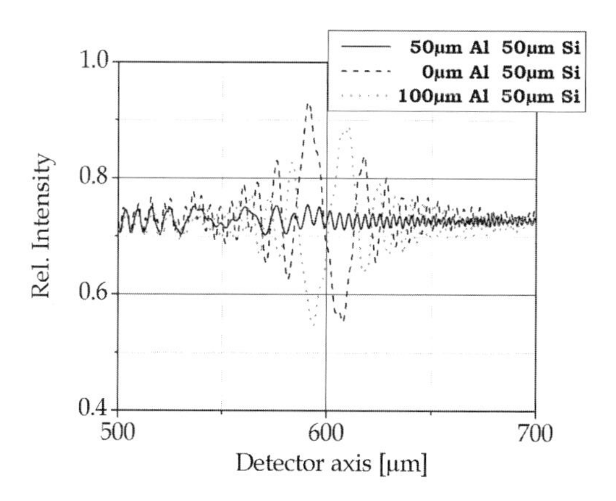


**Figure 15.17** Phase boundary between two slabs (Al and Si) having nearly the same index of refraction, but different thickness [5]. At the position of 600 μm Al and Si stick together.

tudes emerging from the structure which causes the phase gradient that contribute to a measuring point. Figure 15.17 shows the result of such calculations for a phase boundary consisting of Al and Si which have very similar index of refraction. Therefore, for equal thickness no intensity variation occurs (both 50 μm). In the calculation it was assumed that both slabs were illuminated by a plane wave. This explains the small oscillations that appear approximately more than 200 μm off the boundary at 600 μm.

## 15.2 Tomography

### 15.2.1 Mathematical Introduction

The image procedure described in the previous chapter yields the path-integrated information onto a 2D detector (film or electronic device). However, it is often additionally important to have a 2D slice image or a 3D image of the object. To do this in the simplest way, the object can be parted in one direction (e.g., $z$-direction, perpendicular to the $(x, y)$ plane) into small slices. Each slice, having a finite thickness $\Delta z$, is considered as a 2D function $f = f(x, y)$ which describes, e.g., the position-dependent attenuation coefficient $\mu = \mu(x, y)$. This slice is scanned (cf. Figures 15.18 and 15.19) under equidistant angles from 0° to 180° (360°) and so-called projections $P_\theta(t)$ are measured. From these projections 2D images of the slice can be reconstructed and stacked together to a 3D image reconstruction of the object.

It must be pointed out that any function $f$ can be taken into account which is bounded and finite in a given region and zero outside this region. This is easy to realize for solid or liquid samples or gas in a box, but it is not so easy to do for electric and magnetic fields.

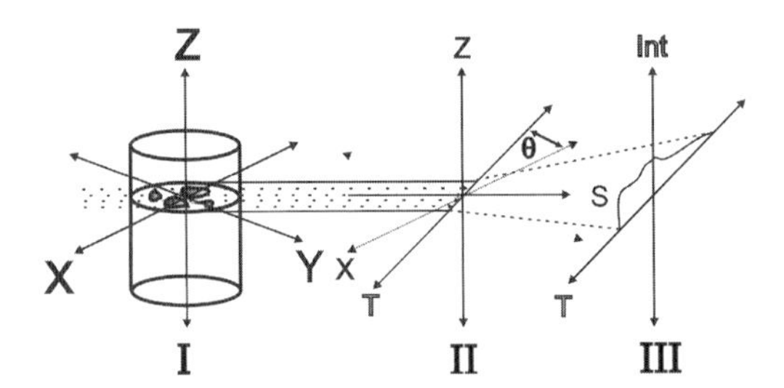


**Figure 15.18** CT geometry to scan an object. Position (I) is the object location with object coordinates $(x, y, z)$, (II) shows the coordinate system for a certain (rotated) orientation of the object with respect to the 2D detector system, and (III) shows the measured projection for this rotated orientation of the sample, T = translation direction.

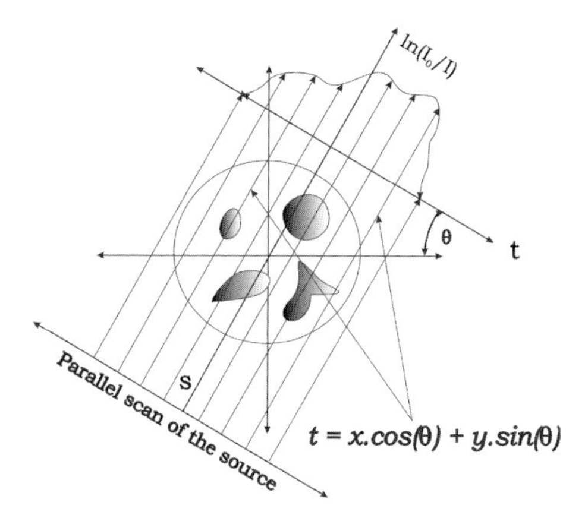


**Figure 15.19** A sample is scanned under the angle $\theta$. The straight line across the slice is given by $t = x\cos(\theta) + y\sin(\theta)$. $t$ is the scanning direction perpendicular to the ray direction $s$.

The sample is scanned under angles $\theta$ with $0 < \theta < 180°$ ($360°$) and the transmitted intensity is registered as a function of the translation $t$. To get a 2D image of the slice, we follow the theoretical description given by Kak and Slaney [11], and Rosenfeld and Kak [12]. A set of pencil rays traverses the sample as shown in Figure 15.19 and the transmitted intensity is measured as a function of the translation $t$. The transmitted intensity is given by Lambert's law as

$$I(x, y) = I_0 \exp\left(-\int_{\text{ray}} \mu(x, y)\,\mathrm{d}s\right) \tag{15.20}$$

where $I_0$ is the intensity of the incident neutron beam. Now one defines a new rectangular coordinate system $(t, s)$ that expresses the rotated detector system $(t, s)$ with respect to the fixed sample system $(x, y)$ (or vice versa, if the sample is rotated as it is done in neutron tomography). The path across the sample is given in terms of $t$ and $\theta$ by eliminating $s$,

$$t = x \cdot \cos(\theta) + y \cdot \sin(\theta) \tag{15.21}$$

From Eq. (15.20) one can define a so-called projection $P_\theta(t)$ as

$$P_\theta(t) = \ln\left(\frac{I_0}{I}\right) = \int_{\text{path}} \mu(x, y)\,\mathrm{d}s \tag{15.22}$$

The set of all projections $P_\theta(t)$ of $\mu(x, y)$ is called the Radon transform of $\mu(x, y)$. From these projections a 2D image can be reconstructed using the "Fourier slice theorem."

### 15.2.2
### Slice Theorem, Shannon Theorem

The basic idea of image reconstruction from projections is contained in the Fourier slice theorem. It states that the Fourier transform of a projection $P_\theta(t)$ of $\mu(x, y)$ is a subset of the Fourier transform of $\mu(x, y)$:

$$\mathrm{FT}\{P_\theta(t)\} = S_\theta(w) \subset \mathrm{FT}\{\mu(x, y)\} = M(u, v) \tag{15.23}$$

To demonstrate this relation we follow again the treatise in [10] and [11]. We assume $\mu(x, y)$ to be bounded and finite, so $\mu(x, y)$ has a Fourier transform $M(u, v)$;

$$M(u, v) = \int_{-\infty}^{\infty}\int_{-\infty}^{\infty} \mu(x, y) \cdot \mathrm{e}^{2\pi \cdot i \cdot (ux+yv)}\, \mathrm{d}x\, \mathrm{d}y \tag{15.24}$$

To prove the Slice theorem one consider the simple case of $\theta = 0$. In this case the two coordinate systems $(x, y)$ and $(u, v)$ coincide, i.e., $x$ is parallel to $u$ and $y$ is parallel to $v$. A projection $P_\theta(t)$ in this case is simply the integral

$$P_{\theta=0}(t) = \int_{-\infty}^{\infty} \mu(x, y)\, \mathrm{d}y \tag{15.25}$$

If we look at the Fourier transform of $\mu(x, y)$ in the case of $\theta = 0$, then the Fourier transform of the object becomes (due to $v = 0$)

$$\begin{aligned} M(u, 0) &= \int_{-\infty}^{\infty}\int_{-\infty}^{\infty} \mu(x, y) \cdot \mathrm{e}^{-2\pi \cdot i \cdot (ux)}\, \mathrm{d}x\, \mathrm{d}y \\ &= \int_{-\infty}^{\infty} \left\{\int_{-\infty}^{\infty} \mu(x, y) \cdot \mathrm{d}y\right\} \mathrm{e}^{-2\pi \cdot i \cdot (ux)}\, \mathrm{d}x \end{aligned} \tag{15.26}$$

Now one can substitute Eq. (15.25) into Eq. (15.26) and get

$$M(u, 0) = \int_{-\infty}^{\infty} P_{\theta=0}(x) \cdot \mathrm{e}^{2\pi \cdot i \cdot ux}\, \mathrm{d}x \tag{15.27}$$

If the Fourier transform of $P_\theta(t)$ is given by

$$S_\theta(w) = \int_{-\infty}^{\infty} P_\theta(t) \cdot \mathrm{e}^{-2\pi \cdot i \cdot w \cdot t}\, \mathrm{d}t \tag{15.28}$$

then $S_{\theta=0}(w)$ equals $M(u, 0)$, the Fourier transform of $\mu(x, y)$ for $\theta = 0$. Because the orientation of the object in the $(u, v)$ system is arbitrary with respect to the $(x, y)$ system this relation enables us to use it for all angles $\theta$ keeping in mind that the Fourier transform conserves the rotation taken in real space. The Fourier transform of $P_\theta(t)$ yields the values of the Fourier transform of $\mu(x, y)$. A dense set of values in the Fourier space (given in polar coordinates due to the polar scanning of the object) can be approximated to rectangular coordinates and back transformed into real space.

An important question must be answered: How many projections yield the best result? This is answered by the so-called Shannon theorem which states that a unique reconstruction is given if the image is sampled with a frequency which is twice higher than the highest frequency in the Fourier transformed image. For the parallel mode scanning using $N$ steps for one projection one can easily derive that $M$ projections (angles $\theta$) are required to fulfill the Shannon theorem.

If $D$ is the diameter of the object that shall be scanned, and $\Delta x$ is the difference between two scanning points, then the number of scanning points/projection is

$$N = \frac{D}{\Delta x} \tag{15.29}$$

assuming that $\Delta x$ fulfills the Shannon condition. The rotation of the sample around 180° means a path length $y$ of an edge point $\pi D/2$. We require that $\Delta x = \Delta y$, so that the number projections $M$ must be

$$M = \frac{\frac{D}{2}\pi}{\Delta y} = \frac{D \cdot \pi}{2 \cdot \Delta y}$$

Using now Eq. (15.27) we get ($\Delta x = \Delta y$)

$$M \geq \frac{\pi}{2} \cdot N \tag{15.30}$$

$N$ and $M$ determine the quality of the reconstruction. Note that Eq. (15.30) is only one of some other requirements. If $M$ does not fulfill the inequality (15.30), i.e., if less than $M$ projections are registered, one can then calculate the reduced $D = D^*$ that fulfills the Shannon condition with Eqs. (15.29) and (15.30).

### 15.2.3
### Image Reconstruction

As mentioned in the previous chapter, a 2D reconstruction of the slice $\mu = \mu(x, y)$ can be yielded by the back transformation (inverse Fourier transformation) of the Fourier transform $S_\theta(w)$. However, the more efficient (and elegant) way is the so-called filtered backprojection (FBP), which is described very detailed in [10] and [11]. The basic idea is embedded in the scanning mode.

The sample (slice) is scanned from 0 to 180°, which involves that $P_\theta(t) = P_{\theta+180^\circ}(-t)$ and suggest to use a polar coordinate system rather than a rectangular one. The projections are registered in polar coordinates, so that the Fourier transformed projections of $\mu(x, y)$, the $S_\theta(w)$ data, are discrete polar function values. So it seems to be apparent to write $\mu(x, y)$ as a Fourier transform representation in polar coordinates. The function $\mu(x, y)$ written as the Fourier representation with $M(u, v)$ in rectangular coordinates is

$$\mu(x, y) = \int_{-\infty}^{\infty} \int_{-\infty}^{\infty} M(u, v) e^{2\pi \cdot i \cdot (ux + yv)} \, dx \, dy \tag{15.31}$$

The same representation in polar coordinates is

$$\mu(x, y) = \int_{0}^{2\pi} \int_{-\infty}^{\infty} M(w, \theta) \cdot e^{+2\pi \cdot i \cdot w \cdot (x\cos(\theta) + y\sin(\theta))} w \cdot dw \cdot d\theta \tag{15.32}$$

The transformation of the $(x, y)$ system into the $(w, \theta)$ system uses the relations

$$\begin{aligned} u &= w \cos(\theta) \\ v &= w \sin(\theta) \\ du \cdot dv &= w \cdot dw \cdot d\theta \end{aligned} \tag{15.33}$$

Substituting these formulas in Eq. (15.32) we get (with the slice theorem)

$$\begin{aligned} \mu(x, y) &= \int_{0}^{\pi} \int_{-\infty}^{\infty} [M(w, \theta) \cdot e^{+2\pi \cdot i \cdot w \cdot t} |w| \cdot dw] \, d\theta \\ &= \int_{0}^{\pi} \left[ \int_{-\infty}^{\infty} S_\theta(w) \cdot e^{+2\pi \cdot i \cdot w \cdot t} |w| \cdot dw \right] d\theta \end{aligned} \tag{15.34}$$

The integral given in brackets can be considered as the Fourier transform of $P_\theta(t)$, however, multiplied by the $|w|$-function which is a special filtering functions of the frequencies in the frequency space:

$$Q_\theta(t) = \int_{-\infty}^{\infty} S_\theta(w) \cdot |w| \cdot e^{+2\pi \cdot i \cdot w \cdot t} \cdot dw \tag{15.35}$$

With $t = x\cos(\theta) + y\sin(\theta)$, $\mu(x, y)$ simply becomes

$$\mu(x, y) = \int_{0}^{\pi} Q_\theta(x\cos(\theta) + y\sin(\theta)) \, d\theta \tag{15.36}$$

A product in Fourier space inverse Fourier transformed is the convolution of the inverse Fourier transformed functions, i.e., we use the relation

$$\mathrm{FT}^{-1}\{S_\theta(w)\cdot|w|\} = P_\theta(t) \otimes \mathrm{FT}^{-1}\{|w|\} \tag{15.37}$$

The convolution operation is denoted by $\otimes$. The function $|w|$ is not a square integrable function, so it has no inverse Fourier transform; however, $\mathrm{FT}^{-1}\{|w|\}$ can be approximated by several filter response functions (convolution kernels, filter functions). The reconstruction of $\mu(x,y)$ from projections $P_\theta(t)$ works now in the following way: each $P_\theta(t)$ is convoluted with a proper filter function (e.g., Shepp–Logan) and the resulting values are "smeared" over the $(x,y)$-planes as demonstrated in Figures 15.20 and 15.21.

In Figure 15.21 four different reconstructions are shown. The $(x,y)$ plane looks like the first image if only one projection is used, then like the second image if two projection, the $\theta=0$ and $\theta=90^\circ$ projections, are added, and so on. Summing up 1000 projection yields the last image (right). Note that the star artifacts are nearly suppressed.

To illustrate one of the filtered projection a part of the filtered function (circle) is extracted and shown in Figure 15.22.

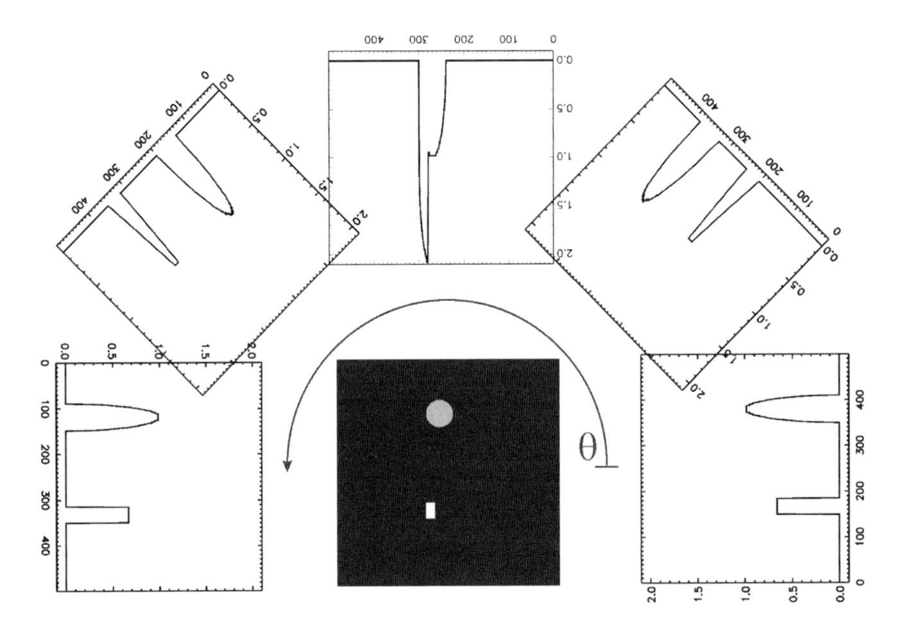


**Figure 15.20** Scanning of a simple 2D object (square and circle) and corresponding projections (from Tilmann Donath, GKSS).

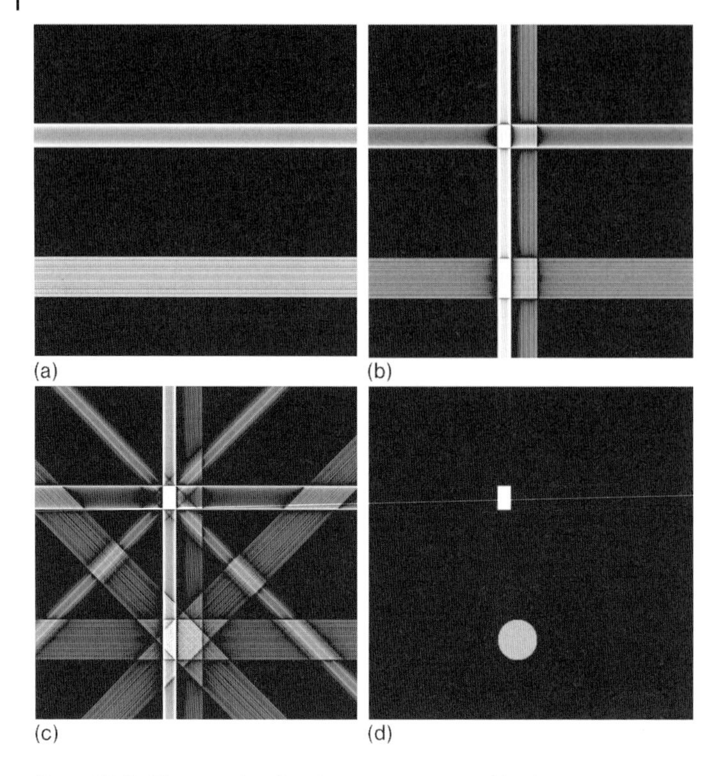


**Figure 15.21** The convoluted projections are smeared back (backprojected) over and summed up in the $(x, y)$ plane (from Tilmann Donath, GKSS).

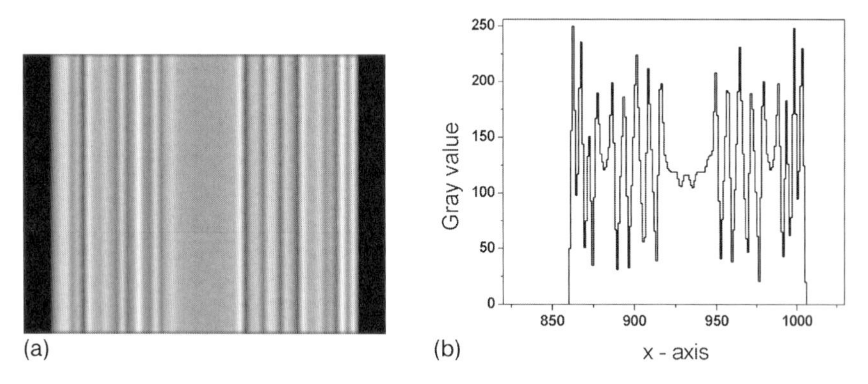


**Figure 15.22** (a) Part of the image "1 projection" (see Figure 15.21) for convenience rotated by 90°; (b) the corresponding filtered function (= convolution of the projection $P$ with a filter function) which is smeared back (backprojected) over the $(x, y)$ plane which yields the left part of the figure.

## 15.3
## New Developments in Neutron Tomography

The examples above identify another question: What can be done if attenuation of neutrons (or X-rays) as an imaging signal do not work? Or to state the physical problem: How can structures in objects be visualized that do not contribute to absorption?

Theorem (without proof): Any interaction that can be measured can be used as imaging signal.

As was shown in Section 15.1.6 the size of the coherence area of the neutron (X-ray) wave determines the effect: if the coherent width (lateral coherence length) is smaller than the object, then refraction occurs. If the object is smaller than the coherent width, small angle scattering occurs.

### Refraction

We consider a function $n(x, y)$ that describes the position-dependent index of refraction: $n(x, y) \neq 1$ inside the sample and $n(x, y) = 1$ outside. From the Snellius law one gets $n_1 \sin(\varepsilon_1) = n_2 \sin(\varepsilon_2)$, with $\varepsilon_1$ the corresponding angle of incidence and $\varepsilon_2$ the refraction angle; $n_1$ and $n_2$ are the corresponding indices of refraction of medium 1 and medium 2. From simple geometrical optics it is also well known that the larger $\varepsilon_1$ the larger is $\varepsilon_2$ up to the case of total reflection. Leaving the sample, neutrons will have a final angle of deflection $\Delta$, which is the sum of all individual deflections $\delta_l (\delta_l = \overline{\varepsilon_l} - \varepsilon_{l+1})$ along the path through the sample:

$$\Delta = \sum_{i=1}^{n} \delta_i \tag{15.38}$$

A deflection $\delta_l$ from the preceding direction is only present at the adjacent volumes with different indices of refraction and if $\overline{k} \cdot \nabla n(x, y) \sim 0$ ($\overline{k}$ is the neutron wave vector). One can determine $\nabla n(x, y)$ and by that $n(x, y)$ by measuring $\Delta$ of each path (line integral, Eq. 15.38) and varying the orientation $\theta$ of the sample from $0^\circ$ to $180^\circ$. If $\overline{k} = (\overline{k}_\parallel, \overline{k}_\perp)$ is represented by a component parallel and perpendicular to $\nabla n(x, y)$, and $\overline{k}_\perp$ is the component of $\overline{k}$ parallel to $\nabla n(x, y)$, then a point $P_\theta(t)$ of a projection is given by

$$P_\theta(t) = P(t, \theta) = \int_{\text{Path}} \nabla n(x, y) \cdot \overline{k}_\perp \, ds \tag{15.39}$$

If the incident direction is known, $\Delta_t$ and $P_\theta(t)$ can be determined from the experiment. The total set of functions $P_\theta(t)$ is then another form of the Radon transform of $n(x, y)$. The reconstruction of $n = n(x, y)$ from projections can follow the well-known mathematical procedures. An estimation of the total angle of deflection $\Delta$ is given by the calculation of the index of refraction $n$, which is given for neutrons by

$$n = 1 - \frac{\lambda^2 N b_c}{2\pi}$$

With $\lambda = 4.76 \times 10^{-10}$ (m), $N$ = number of unit cells $m^{-3} \sim 5 \times 10^{28}$, and $b_c$ = coherent scattering length $\sim 10^{-15}$ (m) of the sample volume of interaction; $1 - n \sim 10^{-6}$ gives the size of the angle of refraction which has to be measured. This can only be done with a special double crystal instrument as it is described in [13, 14]. A simple calculation of $\Delta_t$ for a given orientation $\theta$ for a brass sample (Figure 15.23, ∅ = 10 mm four holes with ∅ = 1, 2, 3, and 4 mm) is shown in Figure 15.23. Figure 15.24 shows the tomograms of a brass cylinder, ∅10 mm, ∅ holes: 1, 2, 3, and 4 mm.

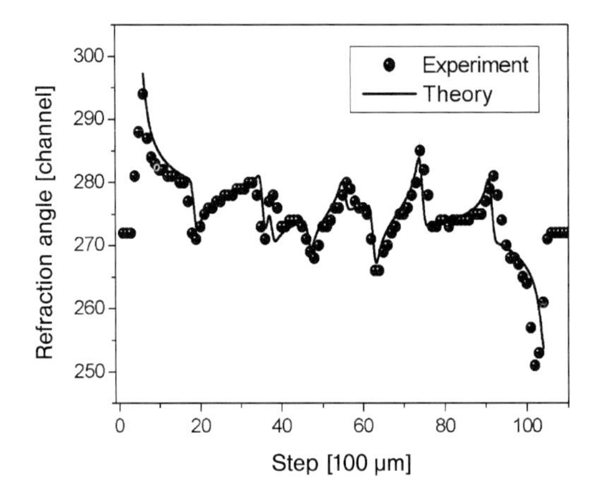


**Figure 15.23** Calculated and measured peak shift for a certain orientation $\theta$ of a brass sample.

A 3D-refraction tomography could also be already realized [15]. In worse cases one yields no absorption contrast; only refraction contrast yields a good reconstruction as demonstrated in Figure 15.25.

#### Ultrasmall Angle Neutron Tomography

In the case of USANS (ultrasmall angle neutron scattering) tomography $\mu(x, y)$ has to be replaced by a suitable function that represents the USANS effects in the sample. The broadening $B$ ($nm^{-1}$) of a scattering function due to (ultra) small angle scattering can be (roughly) approximated by a Gaussian distribution and – involving multiple scattering – written as [16]

$$B_\theta(t) = \sqrt{\int_{\text{path}} \frac{\sigma(x, y) \cdot N(x, y)}{R(x, y)^2} \cdot \mathrm{d}s} \tag{15.40}$$

where $\sigma$ is the scattering cross section, $N$ is the particle density, and $R$ is a parameter with the dimension of a length specifying an average size or correlation

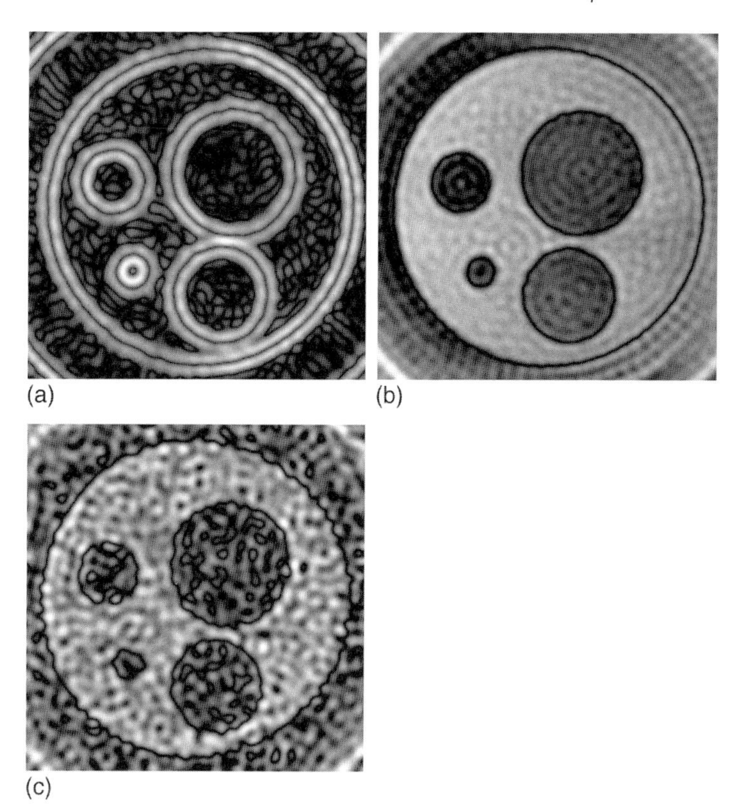


**Figure 15.24** Reconstruction using either (a) refraction data only, (b) integrated refraction data, or (c) absorption data only.

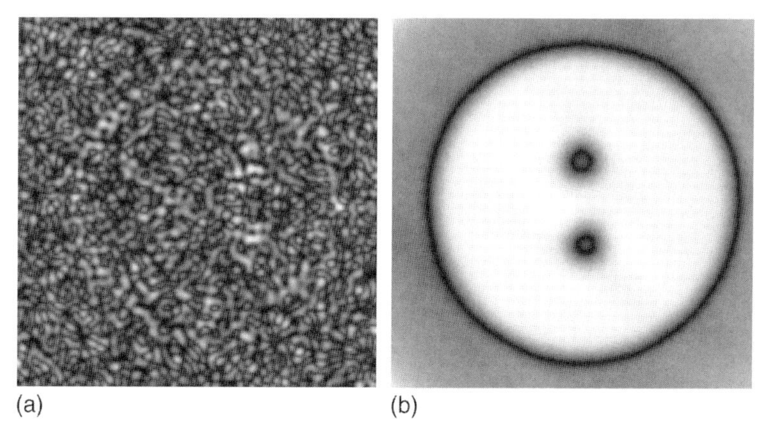


**Figure 15.25** Al-sample: diameter 6 mm; two holes, each 2 mm. (a) Attenuation data, (b) reconstruction with refraction data only.

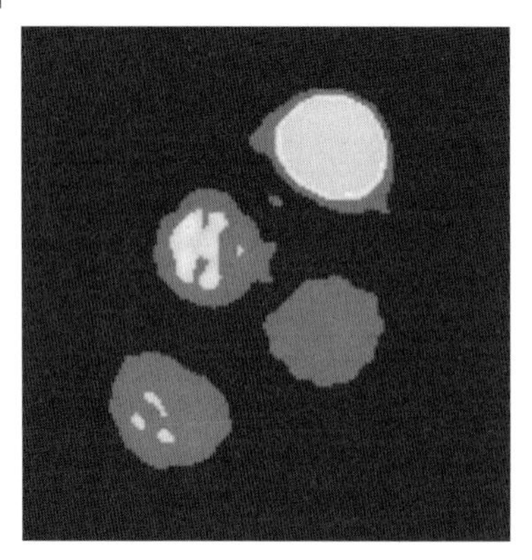

**Figure 15.26** 2D reconstruction from USANS data only; dark grey ~ 3.8 wt.%; grey ~ 5.8 wt.%; light grey (see particle on the top) ~ 12 wt.%; particle size: ~ 150 nm (β-carotene).

length in the scattering object (corresponding to the Gaussian approximation). Now one can use $B$, resp. $B^2$, as imaging signal to use ultrasmall angle scattering for image reconstruction. So the detection of clusters of approximately 150 nm large particles (β-carotene), dissolved in $D_2O$ (~ 5%–12%), could be realized. In this experiment refraction, USANS and absorption were measured simultaneously with the double crystal diffractometer [13, 14]: the shift of the rocking curve yields the final deviation and determines the refraction, the broadening the small angle scattering and the decrease of the integral intensity the absorption of the neutron beam. All these data were derived from a single rocking curve.

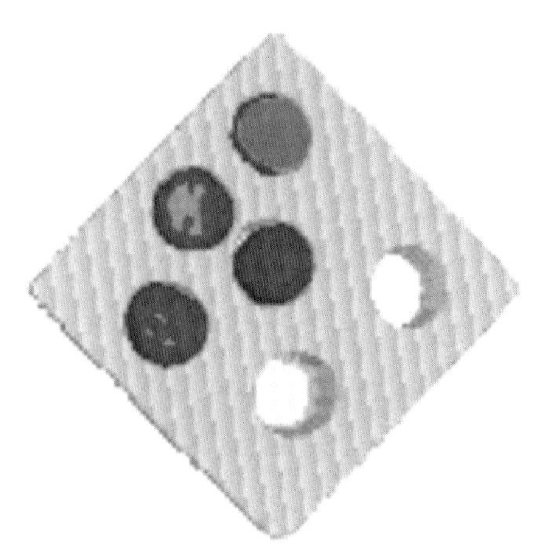

**Figure 15.27** Reconstruction of an aluminum box having six holes, four were filled with different weight percentages (wt.%) of $D_2O$ and β-carotene (see also Figure 15.26).

Figure 15.26 shows the first successful results. Combining USANS and refraction data yields a complete reconstruction (Figure 15.27). With pure attenuation, refraction or USANS data only incomplete, imperfect information is reconstructed.

These methods offer new possibilities for a dedicated computerized tomography, especially for nondestructive testing and material research. Especially the coherence properties, phase-based radiography, and tomography will contribute to a much more details investigation of solid state physics problems. Recently, it was demonstrated [17, 18] that the development of another new techniques and their applications shows the large potential of CT in material research, underlying the large potential of neutron radiography and neutron tomography. Finally, it must be mentioned that tomography with polarised neutrons were already tested sucessfully [5] and that this option of tomography will open a new and large field of applications is science and industrial research.

## References


1 S.G. Glasstone, M.C. Edlund, *Kernreaktortheorie*, Springer-Verlag, Wien, 1961, p. 30.

2 A.A. Harms, D.R. Wyman, *Mathematics and Physics of Neutron Radiography*, Reidel, Dordrecht, Holland, 1986.

3 E. Krestel, *Bildgebende Systeme für die medizinische Diagnostik*, Grundlagen, Berlin München Siemens Aktiengesellschaft, 1980.

4 N. Kardjilov, S. Baechler, M. Bastürk, M. Dierick, J. Jolie, E. Lehmann, T. Materna, B. Schillinger, P. Vontobel, New features in cold neutron radiography and tomography: Part II. Applied energy-selective neutron radiography and tomography, *Nucl. Instrum. Methods Phys. Res. A* 2003, **501**, 536–546.

5 W. Treimer, N. Kardjilov, U. Feye-Treimer, A. Hilger, I. Manke, Markus Strobl, Absorption- and phase-based imaging signals for neutron tomography, *Adv. Solid State Phys.* 2005, **45**, 407.

6 X. Wu, H. Liu, Clinical implementation of x-ray phase-contrast imaging: theoretical foundations and design considerations, *Med. Phys.* 2003, **30**, 8, 2169.

7 W. Treimer, A. Hilger, N. Kardjilov, M. Strobl, Review about old and new imaging signals for neutron computerized tomography, *Nucl. Instrum. Methods A* 2005, **542**, 367.

8 A. Pogany. D. Gao, S.W. Wilkins, Contrast and resolution in imaging with a microfocus x-ray source, *Rev. Sci. Instrum.* 1997, **68**, 2774.

9 K.M. Podurets, V.A. Somenkov, S.Sh. Shil'shtein, Neutron radiography with refraction contrast, *Physica B: Phys. Condens. Matter*, 1989, **156**, 691–693.

10 B.E. Allman, P.J. McMahon, K.A. Nugent, D. Paganin, D.L. Jacobson, M. Arif, S.A. Werner, Phase radiography with neutrons, *Nature* 2000, **408**, 158.

11 A.C. Kak, M. Slaney, *Principles of Computerized Tomographic Imaging*, IEEE Press, New York, 1999.

12 A. Rosenfeld, A.C. Kak, *Digital Picture Processing, Computer Science and Applied Mathematics*, Academic Press, New York, 1982.

13 W. Treimer, M. Strobl, A. Hilger, C. Seifert, U. Feye-Treimer, Refraction as imaging signal for computerized (neutron) tomography, *Appl. Phys. Lett.* 2003, **83**, 2.

14 M. Strobl, W. Treimer, A. Hilger, U. Feye-Treimer, Neutron tomography in double crystal diffractometers, *Physica B* 2004, **350**, 155.

15 M. Strobl, W. Treimer, A. Hilger, First realisation of a three-dimensional refraction contrast computerised neutron

tomography, *Nucl. Instrum. Methods. Phys. Res. B* 2004, **222**, 653.

16 M. Strobl, W. Treimer, A. Hilger, Small angle scattering signals for (neutron) computerized tomography, *Appl. Phys. Lett.* 2004, **85** (3), 488.

17 F. Pfeiffer, T. Weitkamp, O. Bunk, C. David, Phase radiography with neutrons, *Nature Phys.* 2006, **2**, 258.

18 W. Treimer, A. Hilger, M. Strobl, N. Kardjilov, I. Manke, *Appl. Phys. Lett.* 2006, **89**, 203504.

# 16
# Neutron and Synchrotron-Radiation-Based Imaging for Applications in Materials Science – From Macro- to Nanotomography

*Felix Beckmann*

In recent years, 2D and 3D imaging became a standard tool for scientific and industrial applications in materials science. Employing the high-intensity sources at neutron facilities and the highly brilliant and collimated X-ray sources at second and third-generation synchrotron radiation (SR) facilities, specialized contrast techniques were developed. Attenuation-contrast and phase-contrast techniques were optimized to analyze specimens used in basic research studies as well as structural components used in industrial applications.

The image formation at neutron sources was described in Chapter 15. Here, a short introduction to the image formation at synchrotron radiation sources is given. The concepts presented here are also valid for investigations with neutrons. After a short discussion of image creation and quality in tomographic applications, the current instrumentation operated by the GKSS-Research Center Geesthacht, Germany, for computerized tomography using neutrons (NCT) and microtomography using synchrotron radiation (SR$\mu$CT) is presented.

## 16.1 Introduction

The interaction of neutrons with matter is described in details in Chapter 15.1.2. Imaging at current second and third-generation synchrotron radiation facilities makes use of the low divergence and high intensity of the source to record parallel 2D projections of samples using monochromatic X-rays. The principle set-ups and the concept of image formation are presented in Figure 16.1.

### 16.1.1 Attenuation-Contrast Projections

To obtain attenuation-contrast projections radiograms are taken at minimal detector–sample distance. A single projection is obtained by recording the primary intensity $I_0(x, z)$ and the attenuated intensity $I(x, z)$, where $x$ denotes the

*Neutrons and Synchrotron Radiation in Engineering Materials Science: From Fundamentals to Material and Component Characterization*
Edited by Walter Reimers, Anke Rita Pyzalla, Andreas Schreyer, and Helmut Clemens

ISBN: 978-3-527-31533-8

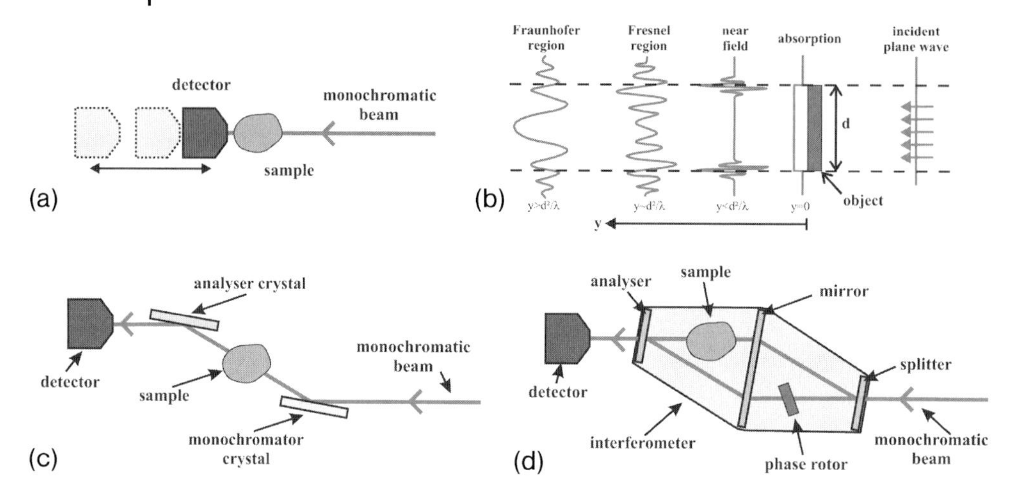


**Figure 16.1** (a) Schematic set-up used for imaging at different sample–detector distances. (b) Coherent image formation with partially coherent beam depending on the object–detector distance $y$, sample dimension $d$ and wavelength $\lambda$. (c) Concept for phase outline contrast applying diffraction-enhanced imaging. A monochromator and analyzer crystal is used to record images at different positions of the rocking curve of the analyzer crystal. (d) Experimental set-up for interferometric phase contrast. A monolithic X-ray interferometer is used to divide the monochromatic beam into two coherent beams. The phase rotor introduces a defined phase shift and the direct phase projection can be obtained.

horizontal and $z$ the vertical dimension. The attenuation projection $P(x,z)$ can be determined by

$$P(x,z) = \int \mu(x, y, z)\, dy = -\ln\left[\frac{I(x,z)}{I_0(x,z)}\right] \tag{16.1}$$

Using the atomic number $Z$, photon energy $E$, and mass density $\rho$, the attenuation coefficient $\mu$ except close to absorption edges can be estimated by

$$\mu \propto \frac{Z^4}{E^3}\rho \tag{16.2}$$

The photon energy $E$ is matched with the sample characteristics in order to obtain sufficient contrast in the radiogram. Therefore, only the structure of the most absorbing material of the specimen can be investigated well. To investigate samples consisting mainly of light elements, the more sensitive phase contrast has to be used [1].

### 16.1.2
### Phase-Contrast Projections

X-rays and neutrons are not only attenuated by the specimen, they are also shifted in phase. This behavior is described by the complex index of refraction

$$n = 1 - \delta - i\beta \tag{16.3}$$

where the real part $\delta$ corresponds to the phase shift due to refraction and the imaginary part $\beta$ belongs to absorption. Depending on the type of the source different phase-contrast imaging techniques were developed, which are suited for different types of sample composition. At incoherent X-ray sources of second-generation synchrotron radiation facilities, multiple crystal reflections or X-ray interferometers have to be used. At third-generation sources, phase-enhancement can be achieved by varying the sample–detector distance due to the partial coherent X-ray beam [2, 3]. In Figure 16.1b, the coherent image formation of a transmission image is shown. Using sample dimension $d$ and wavelength $\lambda$ four regions of sample–detector distances $y$ can be distinguished [4]:

| | | |
|---|---|---|
| $y = 0$ | Absorption | The image is a pure absorption image. |
| $y < d^2/\lambda$ | Near field | Contrast is given by sharp changes in the refractive index, i.e., at interfaces. |
| $y \sim d^2/\lambda$ | Fresnel region | The image loses more and more resemblance with the object. |
| $y > d^2/\lambda$ | Fraunhofer region | The image intensity is the Fourier transform of the object transmission function. |

### 16.1.3 Phase-Enhanced Projections

Using a partially coherent beam, images are taken in the near field. Rotating the sample and performing a tomographical scan emphasizes the sharp changes in the refractive index of the sample. Then, this is called phase outline contrast [5, 6]. A more sensitive technique makes use of multiple crystal reflections. In Figure 16.1c, the schematic set-up is shown. Images at different positions of the rocking curve of the two crystals are recorded. From these images, an absorption image and a phase-enhanced image can be obtained [7, 8]. Using this as the contrast method in tomography again results in mapping of borders of changes in the refractive index.

### 16.1.4 Direct Phase-Contrast Projections

In order to obtain the refractive index in the tomogram, phase projections of the sample have to be measured. A direct technique makes use of an X-ray interferometer which allows for the measurement of a phase projection modulo $2\pi$. The principle set-up is shown in Figure 16.1d. For one phase projection, interference patterns – with and without the sample – are obtained at different overall phase shift introduced by the phase shifter [9–11]. Using the electron density $\sigma$, the phase shift $\phi$ is given by

$$\phi \propto \frac{\sigma}{E} \tag{16.4}$$

### 16.1.5 Indirect Phase-Contrast Projections

An indirect method to measure the phase projection makes use of the partial coherence of third-generation SR beams. By varying the sample–detector distance and recording images, different spatial frequencies in the projected phase images vanish (Talbot effect). The so-called holographic reconstruction procedure then determines the phase projection [12]. More recently, new phase-contrast techniques using gratings were developed and applied at coherent and incoherent sources [13, 14].

## 16.2 Parallel-Beam Tomography

The concept of 3D imaging using parallel-beam geometry is based on recording a set of 2D projections of a specimen at different sample rotations [15]. In Figure 16.2, the principle set-up used for parallel-beam tomography is given. Here, a virtual test object consisting of several spheres and cubes is used to simulate the tomographic measurement, reconstruction data evaluation, and visualization process.

The detector size is $501 \times 501$ pixels. The basic volume of the sample consists of $501 \times 501 \times 501$ voxels. The sample is defined by the combination of the following objects presented in Table 16.1. The position is given in % of the maximal size of 501. Furthermore, two boxes are tilted in the volume. The maximum projected attenuation of the sample is about 2. For the shown simulated measurement projections were calculated from 0 to 180° with an angular increment of 0.5°.

**Table 16.1** Composition of the virtual sample used for simulation.

| Shape | $x$ (%) | $y$ (%) | $z$ (%) | Radius/size (%) | Attenuation | Color |
|---|---|---|---|---|---|---|
| Sphere | 45 | 42 | 50 | 40 | 0.002595 | Yellow |
| Sphere | 20 | 45 | 50 | 10 | 0.006986 | Red |
| Sphere | 60 | 40 | 25 | 11.8 | 0.006986 | Red |
| Sphere | 40 | 50 | 75 | 6.67 | 0.006986 | Red |
| Box | 65 | 22 | 40 | 12.5 | 0.009980 | Blue |
| Box | 58 | 62 | 60 | 11.1 | 0.009980 | Blue |
| Box | 26 | 26 | 37 | 10 | 0.009980 | Blue |

### 16.2.1 Measurement and Reconstruction

In Figure 16.2, the measurement of a single projection is shown. If the rotation axis is perpendicular to the beam, a single slice perpendicular to the axis is projected onto a single line on the 2D detector for all sample rotations. Therefore, measurement and reconstruction can be reduced to a 2D slice with 1D projections. The mathematical description of this tomographic measurement for a single slice $f(x, y)$ is given by the Radon transform [16]

$$P_\theta(t) = \iint f(x, y)\delta(x \cos \theta + y \sin \theta - t)\, dx\, dy \tag{16.5}$$

The tomographic reconstruction is the implementation of the inverse Radon transform. A schematic view of the measurement and reconstruction process is given in Figure 16.3. The complete measurement for a single slice $f(x, y)$ is

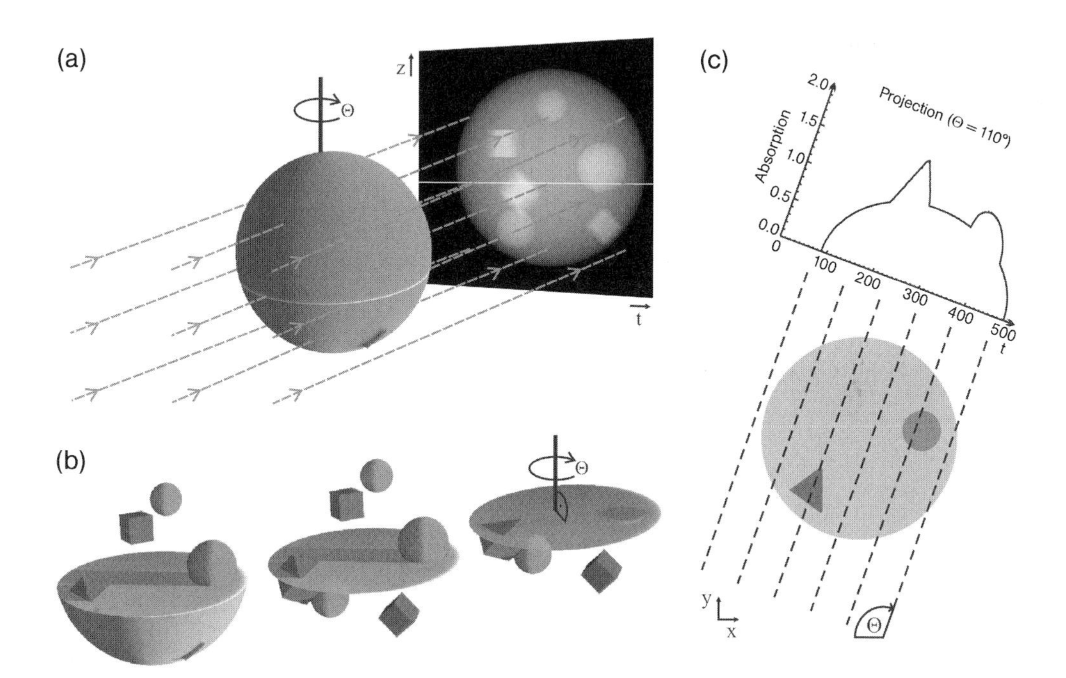


**Figure 16.2** Schematic set-up used for recording of 2D projections in parallel-beam geometry (a). The sample consists of several spheres and cubes. Different colors represent different attenuation (b). If the rotation axis of the sample is perpendicular to the beam, the slice perpendicular to the axis is projected onto the same line of the 2D detector for all sample rotations (c).

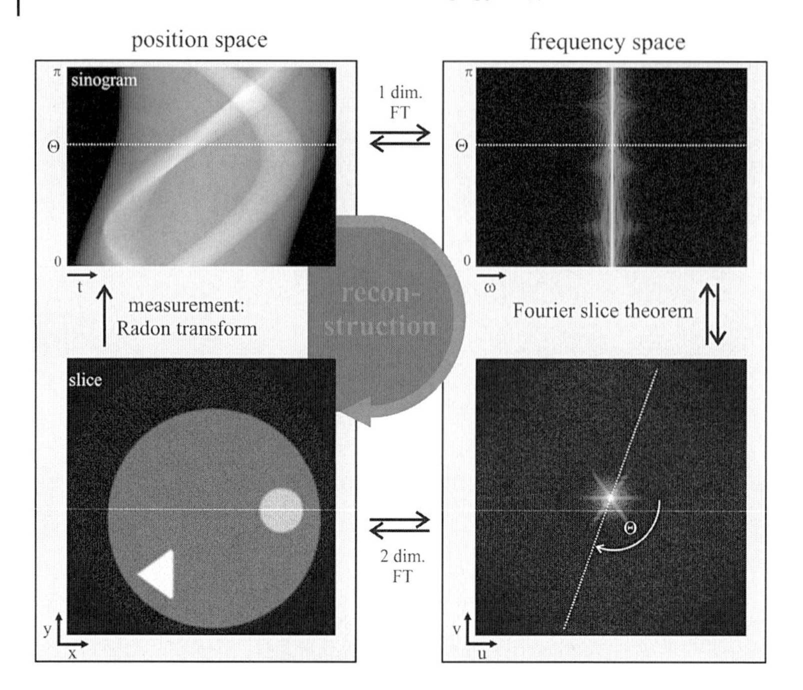


**Figure 16.3** Scheme of the measurement and reconstruction process for parallel-beam tomography. On the left side in the position space the tomographic measurement of a single slice is represented by the sinogram. The tomographic reconstruction can be performed as the sinogram and the slice are connected by the Fourier slice theorem in the frequency space shown on the right side.

the so-called sonogram $P_\theta(t)$. A 1D Fourier transformation of the projection at the angle $\theta$ represents a single line in the 2D Fourier transformation of the slice $f(x, y)$. This is described by the so-called Fourier slice theorem. An implementation of this reconstruction is the back-projection of filtered projection algorithm presented in Chapter 15.2.

## 16.2.2
## Density Resolution and Detector Quality

The spatial resolution in a tomogram can be measured by obtaining the modulation transfer function (MTF) which was discussed in Chapter 15.1.4. Here, special emphasis will be given to the quality of the density resolution in tomography. Assuming an implementation of an optimal measurement, which should, therefore, result in an artifact-free reconstruction, a simulation of a standard tomographic measurement and reconstruction was performed for detector systems,

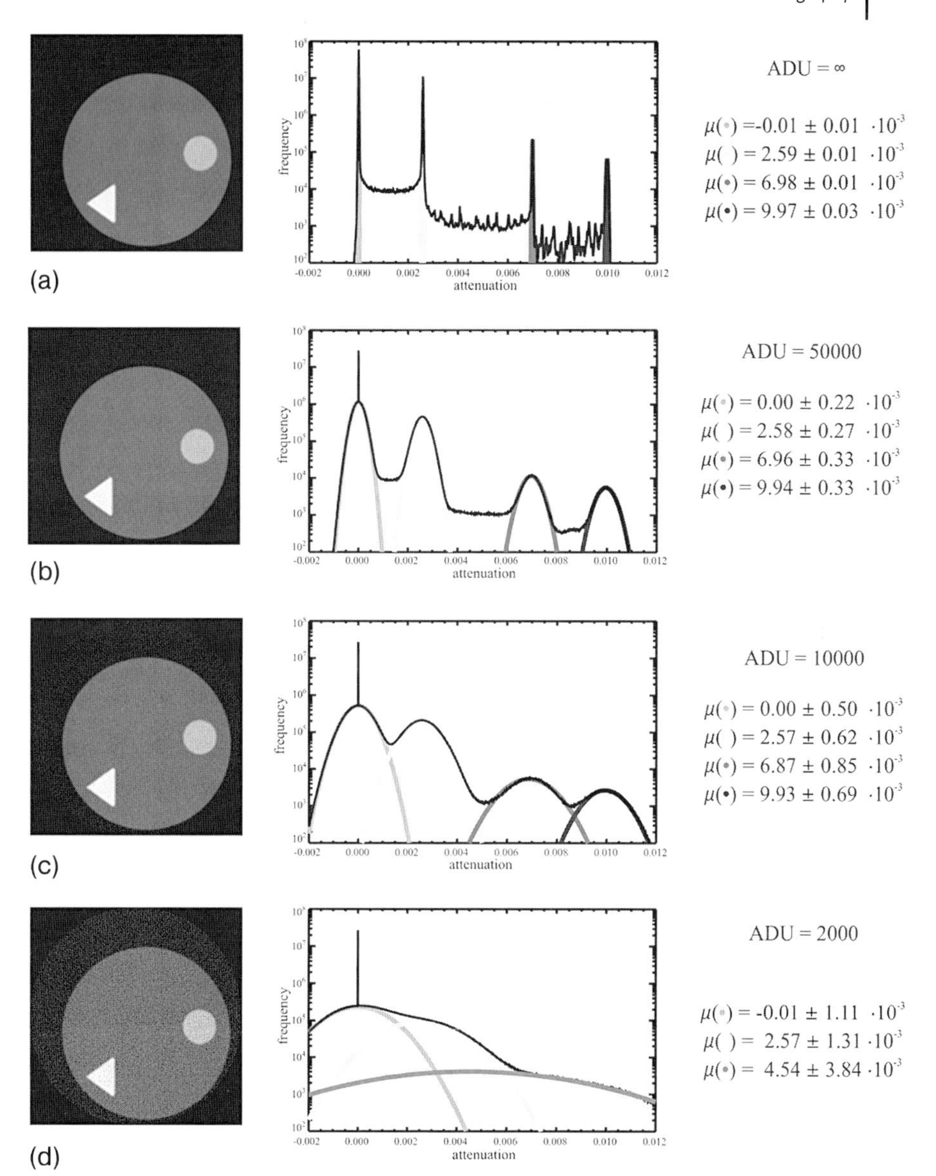


**Figure 16.4** Simulated measurement and reconstruction for the sample presented in Figure 16.1. Left: Reconstructed slice; middle: histogram of all slices representing the frequency of different attenuation values. Right: Obtained parameter for the shown Gauss peaks. The quality of the detector decreases from (a) to (d), resulting in a noisy tomogram where different absorptions used for spheres and cubes cannot be distinguished.

which differ in the number of distinguishable analog digital units (ADU). In Figure 16.4, a resulting single reconstructed slice, the histogram of all slices and the obtained Gauss fits to the histogram are given. It can clearly be seen that for low-statistical qualities of the used detector the different absorbing materials can hardly be separated.

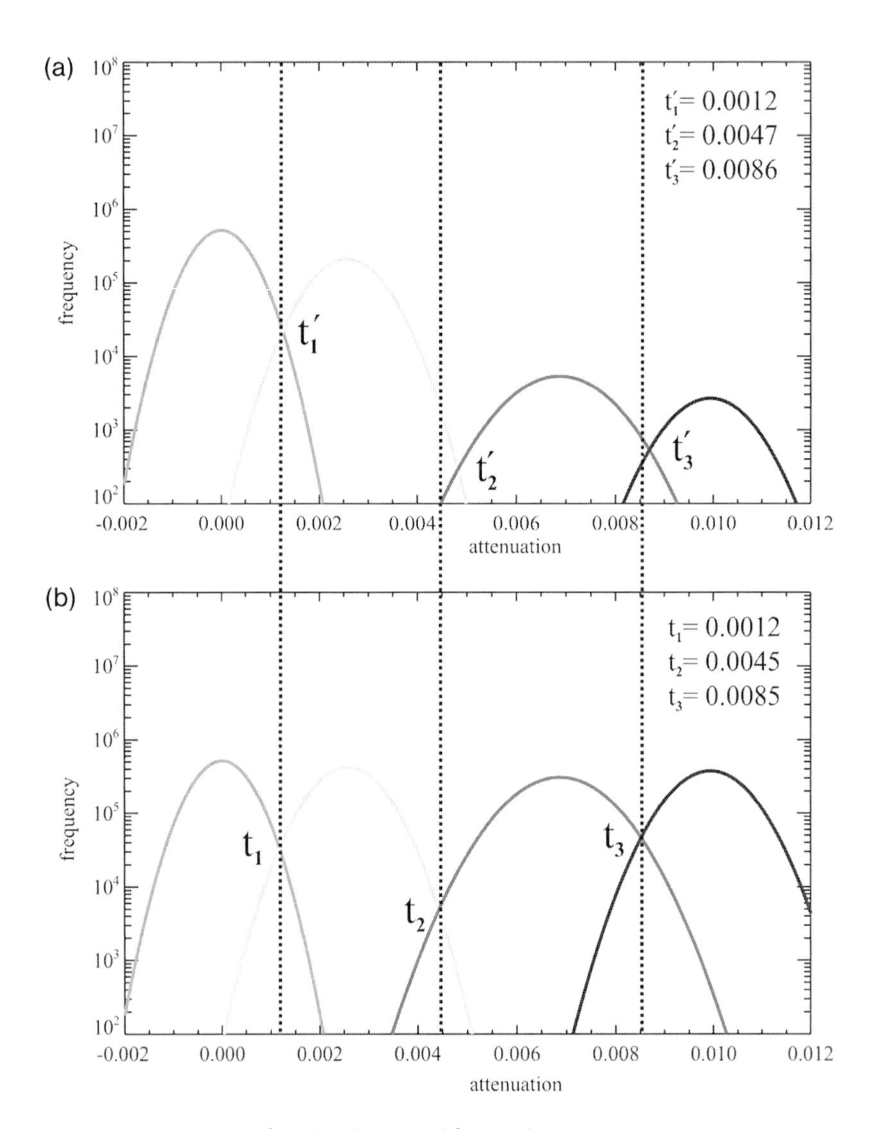


**Figure 16.5** Calculation of the thresholds used for visualization. (a) Distorted thresholds of the original peaks. (b) Correct thresholds of peaks weighted to equal mass. The used simulated scan is presented in Figure 16.4c.

### 16.2.3
### Data Evaluation and Visualization

For further data evaluation very often thresholds are used to distinguish different materials in the tomograms. For this purpose, the intersections of the Gauss peaks are calculated. It is very important to weigh the Gauss peaks to equal mass as otherwise the thresholds will be distorted as shown in Figure 16.5. For 3D visualization of the total reconstructed volume most often the floating point representation is scaled to a byte value. By applying digital filters, the noise in the tomograms can be reduced as shown in Figure 16.6. In Figure 16.7, a special software has been used to virtually slice the volume, to colorize different material, and to perform 3D renderings. The software VGStudioMax of Volume Graphics GmbH Heidelberg was used for the digital filtering and the 3D volume rendering [17].

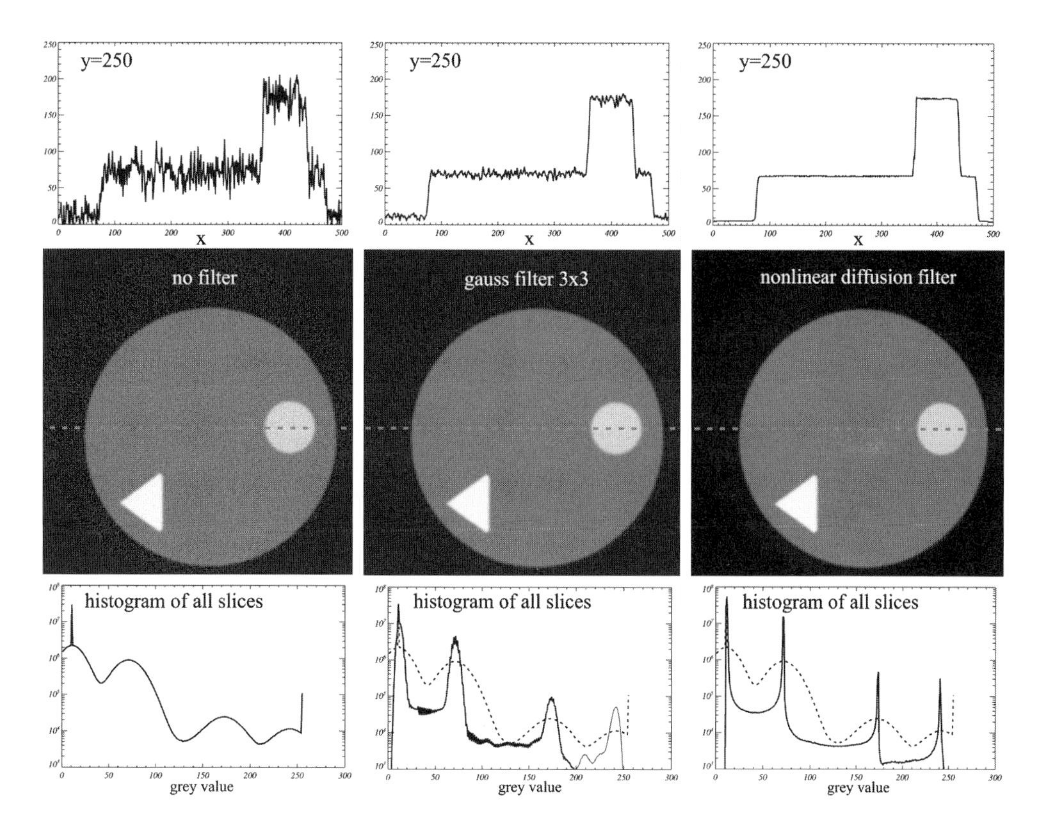


**Figure 16.6** Demonstration of different digital filters to reduce the noise in the tomogram. The used simulated scan is presented in Figure 16.4c. The slices were scaled from −0.0005 to 0.0105 to byte 0 to 255. The resulting thresholds presented in Figure 16.5 are $t_1 = 39.6$, $t_2 = 116.4$ and $t_3 = 209.5$, respectively.

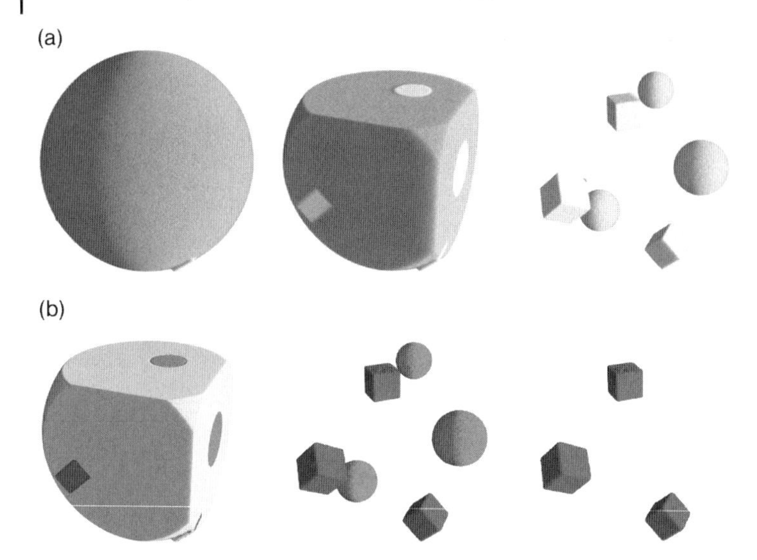


**Figure 16.7** Volume renderings of the simulated scan presented in Figure 16.4c. (a) The unfiltered volume (left), virtually sliced (middle) and the visualization of the inner spheres and boxes (right). By applying the digital filter presented in Figure 16.6c, the different material is segmented and can be separately visualized (b).

## 16.3 Macrotomography Using Neutrons

As an example for a neutron radiography facility, the GENRA-3 facility at the research reactor Geesthacht FRG-1 specialized in nondestructive testing for industrial applications will be shown in detail. Recently, this facility was extended with a tomography set-up for 3D investigations [18].

### 16.3.1 Experimental Set-up

The set-up for neutron tomography can be installed in the irradiation chamber of GENRA-3 at different distances to the source (Figure 16.8a). Therefore, an optimal value of $L/D$ with respect to the size of the sample can be selected. This is required to obtain a good spatial resolution in the tomogram. The neutron beam parameters for different sample–collimator distances $P$ at GENRA-3 are given in Table 16.2.

In Figure 16.9, a new set-up for tomography installed at GENRA-3 is shown. It consists of a 2D neutron detector and a high-precision sample manipulator. The principle set-up is given in the sketch in Figure 16.8b. The components used are summarized in Table 16.3.

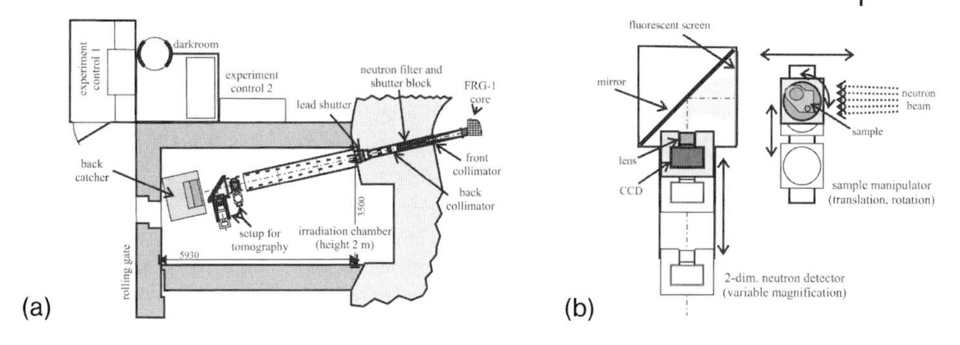


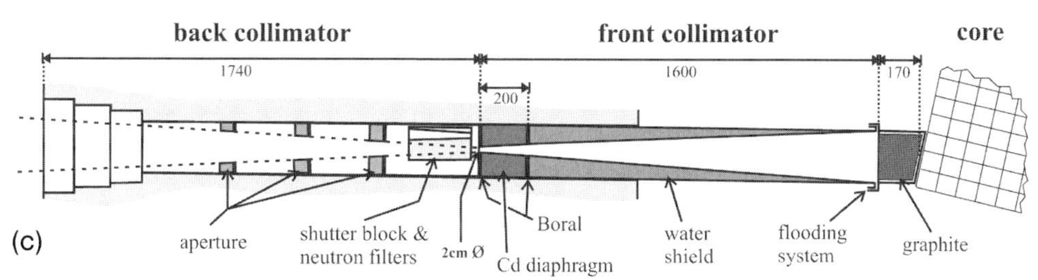


**Figure 16.8** (a) Schematic view of the neutron radiography facility GENRA-3 at the research reactor Geesthacht FRG-1. The tomographic apparatus can be installed at different distances to the source resulting in $L/D$ from 100 up to 300. The distance $L$ is the distance to the aperture installed in the collimator with diameter $D = 2$ cm. The $L/D$ fraction represents the divergence of the beam and is used as a quality value in neutron radiography. (b) Experimental set-up for neutron tomography. The optical magnification of the detector can be varied from a field of view of 10 cm × 10 cm to 40 cm × 40 cm. (c) Collimator for GENRA-3 installed at FRG-1. The shutter block can cover different neutron filters to characterize the spectrum of the neutron beam. A 4 cm Bi filter is used for a thermal and a 0.2 Cd filter for an epithermal neutron beam.

**Table 16.2** Neutron beam parameters at GENRA-3.

| | | | | |
|---|---|---|---|---|
| $P$ (m) | 0.5 | 1.5 | 2.5 | 4.0 |
| Size (cm$^2$) | $16 \times 16$ | $24 \times 24$ | $32 \times 32$ | $47 \times 47$ |
| Flux (n cm$^{-2}$ s$^{-1}$) | $7.4 \times 10^6$ | $3.2 \times 10^6$ | $1.7 \times 10^6$ | $6.8 \times 10^5$ |
| $L/D$ | 100 | 150 | 200 | 300 |

### 16.3.2
### Measurements and Results

In order to test the tomographical experiment at GENRA-3, we chose as a sample a water tap consisting of different types of metal and high-absorbing rubber components (see Figure 16.9b). We set-up the sample axis at 0.5 m distance from the collimator and sample–detector distance to 0.13 m (see Figure 16.3). The tomographical scan was performed by taking 30 flat field images (images without

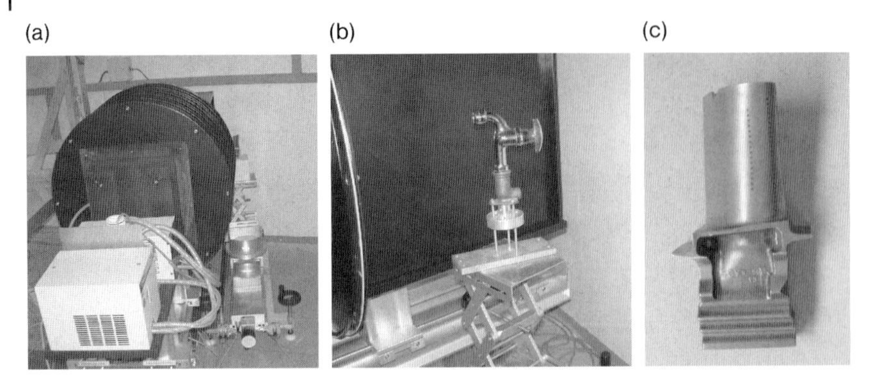


**Figure 16.9** Experimental set-up used for NCT (a); investigation of a water tap (b); and a viewgraph of the turbine blade made of nickel-based alloy (c).

**Table 16.3** Components used for NCT set-up at GENRA-3.

| |
|---|
| *CCD camera*: Kodak KAF-1400, 1320 × 1035 pixels, 6.8 μm × 6.8 μm, 12 bit digitization. |
| *Optical lens*: Nikkor 35 mm f/1.4, Nikon, Tokyo, Japan. |
| *Optical mirror*: Plane mirror, optical surface Al covered by SiO, LINOS Photonics, Göttingen, Germany. |
| *Fluorescent screen*: $^6$LiF/ZnS:Ag, 40 cm × 40 cm, 250 μm thick, ND, green emitting, substrate 1 mm aluminum, Applied Scintillation Technologies, Harlow, UK. |

sample) and 720 projections at different sample rotations equally stepped between 0 and $2\pi$. The exposure time of each image was 10 s. The total scan lasted 4.5 h. To reconstruct the sample the axis of rotation was carefully aligned by software to be perpendicular to the lines of the projections. These aligned projections then were reconstructed, resulting in a 3D dataset consisting of 498 × 498 × 611 voxels representing the size of 130 mm × 130 mm × 160 mm. The visualization of the 3D dataset is presented in Figure 16.10. By using the software VGStudioMax 3D volume renderings and virtual cuts are created. Furthermore, the software is able to extract and colorize the high absorbing sealing rings of the water tap.

A further tomographical scan was performed on a turbine blade made out of a nickel-based alloy (Inconel). The aim of the study is the investigation of the internal cooling system. We recorded 720 projections with an exposure time of 15 s. The total scan lasts 5 h. In Figure 16.9c, an optical viewgraph of the sample is given. The reconstructed data set consists of 437 × 437 × 433 voxel representing a volume of 114 mm × 114 mm × 113 mm. Renderings of the total and sliced volume are presented in Figures 16.11a–c. By using the software VGStudioMax, the voids in the blade representing the cooling canal system are segmented and visualized in dark grey in Figure 16.11d.

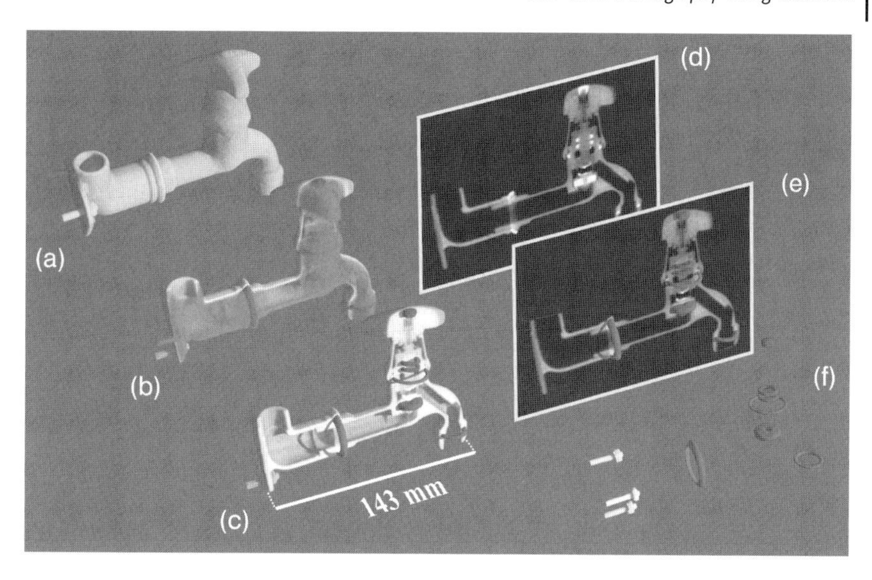


**Figure 16.10** Volume rendering of the total reconstructed volume of the water tap (a). The high-absorbing sealing rings out of rubber material are shown as white regions in the vertical cut in (d). Using the software VGStudioMax these regions are segmented and colored in red in the 3D volume. At the position of the vertical cut, half of the metal material is visualized semitransparent (b) and transparent (c). The 3D arrangement of the sealing rings is shown in combination with the vertical cut (e) and the segmented screws (f).

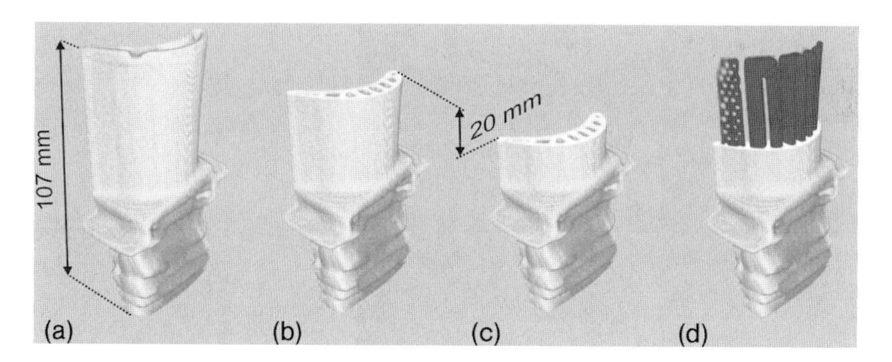


**Figure 16.11** Volume renderings of the reconstructed data set of a turbine blade made of a nickel-based alloy (Inconel). A volume rendering of the total data set is given in (a). Then the volume is sliced at 20 mm (b) and 40 mm (c) from the top. The internal canal system is clearly visible. These voids then are segmented and colored in dark grey. A combination of the cooling system (dark grey) and the remaining nickel-based alloy matrix (light grey) is visualized in (d).

## 16.4 Microtomography Using Synchrotron Radiation

As an example for a microtomography beamline, the recently rebuilt beamline W2 of the storage ring DORIS III at DESY is presented. This beamline HARWI-II (rebuilt and operated by the GKSS-Research Center Geesthacht, Germany, in cooperation with the GFZ Potsdam, Germany, and DESY, Hamburg) is specialized for the use of high-energy X-rays from 20 to 250 keV dedicated to texture, strain, and imaging measurements for materials science applications [19]. During the construction phase of the HARWI-II beamline, special attention was given to the optics concept to obtain a high-energy high-flux X-ray beam with a large field of view optimized for imaging materials science applications [20].

### 16.4.1 Beamline Optics

The main components for the beamline optics are the HARWI-II wiggler, the new front-end filter, the monochromator, the beam stop, and the diagnostics table (Figure 16.12). The design and features of the wiggler installed in the DORIS-III storage ring are presented in detail elsewhere [21]. The main parameters of the wiggler are given in Table 16.4.

To reduce the heat load on the first crystal of the monochromator 3 mm carbon is permanently installed as a high-pass filter. For the use of high-photon ener-

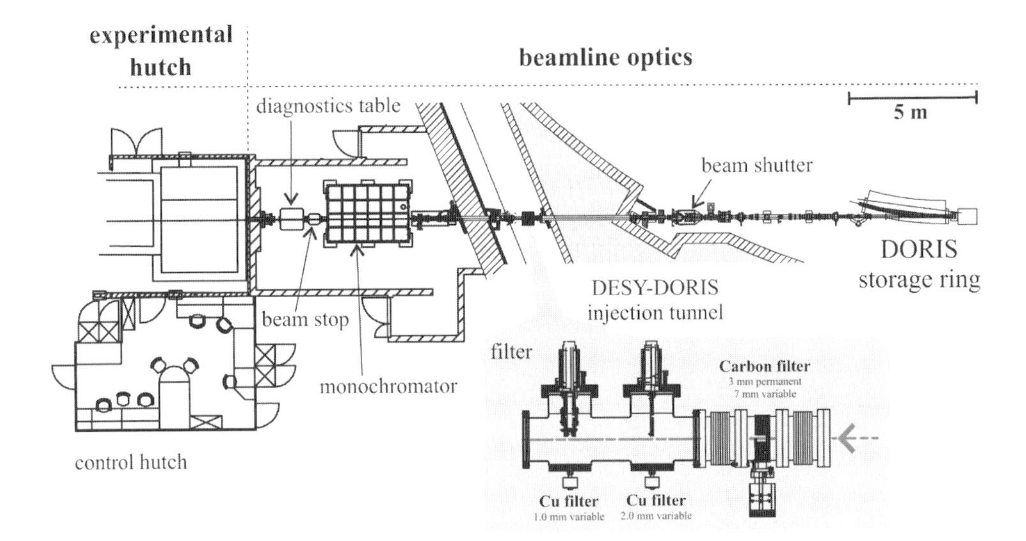


**Figure 16.12** Sketch of the HARWI-II beamline showing the beamline optics. The permanent and optional high-pass filter installed in the DORIS-DESY injection tunnel is shown in the inset. A permanent 3 mm thick carbon filter can be extended by an optional filter (7 mm carbon and 1 mm or 2 mm copper).

**Table 16.4** Parameters of the HARWI-II wiggler.

| | | | |
|---|---|---|---|
| Total length | 4 m | Peak field: | 1.91 T |
| Period length | 110 mm | *K*: | 20.3 |
| Minimal gap | 14 mm | Critical energy: | 26.7 keV |

gies, a 7 mm carbon filter and a 1 mm or 2 mm copper filter can additionally be placed into the beam. The sketch of the filter system is given in the inset of Figure 16.12. A detailed description of the system located in the DORIS-DESY injection tunnel 36 m downstream of the HARWI-II wiggler is given in [22].

The resulting flux behind the filter system is presented in Figure 16.13. The remaining total power of the white beam (70 mm × 10 mm) at minimum gap and minimum filter (3 mm carbon) is about 12.6 kW. To further reduce the heat load on the first crystal the wiggler gap can be increased. By use of different combinations of the filters and wiggler gaps the optimized conditions with respect to photon energy and beam size can be found.

The concept of the monochromator set-up is presented in Figure 16.13. Two different monochromator systems, one reflecting in the horizontal plane, and one reflecting in the vertical plane, are installed. The horizontal monochromator is mainly used for diffraction applications and provides an X-ray beam with a size of 10 mm × 10 mm from 50 to 250 keV. The vertical monochromator is optimized for imaging applications and will deliver an X-ray beam with a size of 70 mm × 10 mm and a photon energy range of 15 to 200 keV. Different pairs of crystals in Laue–Laue and Laue–Bragg geometry will be installed. Furthermore, bent-Laue crystals are used to increase the energy band path of the reflected X-ray beam.

In June/July 2006, the vertical monochromator was successfully set-up. For the first commissioning phase, a fluorescent screen, a Si-111 (Bragg), and Si-111 (bent-Laue) crystal were installed on the secondary crystal stage. The primary crystal stage was equipped with a calorimeter and a water-cooled bent-Laue crystal. Figure 16.13 shows a view into the monochromator tank revealing the installed monochromator system. Both, sketch and viewgraph show the same settings used for 36 keV photon energy.

### 16.4.2 Experimental Set-up

The set-up used for absorption-contrast microtomography is presented in Figures 16.1 and 16.14. It consists of a 2D X-ray detector and a sample manipulator stage. The sample manipulator provides both for the rotation and for the lateral positioning of the specimen. The incident X-rays are converted into visible light which then is projected onto a CCD camera by an optical lens system. The used components are given in Table 16.5.

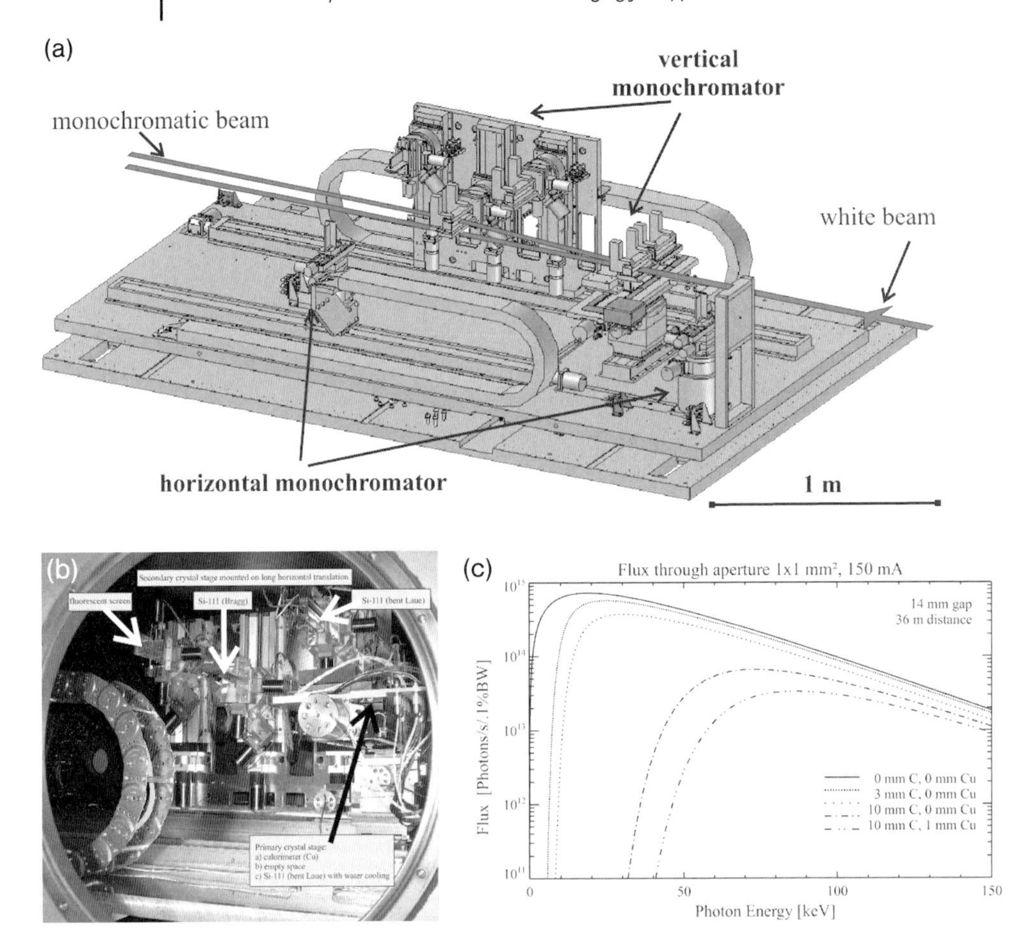


**Figure 16.13** (a) Sketch of the monochromator set-up. The user can switch between two different monochromators. The horizontal monochromator in the foreground and the vertical monochromator in the background are placed into the white X-ray beam by moving the upper plate perpendicular to the beam. The monochromator is installed in the monochromator tank in the optics hutch. The tank is of dimension 3 m $\times$ 2 m $\times$ 1 m ($L \times W \times H$) and can be evacuated to below $10^{-6}$ mbar. (b) View on the installed vertical monochromator as shown in the sketch (top). The primary crystal stage (right foreground) carries a calorimeter and the first Si crystal. The secondary stage (left background) consists of three stages which allow for combinations of different secondary crystal stages with the first crystal. Here the middle secondary crystal stage is set to 36 keV photon energy (Si-111). The fixed exit monochromator has an offset of 40 mm. (c) Comparison of the flux of the HARWI-II wiggler at minimal gap of different filter settings with the unfiltered spectrum at the position of the high-pass filter (36 m distance from the source).

(a)

(b)

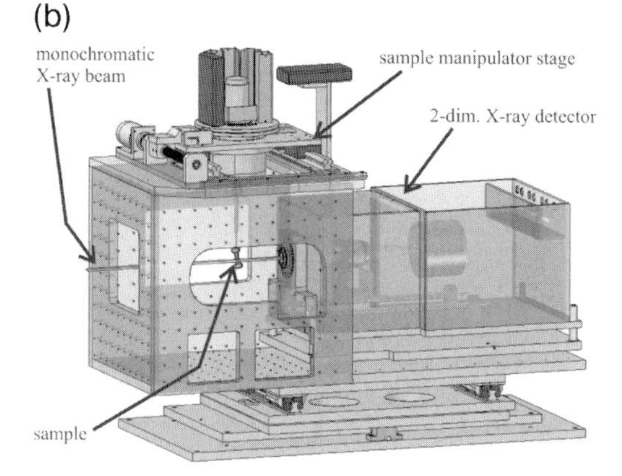


**Figure 16.14** (a) Set-up for microtomography installed at the HARWI-II beamline. The experiment is mounted on a lift table, which can be lowered by 1.5 m. The pit will then be covered by the shown roof. (b) Sketch of the microtomography system. The main components are a high-precision sample manipulator and a 2D X-ray detector. The sample manipulator consists of a rotational axis with integrated *x*–*y*–*z* translation of the sample. The 2D X-ray detector is built from a single crystal luminescent screen, a lens and a CCD camera. The optical magnification can be adjusted to the size of the sample. The length of the total system is about 1 m.

**Table 16.5** Components used for SR$\mu$CT.

| | |
|---|---|
| CCD camera | KX2, Apogee Instruments, Inc.; 14 bit digitalization at 1.25 MHz, 1536 × 1024 pixels; each 9 μm × 9 μm |
| Optical lens | Camera lens, Nikon Inc., 35 mm or 50 mm focal length |
| Fluorescent screen | $CdWO_4$ single crystal, thickness 200–1000 μm |

**Table 16.6** Parameters of SR$\mu$CT at beamlines BW2 and HARWI-II at DESY.

| | BW2 | HARWI-II |
|---|---|---|
| Photon energy | 7–24 keV | 15–200 keV |
| Sample height | 3.5 mm | 8 mm |
| Sample width | 10 mm | 70 mm |
| Spatial resolution | 2 μm | 3 μm |

This set-up is operated as a user experiment for absorption-contrast microtomography at beamline BW2 using photon energies from 7 to 24 keV. The apparatus is furthermore designed to include an X-ray interferometer for performing phase-contrast microtomography [10]. As shown in Figure 16.1d, the monochromatic incoherent X-ray beam is split into two coherent beams. A single phase-contrast projection is obtained by recording interference patterns at different overall phase shifts introduced by the phase rotor. An attenuation projection is calculated from an image with and without the sample. For microtomography

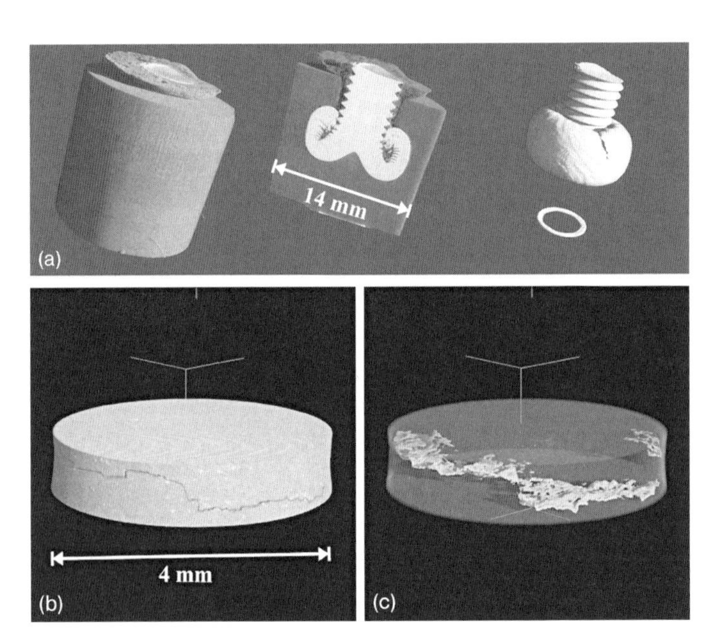


**Figure 16.15** (a) Investigation of an Al-sample studying the material flow in a friction rivet sample. This new joining technique is used for polymers and lightweight metal alloys (here Al). The sample is scanned using 36 keV photon energy. Volume renderings are presented of the total sample (a, left), sliced sample (a, middle), and the Al-rivet, i.e. transparent polymer (a, right). (b, c) Investigation of damage mechanisms and crack extension in an Al2024 aluminum alloy using 28 keV photon energy at HARWI-II. Cut of one specimen, as a full solid (b), semivisible with the cracks appearing in white (c).

typically a set of 720 radiograms at different sample rotations, equally stepped between 0 and $\pi$, were taken. The tomographical reconstructions were calculated by using "back-projection of filtered projection" algorithm. For this reconstruction method, the center of rotation has to be determined within a fraction of pixel. Recently, our reconstruction chain could be extended by an automatized algorithm to find this reconstruction center [23].

For performing microtomography at HARWI-II, the microtomography apparatus was installed onto a lift table in the experimental hutch (Figure 16.14). The installation on the shown lift table allows for fast reinstallation (about 6 min) and operation of the experiment, providing for a fast switching between experiments at HARWI II. The first tomographical scans using the vertical monochromator were successfully performed in July 2006. Within 5 weeks of user operation, 280 tomographical scans were successfully performed in the second-half of 2006 using the photon energy range from 16 to 64 keV.

A few selected examples of materials science applications obtained at the new HARWI-II beamline are presented in Figures 16.15 and 16.16 [24–26]. Different types and sizes of aluminum samples were investigated using photon energies of 28, 36, and 64 keV, respectively.

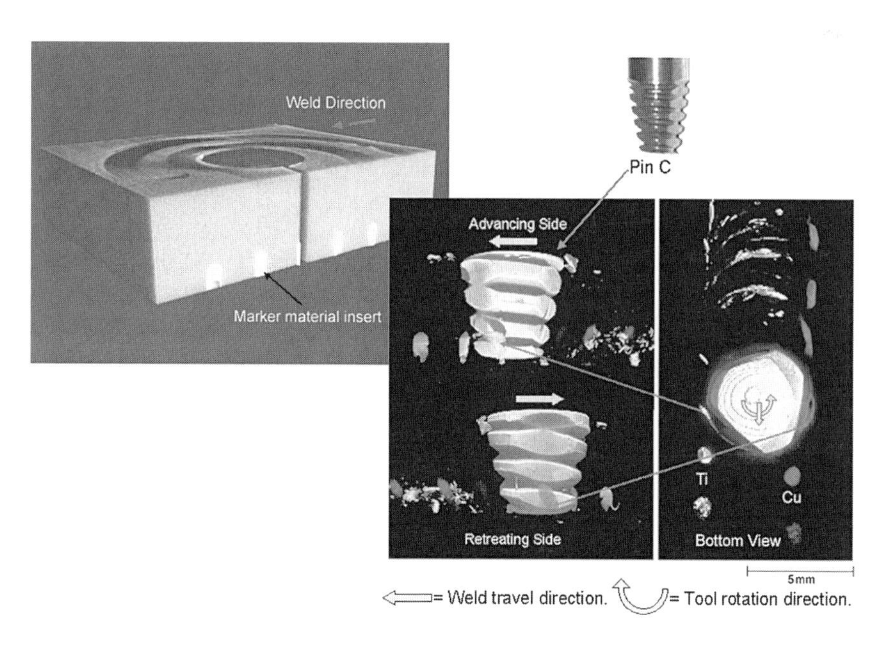


**Figure 16.16** Simultaneous marker material displacement (Cu and Ti) around the aluminum workpiece and the friction-stir welding tool (Pin C – conical and threaded with three equally spaced flats along entire thread length). Left: Volume rendering showing the aluminum sample (5 mm thickness) and the introduced marker material. On the right the aluminum is set transparent to reveal the 3D distribution of the marker material around the pin. The tomographical scans were performed at beamline HARWI-II using photon energy of about 64 keV. More information on friction welding can be found in Chapter 7.4.1.

## 16.5
## Summary and Outlook

Different set-ups for tomography and microtomography at a neutron source and at a second-generation synchrotron radiation facility are presented. The experiments are designed to operate in user mode. Complementary access to both the techniques, neutron tomography and synchrotron radiation tomography, allows the development of new techniques and will give new insight in the 3D behavior of samples especially in materials science.

Exploiting the source characteristics of the new third-generation synchrotron radiation facility PETRA-3 at DESY will allow extending the spatial resolution into the nanometer regime.

## References


1 U. Bonse, F. Busch, X-Ray computed microtomography (CT) using synchrotron radiation (SR), *Prog. Biophys. Molec. Biol.* 1996, **65**, 133–169.

2 A. Snigirev, I. Snigireva, V. Kohn, S. Kuznetsov, I. SchelokovSnigirev, On the possibilities of X-ray phase contrast microimaging by coherent high-energy synchrotron radiation, *Rev. Sci. Instrum.* 1995, **66**, 12, 5486–5492.

3 P. Cloetens, R. Barrett, J. Baruchel, J.-P. Guigay, M. Schlenker, Phase objects in synchrotron radiation hard X-ray imaging, *J. Phys.* 1996, **D 29**, 133–146.

4 http://www.esrf.eu/UsersAndScience/Experiments/Imaging/ID22/Applications/Imaging/CoherentImage.

5 C. Raven, A. Snigirev, I. Snigireva, P. Spanne, A. Souvorov, V. Kohn, Phase-contrast microtomography with coherent high-energy synchrotron X-rays, *Appl. Phys. Lett.* 1996, **69**, 13, 1826–1828.

6 P. Spanne, C. Raven, I. Snigireva, A. Snigirev, In-line holography and phase-contrast microtomography with high energy X-rays, *Phys. Med. Biol.* 1999, **44**, 741–749.

7 Alberto Bravin, Exploiting the X-ray refraction contrast with an analyser: the state of the art, *J. Phys. D: Appl. Phys.* 2003, **36**, A24–A29.

8 E. Pagot, P. Cloetens, S. Fiedler, A. Bravin, P. Coan, J. Baruchel, J. Härtwig, A method to extract quantitative information in analyser-based X-ray phase contrast imaging, *Appl. Phys. Lett.* 2003, **82**, 20, 3421–3423.

9 A. Momose, Demonstration of phase-contrast X-ray computed tomography using an X-ray interferometer, *Nucl. Instrum. Methods* 1995, **A 352**, 622–628.

10 F. Beckmann, U. Bonse, F. Busch, O. Günnewig, X-Ray microtomography ($\mu$CT) using phase contrast for the investigation of organic matter, *J. Comp. Assist. Tomogr.* 1997, **21**, 4, 539–553.

11 A. Momose, Phase-contrast X-ray imaging based on interferometry, *J. Synchrotron Rad.* 2002, **9**, 136–142.

12 P. Cloetens, W. Ludwig, J. Baruchel, D. Van Dyck, J. Van Landuyt, J.P. Guigay, M. Schlenker, Holotomography: Quantitative phase tomography with micrometer resolution using hard synchrotron radiation X-rays, *Appl. Phys. Lett.* 1999, **75**, 19, 2912–2914.

13 T. Weitkamp, A. Diaz, C. David, F. Pfeiffer, M. Stampanoni, P. Cloetens, E. Ziegler, X-ray pase imaging with grating interferometer, *Optics Express* 2005, **12**, 16, 6296–6304.

14 F. Pfeiffer, T. Weitkamp, O. Bunk, C. David, Phase retrieval and differential phase-contrast imaging with low-brilliance X-ray sources, *Nature Phys.* 2006, **2**, 258–261.

15 A.C. Kak, M. Slaney, *Principles of Computerized Tomographic Imaging*, IEEE Press, New York, 1988.

16 J. Radon, Über die Bestimmung von Funktionen durch ihre Integralwerte längs gewisser Mannigfaltigkeiten, *Ber. Ver. Sächs. Akad. Wiss. Leipzig, Math-Phys. Kl.* 1927, **69**, 262–277.
17 http://www.volumegraphics.com.
18 F. Beckmann, J. Vollbrandt, T. Donath, H.W. Schmitz, A. Schreyer, Neutron and synchrotron radiation tomography: New tools for materials science at the GKSS-Research Center, *Nucl. Instrum. Methods Phys. Res.* 2005, **A 542**, 279–282.
19 F. Beckmann, T. Dose, T. Lippmann, R.V. Martins, A. Schreyer, *The new materials science beamline HARWI-II at DESY, AIP Conference Proceedings*, SRI 2006, Daegu, Korea, 2007, CP879, pp. 746–749.
20 F. Beckmann, T. Donath, J. Fischer, T. Dose, T. Lippmann, L. Lottermoser, R.V. Martins, A. Schreyer, *New developments for synchrotron radiation-based microtomography at DESY, Proceedings of SPIE 2006*, 6318, 631810.
21 M. Tischer, L. Gumprecht, J. Pflüger, T. Vielitz, *A New Hard X-ray Wiggler for DORIS III, AIP Conference Proceedings*, SRI 2006, Daegu, Korea 2007, CP879, pp. 339–342.
22 H. Schulte-Schrepping, U. Hahn, *Hard X-ray Wiggler Front End Filter Design, AIP Conference Proceedings*, SRI 2006, Daegu, Korea 2007, CP879, pp. 1042–1045.
23 T. Donath, F. Beckmann, A. Schreyer, Automated determination of the center of rotation in tomography data, *J. Opt. Soc. Am.* 2006, **A 23**, 5, 1048–1057.
24 S. Amancio, J.F. dos Santos, F. Beckmann, A. Schreyer, Computer Microtomographic Investigation of Polyetherimide/Al 2024-T351 Friction Riveted Joints, *DESY Annual Report*, 2006, pp. 485–486.
25 R. Braun, D. Steglich, F. Beckmann, Damage Mechanisms and LCF Crack Extension in Al2024, *DESY Annual Report*, 2006, pp. 497–498.
26 R. Zettler, J.F. dos Santos, T. Donath, J. Herzen, F. Beckmann, D. Lohwasser, Removing the guess work from material flow and deformation mechanics associated with the FSW of Aluminium and its alloys through application of micro-computer tomography and dedicated metallurgical techniques, *DESY Annual Report*, 2006, pp. 667–668.

# 17
# $\mu$-Tomography of Engineering Materials

Astrid Haibel

## 17.1
## Advantage of Synchrotron Tomography

Synchrotron radiation is electromagnetic radiation, generated by the acceleration of ultrarelativistic electrons through magnetic fields. It was discovered in a General Electric synchrotron accelerator in 1947 [1]. The wavelength range of this radiation includes the infrared, optical, ultraviolet, and X-ray light spectrum.

The high brilliance, exceeding other natural and artificial light sources by many orders of magnitude, and the continuous wavelength spectrum of this radiation combined with suitable narrow band pass monochromators make synchrotron radiation unique for experiments requiring continuously tunable sources from the infrared up to the X-ray regions. Further unique properties of synchrotron radiation are the high collimation, the low emittance, and the high level of polarization.

For the synchrotron tomography the monochromatic radiation allows us to resolve and visualize different sample components even if they have similar absorption coefficients with a spatial resolution in the submicrometer range within short measurement times. It is possible to adjust the X-ray energy to the absorption properties of the samples. Further the possibility of monochromizing the beam can be used, e.g., for absorption edge tomography. Due to the monochromatic and almost parallel beam, the images are widely free of artifacts known from laboratory tomography.

*Neutrons and Synchrotron Radiation in Engineering Materials Science: From Fundamentals to Material and Component Characterization*
Edited by Walter Reimers, Anke Rita Pyzalla, Andreas Schreyer, and Helmut Clemens

ISBN: 978-3-527-31533-8

## 17.2
## Applications and 3D Image Analysis

### 17.2.1
### Discharging Processes in Alkaline Cells

Alkaline batteries have been used since about 100 years. They show an excellent performance characteristic, a favorable cost structure, and are extremely reliable in many industrial and household applications [2–5].

The lifetime of such batteries is strongly correlated with the dissolution process of the zinc particles in the anode material [6]. The zinc utilization in alkaline batteries is still an important topic in order to continue the improvement of the cell performance [7].

In this application example for synchrotron tomography the visualization and analysis of the discharging processes of a battery by an *in situ* experiment is shown. The complete process including shrinking and disintegrating of zinc particles in the anode electrolyte material, i.e., the morphology and spatial distribution of zinc and zinc oxide during discharging, is observed [8].

For the measurement a battery was located on a sample holder in the synchrotron beam during the whole experiment, i.e., during the tomographic measure-

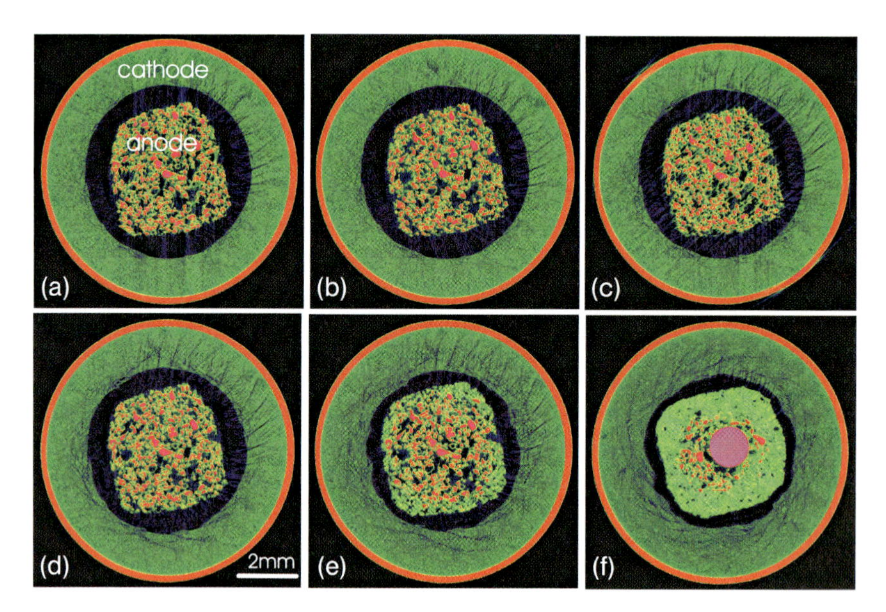


**Figure 17.1** Six different discharging states of the investigated battery. (a) Fully charged state with voltage $U = 1.59$ V. (b) State after 15 min discharging time with voltage $U = 1.47$ V. (c) State after 45 min. discharging time with voltage $U = 1.42$ V. (d) State after 90 min discharging time with voltage $U = 1.37$ V. (e) State after 3 h discharging time with voltage $U = 1.10$ V. (f) A second, structurally identical battery in the almost discharged state (9 h discharging time, $U = 0.5$ V; here the arrester is visible in the middle of the image).

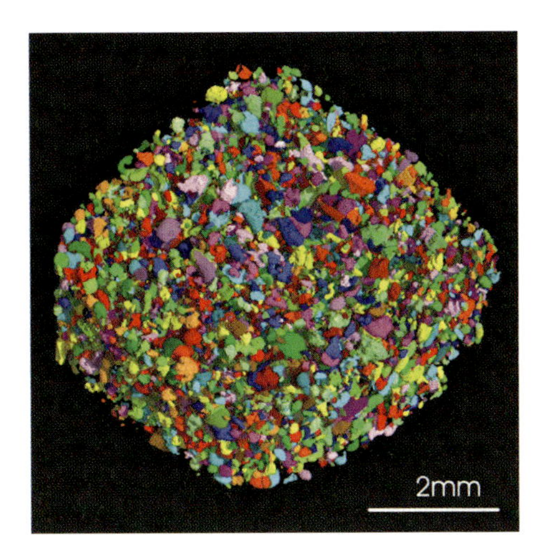


**Figure 17.2** Labeled and color coded zinc particles. They act as anode in alkaline batteries.

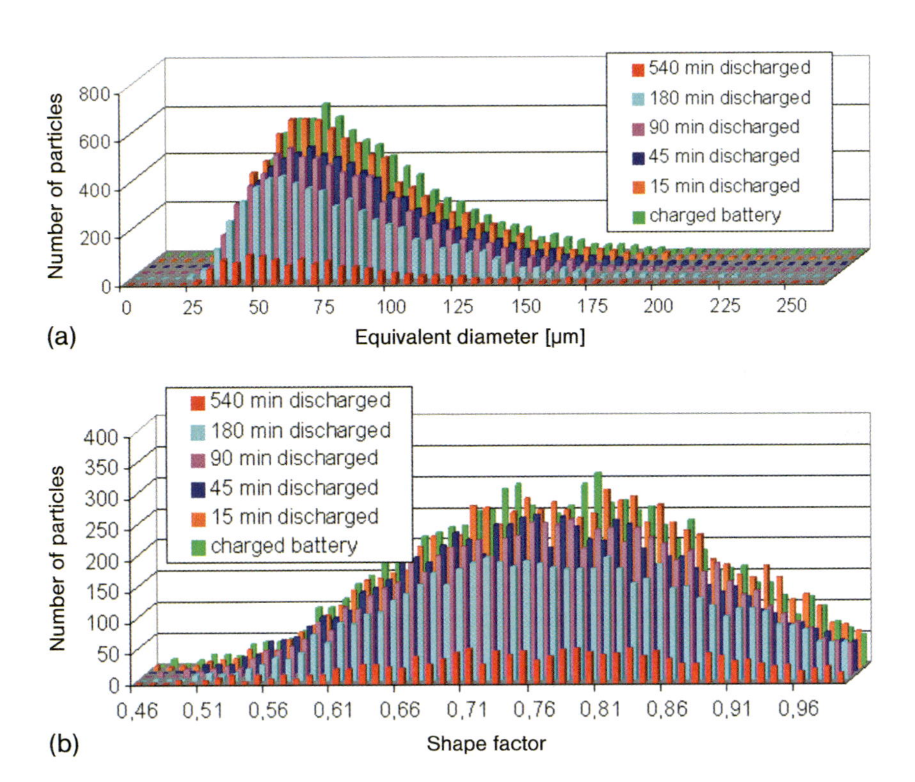


**Figure 17.3** *In situ* tomography on liquid aluminum alloy foams:
(a) Zinc particle number versus diameter for each discharging step.
(b) Zinc particle number versus particle shape for each discharging step (from 0 (planes) to 1 (spheres)).

ments as well as during the discharging. Always the same part of the battery was observed for five different discharging states (see Figures 17.1a–e). In addition, a slice of a fully discharged battery is shown (Figure 17.1f). The discharging times between the tomographic measurements were 15 min, 45 min, 90 min, 180 min, and 540 min.

The battery consists of zinc gel as anode (inner granular structure, 67% zinc and 33% electrolyte) and manganese oxide as cathode (outer green ring) isolated by a separation fleece (blue ring between anode and cathode) and covered by a steel tube (outer red ring). The electrolyte is made of 36% potassium hydroxide, 60% water, and 4% zinc oxide.

As a first qualitative result we see that the amount of small particles in the middle of the pictures in Figure 17.1 decreases in favor of the surrounding areas, i.e., the zinc particles dissolve from outside and zinc oxide accrues. Also the outer ring in the figures increases, i.e., the manganese oxide layer swells. As a consequence the separation fleece is being impacted. Equation (1) describes the chemical reaction at the cathode and at the anode:

$$\begin{aligned} &\text{Anode:} && Zn + 2OH^- \rightarrow ZnO + H_2O + 2e^- \\ &\text{Cathode:} && 2MnO_2 + 2H_2O + 2e^- \rightarrow 2MnOOH + 2OH^- \end{aligned} \tag{17.1}$$

For the quantitative 3D image analysis of the zinc particle dissolving process the particles were separated from the other components in Boolean images by choosing suitable grey value thresholds. The closeness of zinc particles in the battery inhibits a spatial separation of each particle in the Boolean records. Therefore, a 3D watershed algorithm combined with a Euclidian distance transformation was used to separate the particles in the Boolean image. Then the particles were identified by a labeling algorithm and color coded [9, 10]. The result for the zinc particles in the fully charged state of the battery is displayed in Figure 17.2.

Based on this particle separation algorithm the decrease of the zinc particle fraction for each discharging step and the shape of each of these particles could be quantified. The results are shown in Figure 17.3a and b.

The diagrams show that during discharging the mean diameter and the shape of the particles are widely unchanged; only the particle number decreases. This confirms the visual impression of Figure 17.1, which indicates that the particle dissolution takes place from the outside of the zinc gel. Only the outer particles disappear whereas the inner particles remain unchanged.

### 17.2.2
### Microstructural Investigations of $Nb_3Sn$ Multifilamentary Superconductor Wires

The low-$T_C$ superconductor $Nb_3Sn$ is most commonly used to achieve huge magnetic fields higher than 10 T [12, 13]. But, as for the high-$T_C$ materials, $Nb_3Sn$ is very brittle, too. Thus the superconducting wires cannot be bent to form the magnet coils in a simple production step.

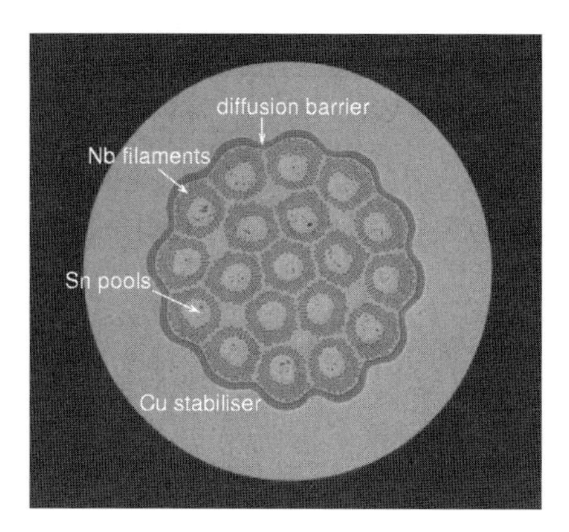


**Figure 17.4** Light microscopic image of a $Nb_3Sn$ precursor wire cross section. The diameter of the wire is 0.7 mm.

To enable coil forming the synthesis of the superconducting $Nb_3Sn$ phase within the wires has to be done after coiling. In a first step the Nb, Cu, and Sn subelements which are assembled inside a tantalum diffusion barrier and a copper can are tubular extruded (see Figure 17.4). Then the strand size is reduced to a cross section less than 1 mm by swaging and drawing.

Afterward, the strands are wired to magnet solenoids. Finally, a heat treatment is applied to transform the constituents into the superconductive phase. The formation reaction occurs through a solid state diffusion reaction at high temperatures in a protecting atmosphere. Tomography has been applied to observe the microstructure during the heat treatment [11].

In Figure 17.5 the cross sections of four internal $Nb_3Sn$ strands are shown after different thermal treatments reconstructed from the synchrotron absorption contrast. It is obvious that the pure Sn in the nonheat treated strand section (Figure 17.5a) is successively transformed into the superconducting phase $Nb_3Sn$ (Figure 17.5d). The linear absorption coefficients $\mu$ at the used X-ray energy $E = 50$ keV are for pure Sn $\mu = 78.12\ \text{cm}^{-1}$, for pure Nb $\mu = 56.81\ \text{cm}^{-1}$, for pure Cu $\mu = 23.42\ \text{cm}^{-1}$, and for $Nb_3Sn$ $\mu = 64.85\ \text{cm}^{-1}$. Hence, in the original state of the wire the grey values in the image for the Sn pools are much brighter than the grey values of the surrounding Nb. After the reaction treatment the grey values of the new formed $Nb_3Sn$ are darker than the remaining Nb.

During the thermal treatment the formation of voids in the strand matrix can be observed [14]. These voids may cause a degradation of the superconductor critical current. Therefore, it is desirable to avoid or at least to better control the void growth during the strand heat treatment, e.g., through optimized strand designs or heat treatment cycles. Taking a glance at Figure 17.5 two types of voids are visible. One void type is formed below the melting temperature of tin in the Sn diffusion centers of the strands (see Figure 17.5b). The cross section of these voids can approach that of an entire Sn pool. The formation of these

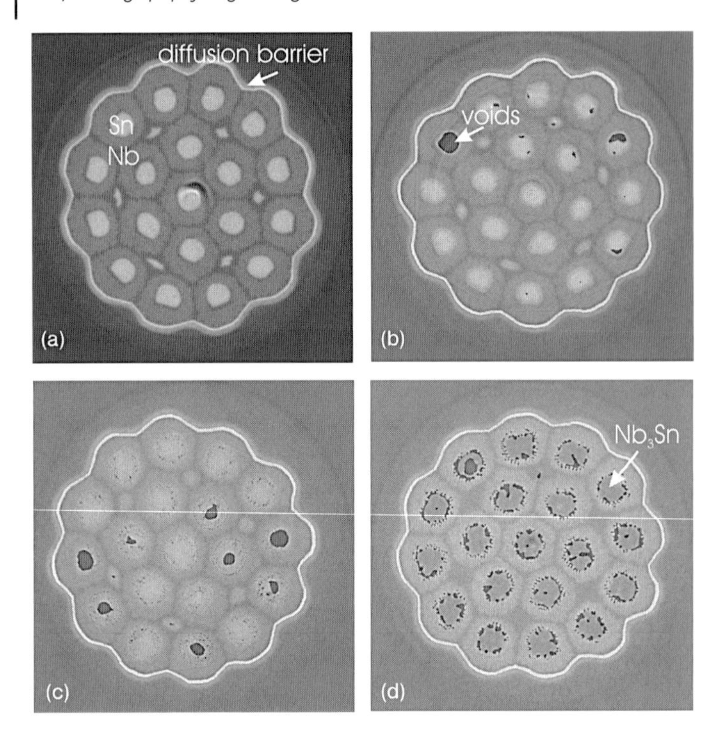


**Figure 17.5** 2D slices of tomograms of $Nb_3Sn$ wires (diameter 0.7 mm): (a) nonheat treated, ductile state; (b) 24 h heat treated at $T = 220$ °C; (c) 24 h heat treated at $T = 310$ °C; (d) 24 h heat treated at $T = 760$ °C.

voids can be explained by a Sn volume increase upon β-Sn ($\rho = 7.26$ g cm$^{-3}$ [15]) to γ-Sn transformation ($\rho = 6.6$ g cm$^{-3}$ [16]) which starts at $T = 162$ °C [16, 17]. Tetragonal β-Sn is transformed into the rhombic γ-Sn which causes a strong volume expansion of about 10% and therefore a plastic deformation of the wire. After the cooling, when γ-Sn is transformed back to β-Sn, voids evolve.

A second type of voids arises in the reacted $Nb_3Sn$ strands above the melting temperature of Sn which has comparatively small (about 1 μm$^2$) cross sections and is distributed around the superconductor filaments. The formation of these voids has been attributed to differences in the diffusion rates of Cu in Sn and Sn in Cu and to the different densities of the formed phases [18, 19].

### 17.2.3
### Influence of the Foaming Agent on Metallic Foam Structures

Metallic foams are highly porous materials with unique properties. They distinguish themselves from other metallic materials by low density, high specific stiffness, and high energy absorption capability [20, 21]. Therefore they become more and more popular for industrial applications. In order to produce metallic

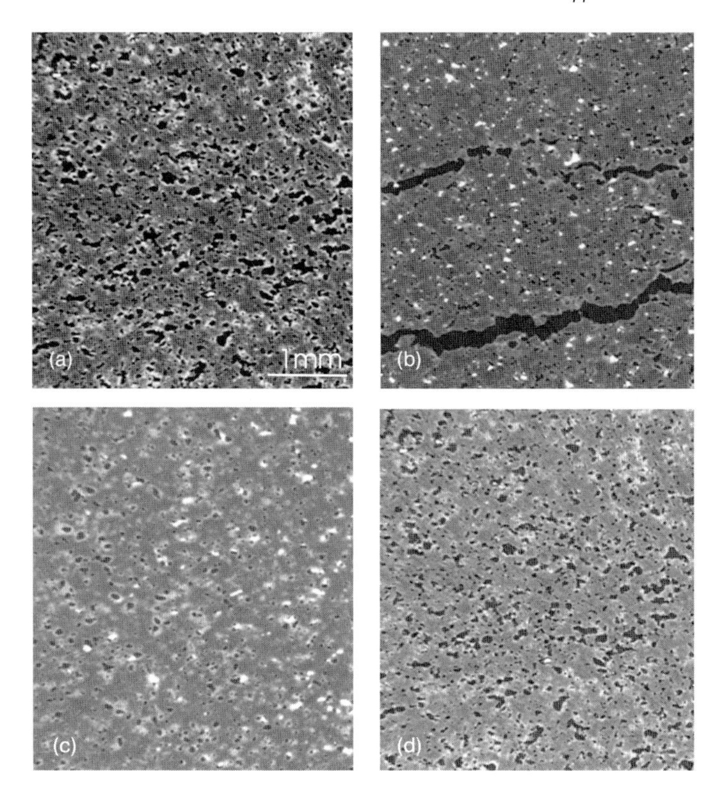


**Figure 17.6** Tomographic slices of four differently prepared aluminum foams in the early foaming stage. (a) Sample prepared with pre-heat-treated $TiH_2$ (180 min/480 °C, decomposition temperature $T = 520$ °C). (b) Sample prepared with untreated $TiH_2$ (decomposition temperature $T = 400$ °C). (c) Sample prepared with sieved $TiH_2$ (particle size $< 20$ μm). (d) Sample prepared with unsieved $TiH_2$ (particle size 20–180 μm).

foams with reproducible macroscopic properties, the precise knowledge of their microscopic structure is needed, i.e., the knowledge of the physical processes during foaming is important for the optimization of the process parameter.

The metal foam presented here was manufactured by admixing 0.5 wt.% titanium hydride powder acting as blowing agent to aluminum alloy powder [22]. Afterward, the powder was pressed to the precursor material. By heating up the precursor to above both the decomposition temperature of the blowing agent and the melting temperature of the metal, hydrogen is released in the melt and a porous structure is generated. The influence of a blowing agent heat pretreatment and of a varying of the blowing agent particle size on the foam structure was investigated by means of synchrotron tomography [23]. Figure 17.6 shows four different prepared foams in the early foaming state.

The treated $TiH_2$ powder (Figure 17.6a) causes a hydrogen release at higher temperatures (decomposition temperature $T = 520$ °C) than the untreated powder ($T = 400$ °C) (Figure 17.6b) which coincides much better with the melting temperature of the alloy ($T = 525$ °C) [24]. Therefore, most hydrogen can be used for forming regular pores whereas the hydrogen release in the sample prepared with untreated foaming agent leads to crack formation in the still solid metal.

The sample prepared with the small $TiH_2$ particles (Figure 17.6c) shows a small pore size distribution of a high number of small pores after foaming. For the sample prepared with a broad $TiH_2$ size distribution (Figure 17.6d) also the pore size distribution is widened.

Hence, the tomographic experiments have shown that by varying the blowing agent parameter (size and/or heat treatment) the metallic foam structure is controllable [25].

## 17.3 Image Artifacts

Tomographic images often contain different kinds of artifacts. A choice of common artifacts is shown in this section.

### 17.3.1 Ring Artifacts

Tomographic images are often superimposed by so-called ring artifacts. Ring artifacts are concentric rings in the images around the center of rotation of the tomographic set-up caused, e.g., by differences in the individual pixel response of the detector or by impurities on the scintillator crystals. They complicate the post-processing of the data, i.e., the segmentation of individual image information.

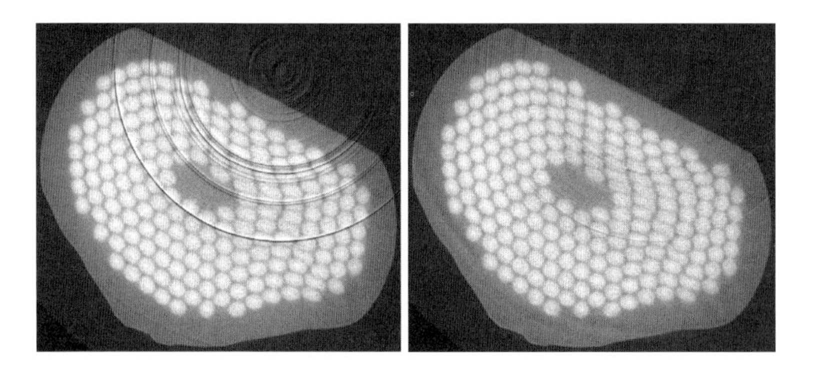

**Figure 17.7** Binary powder-in-tube superconductor strand of $Nb_3Sn$ and copper (sample diameter $\approx 0.7$ mm). Left: Tomographic image superimposed by ring artifacts. Right: The same tomographic slice after using a sinogram correction algorithm [26].

Figure 17.7 (left) shows a tomographic slice of a superconductor strand of $Nb_3Sn$ superimposed by many ring artifacts [26].

A common approach to reduce ring artifacts is the flatfield correction [27]. Thereby images of the background without sample before and after the data acquisition and in certain intervals during the data acquisition are measured. The resulting flatfields include the nonuniform sensitivity of the CCD camera pixels, the nonuniform response of the scintillator screen, as well as the inhomogeneities in the incident X-ray beam. The measurement data are corrected by dividing the normalized radiographic images of the samples and the flatfields ($I_{corr} = I_{measurement}/I_{flatfield}$). However, ring artifacts are not removed completely by this correction when different camera elements have intensity dependent, nonlinear response functions or the incident beam has time-dependent nonuniformities.

In addition, ring artifacts can be avoided by moving the sample or the detector during the measurement in defined horizontal and vertical steps [28, 29]. Then the characteristic of all detector elements is averaged, which leads to significantly reduced ring artifacts.

A third way to reduce ring artifact is the postprocessing of the images, e.g., size and shape filtering after the reconstruction or sinogram processing during the reconstruction [30–32].

### 17.3.2
### Image Noise

The quality of CT images can be degraded by image noise resulting from an unfavorable conversion rate of X-ray photons to visible light photons especially for higher X-ray energies (see Figure 17.8, left) or from high-energy spikes. The use of median filtering in the reconstruction process allows both noise reduction and edge preservation and preserves useful details in the image [33].

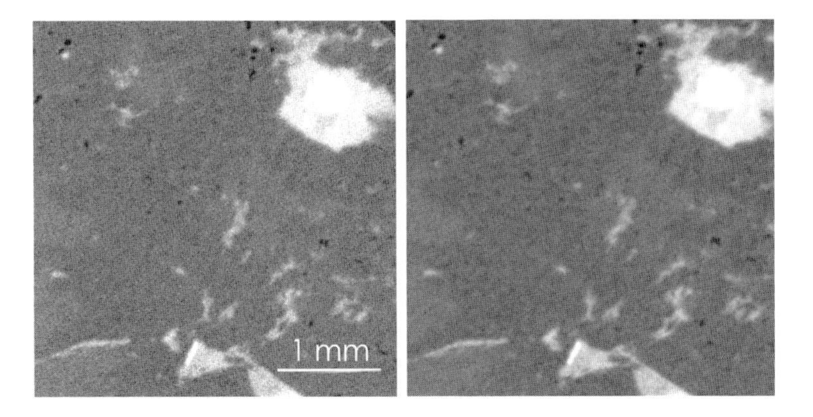


**Figure 17.8** 2D slice of a tomogram of granite rock. Left: Image noise. Right: Median filtered image.

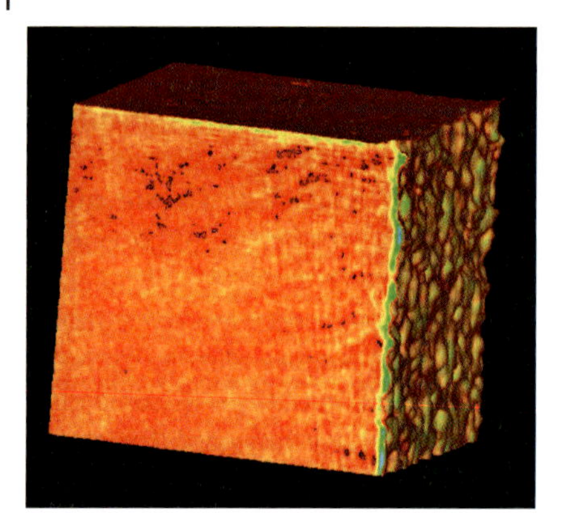

**Figure 17.9** Tomographic slice of a homogeneous aluminum cube with a strong edge enhancement (sample size $2 \times 2 \times 2$ mm$^3$).

### 17.3.3
### Edge Artifacts

Edge enhancement is an adverse artifact for absorption tomography. These artifacts appear as aliasing patterns and as overshoots in the areas of sharp density transitions (see Figure 17.9).

Due to the edge enhancement a precise preprocessing of the data, e.g., the separation of the several components of a sample, is hampered. The effect is caused by scattering of X-rays on materials surfaces where the absorption coefficients change strongly. As shown in Figure 17.9 the effect increases with the finish roughness. The three polished surfaces show a lower effect than the rougher one. To avoid this artifact, the distance between the sample and detector have to be minimized, which reduces the angle of beam spread.

### 17.3.4
### Motion Artifacts

Motion artifacts occur either due to the movement of the sample during the measurement, e.g. tilting, shaking, or shrinking, or due to the inexact movement of the step motors during the data acquisition. The sample in Figure 17.10 was shifted during the date acquisition some horizontal steps intendedly to demonstrate the influence of this motion on the reconstructed 3D image. The left image in the upper part of Figure 17.10 shows one sinogram of the measurement. The shifts during the measurement are clearly discernibly. In the right upper image the associated tomographic slice is displayed.

Whereas the sample movement can be corrected only in a poor way, the correction of the inexact movement of the motors is possible. Here, the discrepancy can be corrected by a translational displacement of each sinogram line at the optimal sinus curve (see Figure 17.10, left lower part) [34].

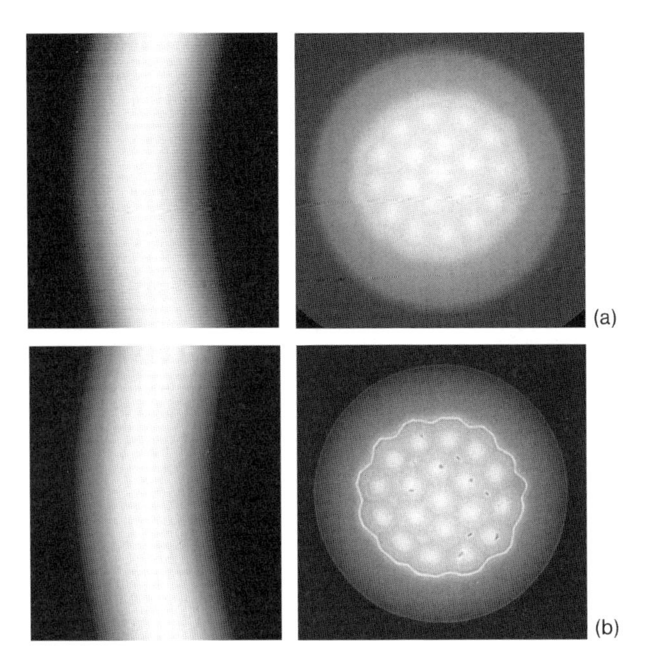


**Figure 17.10** Tomographic slice of a superconductor wire cross section $Nb_3Sn$ (diameter 0.7 mm). (a) Left: Inexact movement of the step motor. Right: Unsharp image of the sample. (b) Left: The corrected sinogram. Right: The same image as above but without motion artifacts.

### 17.3.5
### Centering Errors of the Rotation Axis

A deviation of the rotation axis during the measurements creates semicircle formed artifacts for a 180° tomographic scan as shown in Figure 17.11 (left).

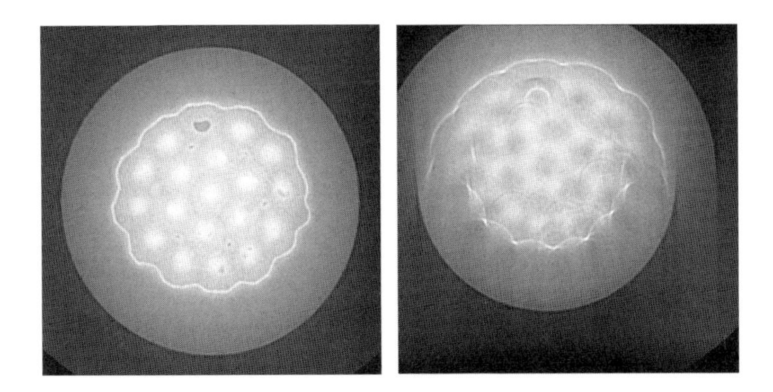

**Figure 17.11** Tomographic slice of a superconductor $Nb_3Sn$ cross section with a diameter of 0.7 mm. Left: The rotation axis of the image differs from the correct axis by 30 pixels. Right: The corrected image.

These artifacts can be corrected easily by an adjustment of the parameter of the wrong rotation axis during the reconstruction calculation [35, 36]. Figure 17.11 shows the result of this correction.

## References


1 F. R. Elder, A. M. Gurewitsch, R. V. Langmuir, H. C. Pollock, Radiation from electrons in a synchrotron, *Phys. Rev.* 1947, **71** (11), 829–830.

2 R. M. Dell, Batteries, fifty years of materials development, *Solid State Ionics* 2000, **134**, 139–158.

3 U. Köhler, C. Antonius, P. Bäuerlein, Advances in alkaline batteries, *J. Power Sources* 2004, **127**, 45–52.

4 C. C. B. M. de Souza, D. C. de Oliveira, J. A. S. Tenorio, Characterization of used alkaline batteries powder and analysis of zinc recovery by acid leaching, *J. Power Sources* 2001, **103**, 120–126.

5 D. Ohms, K. Kohlhase, G. Benczur-Ürmössy, G. Schädlich, New development on high power alkaline batteries for industrial applications, *J. Power Sources* 2002, **105**, 127–133.

6 Q. C. Horn and Y. Shao-Horn, Morphology and spatial distribution of ZnO formed in discharged alkaline $Zn/MnO_2$ AA cells, *J. Electrochem. Soc.* 2003, **150** (5), A652–A658.

7 J.-Y. Huot, M. Malservisi, High-rate of zinc anodes in alkaline primary cells, *J. Power Sources* 2001, **6**, 133–139.

8 I. Manke, J. Banhart, A. Haibel, A. Rack, S. Zabler, N. Kardjilov, A. Hilger, A. Melzer, and H. Riesemeier, *In situ* investigation of the discharge of alkaline $Zn–MO_2$ batteries with synchrotron x-ray and neutron tomographies, *Appl. Phys. Lett.* 2007, **90**, 214102.

9 G. Lohmann, *Volumetric Image Analysis*, Wiley/Teubner, Chichester and Stuttgart, 1998.

10 J. Ohser and F. Mücklich, *Statistical Analysis of Microstructures in Materials Science*, Wiley, New York, 2000.

11 A. Haibel, C. Scheuerlein, Synchrotron tomography for the study of void formation in internal tin $Nb_3Sn$ superconductors, *IEEE Trans. Appl. Supercond.* 2007, **17** (1), 34–39.

12 A. Devred et al., Overview and status of the next European dipole joint research activity, *Supercond. Sci. Technol.* 2006, **19** (3), 67–83.

13 A. Goedeke, *Performance boundaries on $Nb_3Sn$ superconductors*, Ph.D. Thesis, University of Twente, Enschede, The Netherlands, 2005.

14 M. T. Naus, P. J. Lee, D. C. Larbalestier, The interdiffusion of Cu and Sn in internal Sn $Nb_3Sn$ superconductors, *IEEE Trans. Appl. Supercond.* 2000, **10** (1), 983.

15 *Handbook of Chemistry and Physics*, ed. D. R. Lide, 76th edn, CRC Press, Boca Raton, FL, 1995.

16 V. I. Khitrova, Crystal structure of $\gamma$-Sn, *Sov. Phys. Crystallogr.* 1988, **33** (1), 139–141.

17 R. Kubiak, Evidence for the existence of the $\gamma$ form of tin, *J. Less-Common Metals* 1986, **116**, 307–311.

18 A. Paul, *The Kirkendall Effect in Solid State Diffusion*, Ph.D. Thesis, Technische Universiteit Eindhoven.

19 C. Scheuerlein, Ph. Gasser, P. Jacob, D. Leroy, L. Oberli, M. Taborelli, The effect of CuSn intermetallics on the interstrand contact resistance in superconducting cables for the large hadron collider, *J. Appl. Phys.* 2005, **97** (3), 033909-1–033909-7.

20 M. Ashby et al., *Metal Foams: A Design Guide*, Butterworth-Heinemann, Amsterdam, 2000.

21 J. Banhart, N. Fleck, A. Mortensen, eds., Cellular Metals: Manufacture, Properties, Applications, *MetFoam 2003*, MIT-Verlag, Berlin, 2003.

22 J. Banhart, Manufacture, characterisation and application of cellular metal foams, *Prog. Mater. Sci.* 2001, **46**, 559–632.

23 A. Bütow, *Diploma Thesis*, Strukturuntersuchungen an Metallschäumen in verschiedenen Entwicklungsstadien, Hahn-Meitner-Institut Berlin, SF3, 2004.

24 B. Matijasevic, S. Fiechter, I. Zizak, O. Görke, N. Wanderka, P. Schubert-Bischoff, J. Banhart, *Decomposition behaviour of as-received and oxidized* $TiH_2$ *powder, Powder Metallurgy World Congress, Conference Book 2004.*

25 A. Haibel, A. Bütow, A. Rack, J. Banhart, *Quantitative analysis of pore-particle-correlations in metallic foams, WCNDT Congress, Conference Proceedings 2004.*

26 M. Boin, A. Haibel, Compensation of ring artefacts in synchrotron tomographic images, *Opt. Exp.* 2006, **14** (25), 12071–12075.

27 G. T. Herman, *Image reconstruction from projections – the fundamentals of computed tomography*, Academic Press, New York, 1980.

28 R. Davis, J. C. Elliott, X-ray micro tomography scanner using time-delay integration for elimination of ring artefacts in the reconstructed image, *Nucl. Instrum. Methods Phys. Res. A* 1997, **394**, 157–162.

29 W. Görner, M. P. Hentschel, B. R. Müller, H. Riesemeier, M. Krumrey, G. Ulm, W. Diete, U. Klein, R. Frahm, BAMline: the first hard X-ray beamline at BESSY II, *Nucl. Instrum. Methods A* 2001, 467–468, 703–706.

30 J. Sijbers, A. Postnov, Reduction of ring artifacts in high resolution micro-CT reconstructions, *Phys. Med. Biol.* 2004, **49** (14), 247–253.

31 C. Raven, Numerical removal of ring artifacts in microtomography, *Rev. Sci. Instrum.* 1998, **69**, 2978–2980.

32 C. Antoine, Per Nygard, O. Gregersen, Rune Holmstad, T. Weitkamp, C. Rau, 3D images of paper obtained by phase-contrast X-ray microtomography: image quality and binarisation, *Nucl. Instrum. Methods Phys. Res. A* 2002, **490**, 392–402.

33 A. Marion, *An Introduction to Image Processing*, Chapman and Hall, London, 1991, p. 274.

34 W. Lu et al., Tomographic motion detection and correction directly in sinogram space, *Phys. Med. Biol.* 2002, **47**, 1267–1284.

35 http://ftp.esrf.fr/pub/scisoft/ESRF_sw/doc/PyHST/GettingPyHST.html.

36 T. Donath, F. Beckmann, A. Schreyer, Automated determination of the center of rotation in tomography data, *J. Opt. Soc. Am.* 2006, **A 23**, 1048–1057.

# 18
# Diffraction Enhanced Imaging

*Michael Lohmann*

A new X-ray radiography imaging technique, called diffraction enhanced imaging (DEI), enables almost scatter free absorption imaging, as well as the production of the so-called refraction images of a sample [1]. DEI produces images with improved contrast as compared to standard imaging applications.

In the DEI set-up a crystal monochromator is used to select a small energy band from the incident synchrotron radiation. An additional analyzer crystal is installed between sample and detector. By adjusting the analyzer crystal to match the monochromator settings only X-rays passing straight through the sample are selected and detected; any X-ray photon that is scattered by the sample will be rejected. This produces an almost scatter free absorption image, which is called top image. Due to small variations in refractive index of objects inside the sample, X-rays will be refracted at the edges of these objects, and as a result the direction of the X-rays is slightly changed. These X-rays can be recorded by tuning the analyzer crystal to the appropriate angles. With images recorded at the high-angle side and the low-angle side of the top image, the so-called refraction image can be calculated. The refraction image shows contrast in samples even when, due to too small changes of the absorption coefficients, no contrast in absorption images is observable.

In Europe at several beamlines different DEI set-ups are installed (EU project, contract No. HPRI-1999-CT-50008) in order to investigate the possibilities and the limits for beamline parameters and for various materials. At the HASYLAB wiggler beamline W2 at the second-generation storage ring DORIS a 5 cm wide beam with an adjustable energy between 15 and 80 keV is available. Visualization of fossils, detection of internal pearl structures, monitoring of bone, and cartilage and documentation of implant healing in bone as well as material science are given as examples of applications at HASYLAB.

*Neutrons and Synchrotron Radiation in Engineering Materials Science: From Fundamentals to Material and Component Characterization*
Edited by Walter Reimers, Anke Rita Pyzalla, Andreas Schreyer, and Helmut Clemens

ISBN: 978-3-527-31533-8

## 18.1 Introduction

### 18.1.1 Basics

Ever since the discovery by W. C. Röntgen [2] in 1895 material-dependent absorption has been used as a standard application of X-rays. This phenomenon is described by the well-known absorption law

$$I = I_0 \cdot e^{-\mu \cdot x} \tag{18.1}$$

where $I$ is the intensity of a beam after traveling an object with thickness $x$ and absorption coefficient $\mu$, which is a function of the matter, and the energy $E$ of the beam. $I_0$ is the incident intensity. In order to treat the effects of refraction we have to use the more fundamental wave equation:

$$A = A_0 \cdot e^{i\omega(t - s \cdot n/c)} = A_0 \cdot e^{i(\omega \cdot t - k \cdot n \cdot s)} \tag{18.2}$$

where $A_0$ is the amplitude of the electromagnetic wave with the angular velocity $\omega$ and the time $t$ at the position $s$, in the right-hand side of Eq. (18.2) the speed of light $c = \lambda \cdot v = \lambda \cdot \omega/2\pi$ is replaced by the wave vector $\boldsymbol{k} = 2\pi/\lambda$. DEI makes use of the fact that in matter the refractive index $n$ is not exactly one, even not for X-rays in the keV range, and is different for different matter.

The complex refractive index given by [3]:

$$n = 1 - \delta - i\beta \tag{18.3}$$

Inserting Eq. (18.3) into Eq. (18.2) yields

$$A = A_0 \cdot e^{i\omega t} \cdot e^{-ik(1-\delta)s} \cdot e^{-k\beta s} \tag{18.3a}$$

The real part of the refractive index, $1 - \delta$, determinates the phase shift in matter, and the imaginary part, $\beta$, corresponds to the absorption. The intensity $I$ is given by multiplication of the complex wave function $A$ with it conjugated complex $A^*$, hence it follows that the linear absorption coefficient is given by $\mu = 2k\beta$. In Figure 18.1, the effect of the phase shift behind the object is illustrated. Note that the real part of the refractive index for X-rays is smaller than one, and therefore, the wave fronts passing through the sample are in advance of the undistorted wave. The order of magnitude in the keV range of $\delta$ is about $10^{-6}$. Due to the small difference from unity, the deviation of the beam direction, expressed as the wave vector $\boldsymbol{k}$, is in the order of μrad. $\delta$ depends very strongly on the X-ray energy used as well as on nature of the matter traversed.

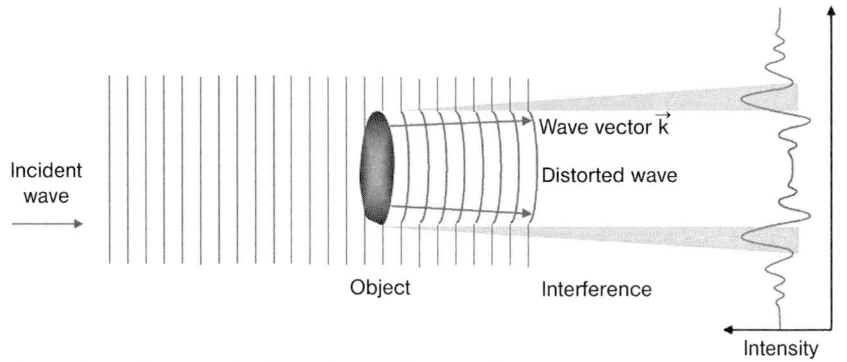


**Figure 18.1** Influence of the phase of waves due to an object.

As indicated in Figure 18.1 the direction $\boldsymbol{k}$ without interaction with the object and the direction $\boldsymbol{k}$ after traversing the object are slightly different, leading to a superposition of the two waves mainly at the margin of the object. This superposition can lead to interference and result in a contrast enhancement, which depends on the distance object–detector. For this so-called in-line imaging to work, the incoming beam must be at least partially coherent [4]. DEI, on the other hand, does not require a coherent beam and is described in more detail in the next section.

### 18.1.2 Extinction Contrast

When objects are illuminated, X-rays are not only absorbed but also scattered into different directions. This scattered beam will be superimposed with the transmitted beam, leading to a decrease in contrast in the recorded image. Therefore, in medical applications a so-called antiscatter grid is installed between object and detector as illustrated in Figure 18.2a, and the increased contrast in the image is

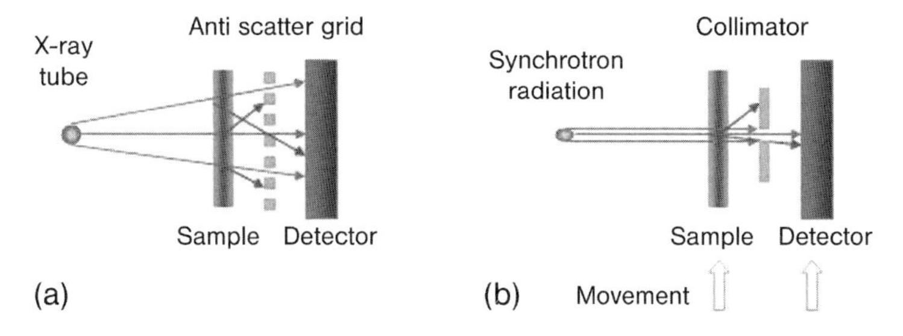


**Figure 18.2** Rejection of scattered photons by an antiscatter grid in classical medical imaging (a) and by a collimator in line scan mode at synchrotrons (b).

called extinction contrast. At synchrotrons wide and parallel beams with a height of the order of sub mm are used, and a collimator after the sample has the same effect as an antiscatter grid. In order to image the complete object, object and the detector are moved synchronously through the beam as illustrated in Figure 18.2b.

### 18.1.3 Principles of DEI

The principle of diffraction enhanced imaging is based on the very narrow angular acceptance of perfect crystals, which serve the same purpose as antiscatter grids or collimators, but with much greater selectivity. A schematic set-up is given in Figure 18.3. A highly perfect (monochromator) crystal is used to select a specific wavelength of the incoming beam. A second (analyzer) crystal is installed behind the monochromator crystal.

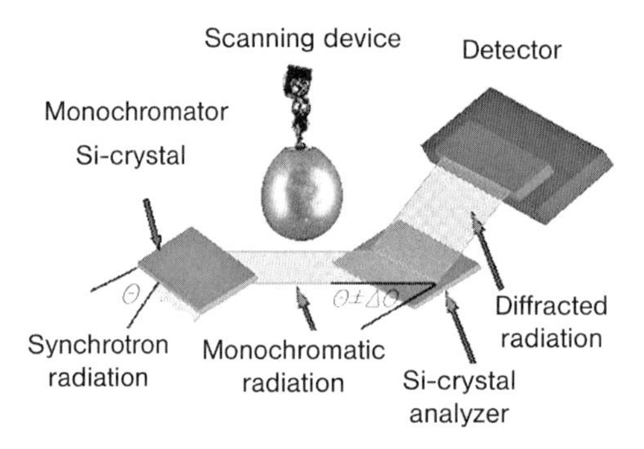


**Figure 18.3** Typical schematic of a DEI setup.

The reflectivity of the analyzer crystal as function of angle is schematically given in Figure 18.4. The width of this reflectivity is called the Darwin width. It is nonzero, even for perfect crystals. Typical Darwin widths are 20 μrad for Si(1 1 1) at 14 keV, which shows the analyzer crystal to be a very selective collimator. If the analyzer crystal is perfectly aligned with the monochromator crystal, we sit on top of the so-called rocking curve, which explains why the images obtained at this position are called top images. The effect of contrast enhancement is clearly visible when comparing the images of a plexiglass bar without analyzer crystal (Figure 18.5a) and with analyzer crystal (Figure 18.5b).

Another modality of using DEI is by adjusting the analyzer crystal to the shoulder of the rocking curve as indicated in Figure 18.4b. Images obtained with this setting are the so-called shoulder images. At this setting the analyzer crystal reflects less of the straight through beam (absorption image) and more of the refracted beam, which has a slightly different angle, therefore enhancing the sensitivity to the edges of the object as explained before. Figure 18.5c gives the corresponding shoulder image of the plexiglass bar.

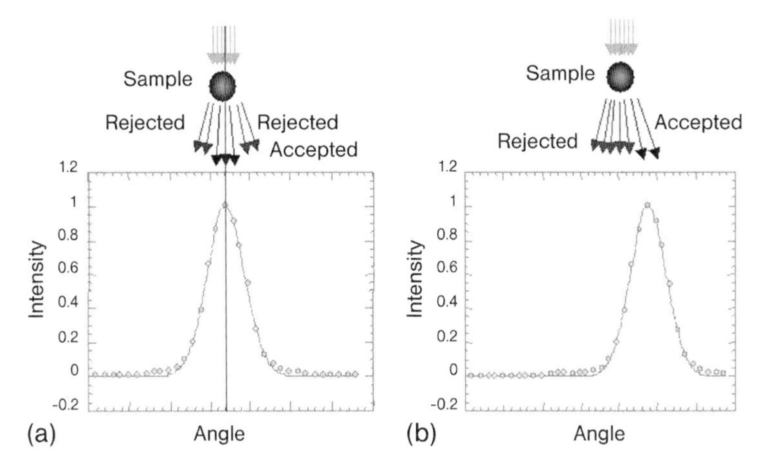


**Figure 18.4** The effect of contrast enhancement by using an analyzer crystal. (a) Analyzer crystal aligned to the angle of the monochromator crystal and (b) analyzer crystal misaligned to the angle of the monochromator crystal.

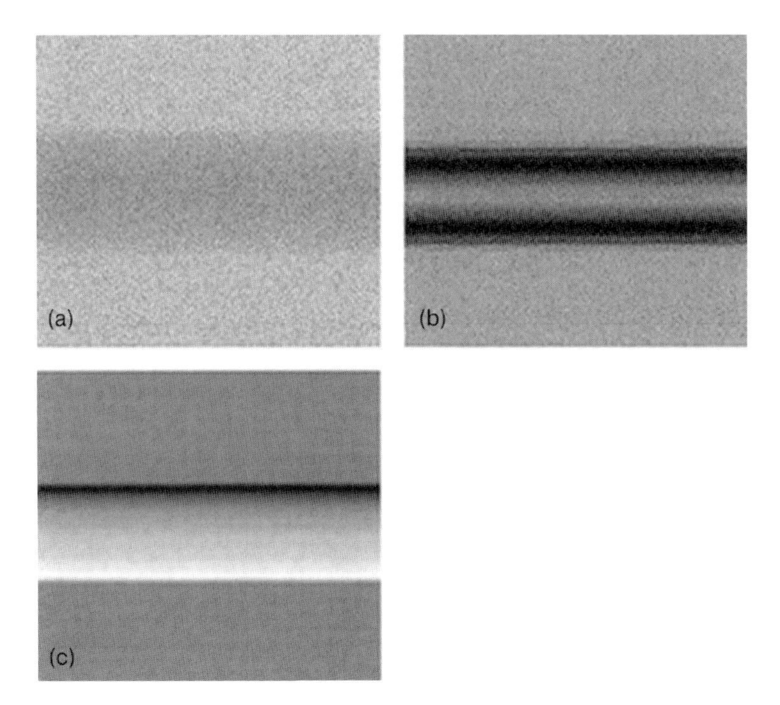


**Figure 18.5** A plexiglass bar with different image modalities: (a) without analyzer crystal, (b) with analyzer crystal adjusted at the top position of the rocking curve, and (c) with analyzer crystal adjusted at the shoulder position of the rocking curve.

The contrast enhancement and improved signal-to-noise ratio in the extinction contrast image (Figure 18.5b), and the shoulder image (Figure 18.5c) over the normal radiograph (Figure 18.5a) is obvious.

The intensity behind the analyzer crystal is given by

$$I_L = I_R \cdot R(\Theta_L + \Delta\Theta_z) \approx I_R \cdot \left( R(\Theta_L) + \frac{\mathrm{d}R}{\mathrm{d}\Theta}(\Theta_L) \cdot \Delta\Theta_z \right) \tag{18.4a}$$

$$I_H = I_R \cdot R(\Theta_H + \Delta\Theta_z) \approx I_R \cdot \left( R(\Theta_H) + \frac{\mathrm{d}R}{\mathrm{d}\Theta}(\Theta_H) \cdot \Delta\Theta_z \right) \tag{18.4b}$$

where $I_R$ is the intensity of the refracted beam in front of the analyzer crystal, $R$ is the analyzer reflectivity at the adjusted angle position $\Theta$ plus the refraction angle in the vertical direction $\Delta\Theta_z$ caused by the sample. The indices $L$ and $H$ denote the high-angle side and the low-angle side of the rocking curve, respectively. The second form of Eq. (18.4) are two-term Taylor series approximations. By using these two equations the refraction image

$$\Delta\Theta_z \sim \frac{I_H - I_L}{I_H + I_L} \tag{18.5}$$

in the vertical direction $z$ can be separated. This type of equation is only valid if the position at the high-angle side and the low-angle side of the rocking curve is the same, which means the reflectivity and the degrees of the slopes have the same values each. In conclusion the intensity distribution of an image obtained by Eq. (18.5) represents a map of the refraction angle in the vertical direction [1]!

## 18.2 Experimental Set-up

A scheme of the diffraction enhanced imaging (DEI) set-up at the W2 beamline at DORIS (HASYLAB) is given in Figure 18.6. In the first part of the wiggler beamline W2, a premonochromator containing Si[1 1 1] crystals is installed (PM-L, PM-B). A monochromator (M-B1, M-B2), ionization chambers (I1, I2, I3 and I4), the analyzer (A–B) and the detector (D) are mounted on an optical table (honey comb table) in order to stabilize the system and to suppress vibrations from the ground (Figure 18.6). The scanning device and rotation device for the sample (O), respectively, is separated from the optical table to avoid vibrations of the system generated by the stepper motor.

The system is used in a line scan mode with two different detectors. Each detector is based on a photodiode array with a phosphor screen and is 5 cm wide with an effective spatial resolution of about 200 μm. Alternatively, a CCD camera can be installed to realize a higher spatial resolution of about 60 μm but a width of 1 cm only.

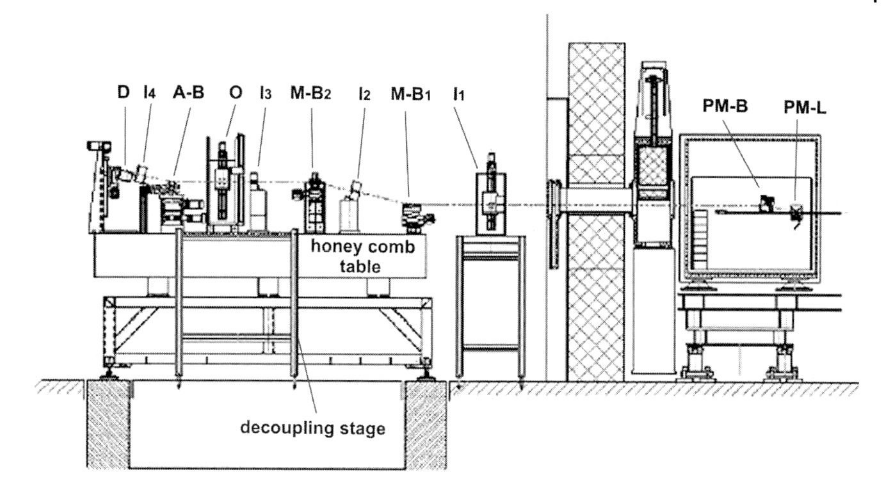


**Figure 18.6** Schematic view of the system.

## 18.3 Examples

### 18.3.1 Complete Set of DEI Images

Figure 18.7 shows as an example different DEI images of a bee. Images are taken without an analyzer crystal (Figure 18.7a), with the analyzer adjusted on the top of the rocking curve (Figure 18.7b), at the low-angle side and at the high-angle side (Figure 18.7c and d) on the shoulder of the rocking curve, respectively, whereby the intensity was 50% compared to the intensity at the top adjustment. The refraction image (Figure 18.7e) finally is calculated from the shoulder images according to Eq. (18.5).

### 18.3.2 Material Science

The study of composite materials and reinforced materials is an important field in material science. An alloy composed of 96% Mg with 3% Al and 1% Zn, is called AZ31, is manufactured by casting. The inner structure of this sample, and therefore, the quality is different at different locations of the cast. In Figure 18.8 DEI images of different parts of the melt are given. The left two images (a) and (b) show small bubbles where the right one shows only a homogenous structure. The sample in the middle shows some additional structures.

Such irregularities are not visible in absorption images, even not by using high-resolution synchrotron radiation.

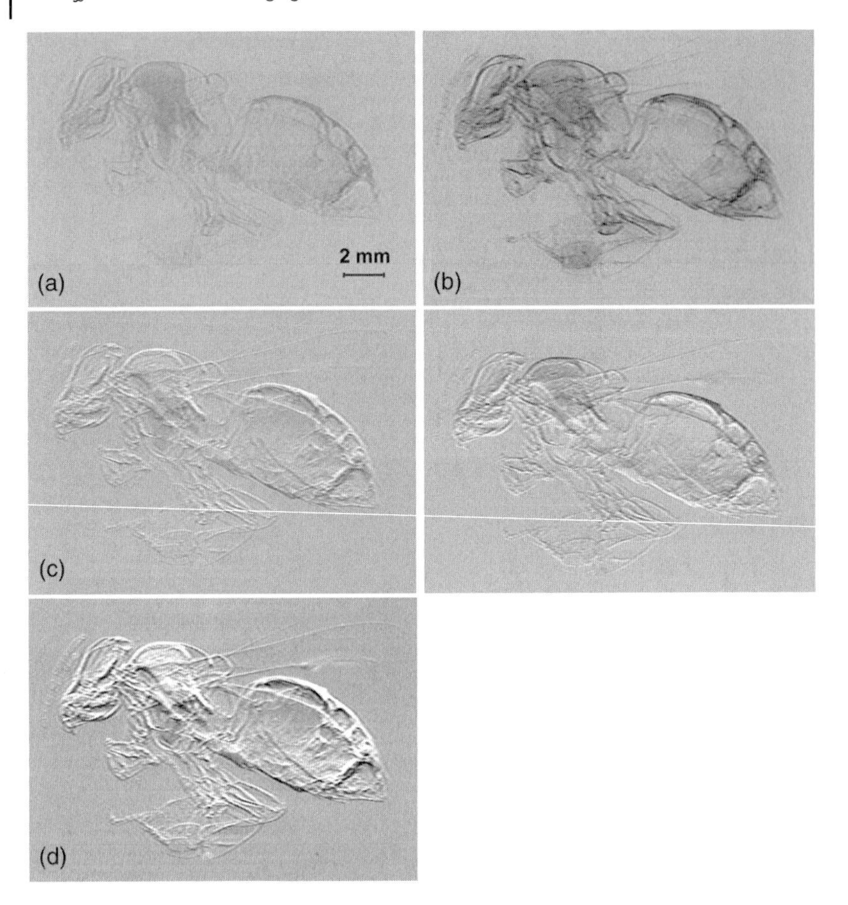


**Figure 18.7** Image of a bee: (a) absorption image without analyzer, (b) image with analyzer on top of the rocking curve (extinction contrast), (c) image at the shoulder of the rocking curve (high-angle side, 50% of intensity $I_H$), (d) as image "c" but low-angle side ($I_L$), and (e) refraction contrast image $\Delta\theta$ (subtraction of images: $(c-d)/(c+d)$).

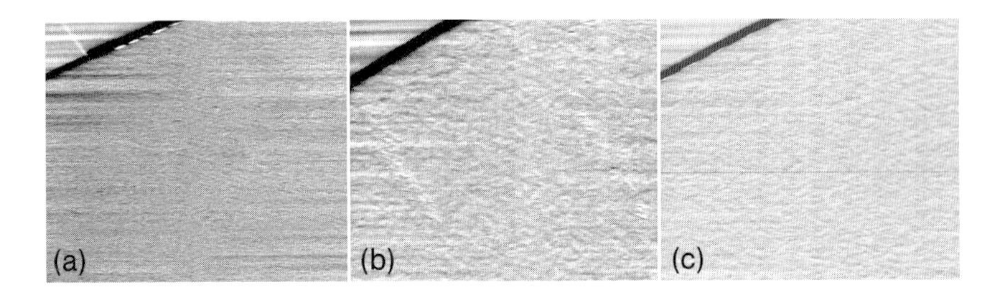


**Figure 18.8** DEI images of AZ31 cubes from different locations of a melt.

### 18.3.3 Example of Mineralogical Investigations

Still the most sophisticated and reliable way to identify the nature of pearls is an examination using various X-ray techniques. The combination of X-radiography, X-ray diffraction, and X-ray luminescence can identify pearls in almost all cases [5].

Most cultured pearls have a characteristic internal structure, the shell bead. This structure is formed when a shell bead is introduced into the mussel together with an epithelium able to produce mother of pearl. Natural pearls have no shell bead. They are rare, and their value exceeds the value of cultured pearls.

Since the second half of the last century a culturing technique without a shell nucleus has been developed for freshwater mussels using only epithelium. These pearls are known as tissue-nucleated freshwater cultured pearls. With the recent appearance of large, high-quality tissue-nucleated freshwater cultured pearls on the markets a new problem arises. Here, the fine central growth irregularity, which proves a pearl to be tissue nucleated, often cannot be made visible by conventional X-radiography (Figures 18.9 and 18.10). The difference in size between

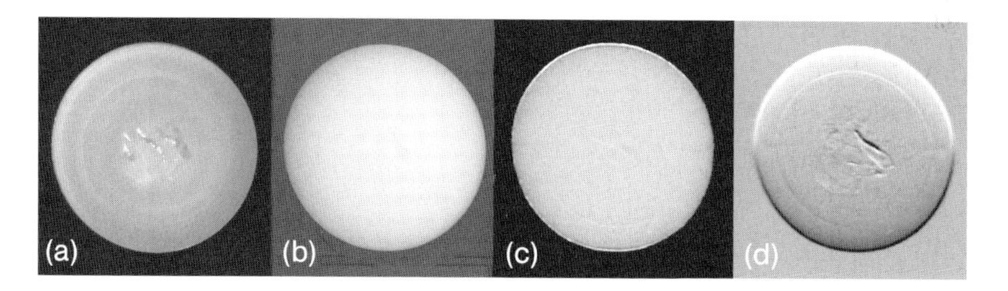


**Figure 18.9** A Chinese tissue-nucleated freshwater pearl. (a) The original pearl cut open (photo by K.-C. Lyncker). (b) Conventional X-radiograph. (c) DEI top image. (d) DEI refraction image.

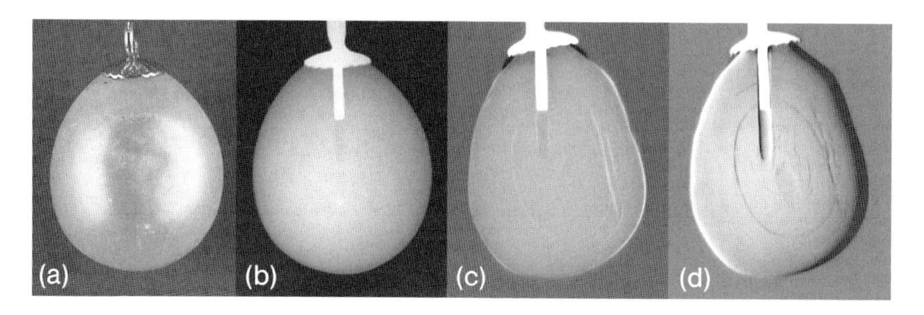


**Figure 18.10** Submitted pearl for an expertise. (a) Photo of the original pearl (photo by K.-C. Lyncker). (b) Conventional X-radiograph. (c) DEI top image, the growth structures are visible. (d) DEI refraction image, additionally the central irregularity aroused by tissue-nucleation is visible.

the tiny irregularity in the center of the pearl and the large pearl body around it, which has grown like a natural pearl, is so large that the differences in the X-ray absorption caused by those irregularities are not visible by using the standard methods.

Pearls of different origin were examined by DEI [6]. These images were compared to those made by conventional X-rays. In some cases additionally a photo of the pearl cut open was taken for comparison.

## 18.4 Conclusions

The diffraction enhanced imaging (DEI) method dramatically enhances contrast for low-absorbing material. Therefore, it can reveal internal structures formerly invisible for conventional X-ray methods. Computational DEI processing allows for the incorporation of partial signals from absorption and diffraction, optimized analysis of the contribution of low-absorption and high-absorption material permit to discriminate those two components by using all kinds of images taken with a DEI set-up.

Furthermore, Dilmanian et al. showed in [7] how to use the main Eq. (18.5) for the refraction images of computed tomography (CT), which description would go beyond the scope of this article.

## Acknowledgments

This work was supported by the European PHASY Project under contract number HPRI-CT-1999-50008 and by local beamtime and travel grants from DESY and ESRF. The author is grateful for all the contributions of the collaborators during measurements and for the sample preparation.

## References

1 D. Chapman, et al., *Phys. Med. Biol.* 1997, **42**, 2015.
2 http://detserv1.dl.ac.uk/herald/xray_history.htm.
3 R. W. James, *The Optical Principles of the Diffraction of X-rays*. Ox Bow Press, Woodbridge, Connecticut, 1962.
4 A. Snigerev, et al., *Rev. Sci. Instrum.*, 1995, **66**, 5486.
5 S. J. Kennedy, Pearl identification. *Aust. Gemolog.*, 1998, **20**, 2–19.
6 J. Schlüter, et al., Diffraction enhanced imaging: a new X-ray method for detecting internal pearl structures, *J. Gemm.*, 2005, **29**, 7/8, 401–406.
7 F. A. Dilmanien, et al., Computed tomography of X-ray index of refraction using the diffraction enhanced imaging method, *Phys. Med. Biol.*, 2000, **45**, 933–946.

# Part III
# New and Emerging Methods

# 19
# 3D X-ray Diffraction Microscope

*Henning Friis Poulsen, Wolfgang Ludwig, and Søren Schmidt*

Three-dimensional X-ray diffraction (3DXRD) is a novel technique aiming at a fast and nondestructive characterization of the individual crystalline elements within mm–cm-sized polycrystalline specimens. It is based on two principles: the use of highly penetrating hard X-rays from a synchrotron source (X-ray energies in the range 30–100 keV) and the application of "tomographic" reconstruction algorithms for the analysis of the diffraction data. In favorable cases, the position, morphology, phase, and crystallographic orientation can be derived for hundreds of elements simultaneously as well as their elastic strains. Furthermore, the dynamics of the individual elements can be monitored during typical processes such as deformation or annealing. Hence, for the first time, information on the interaction between elements can be obtained directly. The provision of such data is vital in order to extend beyond state-of-the-art structural models.

Notably, in a 3DXRD experiment one must prioritize between spatial, angular, and time resolution. This has led to a variety of 3DXRD strategies, which we in the following will summarize as four standard modes of operation. The first two modes enable fast measurements of the *average characteristics* of each grain – such as their center-of-mass position, volume, average orientation, and/or average strain tensor – while the exact location of the grain boundaries is unknown. The two other modes enable full 3D mappings of grains and orientations.

In the following, initially the 3DXRD set-up will be presented and 3DXRD strategies are discussed, and subsequently, the standard modes of operation are presented in more detail with emphasis on reconstruction principles and with selected examples of use. For general reference, a mathematical description of the methodology and additional applications, see [1].

Alternative approaches to provision of 3D maps based on X-ray diffraction exist. These are based on inserting wires [2], slits [3], or collimators [4] between the sample and the detector and scanning the sample with respect to these elements. Not surprisingly such methods will be slower than the tomographic approach of 3DXRD, but they may be associated with other advantages, such as improved options for measuring the local elastic strain. In particular we mention the technique of "differential-aperture X-ray microscopy" [2]. Using a polychro-

*Neutrons and Synchrotron Radiation in Engineering Materials Science: From Fundamentals to Material and Component Characterization*
Edited by Walter Reimers, Anke Rita Pyzalla, Andreas Schreyer, and Helmut Clemens

ISBN: 978-3-527-31533-8

matic microbeam with energies of 8–20 keV and scanning a wire, sub-micrometre resolution in 3D has been demonstrated (see Chapter 20).

## 19.1
## Basic Set-up and Strategy

The basic 3DXRD set-up – sketched in Figure 19.1 – is quite similar to conventional tomography settings at synchrotrons. A (nearly) parallel monochromatic X-ray beam impinges on the sample as a uniform field. The sample is mounted on an $\omega$ rotation stage, where $\omega$ is the rotation around an axis perpendicular to the incoming beam. As an option $x$-, $y$-, and $z$-translations may be added as well as additional rotations.

Any part of the illuminated structure, which fulfills the Bragg condition, will generate a diffracted beam. This beam is transmitted through the sample and probed by a 2D detector. To probe the complete structure, and not just the part that happens to fulfill the Bragg condition, the sample is rotated. Hence, exposures are made for equi-angular settings of $\omega$ with a step of $\Delta\omega$. To provide a uniform sampling the sample is rotated by $\Delta\omega$ during each exposure. Essential to 3DXRD is the idea to mimic a 3D detector by positioning several 2D detectors at different distances to the center-of-rotation, $L$, and exposing these either simultaneously (many detectors are semitransparent to hard X-rays) or subsequently.

Presently, typically two types of detectors are used: near-field detectors with a spatial resolution of ~5 μm in close proximity to the sample ($L$ in the range

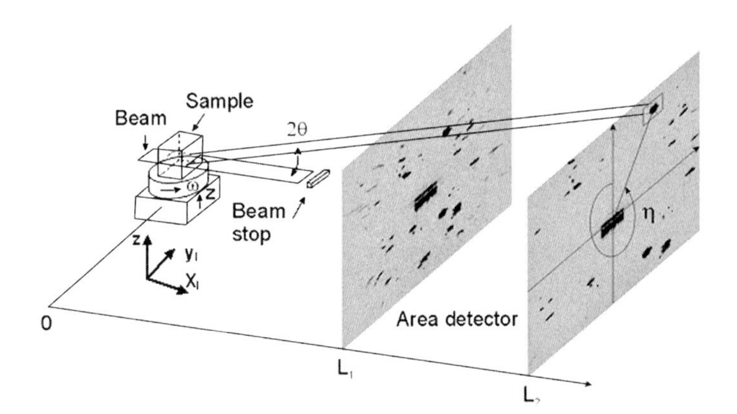


**Figure 19.1** Sketch of the 3DXRD principle for the case of the incoming monochromatic beam being focused in one dimension. Two types of detectors are used: high-spatial-resolution detectors close to the sample and low-resolution detectors far away from the sample (the latter type not shown). The Bragg angle $2\theta$, the rotation angle $\omega$ and the azimuthal angle $\eta$ are indicated for the diffracted beam arising from one grain of a coarse-grained specimen, and for two settings of the high-resolution detector. The axes for the laboratory co-ordinate system are also shown.

2–10 mm) and far-field detectors: conventional CCDs with a resolution of 100–200 μm ($L$ in the range 20–50 cm). The former provide information on position and orientation degrees-of-freedom, while the latter probes strain and orientation. With the near-field detector, images may be acquired at several, typically 3, distances as illustrated in Figure 19.1. This enables ray tracing of the diffracted beam, also known as tracking.

The incoming beam may illuminate the full sample or be focused in one direction to probe only a layer within the material (see Figure 19.1). In the following, we shall refer to these as the 3D and 2D cases respectively. Notably, the 2D case is restricted to use at third-generation synchrotron sources, while the 3D case also applies to second-generation sources and in favorable cases even to work with laboratory sources. In the 2D case, 3D information is generated simply by repeating data acquisition for a set of layers and stacking the resulting reconstructions. An example of a set of exposures obtained in the 2D case is provided in Figure 19.2.

A vital difference between the two types of detectors relates to data acquisition time: with the current technology the far-field detectors provide a time resolution that is several orders of magnitude better than the near-field ones. This makes it necessary to carefully consider the priority of time and space resolution in a 3DXRD experiment. In practice, we may distinguish between four modes of operation, as sketched in Figure 19.3.

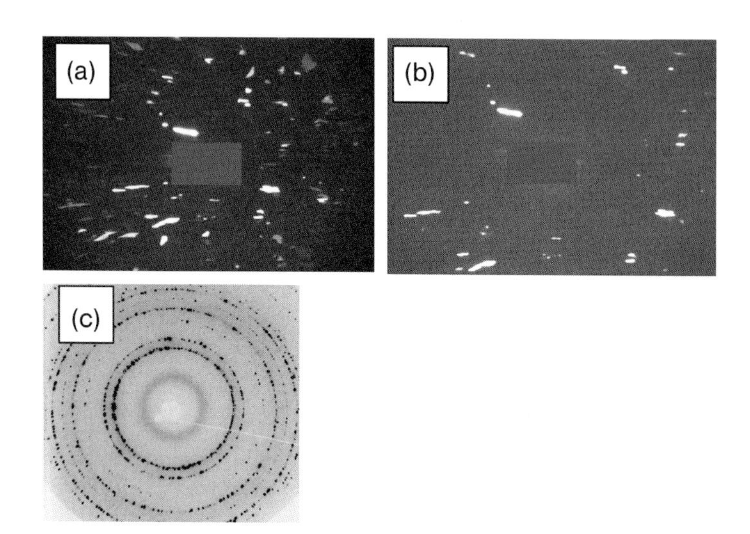


**Figure 19.2** Typical 3DXRD data from a coarse-grained undeformed polycrystal. (a, b) "Near-field" exposures with the high-spatial-resolution detector (see Figure 19.1) at distances of $L = 4$ mm and $L = 8$ mm, respectively. The background is shown as black. (c) A corresponding "far-field" image made with the low-resolution detector at $L = 400$ mm and shown with an inverted grey scale to improve visibility. In the latter case the diffraction spots are positioned on Debye–Scherrer rings.

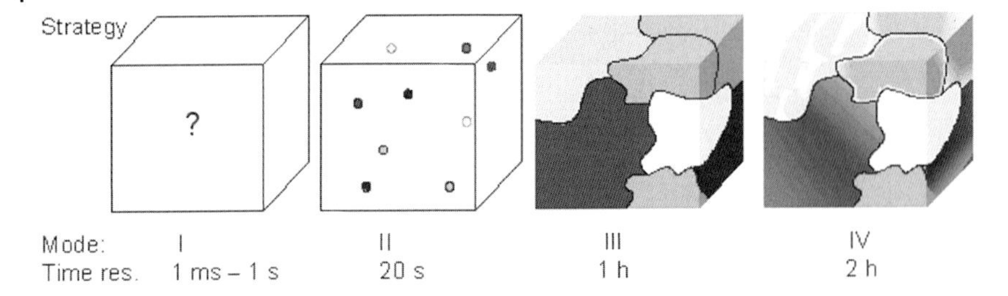


**Figure 19.3** Sketch of four typical 3DXRD modes of operation, listed in direction of increasing spatial information. Mode I: statistical mode, with no spatial information and where not all grains within the illuminated volume are probed. Mode II: information of center-of-mass position, volume, average orientation, and average strain tensor for each grain. Mode III: 3D grain map for undeformed specimens. Mode IV: 3D orientation map for deformed specimens. In all cases, orientations are represented by a grey scale. Typical time resolutions for characterisation of 100 grains are indicated. (Courtesy of E. Nielsen).

Modes I and II are based (mainly) on far-field information. Mode II provides information on the phase, center-of-mass position, volume, as well as average orientation and average strain tensor of each grain in the illuminated part of the specimen. Mode I is an option for very fast measurements, where only a subset of the illuminated grains are characterized and no spatial information is available. Also, the orientation and strain characterization will not be complete. Experimentally the main difference between the two modes is the $\omega$ range covered.

Modes III and IV aim at generation of complete 3D maps. In the case of an undeformed material, the orientation is constant within each grain, and the aim is to provide a *grain map*. In the case of a deformed material, the orientation varies locally. This makes it relevant to measure *orientation maps*, where – similar to EBSD – each voxel in the sample is associated with its own orientation. In both the cases, evidently a high-spatial-resolution detector is needed. The various modes will be detailed below.

In many experiments combinations of these modes are relevant. As an example one may wish to start by mapping extended parts of a sample in order to identify sub-volumes of particular interest. Then one focuses on such parts and performs a fast center-of-mass study on these parts during *in situ* processing. This procedure is then complemented by mapping extended parts again after the processing is completed.

### 19.1.1
### The 3DXRD Microscope

Presently, one dedicated 3DXRD microscope exists. This was developed in collaboration between Risø National Laboratory and the ESRF, and installed at the Materials Science beamline at the ESRF. The optics enable work in the energy range 30–90 keV with a choice between parallel beam, vertical focusing only, and both vertical and horizontal focusing. The energy bandwidth can be varied from $10^{-4}$ to $10^{-2}$, with the most popular setting being $\sim 10^{-3}$. The microscope can carry loads of up to 200 kg. It is equipped with a 25 kN stress rig and several furnaces with operational temperatures of up to 1300 °C. In addition, 3DXRD experiments have been performed at sector 1 at the APS. Currently, this set-up is being converted into a dedicated experimental station.

## 19.2
## Indexing and Characterization of Average Properties of Each Grain

A very important simplification arises in the case where the diffraction pattern is composed of a set of (primarily) nonoverlapping diffraction spots. Three examples of such datasets are shown in Figure 19.2. A large number of problems within polycrystal and powder research can be tailored to apply to this situation.

In this case, the diffraction spots can be sorted with respect to element of origin, say the grain, by means of a polycrystalline indexing scheme [5]. The starting point for such schemes is to apply image analysis routines which identify the spots and determine their center-of-mass positions. Next, the direction of the corresponding diffracted X-rays are determined. In the case of near-field data – that is the tracking set-up illustrated in Figure 19.1 – ray tracing is used. Having identified a set of spots at different detector-to-sample distances which belong to the same reflection, the best fit to a straight line through the center-of-masses of these spots is determined. In the case of far-field data one may simply assume that all grains are positioned at the center-of-rotation and for any spot draw the line from its center on the detector to the rotation axis. In both the cases, once the direction of the ray is known, the corresponding scattering vector is readily determined.

The sorting can be based on three principles: orientation, position, or grain volume (the integrated intensity of any diffraction spot is proportional to the volume of the associated grain of origin). Sorting based on orientation and position has been implemented in the program GRAINDEX [5], which assumes the crystallographic space group to be known. The main limitation is in terms of the probability of spot overlap on the detector. The probability is determined by the number of grains, the texture, and the orientation spread of each grain. Simulations show that for samples with a weak texture and grains with a near-perfect lattice – that is with an orientation spread of order 0.1° or less – several thousand grains can be indexed simultaneously [1]. On the other hand, plastic deformation introduce orientation spread within the grains, which in practice prohibits indexing of embedded grains in materials that are deformed to more than 20%.

Once a grain has been indexed all the tools of the conventional single crystal diffraction analysis is available. Hence, based on the output of GRAINDEX, a comprehensive characterization of each grain immediately follows (mode II analysis). If only the far-field detector is used the grain volume and average orientation can be characterized. Furthermore, as detailed in Chapter 9, one can determine the average values of all the components of the elastic strain tensor

[6, 7]. The time resolution is often ~10 s, limited by the read-out time of the detector. The accuracy is of order 0.3° for orientations and $\Delta\varepsilon/\varepsilon = 10^{-4}$ for the strain. Furthermore, it is possible to perform a full crystallographic analysis of each grain, e.g., a structural refinement [8]. This mode is well suited for *in situ* studies of dynamic phenomena in materials science as it is fast and relatively straightforward.

In a variant of this mode also the center-of-mass position of each grain is determined. At the expense of time resolution, this can be done by tracking using a high-spatial-resolution detector as outlined above. To determine positions with high accuracy, a nonlinear least square fit to the center-of-mass (CMS) positions of the spots on the detector is performed. The accuracy is currently ~5 µm.

For very fast data acquisitions, mode I is relevant. In this case, one simply repeats acquisitions with the far-field detector, oscillating around a given $\omega$ setting. Assuming grains do not rotate, one can then monitor the change in volume of the grains giving rise to diffraction spots at this $\omega$ setting. Several hundred so-called *growth curves* of deeply embedded grains can be obtained simultaneously in this way [9].

### 19.2.1 Application I: Nucleation and Growth Studies

Traditionally, nucleation and growth phenomena have been analyzed using ensemble average properties, such as the volume fraction of transformed material. However, the predictive power of average properties is limited by the neglect of heterogeneities. For example, nucleation may take place preferentially at specific sites, and the growth rate of nuclei may depend strongly on orientation, size, stoichiometry, or relationships with neighboring volumes. 3DXRD is an ideal tool to study the effect of heterogeneities, and as such it has been used for a series of studies related to recrystallization [9], solidification [10], and phase transformations in steel [11], ceramics, and ferroelectrics. In all cases it was demonstrated that the ensemble average "Avrami-type" models are at best gross simplifications.

As an example of mode I work, Offerman et al. studied the phase transformation from the high-temperature austenite to the medium-temperature ferrite phase in carbon steel [11]. The results shown in Figure 19.4 relate to a study where the steel was cooled at a constant rate from 900 to 600 °C over 1 h. During this process the nucleation and growth of ferrite grains was inferred from the appearance and evolution of ~60 distinct diffraction spots originating from the ferrite. By counting the number of new spots appearing in a given time interval, the nucleation rate is determined as function of temperature, a parameter that cannot be determined with high fidelity in any other way. In Figure 19.4a, the nucleation rate is compared to a prediction from a classical nucleation model [12]. From analysis of the data shown in the figure it is found that the predicted and measured activation energies differ by more than two orders of magnitude. Relevant revisions to the model are currently being discussed.

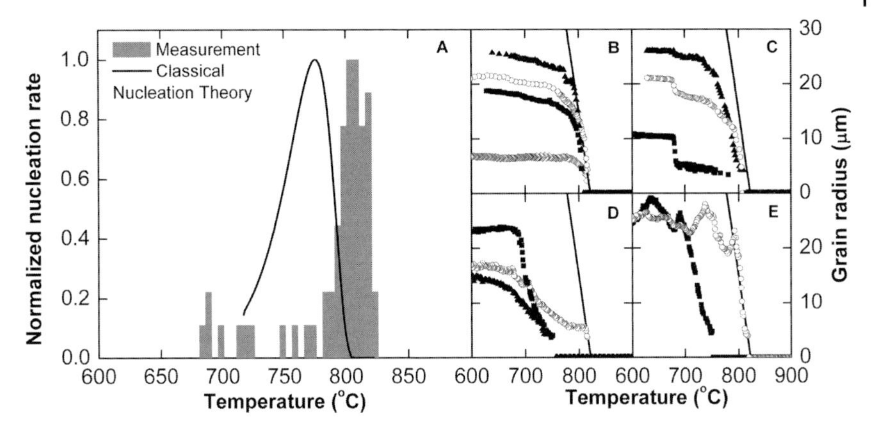


**Figure 19.4** Nucleation of ferrite grains during cooling of a carbon steel. (A) The normalized observed nucleation rate (bars) is compared to theoretical prediction (line) based on a classical model [12]. (B–E) Examples of growth curves for individual ferrite grains. The lines are predictions based on the parabolic growth model of Zener. (Adapted from [11], with permission from Science).

Shown in Figures 19.4b–e are examples of resulting growth curves for the nuclei. It appears that the growth of some grains, those in 'b', followed closely the predictions of Zener, while others displayed large discrepancies. The latter was explained by several effects including local enrichment in the carbon content of the austenite.

### 19.2.2
### Application II: Plastic Deformation

Polycrystal deformation is a topic of prime interest to both metallurgists and geoscientists. However, despite 70 years of effort, there is no consensus on how to approach the modeling. In particular, it is unclear to what extent deformation behavior is determined by the initial grain orientation, the grain–grain interactions or by the emerging dislocation structures. 3DXRD provides *in situ* methods for addressing these issues, for example through real-time observations of grain rotations in bulk materials during the deformation process [13, 14]. As an example of mode II work, Figure 19.5 provides a mapping of the rotation path of 95 grains in a 4-mm thick polycrystalline Al sample strained in tension. These data provide a detailed combinatorial database for critical evaluation of models for polycrystalline deformation. In major parts of the orientation triangle, the paths exhibit a clear dependence on initial orientation. It is found that the grain responses do not match predictions from classical Taylor, Sachs, or self-consistent models, but can be described by subdividing orientation space in four regions with distinctly different rotation behavior [15].

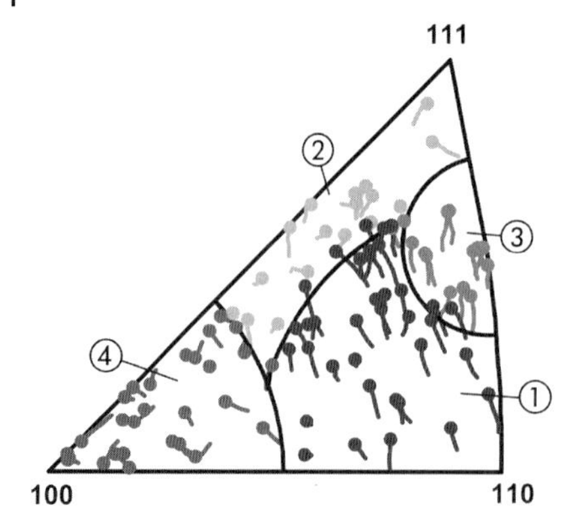


**Figure 19.5** Rotation of the tensile axis of 95 grains within an Al specimen, as expressed in a stereographic triangle. The curves are the observed paths for the average orientation of each grain during tensile deformation from 0 to 6%. The final orientation is marked by a circle. The rotation behavior is interpreted in terms of four regions marked by solid lines. (From [15], with permission from Elsevier).

### 19.2.3
### Application III: Studies of Subgrains and Nanocrystalline Materials

Notably, 3DXRD studies are also relevant for crystalline structures with a grain size smaller than the spatial resolution. Evidently, in such cases the generation of a grain map is not possible. However, by focusing the beam to say 10 μm × 10 μm, diffraction spots from individual grains as small as 70 nm can be detected. By tracing the integrated intensities and positions of such spots as function of time, one can infer changes in the volume, orientation, and strain of the grains of origin.

The main limitation for such studies is spot overlap. To overcome this problem two approaches have been pursued. The first approach is based on reducing the number of illuminated grains by investigating foils [16]. Provided the foil thickness is at least 10 times the grain size, for many annealing processes the grains at the center of the foil may be considered bulk grains. This methodology has been applied to a series of coarsening studies of subgrains in aluminum deformed to both medium and very high strains.

The second approach is to increase the angular resolution of the instrument by 2 orders of magnitude – to 0.004° in $\eta$ and $\omega$ and 0.0005° in $2\theta$. Implemented at sector 1-ID at APS, this improvement is achieved at the expense of flux by use of a six-bounce monochromator and a detector positioned at a distance of 4 m from the sample [17]. This set-up is designed specifically for studies of dislocation patterning. When deforming metals, typically the dislocations assemble into walls separating nearly perfect dislocation-free crystals, the subgrains, with dimensions of 100 nm to 2 μm. With the set-up at APS individual spots from millions of illuminated subgrains in a 300 μm thick Cu sample can be resolved. Full 3D reciprocal space maps are gathered continuously while deforming the specimen. Uniquely, from their time-dependence fundamental questions can be addressed in a direct way, such as "how and when do the dislocation structures form?" and "how do the structures subdivide as function of increasing strain?" In the case of tensile deformation of Cu, it is found that the dislocation structures form just after the plastic on-set and that the structures display intermittent dynamics – a surprising fact not predicted by theory [17].

## 19.3 Mapping of Grains and Orientations

3DXRD microscopy enables a relatively fast generation of large 3D grain maps and 3D orientation maps by use of tomographic reconstruction algorithms. The development of the reconstruction algorithms is nontrivial, as the complexity in terms of the dimensionality and sheer size of the reconstruction space is much larger than, e.g., for absorption contrast tomography. Another difference is that in 3DXRD the number of useful projections is given by the number of observable reflections, which may be limited by experimental geometry to be as few as five.

### 19.3.1 Mode III: Mapping Grains in Undeformed Specimens

At this stage a range of approaches have been presented for the 2D case:

- *Simple backprojection*. In the 2D case, reconstruction algorithms are strictly speaking not needed, as the shape of any grain can be retrieved by back-projecting the periphery of one associated diffraction spot along the direction of the diffracted X-rays. This approach is simple and fast [6], but in practice associated with a spatial resolution which is substantially worse than the approaches below.
- *Forward projection*. For each pixel in the layer of interest in the sample, a forward-projection algorithm scans orientation space to test which orientations match the diffraction patterns. Cross-talk between the pixels in the layer is neglected. Hence, one can only infer the set of *possible* orientations. However in practice, the constraint that all pixels belonging to a given grain must have the same orientation often implies that there is one and only one (possible) orientation for each pixel. If that is not the case, e.g., in cases with substantial spot overlap, the ambiguous result may be used as input for inverse methods presented below. Two forward-projections programs have been developed by Suter and coworkers [18] and Schmidt [19], respectively.

- *Algebraic reconstruction algorithm*. The fact that there is a linear relationship between grain volume and the integrated intensity of an associated diffraction spot implies that algebraic approaches are possible. In [20], the so-called 2D-ART algorithm is detailed. It assumes that the orientations and approximate center-of-mass positions of all grain sections are known *a priori*, e.g., determined by GRAINDEX. The method attempts to reconstruct the boundary of each grain separately. For a specific grain, this is done by associating a "grain-density" to each pixel in the layer. Once the solution has converged, the grain boundary can be defined by setting a threshold. A complete grain map may be obtained by superposing the solutions, the boundaries, of the individual grains. However, such a map will not be space filling as boundaries from neighboring grains may overlap or leave "voids" in the map. To avoid this, one may choose for each pixel the grain where the reconstruction exhibited the largest density.
  First *experimental data* for 2D-ART is reproduced in Figure 19.6. The grain map resulted from an independent reconstruction of 27 grains based on five reflections each. From the overlap between grains and voids in Figure 19.6, the spatial resolution is estimated to be ~5 μm. The limitation is identified to be the instrumental point-spread-function. A few grains are obviously missing from the map – for the particular geometry used the number of independent reflections were too few to enable reconstruction.
- *Monte-Carlo-based reconstruction*. Stochastic approaches are attractive as they easily enable genuine simultaneous reconstructions of all grains rather than grain-by-grain reconstruction as 2D-ART above. On the other hand, they tend to be slow and "get stuck" in local minima if the configuration space is too large. Two approaches have been demonstrated: (a) A *restoration* approach, where a coarse or ambiguous grain map first is generated by a forward-projection algorithm [18] or by ART [21]. A Monte-Carlo-based routine is then used to "restore" the correct pixel affiliations in the grain boundary near regions. (b) An *indexing* approach, in this approach there is no need for an initial grain map; instead the grain map is reconstructed based only on the output from GRAINDEX, that is the number of grains and their orientations [21].

Extensive simulations have been performed in order to estimate the merits of various approaches. The most critical parameter is found to be the number of reflections available for the analysis. Remarkably, for 2D-ART and Monte-Carlo-based reconstructions, reasonable quality maps can be gathered with as little as five projections. In Alpers et al. [21], several algorithms are compared with the

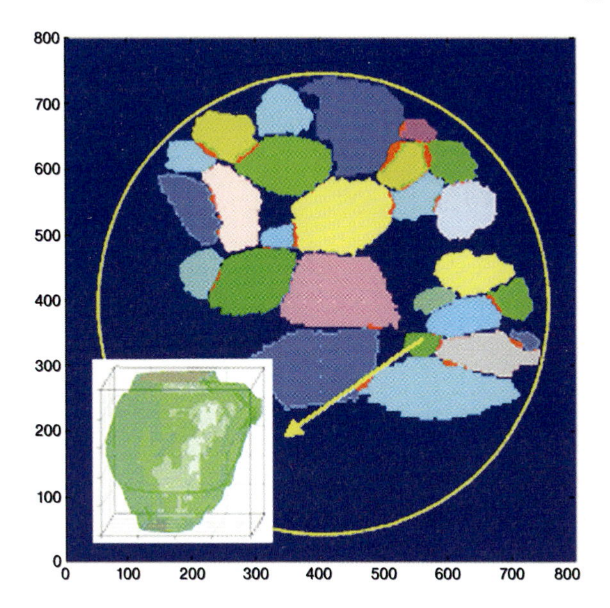


**Figure 19.6** First reported partial grain map of one layer using 2D-ART. The map is in units of μm with the surface of the Al polycrystal indicated by a solid yellow circle. The grain sections were reconstructed independently. This data set was acquired in 10 min. By stacking layers 3D maps are obtained, an example of which is shown in the insert. Orientations are represented by a grey level scheme. (Courtesy of X. Fu).

result that the Monte-Carlo-based reconstructions are superior. Corresponding work on algorithms for the 3D case is currently undergoing (e.g., [22]).

### 19.3.2
### Mode IV: Mapping Orientations in Deformed Specimens

For deformed specimens it is relevant to generate orientation maps, where each voxel in the specimen is associated with each own orientation (see Figure 19.3). Mathematically speaking, the task in this case is to reconstruct a vector-field $\underline{r}(\underline{x})$, where $\underline{r}$ symbolize orientations and $\underline{x}$ position in direct space. To the knowledge of authors, for such fields no transform or algebraic reconstruction algorithm is readily available. Stochastic approaches are possible, but here it is a concern that the configurations space is much larger for deformed case than for the un-deformed.

Recently Rodek et al. [23] presented a Monte-Carlo-based algorithm for moderately deformed specimens, where the size of configuration space has been reduced by orders of magnitude by using known properties of the microstructure. The routine assumes that prior information arising, e.g., from a forward projection makes it possible to identify the number of grains, and for each determine

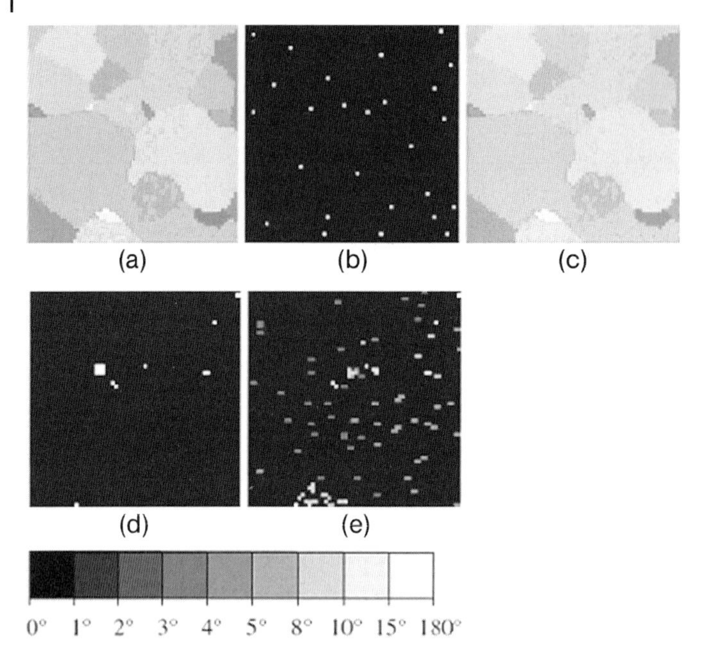


**Figure 19.7** The reconstruction of test case representing a deformed microstructure. (a) Reference orientation map, (b) seeds, (c) reconstructed orientation map. (d) Difference of the reference and the reconstructed grain maps. Black pixels denote identical grain labels, white pixels represent mismatching ones. (e) Difference of the reference and the reconstructed orientation maps. The grey level of the pixels is determined by the misorientation of corresponding orientation pairs, as shown in the key at the bottom of the figure. (Courtesy of L. Rodek).

an approximate average orientation and one pixel in the layer which belongs to this grain. These points function as seeds for the algorithm. This approach was tested on a set phantoms representing microstructures typical of various degrees of deformation. The results are encouraging. As an example in Figure 19.7 the reconstruction of a microstructure with 22 grains and with an internal orientation spread within grains of up to 22°. (In this reproduction of the two maps (a) and (c) on top it is difficult to see the orientation gradient within the grains.) At low noise levels, the reconstructions were nearly perfect. At the time of writing analysis on real data is pending.

### 19.3.3 Application I: Recrystallization

The first 3D movies of grain coarsening related to recrystallization [24], as it here, at least in the initial stage of the process, is relevant to visualize the growth of a single grain (nucleus). More specifically, nucleation in an aluminum single crystal deformed to 42% was stimulated by hardness indents. The crystal was

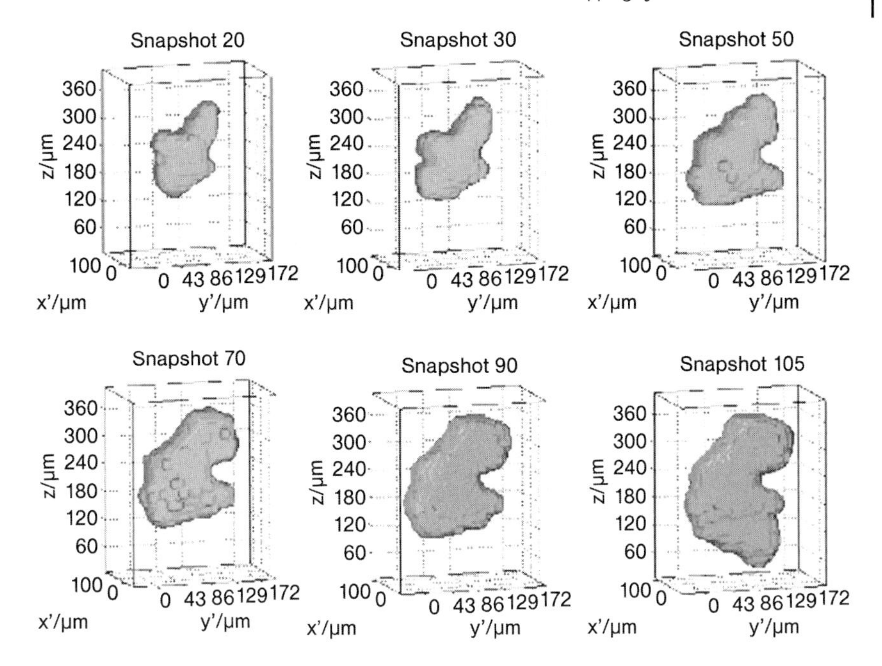


**Figure 19.8** Storyboard showing the growth of one emerging nucleus in an Al single crystal deformed by 42% as function of annealing. (From [24], with permission from Science).

subject to *in situ* annealing for 30 h at temperatures between 270 and 310 °C. The mapping comprised 50 $z$-layers with a spacing of 6 μm. One emerging nucleus was identified at an early stage of the growth. Exposures were made around the $\omega$ setting of a specific reflections belonging to this grain. The nucleus was mapped continuously with a time resolution of at most 10 min. Snapshots of the resulting 3D movie are shown in Figure 19.8. The data quality is sufficient that two general features of the growth are revealed:

1. The growth is very heterogeneous, so the shape is at times remarkably irregular.
2. The growth does not occur smoothly with time, but is often jerky.

These unique observations were explained by the heterogeneous nature of the deformed microstructure.

### 19.3.4 Application II: Grain Growth

In this study, the evolution of the morphology of several hundred grains was monitored simultaneously for the first time [25]. The base material Al–0.1%Mn

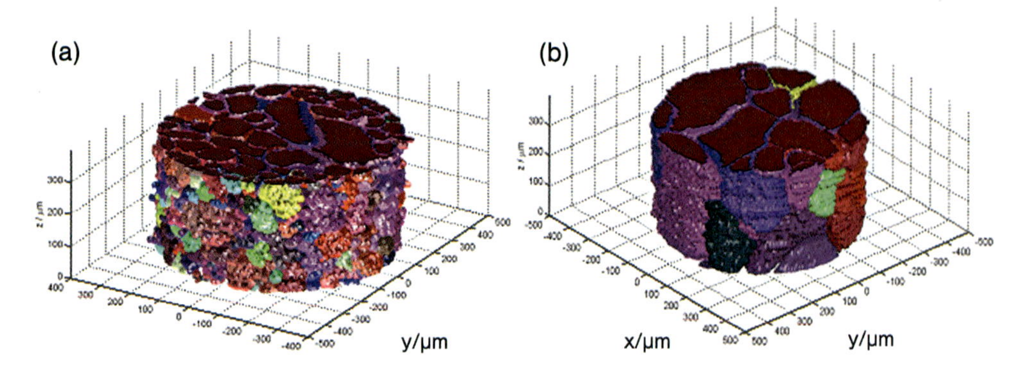


**Figure 19.9** The fully reconstructed grain volumes: (a) the initial grain volume containing 491 grains and (b) final grain volume containing 49 grains. Colors are related to the crystallographic orientation of the individual grains.

was cold rolled 80% and annealed for 8 min at 300 °C producing a fully recrystallized sample with grain sizes between 30 and 100 μm. Afterward, a cylindrical sample with dimensions of 700 μm in diameter and 2 mm in height was selected for further investigations at the 3DXRD microscope. The sample was annealed in an external furnace. Before the first annealing and following each annealing step a cylindrical volume with a diameter of 700 μm and height of 350 μm was fully characterized by mapping out the sample in layers using a planar beam shape at 50 keV. The grain volumes were reconstructed using the GrainSweeper [19] algorithm. In total, five annealing steps were made at annealing temperatures between 400 and 450 °C. Initially, 491 grains were present leaving only 49 grains after the final annealing step. Figure 19.9 shows the fully reconstructed initial and final grain volumes.

The information present in the reconstructed volumes, such as number of grain edges, etc., is to be extracted. Such information may serve as a crucial constraint to any grain growth model.

## 19.4 Combining 3DXRD and Tomography

The 3DXRD set-up sketched in Figure 19.1 is very similar to the set-up typically used for parallel beam tomography, cf. Chapter 15. Hence, it is possible to probe the same specimen with both techniques either simultaneously or consecutively. The relevance of combining tomography and diffraction type information in general is discussed in Chapter 22. More specific to 3DXRD, we note that the basic 3DXRD algorithms developed so-far all apply to mono-phase materials only. A combined data set enables the methodology to be extended to multiphase materials, by first segmenting the volume to be reconstructed based on density and then applying the 3DXRD method to each phase at a time.

### 19.4.1
### Grain Mapping by Tomography

Conventional wisdom tells us that attenuation and/or phase contrast tomography is not sensitive to orientations and as such cannot be used to map grains in a mono-phase material. Recently two methods that overcome this basic limitation have been demonstrated by a group comprising people at ESRF and Risø.

(a) *Diffraction contrast tomography (DCT)*. The principle is illustrated in Figure 19.10: assume a grain is in the Bragg condition. Then there will be two contributions to attenuation of the direct beam: the absorption in the sample and the missing intensity due to diffraction. The latter additional contrast is clearly visible if the grains are near-perfect (mosaic spread well below 1°). Performing first a conventional reconstruction, a density map of the sample is obtained. Based on this one can simulate what the absorption data should look like for a noncrystalline sample. Subtracting the actual images from the simulated ones, one may sort the residual "grain contrast" images, based on a combination of real space (grain shape and position) and orientation space scanning procedures that compare simulated and measured angles. Having indexed a grain in this way, the 3D shape can be reconstructed based on the 3D-ART approach.

    DCT and 3DXRD may be seen as the equivalent of "dark-field" and "bright-field" imaging in TEM. Combined data sets have been acquired in several experiments at ID19 at ESRF. The results demonstrate that this combined method produce grain maps of superior resolution to those based on 3DXRD only. The drawback is that the number of projections required needs to be much higher, of the order of a few 1000 images.

(b) *Topo-tomography* is an alternative tomographic scanning procedure, which allows reconstructing a single grain, using a set-up with additional tilt-stages [26]. For a selected grain, the scattering vector of a suited reflection is aligned

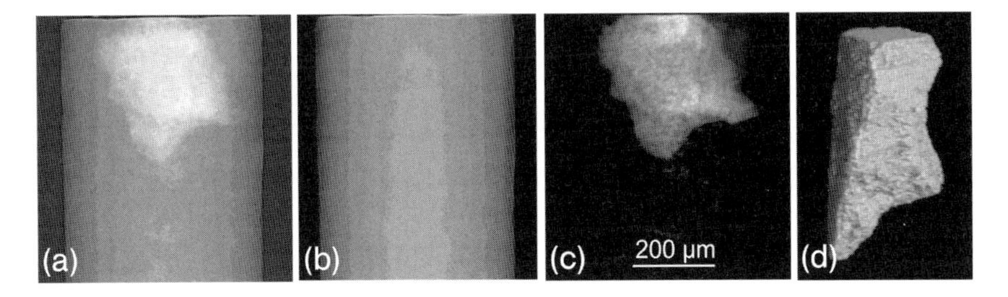


**Figure 19.10** Diffraction contrast tomography: (a) experimental projection with one grain fulfilling the Bragg condition; (b) calculated absorption background; (c) extracted grain projection, in all cases with brightness in image proportional to attenuation; and (d) reconstruction of 3D grain shape by means of algebraic reconstruction techniques (different scale).

parallel to the rotation axis of the tomographic set-up. This particular setting assures that the diffraction condition is maintained while the sample is turned 360° around the rotation axis. Tomographic reconstruction can be performed from data acquired in direct and/or in diffracted beam by means of a standard (cone beam) filtered backprojection algorithm. In comparison to mode III reconstructions by 3DXRD above, the spatial resolution can be substantially better. The disadvantage is the restriction to one grain only.

## 19.5 Outlook

At the time of writing the ID11 beamline is undergoing a major refurbishment. A new generation of hard X-ray optics, an advanced feedback system and temperature control will be installed in order to generate a stable, fully energy and bandwidth tuneable beam, with dimensions as small as 50–100 nm. At the same time work is ongoing to develop a new type of fluorescence screen – a structured screen – for the high-resolution detectors. This detector is specified to have a resolution of 1 μm and an efficiency in the excess of 25% (in comparison present detectors have a resolution of 5 μm and an efficiency of a few percentage).

For modes III and IV work, these developments are predicted to enable the generation of 3D maps with a spatial resolution of 1 μm or alternatively complete maps of 1-$mm^3$-sized materials with a resolution of 5 μm in a few minutes. For modes I and II work, it is predicted that crystalline structures as small as 5–10 nm can be studied during *in situ* processing.

## References

1 H.F. Poulsen, *Three-Dimensional X-Ray Diffraction Microscopy*, Springer, Berlin, 2004.

2 B.C. Larson, W. Yang, G.E. Ice, J.D. Budai, T.Z. Tischler, *Nature* 2002, **415**, 887–890.

3 H.J. Bunge, L. Wcislak, H. Klein, U. Garbe, J.R. Schneider. *J. Appl. Cryst.* 2003, **36**, 1240–1255.

4 T. Wroblewski, O. Clauss, H.-A. Crostack, A. Ertel, F. Fandrich, Ch. Genzel, K. Hradil, W. Ternes, E. Woldt, *Nucl. Instr. Meth.* 1999, **A 428**, 570–582.

5 E.M. Lauridsen, S. Schmidt, R.M. Suter, H.F. Poulsen, *J. Appl. Cryst.* 2001, **34**, 744–750.

6 H.F. Poulsen, S.F. Nielsen, E.M. Lauridsen, S. Schmidt, R.M. Suter, U. Lienert, L. Margulies, T. Lorentzen, D. Juul Jensen, *J. Appl. Cryst.* 2001, **34**, 751–756.

7 R.V. Martins, L. Margulies, S. Schmidt, H.F. Poulsen, T. Leffers, *Mater. Sci. Eng.* 2004, **387–389**, 84–88.

8 G.B.M. Vaughan, S. Schmidt, H.F. Poulsen, *Z. Kristall.* 2004, **219**, 813.

9 E.M. Lauridsen, H.F. Poulsen, S.F. Nielsen, D. Juul Jensen, *Acta Mat.* 2003, **51**, 4423–4435.

10 N. Iqbal, N.H. van Dijk, S.E. Offerman, M.P. Moret, L. Katgerman, G.J. Kearley, *Acta Mater.* 2005, **53**, 2875–2880.

11 S.E. Offerman, N.H. van Dijk, J. Sietsma, S. Grigull, E.M. Lauridsen, L. Margulies, H.F. Poulsen, M.T. Rekveldt, S. van der Zwaag, *Science* 2002, **298**, 1003–1005.

12 W.F. Lange III, M. Enornoto, H.I. Aaronson, *Metall. Trans.* 1988, **19A**, 427–440.

**13** L. Margulies, G. Winther, H.F. Poulsen, *Science* 2001, **291**, 2392–2394.

**14** H.F. Poulsen, L. Margulies, S. Schmidt, G. Winther, *Acta Mat.* 2003, **51**, 3821–3830.

**15** G. Winther, L. Margulies, S. Schmidt, H.F. Poulsen, *Acta Mat.* 2004, **52**, 2863.

**16** C. Gundlach, W. Pantleon, E.M. Lauridsen, L. Margulies, R. Doherty, H.F. Poulsen, *Scr. Mater.* 2004, **50**, 477–481.

**17** B. Jakobsen, H.F. Poulsen, U. Lienert, J. Almer, S.D. Shastri, H.O. Sørensen, C. Gundlach, W. Pantleon, *Science* 2006, **312**, 889–892.

**18** R.M. Suter, D. Hennessy, C. Xiao, U. Lienert, *Rev. Sci. Instrum.* 2006, **77**, 123905.

**19** S. Schmidt, preprint.

**20** H.F. Poulsen, X. Fu, *J. Appl. Cryst.* 2003, **36**, 1062–1068.

**21** A. Alpers, H.F. Poulsen, E. Knudsen, G.T. Herman, *J. Appl. Cryst.* 2006, **39**, 582–588.

**22** T. Markussen, X. Fu, L. Margulies, E.M. Lauridsen, S.F. Nielsen, S. Schmidt, H.F. Poulsen, *J. Appl. Cryst.* 2004, **37**, 96–102.

**23** L. Rodek, H.F. Poulsen, E. Knudsen, G.T. Herman, *J. Appl. Cryst.* 2007, **40**, 313–321

**24** S. Schmidt, S.F. Nielsen, C. Gundlach, L. Margulies, X. Huang, D. Juul Jensen, *Science* 2004, **305**, 229–232.

**25** S. Schmidt, et al., In preparation.

**26** W. Ludwig, E.M. Lauridsen, S. Schmidt, H.F. Poulsen, J. Baruchel, *J. Appl. Cryst.* 2007, **40**, 905–911.

# 20
# 3D Micron-Resolution Laue Diffraction

*Gene E. Ice*

## 20.1
## Introduction

A fundamental question of materials science is how atomic-scale interactions self-organize atoms into mesoscopic structures. This simple question is important because the physical behavior of most materials is dominated by mesoscale structure and dynamics. For example, self-organization is essential to understand grain-growth and deformation microstructure and to understand their effects on plasticity, strength, fracture, transport, and other materials properties.

To understand how mesostructures arise, and how they influence materials behavior, it is essential to map *local* elemental composition, crystal/local structure, and geometrical/chemical defect distributions. In materials, this information is mathematically approximated by three-dimensional (3D) tensor fields, which are typically highly heterogeneous. For this reason, 3D quantitative probes are essential. X-ray microdiffraction is particularly interesting as it provides detailed *atomic-resolution* information about local crystalline structure correlated with the *mesoscale* (0.1–10 μm) real-space resolution of the probe. Furthermore, unlike almost any other probe, X-ray microbeams can nondestructively characterize materials properties in three-dimensions and can observe mesoscale evolution as a response to underlying driving forces (e.g., stress).

### 20.1.1
### The Need for *Polychromatic* Microdiffraction

Although diffraction with micron-scaled beams can provide detailed structural information, *monochromatic* microdiffraction studies are complicated. To efficiently diffract microbeams from small volumes, samples must be simultaneously positioned with submicron precision and aligned with arcsecond angular accuracy. In polycrystalline materials, the penetration of X-rays into the sample further complicates data interpretation as more than one (or none) of the grains intercepted by the incident beam can diffract (Figure 20.1a). If the sample is oriented to

*Neutrons and Synchrotron Radiation in Engineering Materials Science: From Fundamentals to Material and Component Characterization*
Edited by Walter Reimers, Anke Rita Pyzalla, Andreas Schreyer, and Helmut Clemens

ISBN: 978-3-527-31533-8

position one grain to satisfy a Bragg condition, the neighbor grains illuminated by the penetrating beam can be rotated into or out of the beam. This problem is exacerbated by the sphere-of-confusion errors in rotations of the sample (Figure 20.1b). High-energy monochromatic beams that flood many grains can be used for mapping of crystal structures as described in the previous chapter. Here the spatial resolution is determined by the spatial resolution of the detector and by corrections for sphere-of-confusion in the sample rotations. Polychromatic beams, however, have the advantage of requiring no sample rotations for diffraction studies. In this case, spatial resolution is determined to a large degree by the beam size, which can be submicron and even nanoscale. Furthermore, issues of strain or deformation on the measured diffraction volume have – at least in principle – a small effect.

With a polychromatic (Laue) approach, however, there is a completely different problem: an over abundance of diffraction with each subgrain volume diffracting into overlapping Laue patterns on the detector (Figure 20.1c). The complexity of the overlapping Laue patterns is greatly reduced by the use of small X-ray beams that limits the number of subgrains illuminated and can be further simplified by an experimental approach called differential aperture microscopy which resolves the single-crystal-like Laue diffraction from subgrain volumes along the illuminating beam. The theories behind differential aperture microscopy and polychromatic microdiffraction are described briefly below.

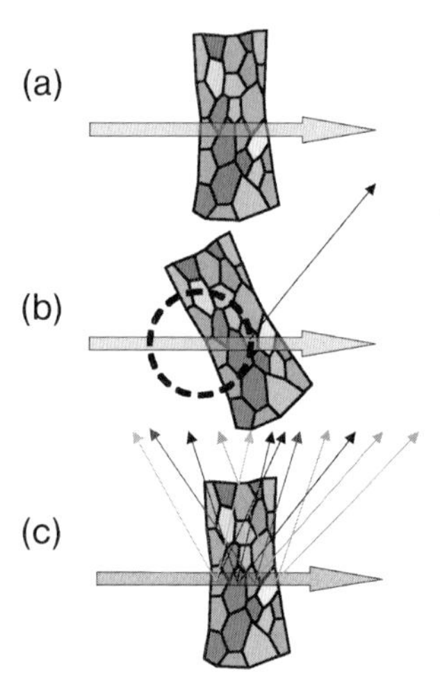


**Figure 20.1** (a) A monochromatic X-ray beam has a low probability of diffracting from an arbitrary crystal grain orientation. (b) If the sample is rotated to orient a grain for Laue diffraction the grain moves within the instrumental sphere-of-confusion even if it is cantered on the goniometer rotation axis. Grains away from the rotation axis are in the incident beam path over an even smaller angular range. (c) With a polychromatic beam, the subgrain volumes intercepted by the incident beam generate Laue patterns that overlap at the detector.

## 20.2
## Theoretical Basis for Advanced Polychromatic Microdiffraction

### 20.2.1
### Modified Ewald's Sphere Description of Laue Diffraction

Monochromatic diffraction is conceptually represented by an Ewald's sphere diagram. *Laue diffraction* can be similarly modeled by a *modified* Ewald's sphere diagram as shown in Figure 20.2. Bragg reflections for a crystal are indicated in reciprocal space by a reciprocal-space lattice that is aligned to the real-space orientation of the crystal. A Bragg reflection is excited when the momentum transfer of an accessible diffraction direction for a particular wavelength matches a reciprocal-space lattice co-ordinate. For a collimated monochromatic beam, possible momentum transfers in reciprocal space lie on the Ewald's sphere, and the sample must be precisely oriented to pass a reciprocal lattice point through the Ewald's sphere.

With Laue diffraction however, a finite bandpass creates a volume in reciprocal space between maximum and minimum Ewald's spheres (note that the centers of the limiting spheres lie along the incident beam direction but are displaced from each other). Bragg reflections that fall within this volume are excited and the directions of the Laue pattern peaks indicate the Bragg plane normals of the crystal. More exactly, the intensity in each pixel in a Laue image is proportional to a line integral through reciprocal space weighted by the incident beam spectral distribution [1].

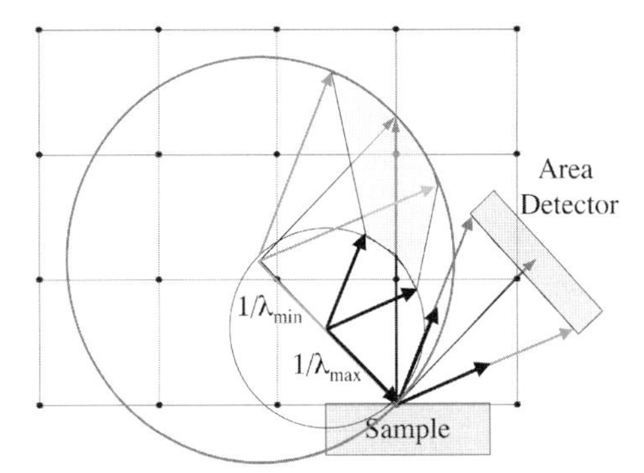


**Figure 20.2** A two-dimensional modified Ewald's sphere diagram. In reciprocal space, the momentum of an X-ray beam depends on its direction and wavelength; for short wavelength (green) the momentum is larger than for large wavelength (black). The momentum transfer is simply the vector difference between the final and incident X-ray momentums. For a fixed wavelength this lies on a sphere. The centers of the spheres for different wavelengths lie along the incident beam direction and are the spheres touch at the 0, 0, 0 position (no momentum transferred). The Laue pattern observed by an area detector is made up of radial line integrals through reciprocal space. Lattice points that lie within the bounds of the minimum and maximum Ewald's spheres contribute to the line integral.

### 20.2.2 Qualitative Information: Phase, Texture, Elastic Strain, Dislocation Density

With the understanding that Laue images represent radial line integrals in reciprocal space, micro-Laue images can offer immediate qualitative information about the sample volume illuminated by the beam.

#### 20.2.2.1 Phase

For example, as shown in Figure 20.3a, the image from a small high-symmetry unit cell looks qualitatively distinct from that of a large or low-symmetry unit cell (Figure 20.3b). Hence, phase distributions can often be determined by mapping the number of reflections emanating from subgrain volumes within the sample.

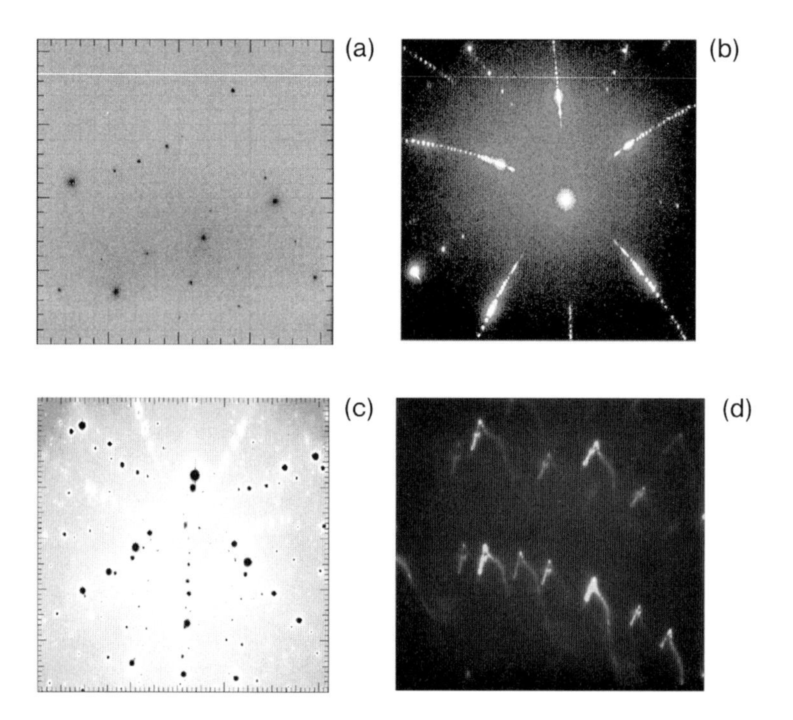


**Figure 20.3** (a) Laue pattern from a $Ni_{80}Fe_{20}$ single crystal. The small unit cell and high-symmetry face-centered-cubic (FCC) real-space lattice results in a sparse body-centered-cubic (BCC) reciprocal-space lattice and consequently only a few Laue spots are detected. (b) Overlapping Laue patterns from a GaN film on a SiC single crystal. The two patterns are superimposed because the GaN film is crystallographically aligned to the SiC substrate. (c) High-temperature superconducting film with buffer layer on a Ni single crystal. There are actually three patterns in the Laue image: one from the substrate and two from the two films. The patterns are at slightly different orientations due to crystallographic tilt between the layers. (d) Deformed Ni crystal shows a dramatically redistribution of intensity near the Bragg peaks. This redistribution is primarily due to long-range rotations of the Bragg planes caused by unpaired dislocations.

#### 20.2.2.2 Texture

Similarly micro-Laue patterns are highly sensitive to orientation and for good quality crystals can measure orientations to 0.001° [2]. Therefore, measurements of local grain orientations (texture) are easily made with 1–3 orders of magnitude better angular resolution than with typical electron probe measurements. For example, in Figure 20.3c, the relative angle between crystal layers in an engineered high-temperature superconductor structure is easily determined. Furthermore, these orientation measurements can be made tens to hundreds of microns below the sample surface so that the local texture in thin films, multilayers, and true 3D samples can be studied.

#### 20.2.2.3 Dislocation Tensor

In addition, dislocations are Krivoglaz defects of the second kind that introduce long-range rotations in the Bragg planes [3] so that sharp Laue spots are smeared (Figure 20.3d). Single images can be used to model the dislocation density and distribution through the illuminated sample volume [4], or micro-Laue images can be resolved by differential aperture microscopy to determine the local Laue pattern within a submicron volume and the 3D distribution of lattice rotations can be used to unambiguously extract the local dislocation tensor [5].

#### 20.2.2.4 Elastic Strain Tensor

Finally, changes in the angles between reflections can be used to determine the local elastic stress. The angles between reflections can be used to determine distortions of the unit cell shape. These distortions are described by the local deviatoric strain tensor [6]. At least four indexed reflections are needed to determine the deviatoric strain tensor (distortion of the unit cell shape). If the energy of one reflection is determined, then not only the unit cell shape but also its volume can be determined and the full strain tensor can be found by comparison to the undeformed unit cell.

## 20.3 Technical Developments for an Automated 3D Probe

Although Laue diffraction is the oldest X-ray diffraction method [7], the recent extension to 3D imaging of crystalline structure has relied on the most modern developments in X-ray sources, optics, and detectors [8]. To go beyond a conceptual tool to an automated 3D probe for mapping crystal structure, five elements have been assembled in the "first-of-its-class" 3D X-ray crystal microscope deployed on the station 34-ID-E at the Advanced Photon Source (Figure 20.4). These are (a) an ultrabrilliant X-ray source, (b) a nondispersive monochromator that can be cycled between monochromatic and polychromatic status, (c) nondispersive focusing optics with submicron resolution, (d) a scanning-wire differential aperture (Figure 20.5), and (e) a high-resolution area detector. The purpose and performance of each element is described briefly below.

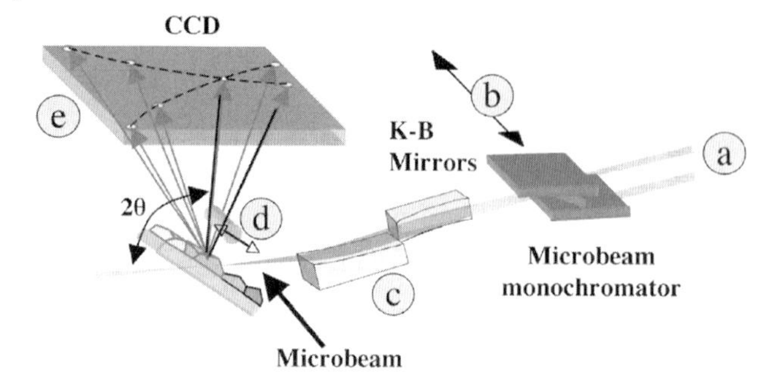


**Figure 20.4** Schematic of the 3D X-ray crystal microscope an instrument optimized for polychromatic microdiffraction experiments. The instrument includes: (a) an ultrabrillant X-ray source; (b) a microbeam monochromator that can switch the beam from polychromatic to monochromatic radiation; (c) a nondispersive focusing Kirkpatrick–Baez mirror system; (d) a differential aperture for deconvolving the Laue patterns generated along the incident beam; (e) an optimized X-ray sensitive area detector.

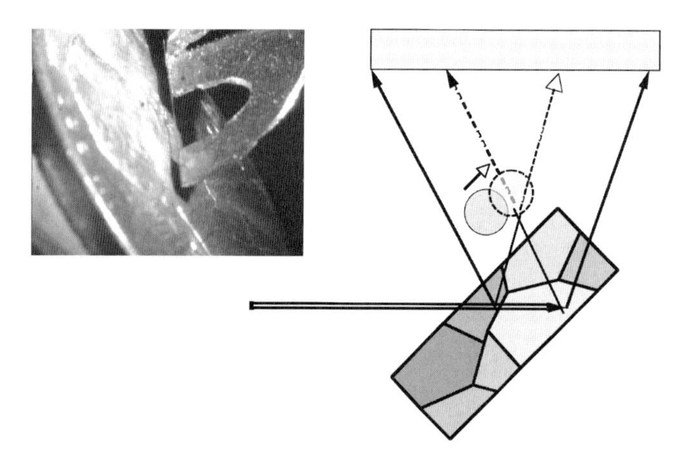

**Figure 20.5** Differential aperture. The intensity is monitored at each pixel as the wire is moved through the diffracted beam. The intensity origin for each pixel is determined by ray tracing back from the pixel position past the edge of the wire and onto the incident beam axis.

### 20.3.1 Source

An ultrabrilliant synchrotron source is the first essential element for automated polychromatic microdiffraction. The combination of small beam size and good collimation needed for simultaneously good real-space and momentum transfer resolution greatly restricts the phase space of the useable beam. This makes it im-

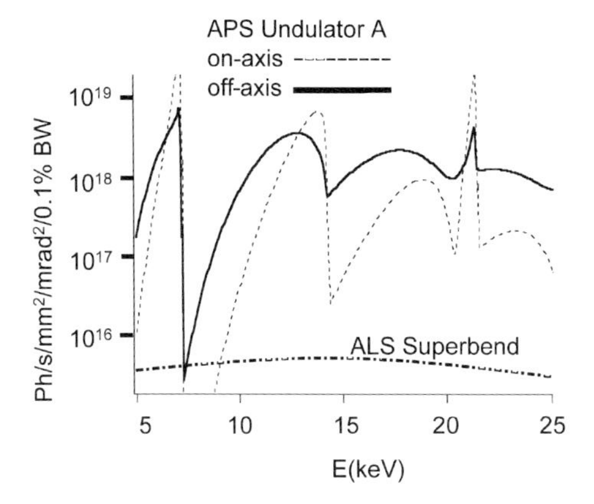


**Figure 20.6** Although a third generation undulator source has about two orders of magnitude greater integrated brilliance than a bend magnet, the spectra have large fluctuations with wavelength that complicate polychromatic microdiffraction. Highest performance is achieved with an optimized undulator, but typically limitations in detector readout make bend-magnet sources competitive for most experiments.

portant that the source have high brilliance (photons/s/$mm^2$/$mrad^2$/eV) over the required bandpass, but also means that the source emittance ($size^2 \times divergence^2$) can be small. A type A undulator at the APS has about 12 (4) orders of magnitude greater X-ray brilliance for monochromatic experiments and about 10 (2) orders of magnitude greater brilliance for wide-bandpass measurements than a laboratory (bend-magnet synchrotron) source (Figure 20.6). This high brilliance allows for micro-Laue images to be collected in milliseconds from many samples. Even though the undulator source is intrinsically highly collimated with a relatively small source, on 34-ID-E only about 0.1% of the total beam emittance is used in the polychromatic microbeam. This makes it possible to field many polychromatic beamlines on one undulator source.

Bend-magnet sources can also be successfully used for polychromatic microdiffraction despite their lower overall brilliance. Currently, the performance of polychromatic microdiffraction is influenced more by the speed of the detector readout than by the counting statistics in the Laue images. Sources that are only a factor of 100 lower in integrated brilliance can therefore do much of the science that is done with an undulator source and can be optimized for particular experiments to achieve very high performance.

### 20.3.2
### Microbeam Monochromator

A microbeam monochromator is the second important element of a polychromatic microdiffraction system. The monochromator allows for measurements of the full strain tensor. As described above, Laue images record radial line integrals

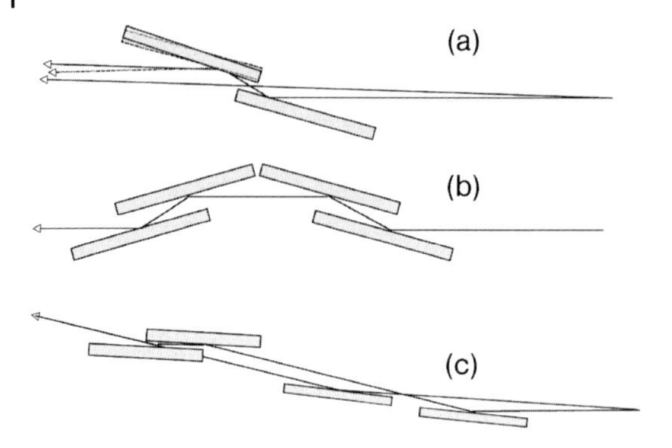


**Figure 20.7** (a) Schematic of the microbeam monochromator on beamline 34-ID-E at the APS. The polychromatic beam is taken ~1 mm above the central beamline axis. This makes the polychromatic beam have a more uniform spectral intensity throughout the useful bandpass of the optics. (b) The micro-monochromator design on beamline 7 at the APS is a four crystal dispersive design that is intrinsically zero offset. (c) The design for the VESPERS beamline at the Canadian light source uses mirrors and a small displacement monochromator to alternate between co-linear polychromatic or monochromatic beams.

through reciprocal space. Strain is determined by comparing the measured unit cell shape to the shape of an undistorted unit cell. Laue images cannot, however, distinguish between unit cells with slightly different volumes but the same shape. Therefore, in order to determine the full strain tensor, it is important to measure the energy of one reflection. Of course the deviatoric strain tensor, which is most sensitive to shear strains can be measured without any energy measurement. On the station 34-ID-E, the microbeam monochromator can be translated into or out of the beam so the beam incident on the focusing optics is either polychromatic or monochromatic. The design also maintains a fixed offset independent of X-ray energy or bandpass [9].

The 34-ID-E monochromator utilizes a two-crystal fixed exit design with a small (~1 mm) offset. The polychromatic beam is taken from ~1 mm above the beam axis and the monochromatic beam is taken from near the beam axis (Figure 20.7a). This optimizes the effective beam brilliance of the undulator source when used for monochromatic measurements and delivers a more uniform energy distribution for polychromatic experiments. Beamline 7 at the ALS uses a four crystal design that is intrinsically zero offset (Figure 20.7b). Beamlines planned for the Canadian Light Source and the Australian light source will use another trick that utilizes mirrors in addition to the crystals to switch between co-axial polychromatic or monochromatic beams (Figure 20.7c). This design combines the simplicity, high-throughput and ease of use of a two-crystal monochromator with the advantages of an intrinsically zero offset dispersion-free design.

### 20.3.3
### Nondispersive Focusing Optics

Nondispersive focusing optics *is the* essential element for polychromatic microdiffraction. On 34-ID-E, total external reflection Kirkpatrick–Baez mirrors are used to focus to an $\sim$300 nm $\times$ 500 nm focal spot. The mirrors are fabricated by profile coating on ultrasmooth substrates [10, 11]. Kirkpatrick–Baez optics are rapidly developing and hard X-ray nanoprobe optics suitable for polychromatic nanodiffraction have been demonstrated [11, 12].

### 20.3.4
### Area Detector

An area detector to record the Laue pattern is also an essential element of a polychromatic microdiffraction station. Unlike fluorescence microscopy it is important that the scattering is out of the ring plane to improve the reflectivity of the polarized incident beam. On 34-ID-E and other (2D) polychromatic microdiffraction beamlines the detector is typically aligned at 90° to the incident beam. This geometry is a compromise that allows for simple sample alignment with reasonable reflectivity of X-rays between 10 and 20 keV (see Figure 20.4). The nominal 90° scattering angle also optimizes the penetration depth of elastically scattered X-rays for a given absorption coefficient.

A detector optimized for polychromatic microdiffraction is quite distinct from area detectors used in protein crystallography and other more common X-ray applications. For example, extremely linear pixel spacing is important to measure strains to 0.01%. Similarly, large dynamic range, low-readout noise, fast readout, and low-point spread functions are all important. Readout speed is particularly important to collect meaningful 3D images in a reasonable time.

### 20.3.5
### Differential Aperture

The final element for a *3D* X-ray crystal microscope is a differential aperture. The differential aperture deconvolves the Laue patterns from subgrain volumes along the incident beam [5]. The principle of differential aperture microscopy is indicated in Figure 20.5. A smooth high Z wire is passed near the surface of the sample (within $\sim$200 μm). The wire is calibrated prior to the measurement so that its position is precisely known relative to the incident beam and the axis of the detector. As the wire moves along the sample surface, it occludes rays that are collected at the detector. The intensity as a function of wire position is observed for each pixel. The derivative of the measured intensity/wire step in each pixel is sum of positive and negative images of the intensity distribution along the wire that reaches that particular pixel separated by the diameter of the wire. By ray

**Figure 20.8** The 3D X-ray crystal microscope on beamline 34-ID-E. The components are mounted on an optical table, which allows new optical approaches to be rapidly tested.

tracing back from the pixel and wire locations, and taking into account small effects due to the angle of the intercepted beam, the origin of the intensity into each pixel can be mapped with submicron precision along the incident beam. A picture of the 3D X-ray crystal microscope on 34-ID-E is shown in Figure 20.8 and a close up of the differential aperture is shown in Figure 20.5. Efforts are now underway to accelerate differential aperture microscopy with multiple wires and more intelligent step scans.

### 20.3.6 Software

Although Laue diffraction is a long-established technique, until recently, no software existed to allow for automated mapping of multiple grains, for differential aperture microscopy or for precision determination of strain and modeling of deformation parameters. This software is still a work in progress, but is now the subject of an international effort to develop a standard package that will foster polychromatic microprobes on beamlines around the world. Existing software packages are described in various references [6, 13] and more advanced user-friendly software is under development.

## 20.4
## Research Examples

The mission of the station 34-ID-E is centered on fundamental investigations of long-standing issues in materials science. These issues include grain growth and 3D organization in polycrystalline materials; 3D dynamics of mesoscale structures; and fundamental assumptions in materials physics modeling. Studies of 3D organization underway include measurements of colony structure in processed materials and studies of fractal misorientation in deformed materials including the development of dislocation hierarchical structures. Other studies of 3D structures include measurements of percolation and diffusion behavior.

Studies of dynamics underway include 3D local growth of grains, measurements of ripening and measurements of anomalous grain growth (i.e., Sn whiskers). Other measurements include studies of nucleation sites, the role of deformation energy and elastic stress in grain nucleation, and grain/subgrain rotations during deformation as a function of global and local environmental conditions.

Fundamental assumptions are being tested by experiments that include: mapping of intra- and intergranular stresses to guide our understanding of the constitutive equations of materials near grain boundaries and other interfaces; measurements of grain boundary shear versus habit; measurement of deformation scaling laws in three-dimensions.

### 20.4.1
### 3D Grain Boundary Networks

How grain boundary/polycrystalline networks interact to applied forces and environmental conditions is a central materials challenge of the 21st century. Modelers want to know the constitutive equations at grain boundaries, and how they change with boundary type. This information impacts efforts to model materials behavior and to engineer ideal microstructures for applications ranging from corrosion resistance to high strength. Similarly, synthesis and processing control of grain boundary distributions is an area of active research with particularly complex issues for nanoscale and advanced layered structures.

Polychromatic microdiffraction allows for unprecedented testing of longstanding assumptions regarding grain boundaries. For example, in coincident site lattice (CSL) theory, grain boundaries are considered to have lower energy if some of the lattice sites of the neighbor grains overlap [14]. The lattice is assigned a number, $\Sigma$, which is the inverse of the fraction of shared sites; if 1/3 of the sites are shared then $\Sigma$ equals 3. According to theory, high-$\Sigma$ boundaries can only have a slight misorientation because the energy minimum for high-$\Sigma$ boundaries is small and the subsequent energy wall minimum is shallow. The good angular precision of polychromatic microdiffraction allows for detailed tests of this and other predictions of CSL theory. Preliminary measurements actually indicate an

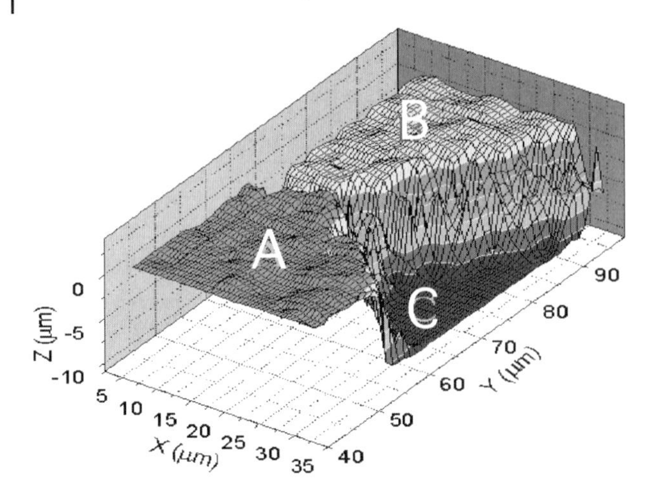


**Figure 20.9** The surface of a Ni triple grain junction color coded to show orientation of grains. The orientations were mapped by polychromatic microdiffraction and show grain boundary grooving, and a step in one of the grains surfaces. Internal grain boundary surfaces are also recovered with these measurements.

opposite trend in the distribution of lattice misorientations near CSL boundary conditions; measured misorientations from coincident site lattice boundaries increase with Σ. Other interesting questions which can be addressed include the role of curvature and faceting in grain boundaries. Grain boundary normals should have lower energy in ideal CSL directions. However, curved surfaces can also lower energy. Here, the role of faceting may allow for energy minimization and studies are underway to study how 3D surfaces accommodate competing processes toward energy minimization. For example, Figure 20.9 shows a step in the surface of a polycrystalline Ni sample at a grain boundary triple junction as well as grain boundary grooving. Polychromatic microdiffraction is certain to shed new light on grain boundaries and on how grain boundaries affect materials behavior.

### 20.4.2 Deformation Behavior and Grain Boundaries

Most theories of deformation are based on the crystallographic orientation of grains with respect to the applied load; the orientation controls the critical-resolved shear stress along slip systems. These theories have now been tested for the first time with the use of high-energy X-ray microdiffraction as described in Chapter 19. However, local environments also affect deformation behavior. For example, Pang [15] has recently studied rotations *across* grain boundaries before and after deformation. Her findings show a strong correlation between grain boundary angle and misorientation magnitude, and finds that coincident site lat-

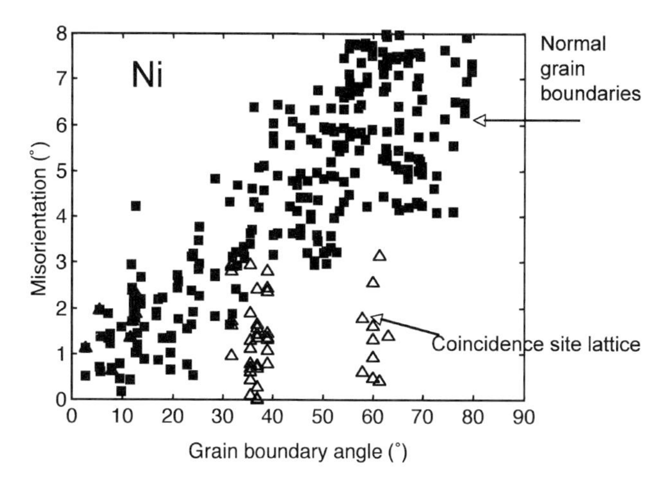


**Figure 20.10** Grain boundary angle before deformation (*x*-axis) compared to the change in misorientation across a grain boundary after deformation (*y*-axis). There is a positive correlation between grain boundary angle and misorientation during deformation. Grain boundary misorientations near coincidence site lattice boundaries however appear to have a distinctly different behavior from other boundaries.

tice boundaries have a much different relationship to deformation than other boundary types (Figure 20.10). Indeed deformation behavior near surfaces and interfaces appears to be quite distinct from the behavior near the center of grains [16].

### 20.4.3 Deformation in Single Crystals

Although the deformation of materials appears to have important complexity introduced by the presence of grain boundaries, there is strong evidence that even in single crystals our understanding of deformation is a work in progress. For example, dislocations tend to organize into mesoscale dislocation walls and cells with scaling laws that extend over many orders of magnitude of deformation. Measurements of single-crystal surface roughening have identified self affine (fractal) behavior in point-to-point surface roughening of single crystals that extends over decades in length [17]. With polychromatic microdiffraction, we can now measure misorientation and the development of 3D deformation structure to test theoretical predictions of deformation theories and to understand how bulk deformation compares to the surface behavior. Pang [18] has tested misorientation *both* at the surface and in the bulk to determine whether the surface observations are indicative of the bulk behavior or are somehow restricted to the near surface region. Her measurements find that the Hurst coefficient is sensitive to both depth and crystallographic orientation/strain load (Figure 20.11).

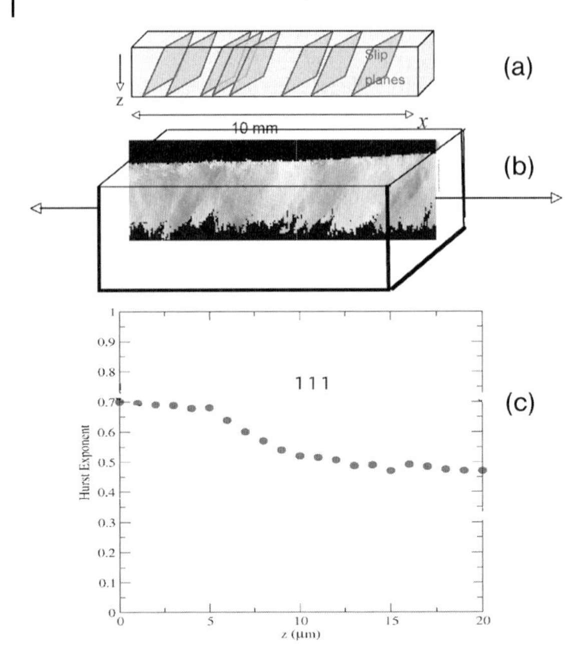


**Figure 20.11** (a) Schematic of single crystal sample used to study deformation structure. Tensile specimens were pulled along various crystallographic orientations. Misorientations in points along the tensile specimen were studied as a function of depth and crystal orientation. A false color map, (b) shows the orientation distribution in an interior surface of the sample. Hurst coefficient as a function of depth, (c) and crystallographic orientation, (d) indicate that both proximity to surface and crystal orientation effect the misorientation distribution of a deformed single crystal.

These studies are certain to provide new information about deformation in single crystals that can be directly compared to finite element and molecular dynamics models.

### 20.4.4 Grain Growth on Surfaces and in Three-Dimensions

The penetrating nature of X-ray microdiffraction allows for precision measurements of thin-film growth, which correlate to the underlying substrate. For example, Budai [19] has explained the transition between "good" and "bad" growth of high-temperature superconducting films on textured Ni substrates by demonstrating how the growth mechanism changes with temperature. Similarly Barabash [20–21] has studied defects in GaN thin films grown on patterned Si and SiC substrates (Figure 20.12). Her results clarify how defects are introduced as a response to thermal and lattice mismatch strains.

More generally, the nondestructive nature of polychromatic microdiffraction enables a fundamentally new approach to the study of grain growth in three di-

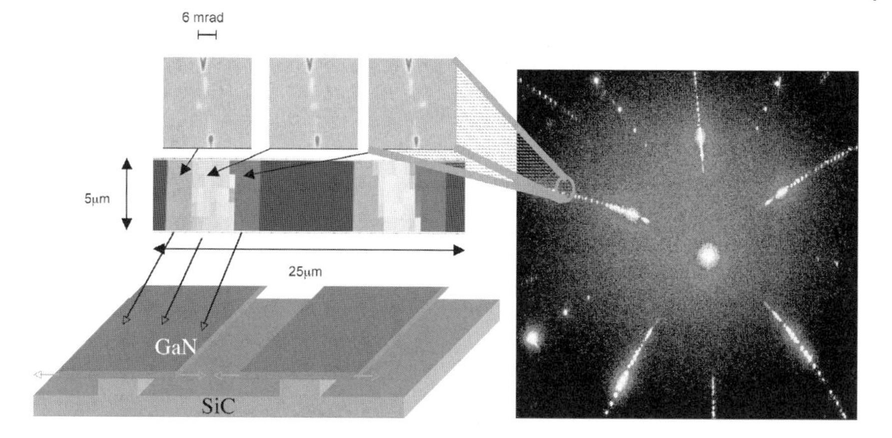


**Figure 20.12** The overall Laue image from this thin GaN film on SiC looks like a single crystal because the film is crystallographically aligned to the SiC substrate. A detailed look at individual reflections find that some are streaked (SiC) due to the finite penetration of the X-ray beam into the substrate. Other reflections change relative position depending on the location of the X-ray beam near pedestals etched into the substrate. These reflections are from the GaN film, which has wings tilted upward on each side of the pedestals.

mensions. Previous studies of grain growth have been limited to 2D growth, have depended on statistical information to understand how grains compete for material during grain growth, or have studied individual grains with little information about the local environment. Budai [22] has recently demonstrated that 3D images of grain structure can be measured and that grain growth can be observed before and after thermal annealing steps. The high-angular precision of the measurements allows for a detailed study of how grain and subgrain boundaries are annealed out. Because the local environment – including the local stress state – can be measured, theories of grain growth can be directly compared to observed local movement of boundaries (Figure 20.13).

### 20.4.5 Anomalous Grain Growth

In addition to normal grain growth, there are very interesting and technologically important cases where anomalous grain growth occurs. For example, Sn films are important as a low toxicity alternative to Pb. However, Sn films can spontaneously grow whiskers that can short circuit electronics. The mechanism behind these phenomena has been the subject of intense speculation and study. Polychromatic microdiffraction allows for detailed measurements of the environment and crystallographic nature of individual Sn whiskers. For example, as shown in Figure 20.14, an orientation map through a Sn film shows individual Sn grains and shows the presence of whiskers extending out from the surface. These whiskers are observed to include a range of crystallographic orientations and to

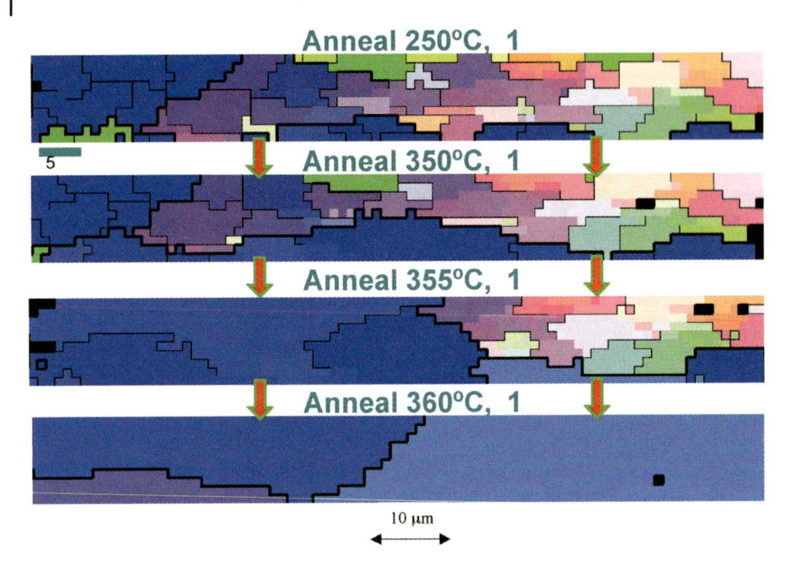


**Figure 20.13** Al grain structure at an interior sample surface before and after thermal annealing steps. The colors indicate crystal orientation and the thickness of the lines indicates the misorientation across the grain boundary; the thicker the line the larger the misorientation.

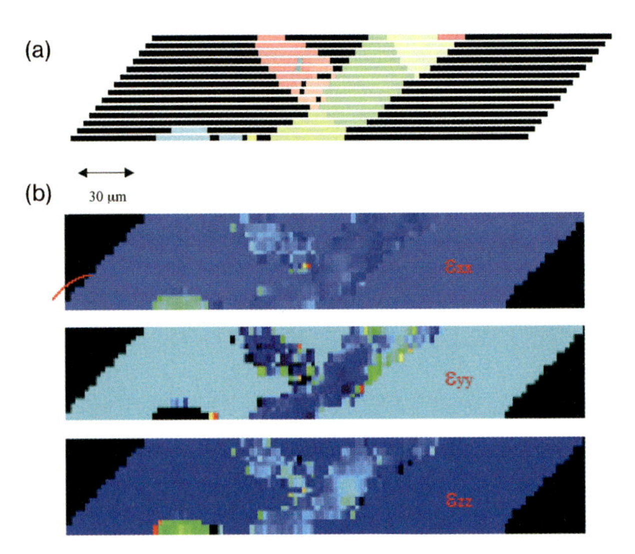


**Figure 20.14** Cross-section through a thin Sn film with a false color representation of grain orientation (a) and with false color representations of the magnitude of the diagonal elastic stress-tensor elements (b). Whiskers sticking out from the film are clearly visible near the lower left and center of the cross-sectional image.

include cases where the whisker is polycrystalline. Furthermore, the elastic strain state in and near the grain can be measured to test theories of grain growth. Results from these experiments are certain to provide essential new information about anomalous grain growth in Sn films and show the potential for addressing anomalous grain growth in a range of materials.

## 20.5 Future Prospects and Opportunities

Because polychromatic microdiffraction is still an emerging new field, research opportunities currently swamp available facilities. Extensive studies of the electromigration response of electronic interconnects have revealed the importance of deformation in the electromigration process [23]. Similarly, measurements of domain wall structures, stresses in implanted materials, fracture, minor phases and precipitates, phase transitions, and interface phases will all contribute to our understanding of materials behavior. Recent measurements to identify tiny microcrystalline grains inside bulk amorphous substrates, demonstrates an ability to test how materials segregate and accommodate metastable equilibrium under rapid annealing [24].

As equipment and techniques improve, the range of possible experiments will also increase. Faster detectors, detectors with energy resolution and smaller beams will all extend experimental possibilities. We envision in the near future, for example, a gain of at least a factor of 100 in data acquisition speed and an improvement of a factor of 10 in spot size (500–50 nm). With these capabilities, studies of nanocrystalline materials and studies of grain boundary morphology and faceting will become routine. In addition, improved control of beam properties will allow for detailed studies of defects in micron-scaled volumes [25]. This will greatly accelerate diffuse measurements of defect properties and will allow point defect concentrations to be associated with mesoscale defect distributions.

## Acknowledgments

This chapter would not have been possible without the efforts of many collaborators. These include, J.-S. Chung, W. Lowe, B.C. Larson, J.D. Budai, J.Z. Tischler, W. Liu, W. Yang, J.W.L. Pang, F. J. Walker, A. Khounsary, L. Assoufid, A. Macrander, and R.I. Barabash. Work sponsored by the U.S. Department of Energy, Division (DOE), Division of Materials Sciences and Engineering. Oak Ridge National Laboratory (ORNL) is operated by UT-Battelle, LLC, for the U.S. Department of Energy under contract DE-AC05-00OR22725. Experiments were performed on Unicat beamline 34-ID at the Advanced Photon Source Argonne Il. Both 34-ID and the APS are supported by the DOE Office of Basic Energy Science.

The submitted manuscript has been authored by a contractor of the U.S. Government under Contract No. DE-AC05-00OR22725. Accordingly, the U.S.

## References


1 R. Barabash, G. E. Ice, B. C. Larson, G. M. Pharr, K.-S. Chung, W. Yang, *Appl. Phys. Lett.* 2001, **79**, 749–751.

2 W. G. Yang, B. C. Larson, G. E. Ice, J. Z. Tischler, J. D. Budai, K. S. Chung, W. Lowe, *Appl. Phys. Lett.* 2003, **82**, 3856–3858.

3 R. I. Barabash, G. E. Ice, J. W. L. Pang, *Mat. Sci. Eng. A* 2005, **400**, 125–131.

4 R. I. Barabash, G. E. Ice, F. J. Walker, *J. Appl. Phys.* 2003, **93**, 1457–1464.

5 B. C. Larson, W. Yang, G. E. Ice, J. D. Budai, J. Z. Tischler, *Nature* 2002, **415**, 887–891.

6 J. S. Chung, G. E. Ice, *J. Appl. Phys.* 1999, **86**, 5249–5256.

7 M. von Laue, "*Nobel Lecture*", 1915, http://nobelprize.org/physics/laureates/1914/laue-lecture.pdf

8 W. Liu, G. E. Ice, B.C. Larson, W. Yang, J. Tischler, *Met. Mat. Trans. A* 2004, **35A**, 1963–1967.

9 G. E. Ice, J.-S. Chung, W. Lowe, E. Williams, J. Edelman, *Rev. Sci. Inst.* 2000, **71**, 2001–2006.

10 C. Liu, L. Assoufid, R. Conley, A. T. Macrander, G. E. Ice, J. Z. Tischler, *Opt. Eng.* 2003, **42**, 3622–3628.

11 W. Liu, G. E. Ice, J. Z. Tischler, A. Khounsary, C. Liu, L. Assoufid, A. T. Macrander, *Rev. Sci. Inst.* 2005, **76**, 113701(1–6).

12 K. Yamauchi, K. Yamamura, H. Mimura, Y. Sano, A. Saito, A. Souvorov, M. Yabashi, K. Tamasaku, T. Ishikawa, Y. Mori, *J. Synch. Rad.* 2002, **9**, 313–316.

13 N. Tamura, A.A. Macdowell, R. Spolenak, B.C. Valek, J.C. Bravman, W.L. Brown, R.S. Celestre, H.A. Padmore, B.W. Batterman, J.R. Patel, *J. Synch. Rad.* 2003, **10**, 137–143.

14 W. Bollman, *Crystal Lattices, Interfaces, Matrices: An Extension of Crystallography*, 1982, published by the author.

15 J. W. L. Pang, G. E. Ice, W. Liu, Inhomogeneous deformation behavior in intercrystalline regions in polycrystalline Ni, accepted Sept. 2007 by *Acta Mater.*

16 J. W. L. Pang, R. Barabash, W. Liu, G. E. Ice, *TMS Lett.* 2004, **1**, 5–6.

17 M. Zaiser, F. M. Grasset, V. Koutsos, E. C. Aifantis, *Phys. Rev. Lett.* 2004, **93**, 195507(1–4).

18 J. W. L. Pang, W. Lin, and G. E. Ice, Self-organized fractal behavior of plastically deformed Cu, *Plasticity Conf.*, Halifax, Canada, 2006.

19 J. D. Budai, W. G. Yang, N. Tamura, J. S. Chung, J. Z. Tischler, B. C. Larson, G. E. Ice, C. Park, D. P. Norton, *Nature Mater.* 2003, **2**, 487–492.

20 R. I. Barabash, O. M. Barabash, G. E. Ice, C. Roder, S. Figge, S. Einfeldt, D. Hommel, T. M. Katona, J. S. Speck, S. P. DenBaars, R. F. Davis, *Physica Status Solidi (a)* 2006, **203**, 142–148.

21 R. I. Barabash, G. E. Ice, W. Liu, S. Einfeldt, A. M. Roskovski, R. F. Davis, *J. Appl. Phys.* 2005, **97**, 013504(1–5).

22 J. D. Budai, *Mat. Sci. Forum* 2004, **467**, 1373–1378.

23 R. I. Barabash, G. E. Ice, N. Tamura, B. C. Valek, J. C. Bravman, R. Spolenak, J. R. Patel, *J. Appl. Phys.* 2003, **93**, 5701–5706.

24 E. Miura, G. E. Ice, E. D. Specht, J. W. L. Pang, H. Kato, K. Hisatsune, A. Inoue, *X-ray Study Of Pd40Cu30Ni10P20 Bulk Metallic Glass Brazing Filler For Ti-6Al-7Nb Alloy, Proceedings THERMEC2006 (International conference on processing & manufacturing of advanced materials)*, July 4–8, 2006, Vancouver, BC, Canada.

25 G. E. Ice, R. I. Barabash, W. Liu, Diffuse X-ray scattering from tiny sample volumes, *Z. Kristallogr.* 2005, **220**, 1076–1081.

# 21
# Quantitative Analysis of Three-Dimensional Plastic Strain Fields Using Markers and X-ray Absorption Tomography

*Kristoffer Haldrup and John A. Wert*†

## 21.1
## Introduction

Measuring the spatial distribution of strain during plastic deformation is of considerable interest in many areas. From some of the early quantitative deformation experiments [1, 2] to the latest investigations involving micrometer and submicrometer [3–5] scale resolution, the standard technique has been decorating a surface of a work piece with features from whose relative displacement the plastic strain can be inferred. While these and related surface methods have contributed significantly to understand the factors that influence plastic flow, the inherent lack of information from the third dimension has been a drawback.

In recent years, broader availability of intense, synchrotron-based X-ray sources able to penetrate millimeter-sized metallic samples has made available new experimental options based on both diffraction and absorption. Tomography, in which the three-dimensional (3D) structure of a specimen is reconstructed based on a large number of X-ray absorption or phase-contrast images, is one such method allowing researchers to probe the bulk of materials [6].

For the measurements of plastic strain, the basic idea [7] is to track the motion of markers in the bulk of a specimen. The markers can be either intentionally introduced or indigenous to the specimen, but they must have an X-ray absorption coefficient significantly different from that of the matrix. From absorption (or phase-contrast) images, the full 3D volume – the tomogram – can be reconstructed allowing identification of individual marker particles. By obtaining tomograms before and after a given type of externally imposed deformation, the 3D displacement field can be reconstructed, from which the 3D displacement gradient and strain fields can be deduced. The method does not in any way depend on the crystallinity of the sample and can thus be used for crystalline as well as amorphous materials. Additionally, the method is applicable up to high strains.

*Neutrons and Synchrotron Radiation in Engineering Materials Science: From Fundamentals to Material and Component Characterization*
Edited by Walter Reimers, Anke Rita Pyzalla, Andreas Schreyer, and Helmut Clemens

ISBN: 978-3-527-31533-8

## 21.2
## Experimental Approach

In order to track the displacement field in the interior of opaque specimens, research groups have primarily pursued two approaches: direct tracking of artificial markers and direct as well as indirect use of indigenous markers.

### 21.2.1
### Markers

Focusing first on samples with intentionally introduced markers, a technique has been developed whereby markers are introduced into an aluminum alloy. This is accomplished by adding W particles with a size of ~10 μm to an aluminum powder, allowing solid aluminum billets to be created by cold-compaction, hot-pressing, and extrusion of the mixed powder. This yields billets of solid aluminum with embedded W markers. Nonmetallic specimens have also been investigated in which glass particles are embedded in an epoxy matrix, and many other matrix–marker systems can be envisioned.

Turning next to the use of indigenous markers, intermetallic particles [8] in a cast aluminum alloy as well as micropores [9] have been used to track crack opening displacements and the associated crack driving forces. Recently, this research has been complemented by work on Al foams subjected to compression loading [10]. Related work has used techniques based on digital image correlation (DIC) of microstructural features to indirectly track the 3D deformation processes in other types of Al foam [11], in bone [12] and in metal powder undergoing compaction [13]. This chapter focuses on the use artificial markers and direct tracking.

### 21.2.2
### Particle Tracking and DGT Calculation

For any marker type, the first step in the data analysis is the segmentation of tomograms into particle and matrix regions. While very advanced techniques exist for this, it has been found sufficient in most cases to use a simple procedure with a single threshold value for each tomogram. Based on these thresholds, image analysis software then identifies and labels particles in all the tomograms and characteristics such as position, size and moments of inertia for each particle are registered. This allows the identification and tracking of individual particles between deformation steps from which the displacement vector, $\vec{U} = (u_1, u_2, u_3)$ can be found for each particle identified in the data sets. These combined make up the displacement field

$$\vec{U}(\vec{x}) = (u_1(\vec{x}), u_2(\vec{x}), u_3(\vec{x})), \quad \vec{x} = (x_1, x_2, x_3)$$

for a given deformation step.

While the visualization of displacement fields gives compelling evidence for (macroscale) heterogeneities in the plastic flow caused by deformation, understanding and quantifying the underlying deformation processes require that the displacement gradient field be calculated as this is directly related to the local stress–strain response of the specimen. The displacement gradient field is defined by the local displacement gradient tensor (DGT)

$$e_{ij} = \frac{\partial u_i}{\partial x_j} = \begin{pmatrix} \frac{\partial u_1}{\partial x_1} & \frac{\partial u_2}{\partial x_1} & \frac{\partial u_3}{\partial x_1} \\ \frac{\partial u_1}{\partial x_2} & \frac{\partial u_2}{\partial x_2} & \frac{\partial u_3}{\partial x_2} \\ \frac{\partial u_1}{\partial x_3} & \frac{\partial u_2}{\partial x_3} & \frac{\partial u_3}{\partial x_3} \end{pmatrix}$$

This measure contains all the information about the local deformation field, but it is also a challenge to visualize, as there are nine quantities tied to each measurement point in 3D space.

### 21.2.3 Spatial Resolution

The spatial resolution for marker-based strain measurements is directly related to the density of marker particles. For a 1 vol% addition of 10 μm W particles to an Al matrix as described here, this gives a spatial resolution on the order of 25 μm. For the DIC-based techniques, ultimate spatial resolution depends on the resolution of the tomograms and results have been reported with a resolution of 2 μm [11].

## 21.3 Results of Investigations

### 21.3.1 Homogeneous Deformation

To validate the method, compression experiments were performed on cylindrical specimens machined from the as-extruded material which can be considered homogenous on the scale of the specimen (1 mm) as the largest feature sizes of the material are the Al-powder particle cross section and the W markers, both on the order of 10 μm. Detailed studies [7] of the central part of the specimen placed bounds on the resolution of the method as well as on the strain inhomogeneities of the active deformation processes, with both being less than 0.02 (2% strain) at a strain level of 0.1.

An important issue regarding this method for assessing the strain throughout opaque specimens is whether the introduction of marker particles affects the mechanical properties of the material. To investigate this, a later study [14] of the data set for homogeneous compression took advantage of the fact that a small volume near the edge of the specimen contained no markers. From the absence of any variations in strain near this region compared to the global response it was concluded that the effect of adding markers was minimal.

## 21.3.2
## Heterogenous Deformation

Having confirmed the feasibility of tracking marker particles by X-ray microtomography, a second experiment [15] focused on detecting and quantifying local

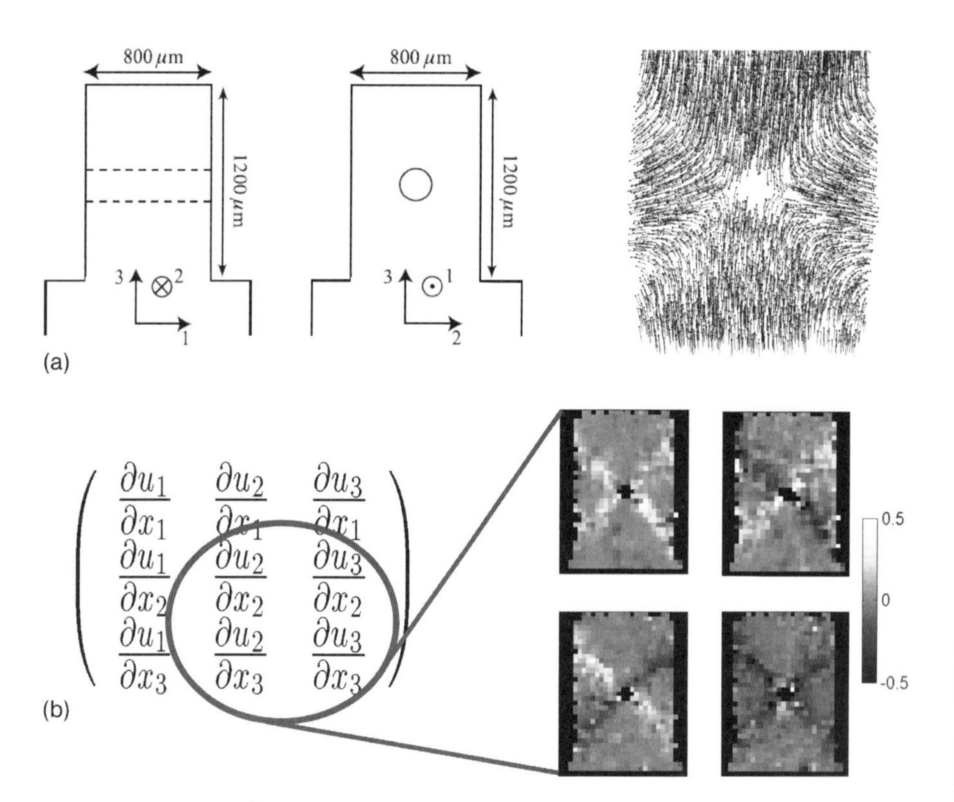


**Figure 21.1** (a) The geometry of the specimen, designed to promote heterogeneous deformation and the associated displacement vectors for 6000 particles after 13% deformation. (b) Diagram showing the four active components of the DGT for this experiment and the magnitude of these four component, color coded and averaged along the hole axis.

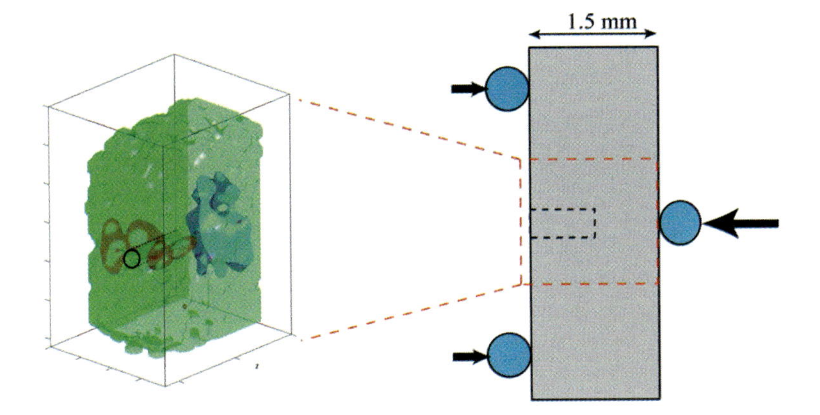


**Figure 21.2** Three-dimensional isosurfaces for $e_{33}$, with the three-direction being along the long axis of a bending specimen with a blind hole. Red colored isosurfaces correspond to elongation while blue is compressive strain. Light green outlines the volume considered for the analysis.

variation in the DGT field by studying a rectangular specimen with a central hole, a structure known from standard deformation mechanics to promote inhomogeneous plastic flow.

Analysis of the displacement field showed an x-shaped flow pattern and quantitative analysis of the DGT field revealed that the x-shaped flow pattern corresponded to two shear planes intersecting along the axis of the hole as shown in Figure 21.1. This in turn allowed the deformation response of the sample to be understood in terms of a mechanical model based on work minimization.

Analysis work is currently in progress to investigate more complicated geometries beyond the scope of macroscale mechanical analysis, i.e., geometries usually investigated only by FE modeling and global stress–strain response. These include a bending specimen with a blind hole as shown in Figure 21.2.

### 21.3.3 Microstructural Effects

Another avenue of research utilizing the technique presented here focuses on the effects of internal properties and structure as compared to the geometry-induced effects presented above. One important class of materials where internal structure plays a crucial role are fiber re-enforced composites where stress–strain concentrations around the re-enforcing phase may lead to debonding, fiber fracture and ultimately failure of the entire component. Figure 21.3 shows two examples of such composites for which data is currently being analyzed, an aluminum compression sample re-enforced with W fibers and a glass-fiber + epoxy compos-

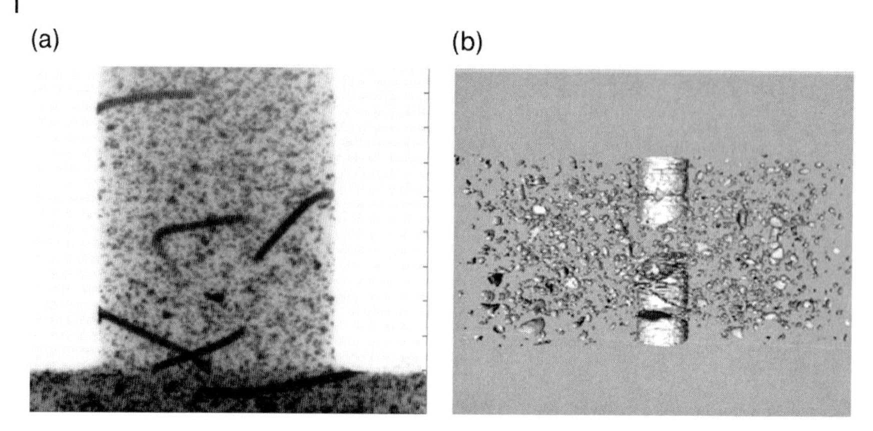


**Figure 21.3** (a) An absorption image of an Al sample containing W wires and marker particles. (b) Reconstructed volume of a glass fiber (light grey)-epoxy(transparent) composite with glass marker particles. The tomogram shown is after straining and fiber breaks are evident. Data for 'B' obtained at the Swiss Light Source.

ite. For the glass-fiber + epoxy sample, both the spatial arrangement of the fiber breaks and the strain field surrounding them is of interest.

Another class of problems where information about the strain distributions related to internal structure is of significant interest on a basic scientific level is in the interaction of crystallographic grains; this work is in progress [16].

The tomographic measurements underlying the results presented in this section have all been performed in collaboration with Dr. F. Beckmann and coworkers at the HASYLAB synchrotron facility in Hamburg, Germany. An exception is the data on fiber re-enforced epoxy, where the measurements were performed with Dr. M. Stampanoni at the Swiss Light Source (SLS) in Villigen, Switzerland.

## 21.4 Outlook

A method has been developed by which plastic strain distributions can be measured throughout the bulk of opaque samples. The method is expected to find application in areas such as deformation and fracture mechanics, where advanced modeling studies have been hard to verify experimentally. For crystalline samples, developments in 3D diffraction techniques [17] in combination with this or similar techniques allow for the first time concurrent, locally resolved measurements of stress, strain, and crystallographic orientation throughout the bulk of specimens. However, as the method is broadly applicable and not limited to the systems presented here several new applications may yet be discovered.

## References


1 G. Sachs, W. Eisbein, Kraftbedarf und Fließvorgänge beim Stangenpressen, *Deutsche Materialprüfungsanstalten Mitteilungen* 1931, **16**, 67–96.

2 M. L. Devenpeck, O. Richmond, Strip-drawing experiments with a sigmoidal die profile, *J. Eng. Ind.* 1965, **87**, 425–428.

3 H.-A. Crostack, G. Fischer, E. Soppa, S. Schmauder, Y.-L. Liu, Localization of strain in metal-matrix composites studied by a scanning electron microscope-based grating method, *J. Microscopy* 2000, **201**, 171–178.

4 D. Raabe, M. Sachtleber, Z. Zhao, F. Roters, S. Zaefferer, Micromechanical and macromechanical effects in grain scale polycrystal plasticity experimentation and simulation, *Acta Materialia* 2001, **49**, 3433–3441.

5 B. M. Schroeter, D. L. McDowell, Measurement of deformation fields in polycrystalline ofhc copper, *Inter. J. Plasticity* 2003, **19**, 1355–1376.

6 L. Salvo, P. Cloetens, E. Maire, S. Zabler, J. J. Blandin, J.-Y. Buffere, W. Ludwig, E. Boller, D. Bellet, C. Josserond, X-ray micro-tomography an attractive characterisation technique in materials science, *Nucl. Instrum. Methods Phys Res B* 2003, **200**, 273–286.

7 S. Nielsen, H. Poulsen, F. Beckmann, C. Thorning, J. Wert, Measurements of plastic displacement gradient components in three dimensions using marker particles and synchrotron X-ray absorption micro to-mography, *Acta Materialia* 2003, **51**, 2407–2415.

8 H. Toda, I. Sinclair, J.-Y. Buffere, E. Maire, T. Connolley, M. Joyce, K. H. Khor, P. Gregson, Assessment of the fatigue crack closure phenomenon in damage-tolerant aluminium alloy by in-situ high-resolution synchrotron X-ray microtomography, *Philosophical Magazine* 2003, **83**, 2429–2448.

9 H. Toda, I. Sinclair, J.-Y. Buffere, E. Maire, K. Khor, P. Gregson, T. Kobayashi, A 3D measurement procedure for internal local crack driving forces via synchrotron X-ray radiation, *Acta Materialia* 2004, **52**, 1305–1317.

10 T. Ohgaki, H. Toda, M. Kobayashi, K. Uesugi, M. Niinomi, T. Akahori, T. Kobayash, K. Makii, Y. Aruga, *In situ* observations of compressive behaviour of aluminium foams by local tomography using high-resolution X-rays, *Philosophical Magazine* 2006, **86**, 4417–4438.

11 E. Verhulp, B. van Rietbergen, R. Huiskes, A three-dimensional image correlation technique for strain measurements in microstructures, *J. Biomech.* 2004, **37**, 1313–1320.

12 B. K. Bay, T. Smith, D. Fyhrie, M. Saad, Digital volume correlation: Three-dimensional strain mapping using X-ray tomography, *Exp. Mech.* 1999, **39**, 217–226.

13 S. A. McDonald, L. C. R. Schneider, A. Cocks, P. J. Withers, Particle movement during the deep penetration of a granular material studied by X-ray microtomography, *Scripta Materialia* 2006, **54**, 191–196.

14 K. Haldrup, S. F. Nielsen, F. Beckmann, J. A. Wert, Plastic strain measurements: From 2D to 3D, *Mater. Sci. Technol.* 2005, **21**, 1428–1431.

15 K. Haldrup, S. F. Nielsen, F. Beckmann, J. A. Wert, Inhomogeneous plastic flow investigated by X-ray absorption microtomography of an aluminium alloy containing marker particles, *J. Microscopy* 2006, **222**, 28–35.

16 K. Haldrup, S. F. Nielsen, F. Beckmann, J. A. Wert, Experimental determination of strain partitioning among individual grains in the bulk of an aluminium multicrystal, *Materials Characterization*, 2007, in press.

17 G. Winther, L. Margulies, S. Schmidt, H. Poulsen, Lattice rotations of individual bulk grains part ii: correlation with initial orientation and model comparison, *Acta Materialia* 2004, **44**, 2863–2872.

# 22
# Combined Diffraction and Tomography

*Anke Rita Pyzalla and Augusta Isaac*

## 22.1
## Introduction

The necessity for obtaining information about microstructure, texture, residual stresses, or damage evolution in three dimensions in the bulk of samples and components has been outlined in Chapters 1 to 3. In many cases microstructure, texture, strain/stress field, and damage evolution are not independent of each other. For instance, the residual stress fields in a material will change in the vicinity of a growing crack. In metal matrix composites (MMCs) particle delamination or particle cracking due to external loading will influence the stress field within the metal matrix. Therefore, subsequent tomography and diffraction experiments have been performed, e.g., in order to follow fatigue damage in fiber-reinforced TiAl6V4 samples [1].

For investigating the evolution of the (residual) strain/stress field and microstructure changes, high-energy synchrotron radiation provides the unique advantage of being non-destructive, providing sufficient penetration depth to allow access to the bulk of samples and small components, and of allowing for resolutions in the order of at least some micrometers in tomography.

In principle, both diffraction and tomography can be performed using monochromatic or white synchrotron X-rays (see Chapters 7, 8, 10, and 16). Due to the higher photon flux compared to monochromatic high-energy X-rays the white beam allows substantially faster data acquisition. However, the resolution at the currently available beamlines is at best 1.5 μm × 1.5 μm × 1.5 μm, and thus an order of magnitude lower than the resolution available in standard experiments using monochromatic radiation, i.e., 0.3 μm × 0.3 μm × 0.3 μm. Tomography experiments using white high-energy synchrotron radiation thus appear advantageous for following the kinetics of processes such as crack growth, sintering processes, or creep damage.

While both diffraction and tomography experiments are well-established techniques, only recently both methods were combined within one experiment [2]

*Neutrons and Synchrotron Radiation in Engineering Materials Science: From Fundamentals to Material and Component Characterization*
Edited by Walter Reimers, Anke Rita Pyzalla, Andreas Schreyer, and Helmut Clemens

ISBN: 978-3-527-31533-8

due to the challenges in choosing an effective compromise, e.g., in the beam properties, in the alignment of the increased number of components of the experimental set-up with the high precision required in using different detection systems simultaneously.

## 22.2
## Experimental Set-up

The experimental set-up for a combined diffraction – tomography – experiment (Figure 22.1) necessitates the use of at least two detection systems, a tomography camera (e.g., a CCD) and a detector for the diffraction part, e.g., a scintillator or a CCD for recording the monochromatic or a Ge semiconductor for recording the energy dispersive diffractogram.

For defining the beam size, e.g., slits can be used. The scattering background can be reduced by limiting the beam by slits in the optical hutch, employing evacuated collimator tubes (e.g., brass tubes).

The optimum wiggler spectrum depends on the absorption coefficient and thickness of the sample. In order to avoid beam-hardening artifacts in the tomography reconstructions, absorbers may be necessary in the primary beam.

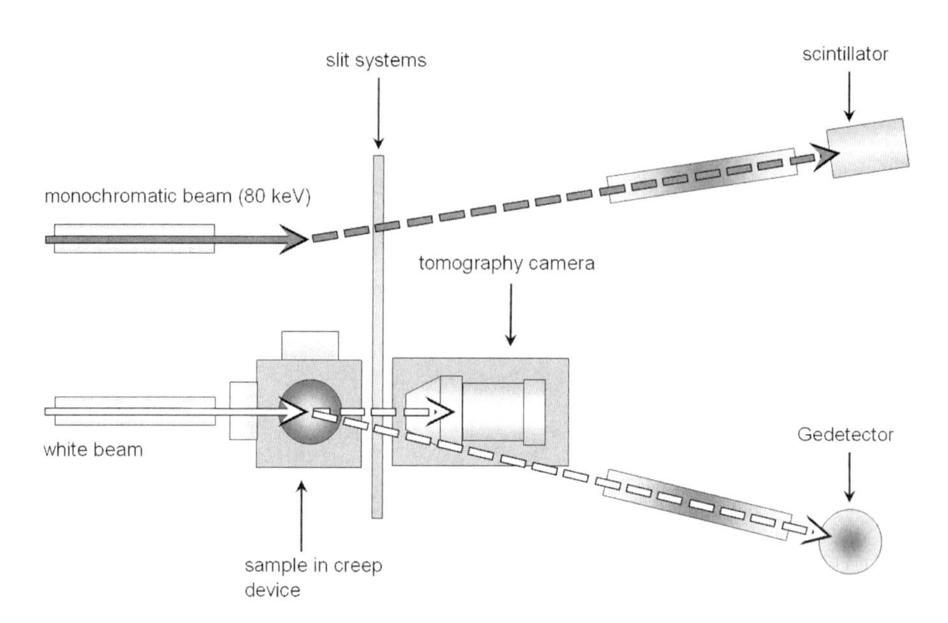


**Figure 22.1** Experimental set-up of a combined diffraction/tomography experiment using both the white and the monochromatic beam for the diffraction part and the white beam for tomography.

## 22.3 Example: Combined Diffraction and Tomography for Investigating Creep Damage Evolution

### 22.3.1 Scientific Background

The efficiency of electricity-generating plants and gas turbines depends strongly on their components' sustainability of loading at high temperatures. The service lifetime of these components in most cases is controlled by creep-induced cavity growth. In many technical applications such as stationary power plants and aero-engines knowledge about the remnant lifetime of creep loaded components is crucial for a safe and cost-effective operation.

So far creep damage could only be determined from two-dimensional microscopy images of the microstructure. Since metallography necessitates cutting the sample, only a snapshot of the damage evolution was available. Compared to microscopy, synchrotron X-ray tomography has the advantages of providing data of damage in the bulk, which often differs substantially from damage at the surface [3], and also of revealing the chronology of damage events and their location within the sample.

### 22.3.2 Experimental Details

#### 22.3.2.1 Miniature Creep Device

In order to investigate the creep damage evolution *in situ* by both synchrotron X-ray diffraction and tomography, a miniaturized creep device (Figure 22.2) was developed at TU Berlin.

Loading is exerted by a spring, which was adjusted to provide a tensile stress of 25 MPa in the sample cross-section. The sample was heated to $375 \pm 2$ °C by an induction-heated loop around the bottom of the sample and the temperature was controlled by a thermocouple. The temperature variation along the active length of the sample is about $\pm 10$ K (dependent on the sample material).

#### 22.3.2.2 Tomography and Diffraction

The experiment was carried out at beamline ID15A of the ESRF [4]. The experimental set-up consisted of three detection systems, which allow to sequentially perform tomography, energy-dispersive diffraction and angle-dispersive diffraction without further alignment or calibration procedures during the experiment. Tungsten carbide slits defined the gauge volume within the sample.

A Si monochromator provided 80 keV radiation with 50 mm offset to the white beam. For fast-tomography and energy-dispersive diffraction a white beam was used which penetrated the monochromator crystal. In order to reduce coarse grain effects, the sample was rotated continuously.

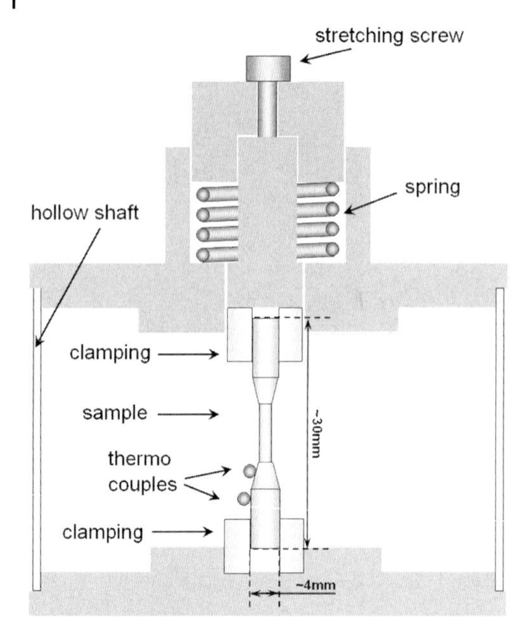


**Figure 22.2** Miniaturized creep device developed at TU Berlin, Germany.

The tomograms were collected using a fast CCD camera, each taking about 3 min, which allows to study the dynamics of the creep process. Data processing was performed by image preprocessing and tomographic reconstruction.

#### 22.3.2.3 Material

As model material, the brass alloy CuZn40Pb2 (~58 mass% Cu, 40 mass% Zn, and 2 mass% Pb) was chosen, which contains three phases: $\alpha$-brass, $\beta$-brass, and Pb. The Pb shows a strong absorption contrast to both brass phases; thus identifying the location of certain Pb particles within axial slices of the sample, we could follow the damage evolution could be followed in a defined sub-volume of the sample with increasing creep time. In order to provide creep conditions as realistic as possible during the limited beamtime available, the sample was tensile deformed (4.47% in total) prior to the creep experiment according to the results given in [5].

### 22.3.3 Results

In the initial state, after 4.47% prestraining – but before the creep test – the hot extruded samples contain small mostly spherical or ellipsoidal shaped voids. The longitudinal axis of the voids is oriented in the longitudinal direction of the samples [2]. During creep, voids nucleate and their number and size increase with increasing creep time (Figure 22.3), see [2] for more quantitative results.

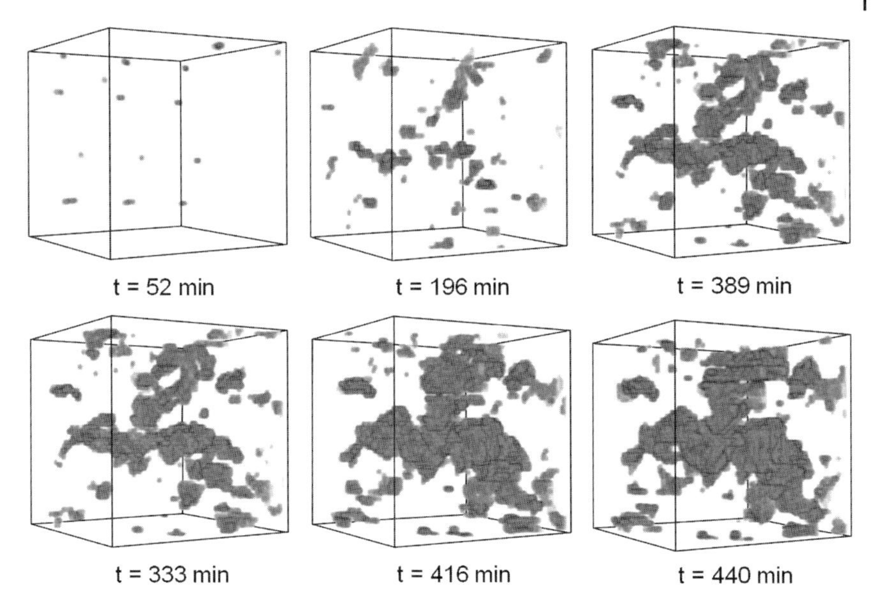


**Figure 22.3** Void evolution (black) during creep of CuZn40Pb2, $T = 375\ ^\circ\mathrm{C}$, $\sigma = 25$ MPa. Size of the sub-volume shown is 80 μm × 80 μm × 80 μm. Spatial resolution: 2.0 μm × 2.0 μm × 2.0 μm [2].

The morphology of the voids changes from spherical, ellipsoid, or rod-like to irregular shapes (Figure 22.4). The change in morphology is due to void growth and void coalescence at grain boundaries which mostly occurs perpendicular to the load axis. In order to unambiguously define the orientation of the voids with

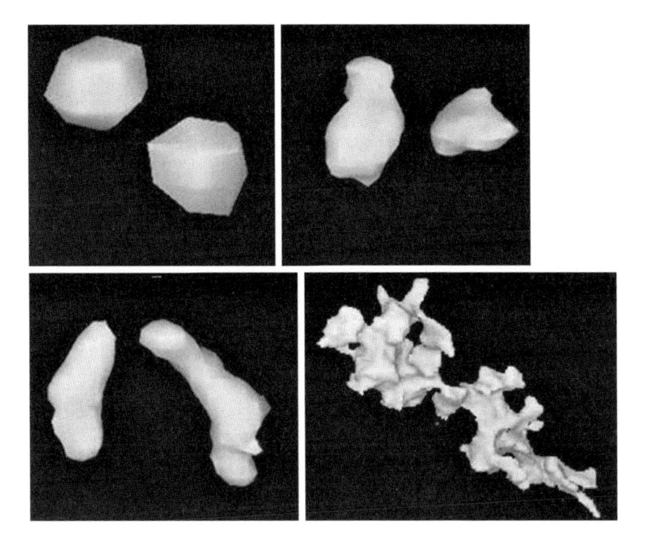

**Figure 22.4** Typical void shapes: sphere-like, ellipsoid-like, rod-like, and irregular.

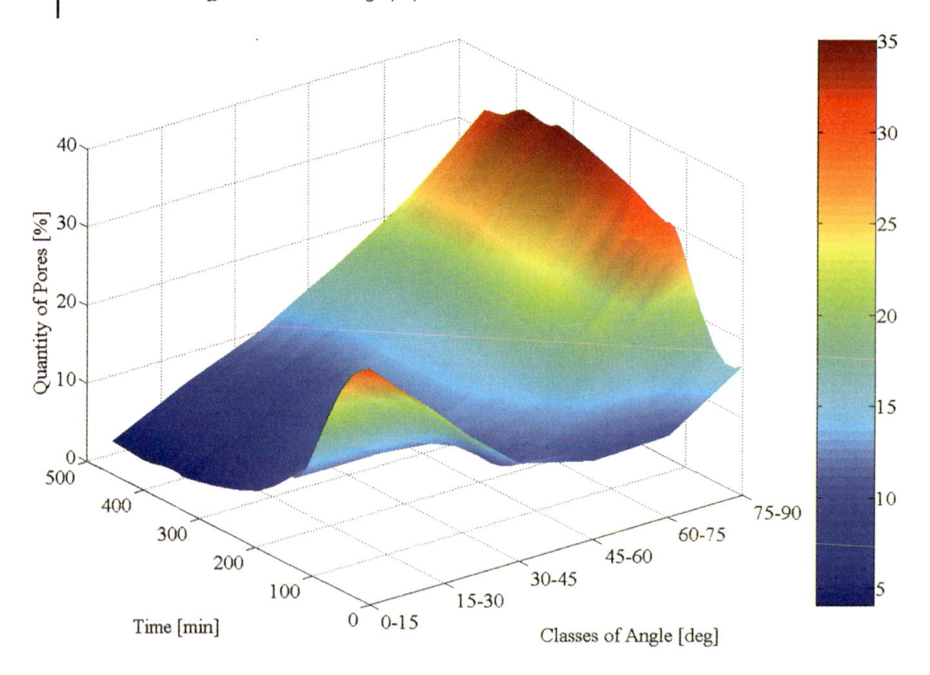


**Figure 22.5** Orientation of ellipsoid-like, rod-like, and irregular voids with respect to sample longitudinal axis (0°: parallel to sample longitudinal axis, 90°: perpendicular to sample longitudinal axis).

respect to the sample longitudinal axis, the voids were represented by equivalent ellipsoids of equal volume and moments of inertia.

The orientations of the longest axis of these ellipsoids with respect to the sample longitudinal axis change from parallel toward perpendicular to the longitudinal axis during the creep test (Figure 22.5). This is presumably due to voids

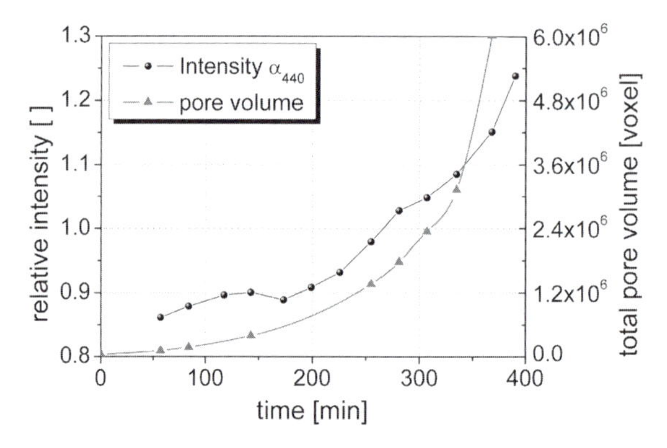


**Figure 22.6** Texture and pore volume development during creep of CuZn40Pb2, $T = 400$ °X, $\sigma = 25$ MPa.

growth and coalescence along the grain boundaries in the later stage of the creep test.

Diffraction revealed the formation of texture complying with the typical tensile test ⟨1 1 1⟩/⟨1 0 0⟩-fiber texture of the fcc $\alpha$-phase of the brass after about 296-min creep time. This fiber texture development presumably is due to the local load increase and due to the strong decrease in supporting cross-section when large voids are formed (Figure 22.6).

## 22.4
## Conclusions and Outlook

Combined tomography and diffraction experiments using either white or monochromatic synchrotron radiation – or even both – during one experiment can give simultaneous access to microstructure, texture, strain/stress field, and damage evolution. The experiments are nondestructive, thus allowing studies of defined sample sub-volumes, and they reveal bulk properties. Using white high-energy synchrotron radiation the speed of data acquisition can be significantly increased thus allowing the investigation of process kinetics.

By exchanging the scintillation counter in the diffraction with the monochromatic beam by a 2D Detector, the determination of texture development can be significantly improved. By moving the slits in front of the sample and behind the sample stress analyses with spatial resolution in future will allow for coupling information about local stresses and the occurrence and development of failure under thermal and/or mechanical loads. Currently, a new instrument DITO at HARWI of GKSS at DESY is developed, which is optimized for simultaneous diffraction and tomography [6]. Methods and optics developments are underway for optimizing simultaneous diffraction and tomography in bulk samples with respect to high local resolution and speed.

## Acknowledgments

The authors acknowledge discussions with Prof. Dr. W. Reimers, TU Berlin and Prof. Dr. G. Sauthoff, MPI für Eisenforschung. We also acknowledge the support of Ms. Dipl.-Ing. B. Camin and Ms. A. Pernack, TU Berlin, Germany. Dr. T. Buslaps and Dr. M. di Michiel, both ESRF Grenoble, France, are also acknowledged for performing the combined diffraction/tomography experiment and the support of Mr. Dipl.-Ing. A. Kottar, TU Wien, in software development for creep damage quantification is also acknowledged. The authors thank Dr. H.M. Mayer for the tensile deformation of the samples prior to the experiment, Mr. Dipl.-Ing. B. Breitbach, MPIE, for drawing the figures of this chapter, D. Fernandez-Carreiras of ESRF for the imaging detector control system and the ESRF for beamtime at ID15A and financial support of traveling expenses. A.P. also gratefully acknowledges the Deutsche Forschungsgemeinschaft (DFG) for financial support of the Project Py9/1-2.

## References

1 S.A. McDonald, M. Preuss, E. Maire, J.-Y. Buffiere, P.M. Mummery, P.J. Withers, *J. Microscopy* 2003, **209** (2), 102–112.

2 A. Pyzalla, B. Camin, T. Buslaps, M. di Michiel, H. Kaminski, A. Kottar, A. Pernack, W. Reimers, *Science* 2005, **308** (5718), 92–95.

3 J.-Y. Buffière, E. Maire, P. Cloetens, G. Lormand, R. Fougères, *Acta Mat.* 1999, **47** (5), 1613–1625.

4 Information about the ESRF ID15A high-energy scattering beamline can be found at www.esrf.fr/UsersAndScience/Experiments/MaterialsScience/ID15.

5 M.R. Willis, J.P. Jones, *Scripta Materialia* 2001, **44** (1), 31–36.

6 http://www.gkss.de/pages.php?page=w_abt_genesys_harwi.htmllanguage=d&version=g (23.10.2006).

# Part IV
# Industrial Applications

# 23
# Diffraction-Based Residual Stress Analysis Applied to Problems in the Aircraft Industry

*Peter Staron, Funda S. Bayraktar, Mustafa Koçak, Andreas Schreyer, Ulrike Cihak, Helmut Clemens, and Martin Stockinger*

## 23.1
## Motivation

Apart from safety aspects, cost reduction and weight saving are most important principles governing design and construction of aircrafts. Today, also reduction of $CO_2$ emission is an important topic. For meeting the various demands, advanced alloys and thermomechanical treatments as well as new and optimized production processes are being developed. Materials and processes are tailored for specific purposes within an aircraft; desired properties concern, for example, formability, age hardenability, weldability, corrosion resistance, or high-temperature properties. For example, welding of fuselage components like stringers or clips made of Al alloys can have several advantages over riveting: weight reduction, cost reduction, and reduction of corrosion problems. Retention of optimum weld microstructure and properties as well as control of welding-related residual stresses and distortion is essential.

However, new materials and techniques only slowly enter aircraft-related production processes because safety issues play a dominant role. Estimates of risks are normally done in a most conservative way. Therefore, the concept of damage tolerance plays an important role, which means, for example, that the question is not whether there is a crack in an aircraft fuselage (because most likely there are some), but how fast does it grow and how can it be stopped. In this context, it will be of growing importance to study residual stress fields in weld configurations and their influence on fatigue crack propagation. Laser beam welding (LBW) and the recently developed solid state joining technique known as friction stir welding (FSW) are two most promising techniques for joining of advanced airframe Al alloys to fabricate weight and cost-efficient novel metallic structural components. Being a solid state joining process, FSW is capable to join even those conventional Al alloys (e.g., 2024) that are usually not weldable with fusion welding techniques. The LBW technique has proved to be very suitable for a

*Neutrons and Synchrotron Radiation in Engineering Materials Science: From Fundamentals to Material and Component Characterization*
Edited by Walter Reimers, Anke Rita Pyzalla, Andreas Schreyer, and Helmut Clemens

ISBN: 978-3-527-31533-8

number of reasons, for example, low distortion while processing at high speeds (i.e., low heat input and smaller weld deposit). In Section 23.3, two examples are given for the analysis of residual stress distributions in laser beam welds of aircraft Al alloys.

An important tool for optimization of production processes is finite element modeling (FEM). Detailed predictions of three-dimensional residual stress distributions can be obtained and the influence of various parameters on the final stress state can easily be tested using an appropriate model. However, the developed models have to be evaluated by measurements before their results can be fully trusted. The following example, which deals with residual stresses in turbine disks made of a Ni-based superalloy, is concerned with such an evaluation.

## 23.2 Residual Stresses in Turbine Disks

### 23.2.1 Introduction

Nickel-based superalloys are being employed for aerospace components due to their unique combination of high-temperature strength and considerably high-fatigue strength [1, 2]. The so-called "direct age process" has been newly developed as a special thermo-mechanical processing route for the Inconel™ alloy 718 (IN 718) in order to obtain enhanced high-temperature strength in turbine disks used for aero engines (Figure 23.1). Direct aging requires water quenching directly after forging [3, 4], contrary to the conventional heat treatment where forging is followed by slow air cooling [5]. A high number of dislocations are frozen

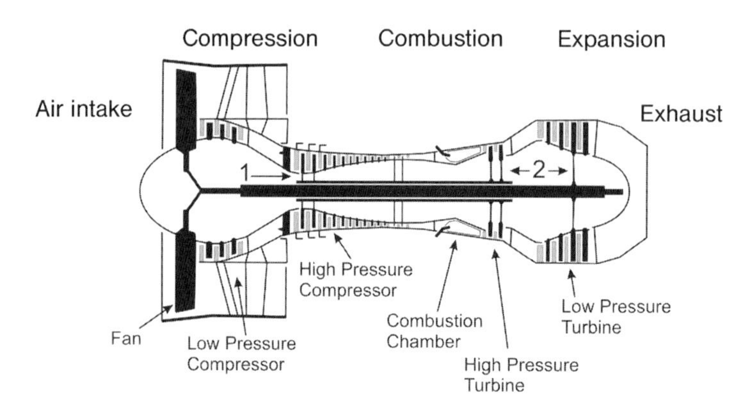


**Figure 23.1** Schematic drawing of a modern jet engine. The working cycle is indicated above, whereas the most important engine parts are named below. (1) and (2) denote the position of compressor and turbine disks, respectively.

in by quenching, acting as potential nucleation sites for the strengthening precipitates that are formed in the subsequent annealing treatment. These precipitates are fine semicoherent $\gamma''$-$Ni_3(Nb, Ti)$ and coherent $\gamma'$-$Ni_3(Al, Ti)$ particles.

However, the high cooling rate connected with water quenching leads to much higher residual stresses than the conventional slow air cooling. Therefore, problems can arise during machining of a disk to its final shape when the residual stresses are not completely relieved during the subsequent annealing treatment, which often seems to be the case. Residual stresses can cause distortion during turning of a disk; in the worst case, the final geometry of the component cannot be machined out of the forging. Thus, the knowledge of the stress gradient through the thickness of the disk is especially important with respect to distortion caused by material removal during machining. If the residual stress state within the disk is known prior to turning, the process can be adjusted to minimize distortion considerably.

For the design of an industrial forging process the special purpose finite element (FE) code DEFORM2D™ is often used. It was also employed to simulate the forging process of the studied turbine disk; consequently, it was also used for modeling the water-quenching step. For this cooling simulation an axisymmetric stress state was assumed and, consequently, a two-dimensional cross section was simulated. In order to verify the residual stress predictions of the FE simulation, residual stresses in a turbine disk that was taken out of the commercial production after forging and water quenching were studied by neutron diffraction [6, 7]. Due to the relatively large dimensions of the part (∅ 320 mm, thickness up to 25 mm), neutron diffraction is the only experimental technique to study bulk residual stresses within the disks.

In addition to the disk, a thin plate of the same material was examined. Due to the small thickness of the plate (6 mm) much less beam time was required for strain measurements due to less absorption. The thin plate also allowed testing the influence of geometry on residual stress modeling.

The results confirmed that the heat transfer rate $h(T)$ is a crucial input parameter for this simulation. Temperature-dependent heat transfer coefficients have been estimated from temperature measurements in a disk during water quenching [8]. While the simulation produced good residual stress predictions for the disk, the results were not as good for the thin plate. Therefore, optimization of $h(T)$ for the thin plate can still lead to a significant improvement of the simulation.

### 23.2.2 Material

The Ni-based superalloy IN 718 is an alloy developed and optimized for high-temperature application. Its strength is achieved by partially coherent $\gamma''$-$Ni_3(Nb, Ti)$ precipitates as well as coherent $\gamma'$-$Ni_3(Al, Ti)$ precipitates in the face-centered cubic (fcc) $\gamma$-matrix. The corresponding elements are in solution at the forging temperature. The precipitates are formed during an appropriate heat treatment (annealing for 8 h at 718 °C, followed by furnace cooling to 621 °C within 2 h, holding for 8 h, and finally cooling in air) [9]. The main strengthening phase is $\gamma''$, which can also be present in its thermodynamically stable form, referred to as $\delta$-phase, having the same nominal composition as $\gamma''$. Furthermore, M(C, N) and $M_{23}C_6$ carbides can be found in the matrix (M stands for alloying elements such as Ti and Mo), but due to the small carbon content of only 0.004 wt% their volume fraction is very small.

The sluggishness of the $\gamma''$-precipitation is well recognized in literature [10]. As a consequence, in the as-quenched condition merely some clusters with a few nanometer in size were found by transmission electron microscopy [11]. Therefore, it is assumed that the amount of $\gamma''$ can be neglected for the study of the water-quenched condition. The same applies for the $\gamma'$-phase. Unlike the $\gamma''$-particles, $\delta$-phase is not completely dissolved at forging temperature. A quantitative phase analysis performed on SEM images of the as-quenched condition revealed a small volume fraction of $\delta$-phase between 3% and 4% (Figure 23.2). Furthermore, no significant peaks referring to the orthorhombic $\delta$-phase could be identified in neutron or X-ray diffraction. Therefore, considering the material as a single phase is a good approximation and, thus, the strains determined from the matrix reflections are regarded as representative for the whole sample.

The studied disks were compressor disks with a diameter of 320 mm and a thickness varying from 15 to 25 mm. They were taken from the standard production process conducted at Böhler Schmiedetechnik GmbH & Co KG, Austria, comprising a two-step closed-die forging process on a screw press followed by water quenching. The large forging deformation together with forging below the $\delta$-phase solvus temperature leads to a fine grained microstructure with a mean grain size of 5 μm [3]. Therefore, the number of grains within the gauge volumes used for neutron measurements should be high enough for representing the bulk. The crystallographic texture of a disk was measured using neutron diffraction. The maximum orientation density was about 1.5 times that of the

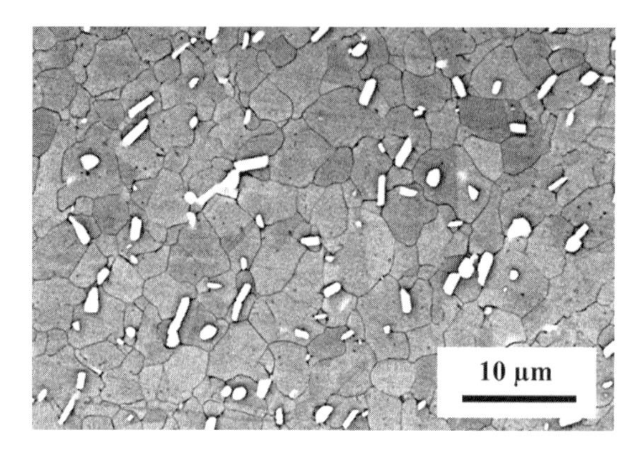


**Figure 23.2** Microstructure of the forged and subsequently water-quenched IN 718 alloy. The $\delta$-phase appears bright; the $\gamma$-matrix appears gray (SEM image).

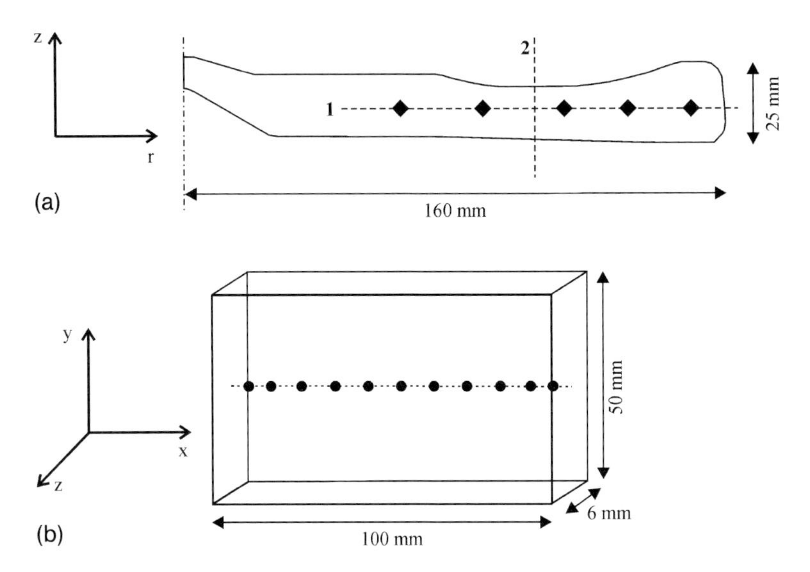


**Figure 23.3** (a) Cross section of the IN 718 turbine disk with a rotational symmetry, showing the locations of strain measurements (dashed lines) as well as the approximate sizes of the gauge volumes. (b) Shape of the IN 718 plate with the locations of the strain measurements.

random distribution, which was considered small enough not to affect the residual stress analysis.

In addition, a thin plate with the size $6 \times 50 \times 100$ mm$^3$ was cut from a turbine disk. It was subjected to an additional annealing treatment at forging temperature and subsequently quenched in water. The shape of the samples and the positions of the scan lines are shown in Figure 23.3.

### 23.2.3 Finite Element Modeling

For the optimization of the forging process of the investigated turbine disks the FE code DEFORM2D™ was used, which is a special code for hot massive deformation. It was also used for modeling the water-quenching step. An axisymmetric stress state was assumed according to the symmetry of the disk and, consequently, a two-dimensional model was established. The FE net consisted of a relatively high number of 10,000 almost rectangular elements in order to be able to extract the stress values at various locations within the cross section with high accuracy.

Material input data like thermal conductivity, thermal expansion, or heat capacity were taken from literature [12, 13]. Temperature- and strain rate-dependent flow stress behavior was determined experimentally as those data were not available in literature for the investigated material. For instance, the data sets describing the yielding behavior in the elastoplastic simulation were determined in conventional compression tests at various temperatures, whereas Young's modulus and Poisson ratio were determined dynamically (ultrasonic pulse-echo overlap measurements [14], Table 23.1). The measured thermophysical data (Table 23.2) are similar to the values given by Dye et al. [15]. For proprietary reasons

**Table 23.1** Dynamically determined Young's modulus and Poisson ratio of IN 718 as a function of temperature.

| $T$ (°C) | $E$ (GPa) | $\nu$ |
|---|---|---|
| 28 | 199 | 0.30 |
| 300 | 177 | 0.32 |
| 500 | 174 | 0.33 |
| 700 | 169 | 0.35 |
| 800 | 162 | 0.36 |
| 900 | 153 | 0.36 |
| 1000 | 143 | 0.37 |
| 1075 | 134 | 0.37 |
| 1150 | 126 | 0.37 |

These values were used in the FE simulation.

**Table 23.2** Experimentally determined heat capacity $c_P$ as a function of temperature.

| $T$ (°C) | $c_P$ (J kg$^{-1}$ K$^{-1}$) |
|---|---|
| 20 | 349 |
| 100 | 359 |
| 200 | 373 |
| 300 | 386 |
| 400 | 399 |
| 500 | 413 |
| 550 | 419 |
| 600 | 450 |
| 650 | 464 |
| 700 | 456 |
| 750 | 482 |
| 800 | 570 |
| 850 | 593 |
| 900 | 541 |
| 950 | 513 |
| 1000 | 497 |
| 1100 | 504 |
| 1200 | 511 |

These values were used in the FE simulation.

the applied yielding data as well as the thermal conductivity data cannot be disclosed.

The heat transfer coefficient $h(T)$ is a crucial parameter for the development of the residual stress state. It cannot be measured directly; therefore, it was deduced from temperature measurements in a disk during water quenching. An example for such a determination can be found in [16] for a cylinder made of brass.

### 23.2.4 Neutron Diffraction

Strain measurements [17] were performed with the neutron diffractometer ARES at the GKSS Research Centre in Geesthacht, Germany, using a constant wavelength of $\lambda = 0.164$ nm [18]. For the measurement of IN 718, which is dominated by the fcc Ni-matrix, the (3 1 1) reflection was used because it is generally recommended in literature for the evaluation of macrostrains in fcc lattices [19, 20]. The influence of intergranular stresses on (3 1 1) lattice planes is small and thus it should be best to reflect macrostresses. The shifts of the Bragg peaks $\Delta\theta$ were determined using an area detector with an active area of $300 \times 300$ mm$^2$.

In case of the disk the symmetry axes (radial, tangential, axial) were used as coordinate system, in case of the plate the longitudinal ($x$), transverse ($y$), and normal ($z$) directions were used as defined in Figure 23.3. The diffraction elastic constants (DEC) $E_{hkl} = 195$ GPa and $\nu_{hkl} = 0.31$ for the (3 1 1) reflection were calculated using the Kröner model [21, 22] with the following single crystal values: $C_{11} = 248$ GPa, $C_{22} = 152$ GPa, $C_{44} = 125$ GPa [23].

The locations of strain measurements in the turbine disk are indicated by dashed lines in Figure 23.3. A matchstick-like gauge volume with a cross section of $2 \times 2$ mm$^2$ was chosen in order to increase intensity; its length was 30 mm for measurements of axial and radial strains and 10–15 mm for tangential strains. It is thus assumed that the stresses are constant in the hoop direction over the length of the gauge volume [24].

For the disks, the value for the unstrained lattice parameter $d_0$ was determined by measuring small cubes (with 4 to 5 mm edge length) using a cubic gauge volume with 2 to 3 mm edge length. The cubes were cut out of the disks by electrodischarge machining with their faces parallel to the axial, radial, and hoop directions. Measurements of cubes cut from different locations within a disk had shown that the disks are chemically very homogeneous so that a single reference cube should be representative for the whole disk. Due to the small size of the cubes they were considered to be free of macrostresses [25]. The cubes were measured in the three orthogonal directions, that is, along the three principal axes of the disk. Since there were only small differences in the lattice parameters for the three directions, the mean value was used as $d_0$.

For the plate, the stress differences $\sigma_x - \sigma_z$ and $\sigma_y - \sigma_z$ were analyzed for which cutting of a reference sample is not required. A cubic gauge volume with an edge length of 3 mm was used.

## 23.2.5
## Results

### 23.2.5.1 *In Situ* Tensile Test

In order to understand the influence of plastic deformation history on the evolution of residual stresses, an *in situ* tensile test was performed at the beam line BW5 at the HASYLAB at DESY in Hamburg, Germany. The tensile test specimen was irradiated with a parallel, monochromatic photon beam with an energy of ~100 keV, a wavelength of $\lambda = 0.12$ Å, and a cross section of $1 \times 1$ mm$^2$. As the average grain size of the tensile samples was the same as depicted in Figure 23.2, the irradiated volume enclosed more than $10^6$ grains, which guarantees good grain statistics. Eight complete diffraction rings could be recorded on an image plate, which enabled the simultaneous measurement of the strain parallel and one of the strain components perpendicular to the loading direction. The tensile test was performed in a strain control mode by increasing the macrostrain step-

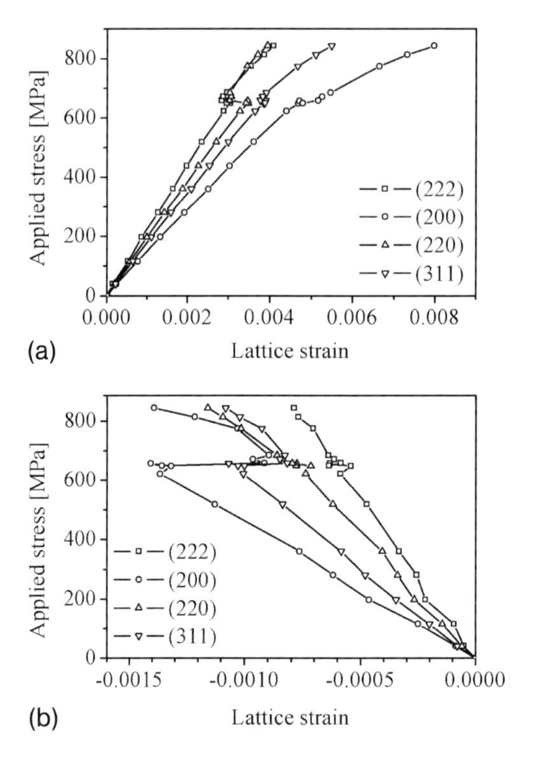


**Figure 23.4** Lattice strains of (3 1 1), (2 2 0), (2 0 0), and (1 1 1) reflections during loading in an *in situ* tensile test measured with high-energy X-rays. For the sake of clarity, error bars are omitted. (a) Parallel to the loading direction; (b) perpendicular to the loading direction. The average errors in strains are $30 \times 10^{-6}$ for (3 1 1), $50 \times 10^{-6}$ for (2 2 0), $75 \times 10^{-6}$ for (2 0 0), and $50 \times 10^{-6}$ for (1 1 1).

**Table 23.3** Young's moduli and Poisson ratio as determined from *in situ* tensile tests and corresponding values calculated via the Kröner model.

| *h k l* | Calculated | | Measured | |
|---|---|---|---|---|
| | *E* (GPa) | $\nu$ | *E* (GPa) | $\nu$ |
| 3 1 1 | 195 | 0.309 | 172 | 0.28 |
| 2 2 0 | 215 | 0.289 | 192 | 0.24 |
| 2 0 0 | 169 | 0.334 | 143 | 0.32 |
| 1 1 1 | 237 | 0.268 | 214 | 0.20 |

wise and holding it constant during the X-ray measurements. The macroscopic load versus elongation curve was recorded by a load cell and an extensometer was mounted directly on the sample.

Figure 23.4 shows the lattice strains correlated with the macroscopic applied stress. In the elastic region all reflections show a linear behavior. Table 23.3 gives the Young's moduli $E_{hkl}$ determined from the slope of a linear fit in the elastic region and the values calculated via the Kröner model. Using the measured $E_{hkl}$, the residual stresses determined by neutron diffraction are shifted by only 10 MPa. So the qualitative agreement between the measurement and the Kröner model is good, but there is a deviation from the calculated $E_{hkl}$ values of 10–15%, depending on the reflection. In the plastic region, the load is no longer evenly distributed between the grains, which results in a nonlinear behavior for all sets of grains. It is most pronounced for the (2 0 0) peak, which is in accordance to measurements as well as to model calculations reported in literature [26–29].

The results of this tensile test confirmed that the 3 1 1 reflection gives a good measure for macrostresses in the studied material.

#### 23.2.5.2 Stresses in a Turbine Disk

The FE model predicts tensile radial and tangential stresses up to about 600 MPa in the interior of the water-quenched disk (Figure 23.5). These tensile stresses are balanced by compressive stresses near the surfaces of the disk. The distributions of radial and tangential stresses are very similar except for the outer rim of the disk. Axial (through-thickness) stresses are zero except for a region near the circumference of the disk. Neutron diffraction measurements were conducted along the lines indicated in Figure 23.3. Repeated measurements along three different radii that are 120° apart from each other showed a rotational symmetry of the stress state.

A comparison shows that the FE model describes the measured stresses quite accurately (Figure 23.5). Especially, the maximum stress levels are well reproduced. However, there is a significant difference in the thin section of the disk (between 80 and 120 mm radius), where the measured stresses are significantly

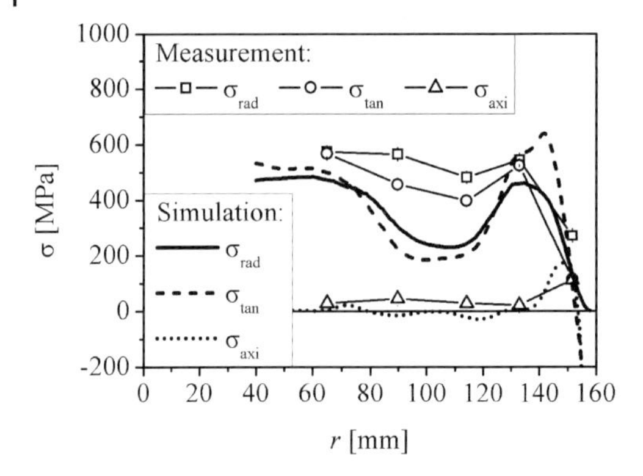


Figure 23.5 Comparison of measured stresses (symbols) and predictions of the FE model (lines) along the radius of a disk (cf. Figure 23.3).

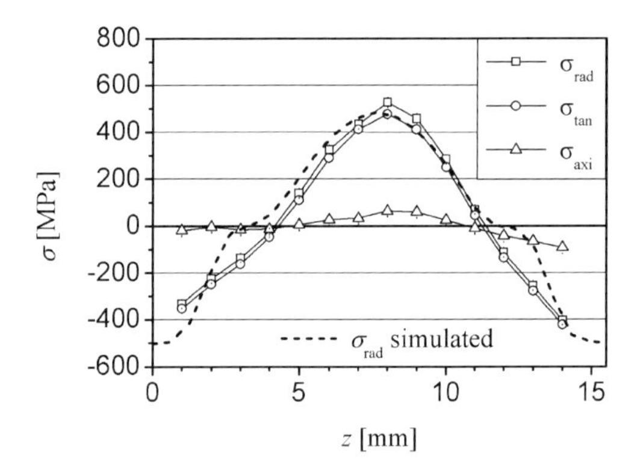


Figure 23.6 Stresses through the thickness of a disk (cf. Figure 23.3). Symbols: as measured; dashed line: FE result.

larger than the FE predictions. To check this in more detail, a through-thickness scan was conducted at the thin section at a radius of 110 mm. The results show that the agreement between simulation and measurement in this case is very good (Figure 23.6). As a consequence, the differences observed in Figure 23.5 most likely result from a positioning error.

The simulation of the disk was initially performed using a constant heat transfer coefficient $h = 4000\ \mathrm{W\ m^{-2}\ K^{-1}}$ [30], which is based on data for water as reported in literature [31]. Variations of this constant value within reasonable boundaries did not lead to a significant improvement of the model in terms of residual stress predictions for a disk [32].

#### 23.2.5.3 Stresses in a Thin Plate

While modeling the thin plate using a constant $h$, satisfying results were not achieved. Figure 23.7 shows the poor agreement between measured and calcu-

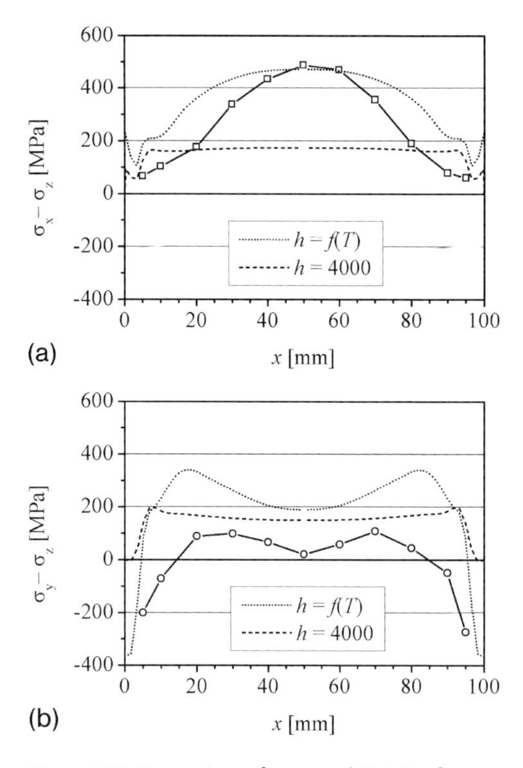


**Figure 23.7** Comparison of measured (3 1 1 reflection; symbols) and simulated (dotted and dashed lines) stresses for the thin plate, (a) difference $\sigma_x - \sigma_z$, (b) difference $\sigma_y - \sigma_z$. (The unit of $h$ is W m$^{-2}$ K$^{-1}$).

lated stresses for $h = 4000$ W m$^{-2}$ K$^{-1}$. The simulation gives flat stress distributions while measured stresses vary significantly over the length of the plate. A systematic variation of various input parameters of the simulation revealed that a more detailed knowledge of the heat transfer coefficient $h = f(T)$ as a function of temperature seemed to be required for improving the model in case of the thin plate [33]. The smaller thickness, leading to a faster cooling of the plate, seems to be responsible for a stronger influence of the exact function of $h(T)$ on the predicted stress distribution.

The simulation of the plate was repeated with the temperature-dependent $h(T)$. The results show a much better agreement of simulation and measurement, although some details could still be further improved (Figure 23.7). The predicted transverse stresses, for example, are significantly larger than measured ones. It must be noted that discrepancies between simulation and measurement can partly be due to a resolution effect. While the measurement gauge volume in the plate had a size of $\sim 3 \times 3 \times 3$ mm$^3$, the calculated data could not yet be integrated over a corresponding volume. However, this influence is considered to be relatively small and therefore will not lead to major changes in the presented results.

### 23.2.6 Summary

The residual stress states in a water-quenched IN 718 turbine disk and a thin, water-quenched plate of the same material were studied using FE modeling and neutron diffraction. The results have shown that the initial model using constant heat transfer coefficients found in literature gives a good overall description of the residual stress state in a turbine disk after water quenching. However, for the prediction of the residual stress state in the thin plate a temperature-dependent heat transfer coefficient has to be used. With this improvement, the stress distribution in the thin water-quenched plate could be described much better, although further improvements are still possible. This means that the influence of variations of the processing steps on the resulting stress state in the turbine disks can be predicted with good accuracy, thus allowing an optimization of subsequent treatments such as turning to final shape.

## 23.3 Residual Stresses in Laser-Welded Al Joints

### 23.3.1 Introduction

Use of advanced welding techniques for Al alloys is considered as an essential long-term alternative to the conventional riveted airframe structures (Figure 23.8) because of its potential in weight and cost reduction [34]. However, the chemical and physical characteristics of Al may pose challenges associated with the quality of welds. These challenges knowingly include hot cracking, porosity formation, and susceptibility to corrosion. In addition, the strength of the weld material often is less than that of the base material (so-called undermatching case). However, successful developments both in advanced welding techniques and Al alloys combined with engineering strategies for optimum joint designs have produced industrially applicable solutions [35]. The Al alloy AA6056 is such an alloy, being considered as one of the candidate alloys for airframes [34]. Essentially, this is a weldable AlMgSiCu alloy that can be precipitation hardened to different strength levels. In the T4 temper (underaged), the strength is low but can be improved by suitable heat treatments such as T6 (peak-aging) or T78 (overaging). The LBW process successfully provides skin-stringer T-joints for integral airframe structures made of Al alloys of the AA6xxx series [36–39].

In future aircrafts, welded T-joints provide a possible joint design for "clip-skin" joints of an airframe. Currently, all clips (which join skin, frame, and stringers) are riveted but in principle a one-sided LBW process offers a weight- and cost-effective next step to the rivet-free aircraft fuselage. Since residual stresses are sensitive to joint geometry and restraint conditions, care has to be taken for residual stresses in welded clips. Higher degrees of restraint against shrinkage may lead to higher residual stresses. The magnitudes of welded resid-

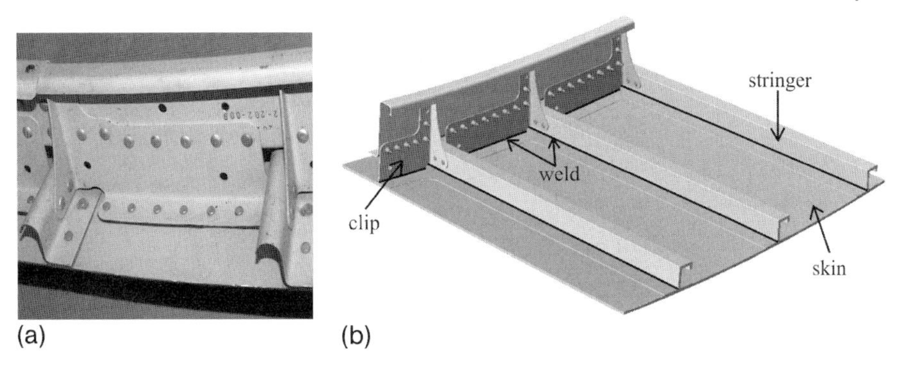


**Figure 23.8** Part of an airframe structure joined with rivets (a). Replacement of skin-stringer and skin-clip riveted joints by welds (b).

ual stresses can often be a high proportion of the yield or proof strength [36, 37, 40–42] depending on joint configuration and welding process. For studying fundamental aspects of the residual stress field around welded clips, a model clip-skin configuration was produced. Strain measurements were conducted at various locations both in the weld and in the clip involving also run-in and run-out locations [43].

A further step toward a future fully rivet-free aircraft is the possibility of skin–skin joints (butt joints) by FSW or LBW, joining larger panels configured with stringers and clips. However, skin–skin joints are critical joints that have to be most reliable. When operational stress patterns are superimposed on residual stresses, the internal stress state will be modified. Damage processes are influenced depending on the presence of either tensile or compressive residual stresses at the given location. Thus, residual stresses may have a detrimental influence on the service performance of the weld. This can be counterbalanced at least in part by a suitable postweld heat treatment (PWHT). Thereby, not only softened regions such as the heat-affected zone (HAZ) can be hardened through PWHT and strength restored, but also residual stresses due to welding can be reduced. Thus, for a comprehensive assessment of the performance of structural welds, a careful evaluation of the state of residual stress in various sheet thicknesses and heat conditions is essential. Although significant amount of information on conventional steel welds is available, lack of data still exists for the laser beam-welded Al alloys of aerospace grade. Therefore, in this work different $CO_2$ laser beam-welded AA6056 butt joints were investigated using neutron and high-energy X-ray diffraction [44].

### 23.3.2
### Materials and Welding

The investigated laser beam-welded T-joint specimen consists of a 2-mm-thick AA6013-T6 sheet as a clip part and a 6-mm-thick AA6056-T78 base plate of size $160 \times 275$ mm$^2$ (Figure 23.9a). The total length of the weld was 120 mm and it

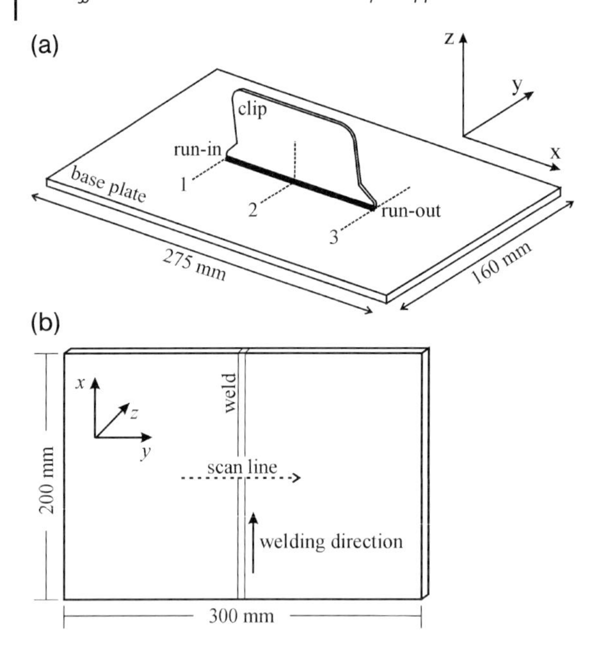


**Figure 23.9** Drawing of the investigated Al samples with laser-beam welds. (a) T-joint configuration with 6-mm-thick base plate and 2-mm-thick clip and (b) butt joint configuration.

was welded from one side only. The welding was carried out using AA4047 grade (12% Si) Al alloy filler wire of diameter 1.2 mm with a welding speed of 1.8 m $min^{-1}$ employing the 3.3 kW Nd:YAG LBW unit at GKSS. Both base sheet and clip were fixed to the welding table using a vacuum system. No PWHT was carried out after the welding.

The cross section of the weld joint is shown in Figure 23.10a. The one-sided LBW process produced a rather narrow fusion zone with ~2-mm width at the midsection and the root of the weld, while at the top the fusion zone was widened to about 3 mm. Microhardness measurements have been carried out along the dashed line indicated in Figure 23.10a. The hardness profile shows a drop of hardness in the weld zone to a minimum value, which indicates a lower strength (undermatching) of the weld deposit (Figure 23.11). The decrease in hardness in the weld can partly be attributed to the dissolution of strengthening precipitates.

For the butt joints, sheets of the airframe Al alloy AA6056 with thicknesses of 3.2 and 6 mm were studied. Sheets with dimensions 250 mm in width and 500 mm in length (i.e., half the weld coupon size) were cut in the direction perpendicular to the rolling direction and welded in the butt joint configuration by using the filler wire AlSi12 with a $CO_2$ laser. The produced welded coupons were $500 \times 500$ $mm^2$ in dimensions. Welding was done in the T4 and the T6 states.

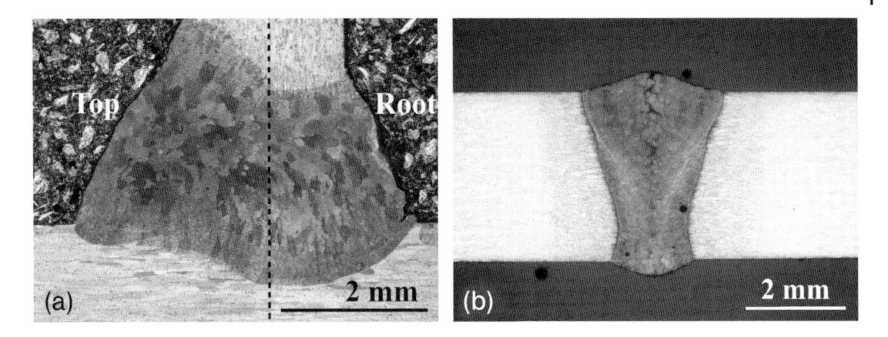


**Figure 23.10** Optical micrographs of weld cross sections. (a) T-joint with base plate at the bottom and (b) butt joint. Hardness measurements were performed along the dashed line indicated in panel (a).

Sheets welded in T4 were given PWHT to T6 and T78. The maximum temperature used in both cases was the same (190 °C) but aging times differed, being longer in the case of T78. The welds were relatively slim (Figure 23.10b); the width of the weld increased with the increase in sheet thickness and varied from about 3 mm for 3.2-mm sheet thickness to about 5 mm for 6-mm sheet thickness. The HAZ in these cases were symmetrical on either sides of the weld and extended up to 0.4 and 2.5 mm, respectively. Furthermore, with the constancy of microhardness in the base metal as the criterion, the extent of the hardness dip as measured from the weld center varied from about $\pm 2.5$ to $\pm 8$ mm, respectively.

For residual stress measurements samples having the dimensions $200 \times 300\ mm^2$ were extracted from the welded coupons by using water jet cutting with the weld placed in the center (Figure 23.9b). Internal strains were measured at

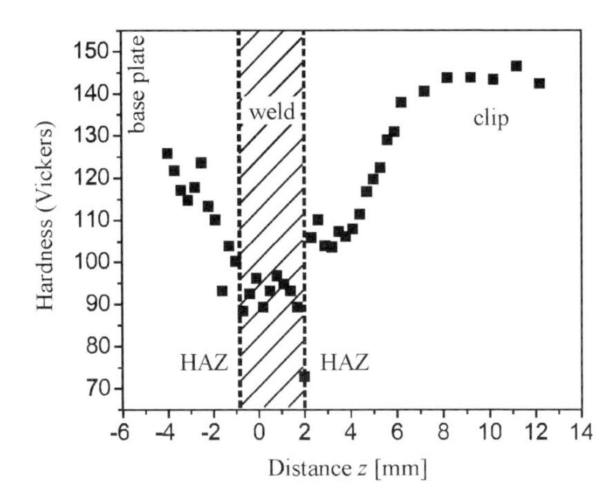


**Figure 23.11** Microhardness along the line shown in Fig. 23.10a. Zero is at the base plate surface.

19 points along a 100 mm long scan line in the middle of each sheet using neutron diffraction.

### 23.3.3
### Neutron Diffraction

Residual strain measurements were performed with the neutron diffractometer ARES at the Geesthacht Neutron Facility (GeNF) at GKSS [18]. Neutrons of 0.164-nm wavelength from an elastically bent perfect Si (3 1 1) monochromator were used. The Al (3 1 1) diffraction peak was recorded with an area detector at an angle of about 84°. This peak is recommended because the influence of microstresses on the stress results should be small. As explained in the previous section, the (3 1 1) peak gives a good measure for macrostresses [20]. Measured diffraction peaks were fitted with a Gaussian function. The principal sample axes $x$, $y$, $z$ are indicated in Figure 23.9. The gauge volumes in which the strains are measured were defined by Cd slits. Residual stresses were calculated from measured strains [24]. In general, DEC have to be used for $E$ and $\nu$ that depend on the Miller indices ($h$ $k$ $l$) of the reflection that is used for the strain measurement [45]. However, the crystallographic anisotropy of the elastic constants is small for Al and, therefore, for the Al (3 1 1) reflection the macroscopic values of $E=73$ GPa and $\nu=0.33$ were used [20].

For the determination of a triaxial stress state, the knowledge of the unstrained lattice parameter $d_0$ is required. Both Al alloys AA6056 and AA6013 are heat treatable alloys in which the solute (mainly Mg, Si, Cu) concentration in the matrix can depend on the heat treatment. As the lattice parameter can change considerably with the solute content, the $d_0$-values can also depend on the heat treatment the materials were subjected to during welding [46]. One way to solve this problem is to cut a reference sample ("comb") with thin teeth out of the weld region by spark erosion in which most of the macroscopic stresses are known to relax [47]. Such a comb can, therefore, serve as stress-free reference material. However, so far another approach was used: Assuming a plane stress state in a thin specimen from the condition $\sigma_z=0$, $d_0$ can be calculated for each point; the stresses $\sigma_x$ and $\sigma_y$ can be evaluated using the biaxial formula [24]. The plane stress assumption was used for both butt joints and T-joints. For the T-joint, it was used for base plate and clip separately. The resulting $d_0$ values for both materials of the T-joint were very similar. Values for the weld material of the T-joint were simply interpolated, being aware of the fact that the error in the resulting stresses in the T-joint weld can be large due to a possible influence of $d_0$ variations. Cutting of reference samples, in which macrostresses can relax, will be the next step in the investigation of the T-joints.

For the T-joints, strain measurements were conducted on both base plate (scan lines 1–3) and clip (scan line 2, Figure 23.9b). A matchstick-like gauge volume of nominal size $1 \times 1 \times 30$ mm$^3$ was used for scanning strains $\varepsilon_y$ and $\varepsilon_z$ at the mid-clip position in the base metal and the clip for improving the intensity [48], assuming that the stresses are constant along $x$ over the length of the volume element. A smaller gauge volume of size $1 \times 1 \times 2\ mm^3$ was used for scanning $\varepsilon_x$ because large stress gradients are present in the $y$-direction. For the measurements at the run-in and run-out positions a volume of $2 \times 2 \times 10\ mm^3$ was used for measurements of $y$ and $z$ strains and a cubic volume of $2 \times 2 \times 2\ mm^3$ for measurements of $x$ strains. In order to compare the results obtained from run-in and run-out to the results from mid-position of the base plate, one measurement at the mid-position was performed with the same gauge volume as for the measurement on run-in and run-out. However, no significant differences were observed in the results.

For the butt joints a matchstick-like gauge volume of nominal size $2 \times 2 \times 30\ mm^3$ with the long edge parallel to $x$ was used for measurements of transverse ($y$) and normal ($z$) strains for improving the intensity [48]. A cubic gauge volume of approximate size $2 \times 2 \times 2\ mm^3$ was used for measurements of longitudinal strains because large stress gradients are expected in the transverse direction. In order to increase the number of diffracting grains for the small gauge volume, it was shifted by 20 mm in the longitudinal ($x$) direction during the measurement of longitudinal strains. Thus, a similar volume as for the two other directions was analyzed. Both material versions tested had pancake-shaped grains, for example the HDT version contained grains of average length of about 108 μm and width of about 23 μm, which gives an aspect ratio of about 4.7. The grain size proved to be small enough to ensure a good grain statistics even with the small gauge volume.

### 23.3.4 Stresses in Laser-Welded T-Joints

The residual stress distributions in base plate and clip at the mid-clip position (scan line 2, see Figure 23.9a) are shown in Figure 23.12. Several measurements were done to check the repeatability. In Figure 23.12 the mean value of all measurements is shown. There are tensile residual stresses in the $x$-direction (longitudinal) both in the base plate and in the clip. The maximum longitudinal tensile residual stress of 200 MPa is observed in the clip at a distance of 8 mm from the weld center (Figure 23.12b). Compressive transverse ($y$) residual stresses appear in the range about $\pm 8$ mm in the base plate (Figure 23.12a).

The level of longitudinal tensile residual stresses is significantly higher at the run-out (scan line 3) compared to the run-in (scan line 1) location (Figure 23.13). On the other hand, no significant difference exists between compressive stress components of these two locations. As expected, tensile residual stresses were measured to be lower at both run-out (70 MPa) and run-in (20 MPa) locations than at the mid-clip position (140 MPa) of the welds due to the differences in constraint and thermal cycle conditions. Compressive transverse stresses are around 65 MPa at run-in and run-out, which are significantly lower than those at the mid-clip position.

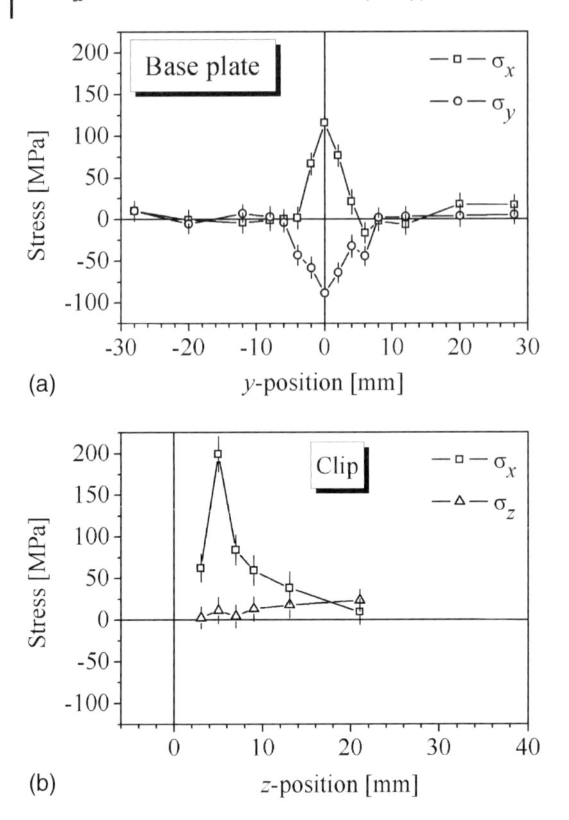


**Figure 23.12** Residual stress distributions at the mid-clip position (scan line 2 in Figure 23.9a), (a) in the base plate and (b) in the clip. $x$ is parallel to the weld.

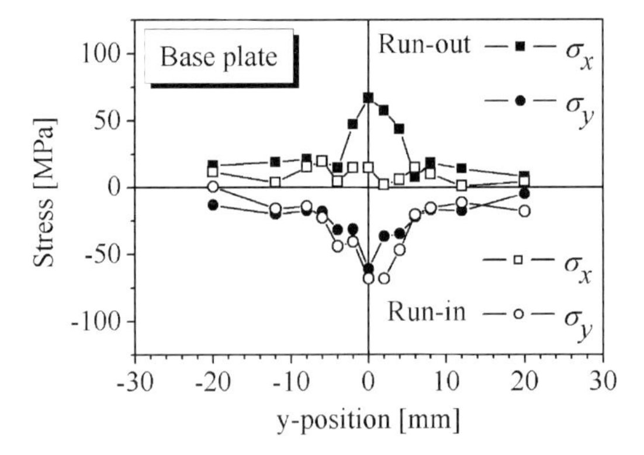


**Figure 23.13** Residual stress distributions at run-in and run-out positions (scan lines 1 and 3 in Figure 23.9a). $x$ is parallel to the weld.

### 23.3.5
### Stresses in Laser-Welded Butt Joints

From practical view points the effect of following three parameters on the development of residual stress was of immediate interest: (1) the sheet thickness at a given temper, (2) the initial temper in which welding was undertaken, and (3) the possibility of stress relaxation through a PWHT. For damage tolerance under tensile loading tensile residual stresses are of immediate concern. In all the cases investigated there were longitudinal tensile residual stresses, with a minimum in the weld center, extending well outside the weld into the base metal.

PWHT is one of the methods for stress relaxation well known from steel welds. PWHT in the present case was supposed to be beneficial in improving strength coupled simultaneously with reduction of residual stresses. This should contribute not only to the structural stability but also to economy. It turned out, however, that while the strength was improved as expected, the initial distribution as well as the level of residual stress was more or less retained (Figure 23.14). Con-

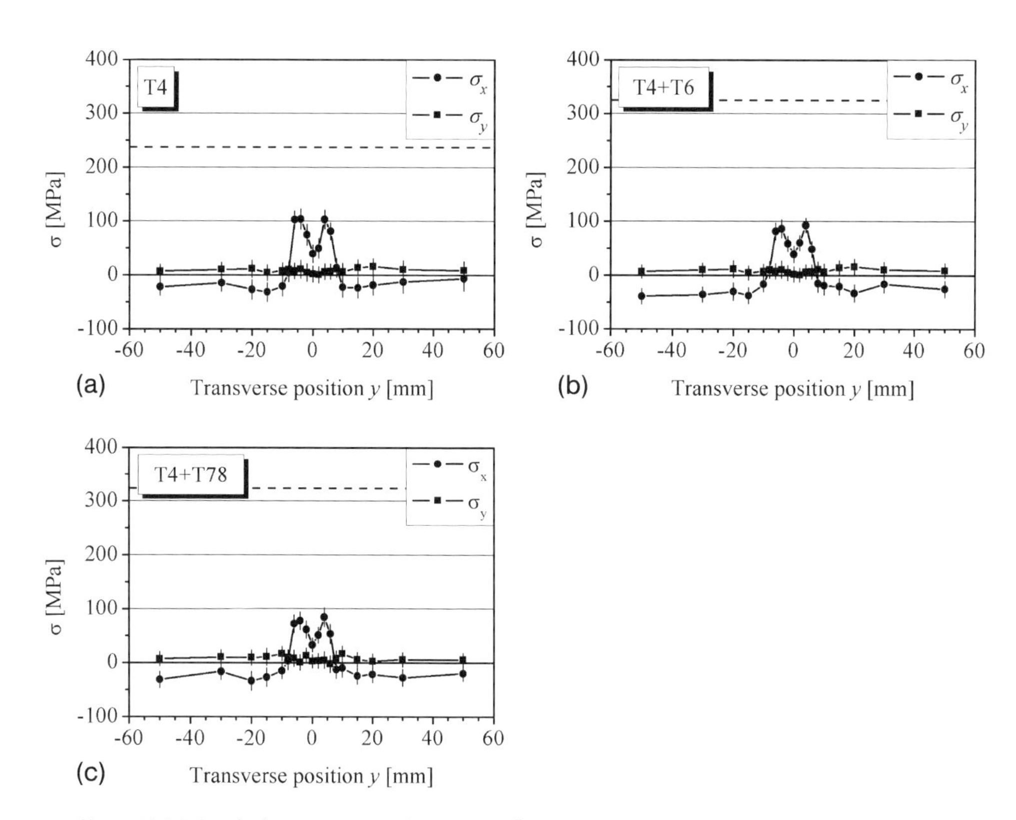


**Figure 23.14** Residual stresses in AA6056/3.2-mm butt joints (HDT version) welded in T4 (a), after additional PWHT T6 (b), and T78 (c). The dashed line denotes the yield stress $R_{p0.2}$ for the weld.

sequently, in this context the outcome in stress relaxation was marginal applying standard PWHT tempers. Since higher temperatures than that employed ($\geq$190 °C) are not advisable due to loss of strength, further methods (such as shot peening, modified temper cycle) must be sought for reduction of residual stress, assuming that the level of residual stress is not tolerable or is detrimental. Yield stresses within the weld material (dashed lines in Figure 23.14) were determined with small specimens cut out of the welds.

Further measurements showed that with an increase in the sheet thickness from 3.2 to 6 mm, the residual stress profile changed more than the stress level [44]. A wider zone of tensile stresses can be attributed in the first instance to the increase in the heat input required for welding, which increases with the increase in the sheet thickness.

Furthermore, in contrast to T4 as the starting temper for LBW, residual stresses in T6 as the starting temper for LBW were found to be higher and close to the yield strength of the weld [44]. A significant difference was observed in the stress profile, which was broader in T4 than that in T6 for the same sheet thickness. Such a difference may be resulting from the difference in the initial temper and the modification in chemical composition. Welding in T4 appears to be beneficial since the residual stress level is reduced nearly to half of that when welded in T6. Thus, welding in the underaged condition may be an approach to reduce residual stresses. However, this needs to be confirmed with further measurements.

### 23.3.6 Summary

Residual stresses in a laser beam-welded Al alloy T-joint resembling a simplified clip-skin configuration of a welded airframe were investigated to provide basic understanding of residual stress development at various locations. Following observations have been made: first, the presence of a maximum of longitudinal tensile stresses within the clip at a distance of about 8 mm from the weld. Second, tensile residual stresses are much higher at the mid-location of the weld length than at the run-in and run-out locations, which was attributed to a less constrained condition at the weld ends compared to the middle of the weld. Third, tensile stresses at the run-out location are higher compared to tensile stresses at the run-in. This can be explained by an increasing stiffness via solidification of the weld seam with increasing weld length.

Based on these results, welding procedure and joint geometry, including thickness increase of the base plate at the weld area ("socket"), will be modified. Residual stresses in sheets with areas of reduced thickness ("pockets") as well as new Al alloys in combination with numerical simulation will be subject of further investigations. Furthermore, an investigation of the fatigue properties of these welds will be conducted in order to assess the effects of the residual stress distributions.

Residual stresses in laser beam butt-welded AA6056 Al sheets were determined to study the influence of initial temper, PWHTs, and sheet thickness on the residual stress state. Regarding the reduction of residual stresses, LBW in the underaged condition (T4) was found to be beneficial. A further reduction through a PWHT was, however, not possible at the temperature employed (190 °C). Nonetheless, irrespective whether the sheets are welded in underaged or peak aged (T6) condition, residual stresses are found to be well below the yield stress of the weld in 3.2-mm sheets. In this sense, LBW can be applied safely to thin sheets. In all cases, residual stresses are found to extend well beyond the weld. Hence, the reduction of residual stresses in general and in thick sheets in particular through proper measures remains to be a challenge.

## 23.4 Conclusions

Diffraction-based residual stress analysis is a powerful tool for the engineer interested in optimization of materials and production processes such as forging of components or welding of sheet structures. Residual stress analysis is possible in the interior of thick material (up to several centimeters) and in many cases it can be done in a completely nondestructive way. The spatial resolution achieved by neutron diffraction is sufficient for a large number of problems. For example, validation of stress predictions gained from FE models are a broad field of application as it was shown for forged turbine disks. As modern joining techniques make their entrance into aircraft construction, the inclusion of residual stresses in damage concepts, like crack initiation and growth models, will be of growing importance. Diffraction-based residual stress analysis can contribute largely to such investigations as today it is a routine technique that can deliver reliable results. Finally, the costs of such measurements are comparable to that of other sophisticated experimental methods.

## Acknowledgments

The authors are grateful to J. Homeyer, S. Yi, M. Panzenböck, and H.-U. Ruhnau for assistance with the tensile test of IN 718 samples. The IN 718 measurements were supported by the European Commission under the Sixth Framework Program through the key action: strengthening the European research area, research infrastructures. Contract number: RII3-CT-2003-505925. The investigation on the T-joint was conducted within the framework of the HGF Virtual Institute PNAM (Photon and Neutron Research on Advanced Engineering Materials) and EU-Project WEL-AIR (development of short distance welding concepts for airframes). Investigations on the butt joints were conducted within the framework of the EU-project IDA (investigation on damage tolerance behavior of Al alloys). The authors wish to thank S. Riekehr, GKSS, for conducting the T-joint laser beam welds, and W. V. Vaidya, GKSS, for providing the butt joints.

## References


1 E.A. Loria (ed.), Superalloys 718, *The Minerals, Metals & Materials Society (TMS)*, Warrendale, PA, USA, 2005.

2 W. Betteridge, *Materials Science and Technology*, Vol. 7, Wiley-VCH, Weinheim, New York, Basel, Cambridge, 1992, 641.

3 W. Horvath, W. Zechner, J. Tockner, M. Berchthaler, G. Weber, E.A. Werner, Superalloys 718, 625, 706 and Derivates, *The Minerals, Metals & Materials Society (TMS)*, Warrendale, PA, USA, 2001, 223–228.

4 G.A. Rao, M. Kumar, M. Srinivas, D.S. Sarma, *Materials Science and Engineering A*, 2003, **355**, 114.

5 S.J. Hong, W.P. Chen, T.W. Wang, *Materials Science and Transactions A*, 2001, **32**, 1887–1901.

6 U. Cihak, P. Staron, W. Marketz, H. Leitner, J. Tockner, H. Clemens, *Zeitschrift für Metallkunde*, 2004, **95**, 663–667.

7 U. Cihak, P. Staron, H. Clemens, J. Homeyer, M. Stockinger, J. Tockner, *Materials Science and Engineering A*, 2006, **437**, 75–82.

8 U. Cihak, P. Staron, M. Stockinger, H. Clemens, *Advanced Engineering Materials*, 2006, **8**, 1088–1092.

9 D.D. Krueger, Superalloy 718 – Metallurgy and Applications, *The Minerals, Metals & Materials Society (TMS)*, Warrendale, PA, USA, 1989, 279–296.

10 S. Azadian, L.-Y. Wei, R. Warren, Delta phase precipitation in Inconel 718, *Materials Characterization*, 2004, **53**, 7–16.

11 L. Geng, Y.-S. Na, N.-K. Park, Continuous cooling transformation behavior of alloy 718, *Material Letters*, 1997, **30**, 401–405.

12 K.C. Mills, Recommended values of thermophysical properties for selected commercial alloys, *National Physical Laboratory (NPL)*, ASM International, Cambridge, UK, 2002.

13 G. Pottlacher et al., Thermophysikalische Eigenschaften von festem und flüssigem INCONEL718, *Thermochimica Acta*, 2002, **382**, 255–267.

14 A.G. Hellier, S.B. Palmer, D.G. Whitehead, *Journal of Physics E: Scientific Instruments*, 1975, **8**, 352–354.

15 D. Dye, K.T. Conlon, R.C. Reed, Characterization and modelling of quenching induced residual stresses in the nickel-based superalloy IN 718, *Metallurgical and Materials Transactions A*, 2004, **35**, 1703–1713.

16 A. Buczek, T. Telejko, Inverse determination of boundary conditions during boiling water heat transfer in quenching operations, *Journal of Materials Processing Technology*, 2004, **155–156**, 1324–1329.

17 M.T. Hutchings et al., *Introduction to the Characterization of Residual Stress by Neutron Diffraction*, Taylor & Francis Group, Boca Raton, USA, 2005.

18 P. Staron, H.-U. Ruhnau, P. Mikula, R. Kampmann, The new diffractometer ARES for the analysis of residual stresses, *Physica B*, 2000, **276–278**, 158–159.

19 T.M. Holden, C.N. Tomé, R.A. Holt, Experimental and theoretical studies of the superposition of intergranular and macroscopic strains in Ni-based industrial alloys, *Metallurgical and Materials Transactions A*, 1998, **29**, 2967–2973.

20 B. Clausen, T. Lorentzen, T. Leffers, Self-consistent modelling of the plastic deformation of fcc polycrystals and Ist implications for diffraction measurements of internal stresses, *Acta Materialia*, 1998, **46**(9), 3087–3098.

21 E. Kröner, Berechnung der elastischen Konstanten des Vielkristalls aus den Konstanten des Einkristalls, *Zeitschrift für Physik*, 1958, **151**, 504–518.

22 F. Bollenrath, V. Hauk, E.H. Müller, Zur Berechnung der vielkristallinen Elastizitätskonstanten aus den Werten der Einkristalle, *Zeitschrift für Metallkunde*, 1967, **58**, 76–82.

23 D. Dye, S.M. Roberts, P.J. Withers, R.C. Reed, The determination of the residual strains and stresses in a tungsten inert gas welded sheet of IN 718 superalloy using neutron diffraction, *Journal of Strain Analysis*, 2000, **35**, 247–259.

24 A.J. Allen, M.T. Hutchings, C.G. Windsor, C. Andreani, Neutron diffraction methods for the study of residual stress fields, *Advances in Physics*, 1985, **34**, 445–473.
25 A.D. Kravitz, R.A. Winholtz, Use of position-dependent stress-free standards for diffraction stress measurements, *Materials Science and Engineering A*, 1994, **185**, 123–130.
26 T.M. Holden, R.A. Holt, A.P. Clarke, *Materials Science and Engineering A*, 1998, **246**, 180–198.
27 B. Clausen, T. Lorentzen, M.A.M. Bourke, M.R. Daymond, *Materials Science and Engineering A*, 1999, **259**, 17–24.
28 B. Clausen, Ph.D, Thesis, Risø National Laboratory, Roskilde, Denmark, Risø-R-985 (EN), 1997.
29 M.R. Daymond, M.A.M. Bourke, R.B. Von Dreele, B. Clausen, T. Lorentzen, *Journal of Applied Physics*, 1997, **82**, 1554–1562.
30 U. Cihak, P. Staron, W. Marketz, H. Leitner, J. Tockner, H. Clemens, Residual stresses in forged IN718 turbine discs, *Zeitschrift für Metallkunde*, 2004, **95**(7), 663–667.
31 H. Chandler, *Heat Treater's Guide*, ASM International, Materials Park, Ohio, USA, 1995, 88.
32 U. Cihak, M. Stockinger, P. Staron, J. Tockner, H. Clemens, Superalloys 718 (ed. E.A. Loria), *The Minerals, Metals & Materials Society (TMS)*, Warrendale, PA, USA, 2005, 517–526.
33 D. Dye, K.T. Conlon, R.C. Reed, Characterization and modelling of quenching induced residual stresses in the nickel-based superalloy IN 718, *Materials Science and Engineering A*, 2004, **35**, 1703–1713.
34 K.-H. Rendigs, *Materials Science Forum*, 1997, **242**, 11–24.
35 P. Lequeu, P. Lassince, T. Warner, G.M. Raynaud, *Aircraft Engineering and Aerospace Technology*, 2001, **73**, 147–159.
36 R.A. Owen, R.V. Preston, P.J. Withers, H.R. Shercliff, P.J. Webster, *Materials Science and Engineering A*, 2003, **346**, 159–167.
37 R.V. Preston, H.R. Shercliff, P.J. Withers, S. Smith, *Acta Materialia*, 2004, **52**, 4973–4983.
38 M. Ya, P. Marquette, F. Belahcene, J. Lu, *Materials Science and Engineering A*, 2004, **382**, 257–264.
39 W. Zink, in *Proceedings of the Nineteenth European Conference on Materials for Aerospace Applications* (eds. M. Peters and W.A. Kaysser, DGLR Report, 2001–2002), December 6–8, 2000, Munich (Germany), 2002, 25–32.
40 Z. Feng, *Woodhead Publishing*, England and CRC Press LLC, USA, 2005.
41 A. Stacey, H.J. Macgillivary, G.A. Webster, P.J. Webster, K.R.A. Ziebeck, *Journal of Strain Analysis*, 1985, **20**(2), 93–100.
42 E. Goncalves, M.A.C. Gonzales, in *Proceedings of Eleventh International Conference on Fracture (ICF), Turin (Italy), March 20–25, 2005.*
43 F.S. Bayraktar, P. Staron, M. Koçak, A. Schreyer, *Materials Science Forum*, 2006, **524–525**, 419–424.
44 P. Staron, W.V. Vaidya, M. Koçak, J. Homeyer, J. Hackius, *Materials Science Forum*, 2006, **424–425**, 413–418.
45 V. Hauk, *Structural and Residual Stress Analysis by Nondestructive Methods*, Elsevier, Amsterdam, 1997.
46 H.G. Priesmeyer, *Measurement of residual and applied stress using neutron diffraction* (eds. M.T. Hutchings and A.D. Krawitz), NATO ASI Series E, 1992, 216, 277–284.
47 E. O'Brien, *Proceedings of the Sixth International Conference on Residual Stresses ICRS-6*, IOM Communications Ltd., London, UK, 2000.
48 L. Pintschovius, V. Jung, E. Macherauch, O. Vöhringer, *Materials Science and Engineering*, 1983, **61**, 43–50.

# 24
# Optimization of Residual Stresses in Crankshafts

*Anke Rita Pyzalla*

## 24.1
## Introduction

Innovation in the automotive industry is driven by strong requirements on costs, automobile quality, safety, convenience aspects, and pollution control. Fuel consumption, which is of both economic and ecologic relevance, can be decreased substantially by a reduction in automobile mass. This reduction in mass can be achieved by using new materials, optimizing geometries of vehicle parts, but, also by controlling manufacturing processes in order to achieve optimized material microstructures and residual stress states.

In automotive engines, crankshafts have received increasing attention with respect to a decrease in mass and an increase in its lifetime. The crankshaft is that part of the engine, which translates the linear piston motion into rotation. Therefore, the geometry of the crankshaft is rather complex (Figure 24.1).

Crankshafts are subject to strong bending moments due to gas loading, which are particularly high in modern diesel engines. These bending moments give rise to alternating stresses with stress concentrations in the fillet connecting the rod bearing and the side walls.

Crankshafts are manufactured either from grey cast iron or steels. Steel lately has made significant inroads because of its higher stiffness, its higher fatigue strength, lower weight, and improved noise vibration harshness compared to the cast iron [1]. Crankshaft fatigue strength is enhanced by deep rolling, e.g., [2], nitriding [1] and deep rolling [1, 3]. For automotive crankshafts deep-rolling is the preferred method. Deep rolling of the relief-groove increases surface hardness, decreases of surface roughness, and introduces compressive residual stresses [3, 4].

In order to quantify and optimize the residual stress state introduced by deep rolling of the relieve groove and/or induction hardening of the bearing residual stress analyses were performed on crankshafts deep rolled with different process parameters such as force and number of rolling cycles.

*Neutrons and Synchrotron Radiation in Engineering Materials Science: From Fundamentals to Material and Component Characterization*
Edited by Walter Reimers, Anke Rita Pyzalla, Andreas Schreyer, and Helmut Clemens

ISBN: 978-3-527-31533-8

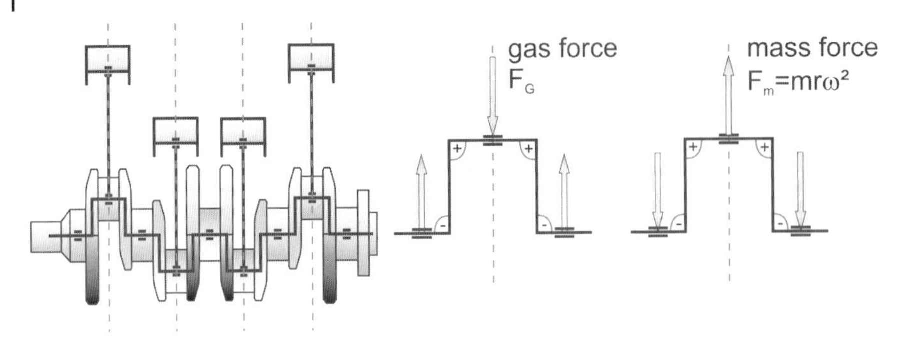


**Figure 24.1** Scheme of a crankshaft design and of the forces acting on a crankshaft.

## 24.2
## Experimental Determination of Residual Stresses in Crankshafts

In order to be efficient with respect to an increase in lifetime the deep rolling of the fillet and/or the induction hardening of the bearing need to introduce compressive residual stresses not only directly at the surface, but within the near-surface layer. The residual stress state introduced, therefore, is three dimensional. X-ray residual stress analyses only give access to the surface of the fillet and the bearing. In order to determine the residual stress distribution also in defined distances from the crankshaft surface using laboratory X-rays, the material needs to be removed stepwise, e.g., by electrochemical etching, after each step a measurement is performed. This procedure is time consuming and yields only the hoop and the longitudinal residual stress component. A further possibility [3] is sampling of the crankshaft and performing X-ray measurements of the residual stresses in radial and hoop direction on the polished surface of the sample (Figure 24.2). This method gives access to the residual stress-depth profile beneath the surface (Figure 24.3) [5], the residual stress values measured, however, are different from those in the original crankshaft, because cutting releases the longitudinal residual stresses. Because of the stress equilibrium conditions (see

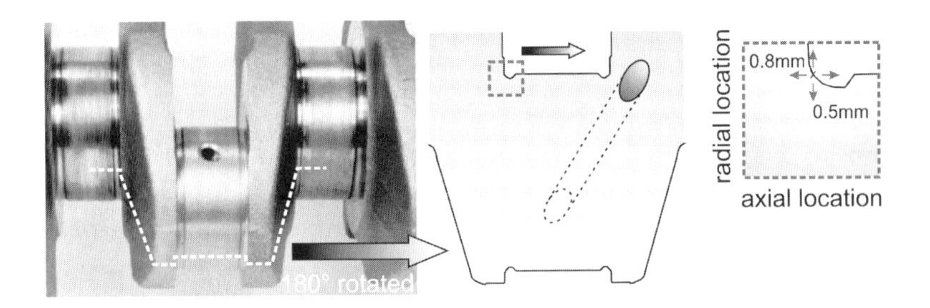


**Figure 24.2** Sampling of the crankshaft.

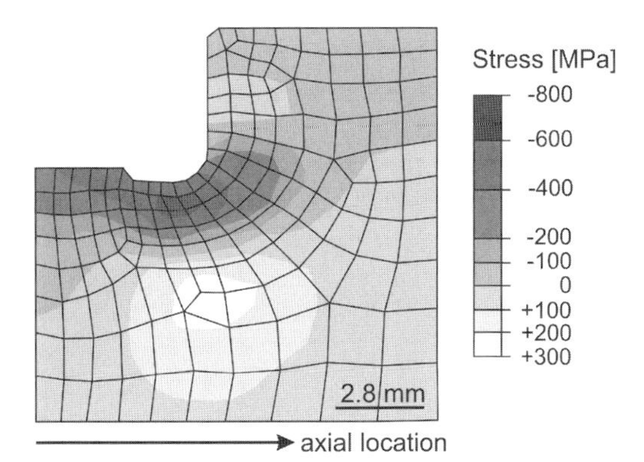


**Figure 24.3** Residual stress distribution in the hoop direction measured using X-rays at Volkswagen AG, Wolfsburg, Germany [5].

Chapter 2), cutting also affects the residual stresses in radial and hoop direction of the crankshaft.

Due to the complex geometry of crankshafts, a nondestructive three-dimensional residual stress analysis only is possible using neutron diffraction. Neutron residual stress analyses were performed at the neutron residual stress diffractometer at the Hahn-Meitner-Institute Berlin GmbH. The sample was mounted in an Eulerian cradle using a sample holder which in combination with a laser system enables a mechanical positioning with an accuracy of at least 100 μm. The location of the surface of the sample in the diffractometer was verified by a through-surface scan [6, 7].

Because of the rotational symmetry of the crankshaft and the rotational symmetry of the deformation introduced by the deep-rolling process and also the rotational symmetry of the temperature field of the induction-hardening process, the principal axes of the residual stress distribution can be assumed to coincide with the axial, hoop, and radial direction of the crankshaft. As a consequence of the large sidewalls of the samples, the beam path in the axial direction becomes long, thus it is not economic to analyze the axial strain components at $\psi = 0^\circ$. Strains therefore were determined in the radial, hoop direction, and at an angle of 50.8° from the radial to the axial direction. In order to minimize the effect of the $d_0$ uncertainty on the residual stress values, further measurements were performed using the $\sin^2 \psi$-method (Figure 24.4), they yield the difference between the residual stress in axial and the residual stress in the radial direction [8], see also Chapter 2 of this book; $\psi$ here denotes the angle between the radial and axial direction of the sample.

Due to the weak texture of the crankshaft the $d$-$\sin^2 \psi$ curve obtained were linear (Figure 24.4) and thus the uncertainty of the residual stress values determined was about ±40 MPa on average. Numerical calculations of the residual stress distribution in deep rolled [3] and induction hardened crankshaft [7] revealed the residual stresses in the radial direction to be small compressive resid-

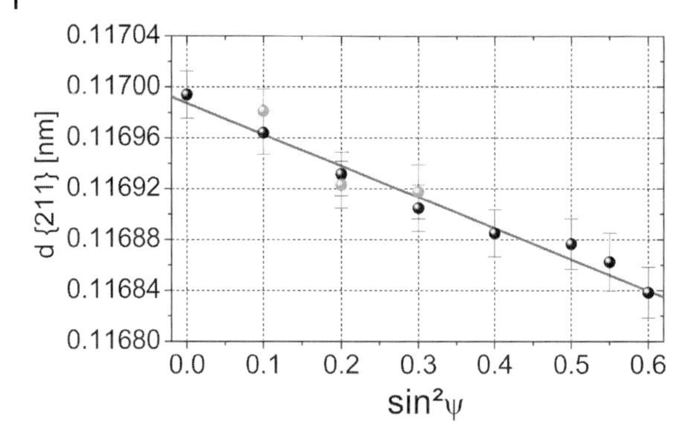


**Figure 24.4** *d*-versus $\sin^2 \psi$ curve obtained in the fillet of a deep-rolled crankshaft (deep-rolling force $F = 14$ kN).

ual stresses; their quantitative value being almost negligible in comparison to the residual stress values in the axial and hoop direction. Hence the results of the analyses using the $d$-$\sin^2 \psi$ method are almost equivalent to the residual stress values in the axial direction.

## 24.3 Experimental Results and Implications

During the deep-rolling process the surface layer of the fillet is expanded beneath the roll. Exceeding the yield strength of the crankshaft material (steel or grey cast iron), once the rolling is finished, results in compressive residual stresses in the near-surface zone and tensile residual stresses below it (Figure 24.5). The depth

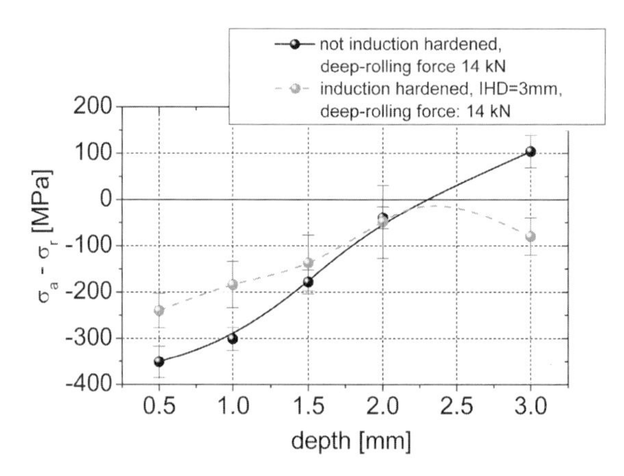


**Figure 24.5** Difference between axial and radial residual stresses in the fillet of a deep rolled and a deep-rolled + induction hardened crankshaft.

of the near-surface zone with compressive residual stresses and the magnitude of the compressive residual stresses depend on the parameters of the deep-rolling process, e.g., the rolling force and the number of rolling cycles and the crankshaft material. In the present investigations a deep-rolling force of 14 kN appeared to be an optimum producing compressive residual stresses within a distance of about 2 mm to the surface of the crankshaft and substantially increasing the lifetime compared to nondeep-rolled crankshafts and crankshafts deep rolled with lower rolling force. The magnitude of the residual stresses that can be introduced by the deep-rolling process is limited by the strength of the crankshaft material; compressive residual stresses introduced into GGG 60 are smaller in magnitude than compressive residual stresses after deep rolling of 38MnVS6 crankshafts [3, 6].

In the induction hardened and deep-rolled crankshafts superposition of the residual stresses caused by the induction hardening of the bearing and by the deep-rolling of the fillet occurs. In the induction hardening process, during heating the surface layer expands more of the bearing. Because the bulk prevents the surface layer from expanding freely, compressive residual stresses form in the near-surface zone, which are balanced by tensile stresses in the bulk. The compressive stresses in the surface layer are reduced with progressive austenitization because the volume of the austenite is smaller than the volume of the originally pearlitic microstructure and because of heat flux from the warmer surface into the cooler bulk, which decreases the difference in thermal expansion. These stresses created during heating, however, are very low in magnitude since the high temperature strength of the austenite is low.

During cooling, austenite contraction in the surface layer results in the formation of tensile stresses in the surface layer and compressive stresses in the bulk. When the temperature reaches the martensite start temperature, the austenite in the surface layer starts to transform into martensite, thus the volume of the surface layer increases and tensile stresses in this layer decrease. Due to the volume increase in the surface layer and volume contraction, further temperature

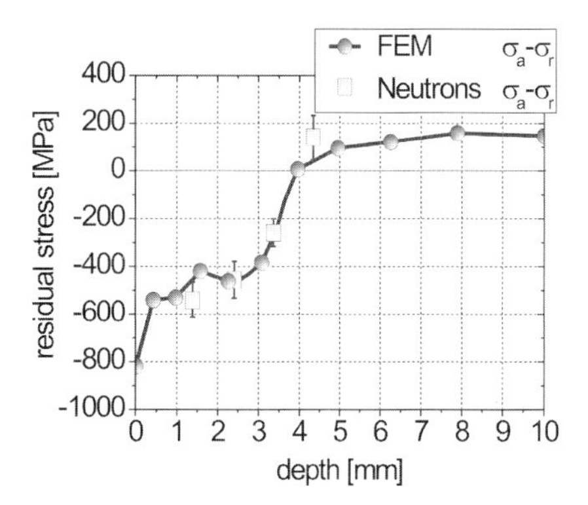


**Figure 24.6** Residual stresses in the fillet of an induction hardened crankshaft [7].

decrease produces increasing compressive stresses in the surface layer and tensile stresses in the bulk of the crankshaft (Figures 24.5 and 24.6). The magnitude and the distribution of the residual stresses depends, e.g., on the dimension of the crankshaft, transformation behavior (CCT diagram), quenching velocity, and high temperature strength of the material.

Neutron diffraction revealed that the maximum of the compressive residual stresses introduced by induction hardening is at the surface of the bearing, while the maximum compressive residual stresses due to deep rolling are beneath the surface of the fillet. The residual stress distribution produced by the induction-hardening process appears more homogeneous in the bulk of the crankshafts compared to the residual stress distribution in crankshafts that were only subjected to deep rolling; induction hardening also increased the depth where compressive residual stresses are present.

A comparison of the three-dimensional residual stress distribution measured using neutron diffraction to the axial and hoop stresses, determined using laboratory X-rays, showed the amount of residual stress relaxation caused by the sampling process. For crankshafts identical in geometry and material, deep rolled using the same process parameters as the crankshaft investigated by Volkswagen AG established a procedure allowing quality control of the crankshafts using laboratory X-ray measurements of the residual stresses [3].

The results of neutron diffraction residual stress analyses on induction hardened crankshafts were used by BMW AG for validating a numerical model (Figure 24.6), which is now used for systematic parameter studies optimizing the induction-hardening process with respect to crankshaft lifetime [7].

## 24.4 Conclusions

The lifetime of crankshafts can be significantly enhanced by induction hardening of the bearing and deep rolling of the fillet, which produce compressive residual stresses at the surface and in the near-surface region. Neutron diffraction revealed the residual stress distribution and residual stress magnitude in the crankshaft and their dependence on the process parameters of induction hardening and deep rolling. The results of the neutron residual stress measurements served for an optimization of the deep-rolling process, a calibration of X-ray residual stress measurements on sectioned pieces of crankshafts in routine quality control and for validating an FEM model of the induction-hardening process.

## References


1 M. Burnett, M. Richards, Improved Forged Crankshaft Performance Utilizing Deep Rolling, http://www.autosteel.org/AM/ (23.11.2006).

2 G. Bernstein, B. Fuchsbauer, *Z. Werkstofftechnik* 1979, **13**, 103–109.

3 C. Achmus, Messung und Berechnung des Randschichtzustandes komplexer Bauteile nach dem Festwalzen, Dissertation, TU Braunschweig (1999).

4 C. Achmus, FEM-Berechnung von Festwalzeigenspannungen. In: *8. Deutschsprachiges ABAQUS-Anwender-Treffen*, Universität Hannover, 1996, pp. 55–66.

5 C. Achmus, J. Betzold, H. Wohlfahrt, *Mat.-wiss. Werkstofftech.* 1997, **28**, 153–157.

6 H.M. Mayer, A. Pyzalla, W. Reimers, C. Achmus, *Mater. Sci. Forum* 2000, Vols. **347–349** and **340–345**.

7 Th. Georges, Ch. Hackmair, H.M. Mayer, A. Pyzalla, H. Porzner, P. Duranton, Modeling Induction Heat Treating of Crankshafts, Heating Industrial Magazine, November 2004, online: http://www.industrialheating.com/CDA/ArticleInformation/features/BNP_Features_Item/0,2832,130998,00.htm and http://www.esi-group.com/Services/Publications/

8 E. Macherauch, P. Müller, *Z. Angew. Physik* 1991, **13**, 305–312.

# Index



*Neutrons and Synchrotron Radiation in Engineering Materials Science: From Fundamentals to Material and Component Characterization*
Edited by Walter Reimers, Anke Rita Pyzalla, Andreas Schreyer, and Helmut Clemens

ISBN: 978-3-527-31533-8

## C

### e

*q*



*r*